Soap Manufacturing Technology

Soap Manufacturing Technology

Editor
Luis Spitz
L. Spitz, Inc.
Highland Park, Illinois, USA

AOCS PRESS

Urbana, Illinois

AOCS Mission Statement

To be a global forum to promote the exchange of ideas, information, and experience, to enhance personal excellence, and to provide high standards of quality among those with a professional interest in the science and technology of fats, oils, surfactants, and related materials.

AOCS Books and Special Publications Committee

M. Mossoba, Chairperson, U.S. Food and Drug Administration, College Park, Maryland
R. Adlof, USDA, ARS, NCAUR-Retired, Peoria, Illinois
M.L. Besemer, Besemer Consulting, Rancho Santa, Margarita, California
P. Dutta, Swedish University of Agricultural Sciences, Uppsala, Sweden
T. Foglia, ARS, USDA, ERRC, Wyndmoor, Pennsylvania
V. Huang, Yuanpei University of Science and Technology, Taiwan
L. Johnson, Iowa State University, Ames, Iowa
H. Knapp, DBC Research Center, Billings, Montana
D. Kodali, Global Agritech Inc., Minneapolis, Minnesota
G.R. List, USDA, NCAUR-Retired, Consulting, Peoria, Illinois
J.V. Makowski, Windsor Laboratories, Mechanicsburg, Pennsylvania
T. McKeon, USDA, ARS, WRRC, Albany, California
R. Moreau, USDA, ARS, ERRC, Wyndoor, Pennsylvania
A. Sinclair, RMIT University, Melbourne, Victoria, Australia
P. White, Iowa State University, Ames, Iowa
R. Wilson, USDA, REE, ARS, NPS, CPPVS-Retired, Beltsville, Maryland

AOCS Press, Urbana, IL 61802

©2009 by AOCS Press. All rights reserved. No part of this book may be reproduced or transmitted in any form or by any means without written permission of the publisher.

ISBN 978-1-893997-61-5

Library of Congress Cataloging-in-Publication Data

Soap manufacturing technology / editor, Luis Spitz.
   p. cm.
  Includes index.
  ISBN 978-1-893997-61-5 (alk. paper)
  1. Soap. I. Spitz, Luis.
  TP991.S685 2009
  668'.12--dc22
                      2009010006

Printed in the United States of America.
13 12 11 10 09   6 5 4 3 2

The paper used in this book is acid-free and falls within the guidelines established to ensure permanence and durability.

Contents

Preface vii

1. **The History of Soaps and Detergents**
   Luis Spitz 1
2. **Implications of Soap Structure for Formulation and User Properties**
   Norman Hall 83
3. **Soap Structure and Phase Behavior**
   Michael Hill and Teanoosh Moaddel 115
4. **Formulation of Traditional Soap Cleansing Systems**
   Edmund D. George and David J. Raymond 135
5. **Chemistry, Formulation, and Performance of Syndet and Combo Bars**
   Marcel Friedman 153
6. **Transparent and Translucent Soaps**
   Teanoosh Moaddel and Michael I. Hill 191
7. **Kettle Saponification: Computer Modeling, Latest Trends, and Innovations**
   Joseph A. Serdakowski 203
8. **Continuous Saponification and Neutralization Systems**
   Timothy Kelly 223
9. **Semi-Boiled Soap Production Systems**
   Boris Radic and Luis Spitz 249
10. **Soap Drying Systems**
    Luis Spitz and Roberto Ferrari 267
11. **Bar Soap Finishing**
    Luis Spitz 303
12. **Manufacture of Multicolored and Multicomponent Soaps**
    Luis Spitz 349
13. **Soap Making Raw Materials: Their Sources, Specifications, Markets, and Handling**
    Michael A. Briggs 377
14. **Analysis of Soap and Related Materials**
    Thomas E. Wood 399
15. **Soap Bar Performance Evaluation Methods**
    Yury Yarovoy and Albert J. Post 417
16. **Soap Calculations, Glossary, and Fats, Oils, and Fatty Acid Specifications**
    Luis Spitz 437

Index 457

Preface

The idea for a new book dedicated exclusively to soaps emerged from looking at material published in three prior books and at unpublished lectures from two previous conferences. I hope this book will be a useful reference guide for those already in the soap industry and for newcomers as well.

The first book I edited and coauthored, entitled *Soap Technology for the 1990's,* published by AOCS press in 1990, is currently out of print. *Soaps and Detergents*, which followed in 1996, is also out of print. My most recently edited and coauthored book, *SODEOPEC*, published in 2004, is still in print.

This new publication, *Soap Manufacturing Technology*, contains revised versions of the most important chapters from previous books. Soap related lectures that were presented in the 2006 and 2008 SODEOPEC (Soap, Detergents, Oleochemicals and Personal Care Products) conferences are included. Additional new material completes the book.

I wish to extend my appreciation to AOCS Press for publishing this updated book dedicated solely to the soap industry. My coauthors and I are pleased that this book will be introduced at the AOCS 100th Anniversary Conference and Expo on May 3, 2009, in Orlando, Florida.

I especially appreciate the work of all the authors who updated their selected chapters from previous books. Thanks also to the new contributors. The writers' sharing of their knowledge will benefit us all.

On a personal note, I feel fortunate to have been involved in the field of soap and to have worked with a product which preserves and enhances people's health, well-being, and beauty. I am grateful for the opportunity the industry has given me to be of service.

Luis Spitz

The History of Soaps and Detergents

Luis Spitz
L. Spitz, Inc., Highland Park, Illinois, USA

Bubbles Since Antiquity: Yesteryears

When was soap, as it is known today, discovered? Who discovered it? Which were the products used by the various ancient civilizations to clean the body and to do laundering? Where does the name soap come from, a name which is very similar in many languages? E.g., *Sapone* (Italian), *Savon* (French), *Seife* (German), *Saippua* (Finnish), *Szappan* (Hungarian).

The origin of the name "soap", as well as the date and circumstances of discovery are not known with precision. Most scholars agree that the discovery was accidental. The name is attributed to a Roman legend.

What is known and proven is that health is directly related to cleanliness and the use of soap and water. Data proves that the infant mortality rate is lower when the consumption of soap is higher in a country.

A brief history of soap follows. Interested parties can find extensive published material on the history of soap—please refer to the list of references.

Mesopotamia, derived from the Greek for "land between the rivers", is an area between the Tigris and Euphrates rivers, presently in southern Iraq. This region, referred to as the "cradle of civilization", was known as Sumer, and Sumerian was the spoken language.

The city of Ur in Sumer is identified in the Bible as the home of the patriarch Abraham.

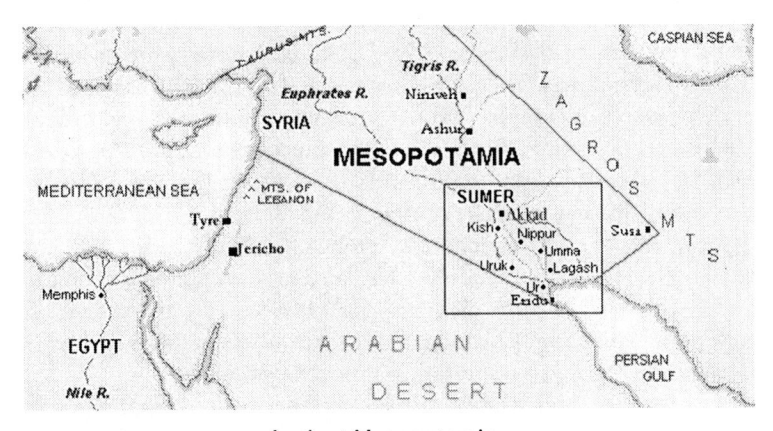

Ancient Mesopotamia.

Dating ≈2500 BCE, the oldest literary reference to soap was found in a Sumerian clay tablet written in cuneiform and relates to the washing of wool. In another Sumerian tablet, from 2200 BCE, there is a formula for soap that consists of water, alkali, and cassia oil.

Sumerian tablet.

The concept of cleanliness has religious beginnings. Among the ancient Hebrews, the importance of cleanliness for health was recognized, and laws were instituted to enforce washing of the hands before and after a meal, and of the hands and feet before entering the temple. The other aspect was spiritual cleanliness. A rabbinical saying states that "Physical cleanliness leads to spiritual purity." Later on, the expression that "cleanliness is next to godliness" was used by George Whitefield, the British religious reformer, and by John Wesley (1703–1791), the celebrated Anglican preacher and founder of the Methodist Church, in his Sermon 93 delivered in 1778.

But cleanliness was not always next to godliness. For centuries, bathing was used for rituals and was unrelated to cleaning and keeping clean. The early church disapproved of the bath. St. Francis of Assisi listed dirtiness as a mark of holiness. There are three passages in the Old Testament in which the word soap appears, but there is no evidence that soap, as we know it now, was used in Biblical times. In the *Good News Bible*, which uses contemporary English, the three passages appear as follows: Jeremiah 2:22, "Even if you washed with the strongest soap, I would still see the stain of your guilt."; Malachi 3:2, "But who will be able to endure the day when he comes? Who will be able to survive when he appears? He will be like strong soap, like a fire that refines metal."; Job 9:30, "No soap can wash away my sins." In the passage from Jeremiah, the Hebrew word *borit* is translated as soap, but *borit* is best translated as a "salt from a plant", "a substance to clean with". *Borit* was probably natural wood or vegetable herbs obtained from burning indigenous plants from Israel, Egypt, or Syria. These passages in the Bible are metaphors for spiritual cleanliness and religious purification via physical cleanliness.

The ancient Egyptians also had religious rituals that demanded cleanliness. Their priests' heads were shaved and they bathed several times a day. They also showered and scrubbed with sand and then anointed the graven images of their gods with oil. The Greeks were the first to bathe for aesthetic reasons. They had no rules for cleanliness on religious grounds. "A sound mind in a sound body" was their idea. There were public baths in Athens at the time of Socrates (469–399 CE). The Romans also built public baths and encouraged cleanliness. The great public baths in Rome became luxurious clubs, and bathing became very popular. The Roman Empire built many aqueducts, which supplied water not only for drinking, but also for washing and cleaning. For the Greeks and Romans, washing consisted of having hot baths and either beating the body with twigs or scraping off dirt with a *strigil* shaped as a shoehorn. By encouraging cleanliness, the Roman Empire suffered very little from the plague and pestilences of the times. When the barbarians overthrew the Roman Empire, all of the aqueducts, baths, and public drains were destroyed. During the Middle Ages, in an era called "a thousand years without a bath", millions died in the cities. The Black Death of 1348 killed 25% of the inhabitants of Italy, Spain, France, Germany, and England.

An often quoted legend tells about animal sacrifices made to the goddess Athena at her temple on Rome's ancient Sapo Hill. When it rained, the animal fat (which remained from the sacrifices) mixed with wood ashes and was washed down the side of the mountain. Roman laundresses washing cloth downstream in the Tiber River found that the yellowish "soapy" waters made their clothes whiter and cleaner. The name for "soap" might have originated on Sapo Hill.

Another theory asserts that the ancient Gauls stumbled upon soap in an effort to extract oil from tallow. Perhaps they experimented by boiling it in water that had been leached through beech tree ashes. The excavations at Pompeii, a city destroyed by an eruption of the volcano Vesuvius in 79 CE, reveal a complete soap factory. The Roman historian Pliny the Elder (23–79 CE) was the first to mention soap in his *Historia Naturalis* around 70 CE. He indicated that the Romans secured soap from the Gauls, and he described in detail the bathing procedure in the Roman Bath: Passing to the baths proper, the citizen entered the *tepidarium*, in this case a warm air room; then he went to the *calidarium*, or hot air room; if he wanted to perspire still more freely, he moved into the *laconium* and gasped in superheated steam; He then took a warm bath and washed himself with a novelty learned from the Gauls, i.e., soap made from tallow and the ashes of the beech or elm. Humans have used soap substitutes or "natural soaps" since primitive times. These were usually plant substances containing "saponins", detergent cleansers naturally produced by some plants. Soap plants are common to the Fertile Crescent, the birthplace of ancient civilizations. The American Indians kept clean without soap. They used roots and soap-like leaves: soap bark, soap root, agave and yucca roots. The Navajo Indians made soap with yucca. The soapy part comes from the root of the plant. It was peeled, sliced, pounded, dropped into water, and churned into suds. It was even good for a foamy shampoo, but it had to be well rinsed to avoid irritation. American Indians also washed with fuchsia leaves and agave, and scrubbed with soapwort leaf washcloths. In South America, Indians still use soapbark and soapberry.

It is believed that the Phoenicians were the first to develop soap making into an art. The Arabs, Turks, Vikings, and Celts all made soap. Soap making was brought to England by the Celts in *c.*1000 CE; from there, its use and manufacture spread throughout Europe. Since the production of soap depends on boiling fats and oils with an alkali, soap making began in countries around the Mediterranean where olive oil and a fleshy plant called Barilla were found in abundance. Barilla is still grown in Spain, Sicily, and the Canary Islands, where its ashes provide the necessary alkali.

In the ninth century CE, Marseilles, France, was already famous for soap making. Then two other great European centers for soap manufacture grew up in Savona, Italy, and Castilla, Spain. England's first soap production began in Bristol in the twelfth century and by the fourteenth century, soap was being widely manufactured in Britain. There was still a reluctance to use soap for washing the body up to the sixteenth century. Like bathing, only the rich could afford fine soaps. Cromwell, in 1712, almost taxed cleanliness into oblivion in England. Soap monopolies, combined with heavy taxes and high prices, kept manufactured soap scarce until well into the nineteenth century. Napoleon paid two francs for a bar of perfumed Brown Windsor, an inflated price for 1808. In 1853, when Gladstone grudgingly repealed the English soap tax, he condemned soap as, "most injurious both to the comfort and health of the people", but soap makers heaved a sigh of relief, and soap making became something of a boom industry. This became a turning point in social attitudes toward personal cleanliness.

Two French and one Belgian chemist made industrial soap production possible.

- 1787: *Nicholas Leblanc* invented the process of obtaining caustic soda from common salt (sodium chloride).

- 1823: *Michel Eugène Chevreul* discovered that the chemical nature of fats and oils were glycerides. When glycerides react with caustic soda or caustic potash, soap is produced and glycerine is liberated.

- 1861: *Ernest Solvay* invented the ammonia-soda process for the production of soda ash (sodium carbonate) to be used widely in soap and glass making. He used common salt, ammonia, carbon dioxide, and lime for his process.

The meat packing business was established in the latter part of the nineteenth century. During meat preparation the meat packers saved large quantities of the inedible fats and oils by-products to make large-scale soap production an economic reality. Leblanc's invention and the meat packing business made it possible to produce soap that was affordable by everyone.

Marseille Soaps

In southeastern France, the Provence is a region of the Camargue in which olive oil, salt, and soda ash were readily available for soap making. In the sixteenthth century, Marseille became the first official soap producing region in France. Jean Batiste Colbert, a minister of Louis XIV, the Sun King, issued The Edict of Colbert on October 5, 1688, prohibiting the use of animal fats for the production of Savon De Marseille (Marseille soap). The soap had to contain 72% of vegetable oils (pure olive oil, copra, and palm oil) to ensure quality. In the nineteenth century, most of the olive oil was replaced by coconut and palm oils. Each cube shaped bar of Marseille soap had to be stamped with the statement, "Contains 72% Extra Pure Oil". Many bars also had the weight in grams stamped on one of the sides.

Being gentle on the hands and the cloth washed, these soaps became so popular that by the 1880s there were about 100 Marseille soap producers in France. During the last decade, Marseille soaps, a name remembered and associated with quality soaps of the past, were rediscovered due to the rise in interest in vegetable-based, natural products. The traditional opaque, cube-shaped soaps are sold mainly as specialty gift soaps. In the last few years, both cube- and regular-shaped opaque and translucent toilet soaps have appeared in Europe and elsewhere, with and without the 72% claim.

Cube shaped Marseille Soap.

Marseille Soap Flakes, Powder, and Liquid Detergents are offered in Europe by several companies. In Italy, it is interesting to note that, Procter & Gamble's *Ace Detersivo Marsiglia,* in powder and liquid versions, show a Marseille Soap Bar, in spite of the products being detergents and not soaps. Other companies also use a Marseille shaped soap illustration for non-non products.

Soap Making in Colonial America

Soap making was one of the first trades in early America. The chore of making soap fell on women. Surplus animal fats and oils were saved to make soap 2-3 times a year. The fats and oils were melted in a large iron kettle, which hung on a rod over a fire, and lye was mixed in using a long-handled iron spoon. A rule-of-thumb formula was used. By experience, it was known that the strength of the lye (alkali) determined the success of good soap making. If an egg or a potato could float to the surface, the lye had the right strength.

Soap makers arrived in 1608 at Jamestown, Virginia, on the second ship from England. At the beginning, the soap they made was used for laundering. Fine toilet soaps were imported from Europe, but few could afford to buy them. The soap maker and the candle maker usually worked together. Both used tallow as the base raw material. Candle making was more profitable than soap making due to its higher demand. Benjamin Franklin's father was a candle maker who wanted his son to be a soap maker, but Benjamin became a printer. The first United States patent was granted in 1790 to Samuel Hopkins, a soap maker, for processing potash by a new method.

The Oldest Living Brands

Yardley (1770)

The young William Yardley paid King Charles I a large sum of money in return for a concession to manufacture soap for all of London. Details of his activities were lost in the 1666 Great Fire of London, but it is known that he used lavender fragrance for his soaps. In 1770, Yardley's original English Lavender soap was introduced. Created for gentlemen to use when shaving, the rich fragrant lather quickly became a favorite with the ladies too. Yardley was the first branded soap in the world, proclaiming its name on every bar.

In 1913, the firm adopted as the trademark for all of its lavender products Sir Francis Wheatley's *Flower Sellers Group,* one of a set of fourteen paintings known collectively today as *Cries of London. The Flower Sellers'* charming, sentimental quality endured and still adorns some of the Yardley Lavender Soap packages. Yardley was established in the United States in 1921, and by 1928, a full factory existed in New Jersey. In 1960, a new factory was built in Totowa, NJ. In 1978 Jovan, the well-known makers of Musk Oil, purchased the U.S. rights for Yardley and started producing the Yardley Old English Lavender Fragranced Soap, advertising it with the slogan, "Most Soaps have a Slogan, Ours has a History." The Lavender bar was followed by five ingredient bars: The Cocoa Butter Soap, The Oatmeal Soap, The Hard Lotion Soap, The Aloe Vera Soap, and The Baby Soap.

Yardley Old English Lavender Fragranced Soap—1920's.

Sir Francis Wheatley's *Flower Sellers Group*—1913.

After Jovan, several companies owned Yardley. In 1979, the British-based Beecham Group acquired Jovan. Jovan ceased operation in 1984. In 1990, Yardley of London, the British parent company, was purchased by the New York-based firm of Wasserstein & Perella. In 1991, the Maybelline Company, owned by Wasserstein & Perella, purchased the right to manufacture and sell Yardley products in North America. In 1996, Maybelline was sold to L'Oreal Cosmetics. In 1997, Yardley of London created an independent organization in the United States. The Wella Company bought the rights to the Yardley name outside of the United States in 1998. In December 2001, Wella purchased Yardley U.S., and Procter & Gamble bought Wella in September 2003. In October 2005, the Lornamead Group from the United Kingdom purchased both the U.S. and English Yardley operations.

Several types of Yardley soaps have been introduced during the last few years. In 2003, six Yardley London Moisturizing Soap bars were launched with a wide range of ingredients under the Secret Cottage product line. These bars were Flowering English Lavender, Natural Oatmeal & Almond, Sweet Summer

Aloe & Cucumber, Natural Performance, Early Morning Rose, and Baby Gentle Natural Moisturizing Soap-Dermatologist Tested.

In 2004, Yardley introduced two specialty bath lines: Nature's Slices Bar Soaps and the Apothecary Line.

The 2008, four Naturally Moisturizing bars were re-positioned: English Lavender with Essential Oils, Fresh Aloe with Cucumber Essence, Oatmeal & Almond with Natural Oats, and Lemon Verbena with Shea Butter.

Pears (1789)

The history of Pears Transparent Soap began in 1789 when Andrew Pears opened a barbershop in London's Gerrard Street, Soho district, a fashionable residential area. He manufactured creams, powders, and other beauty aids. Wealthy clientele used his products to cover up the damage caused by harsh, highly alkaline soaps used in Britain in those days. Mr. Pears recognized the potential for a pure, gentle soap and began to experiment with the production of a fragranced transparent soap for delicate complexions. In 1835, his grandson Francis became a partner. Later, in 1862, Francis's own son Andrew started working with his father and became a partner.

Another partner, Thomas J. Barratt, son-in-law of Andrew Pears, a very creative, enterprising person, was in charge of promotions. Due to his novel approach to advertising, he is considered "The Father of Modern Advertising". At the Paris Exhibition of 1878, Barratt saw a well-known humorous plaster statuette, called *You Dirty Boy,* of an old woman washing the ears of a boy who was very unhappy about it. He purchased it for £500 from G. Focardi, a well-known sculptor, and had it carved in marble. He placed it outside his office. Terra cotta reproductions were offered due to the large demand for both the *You Dirty Boy* and for the advertising materials that followed.

You Dirty Boy by G. Focardi—1878.

Thomas J. Barratt, "The Father of Advertising", was a pioneer in using works of art as sources for advertising. He made advertising history by buying paintings from artists and offering them to the public as poster-reproduced, high-quality chromolithographs.

He acquired an 1886 painting done by Sir John Everett Millais, a very popular painter in Britain, entitled *A Child's World*. The painting depicted a curly-headed little boy–Millais' grandson—blowing a soap bubble through a clay pipe. The painting was retitled *Bubbles* and was issued as a chromolithograph in 1897.

For advertising purposes, a small transparent Pears Soap was added to the corner of the picture, making *Bubbles* the most celebrated and reproduced soap advertising of all time.

The original painting was exhibited in the Art Section of the 1893 World's Columbian Exposition in Chicago. Small colored reproductions of the advertising picture were distributed to visitors.

The *Bubbles* chromolithographs had a companion print, *Cherry Ripe,* which featured a lovely young girl.

Bubbles Poster by Sir John Everett Millais—1886.

The *Pears' Annuals* started in 1891. They contained art and literary items such as Charles Dickens' *Cricket on the Hearth*, followed by *A Christmas Carol* in 1892, *The Battle for Life* in 1893, *The Chimes* in 1894, and *The Haunted Man* in 1895. The *Pears' Annuals* also had illustrations, some in color by famous artists, including Arthur Rackham, Charles Green, Frank Dadd, and others. Pears and many products were advertised on the front or back cover, and at times, large postcard size inserts were included. Each year, two, three, or four (in 1915 six) large-size chromolithographic prints, of excellent quality and strongly resembling the original paintings, were offered for sale by mail at a cost of 1 shilling. A total of 102 Pears chromolithographs were offered from the first issue in 1891 to the last offering in 1924. All of the Pears chromolithographs are highly valued collector's items today. Some of the original Pears paintings are in the Lady Lever Art Gallery in Port Sunlight Village, Wirral, which has one of the most important art collections in England. The *Pears' Shilling Cyclopaedia*, a yearly reference book for everyday use, started in 1897 and is still published today.

Four Pears slogans that became classics are as follows: "Pears Soap Matchless For The Complexion"; "Good morning . . . Have you used Pears Soap? How do you spell Soap dear? Why Ma, Pears of course; and He won't be happy till he gets it!" There were many interesting Pears magazine advertisements, like: "The First Message from Mars - Send up some Pears' Soap"; "Peace, Purity, Pears"; and, "As Pears Soap dissolves Beauty evolves." An often quoted advertisement from 1900 shows the American and British flags and the text reads: "Pears' Soap and an Anglo American Alliance Would Improve the Complexion of the Universe."

"The Transparent Soap" by Pears.

Pears enamel advertising sign.

A Pears Soap advertisement that appeared in 1884 in the British magazine, *Punch*, showed a ragged tramp with a pipe in his mouth stating, "Two years ago I used your soap, since then I have used no other." Two other U.S. companies also used the same tramp to advertise their soaps. An N.K. Fairbank's White Star Soap trade card shows the same tramp sitting at a table with a pipe in his mouth writing a long letter asking for a bar of soap, and on the picture the caption is, "This picture was first used by the N.K. Fairbank Company in 1884." J.S. Kirk's White Russian Laundry Soap, which won the First Prize at the 1893 Chicago World's Columbian Exposition, also used the same tramp, shown sitting at a desk writing this note: "I used your soap two years ago and have not used any other since." Thus, three companies fought for the privilege to have the "friendly tramp" use their soap. It is not known who the first to use the tramp motif was or who copied whom. Unilever records indicate that Pears was the world's first registered commercial brand. A.& F. Pears Ltd. became part of Lever in 1914.

The "Friendly Tramp"—"I used your Soap two years ago since when I have use no other".

Companies of the Past

There were many soap companies at the turn of the century. Some became large and famous with well-known brands. A few of the old ones are still remembered today. The fast growing, big soapers bought the important smaller ones. Many other firms faded away, leaving behind a rich, but difficult to trace history of the many brands and promotional material used in a market that was fiercely competitive from its very beginnings. The United States Bureau of Census listed 238 soap factories in the United States in 1935.

Chicago was the home of many soap companies. Armour Soap Works, now the Dial Corporation, a Henkel Company, is the only company from the past that still exists in Chicago today. Chicago was a preferred location. Railroads, which started in 1850, were operating all over Illinois by 1860. Grains and livestock were shipped into Chicago for the growing meat packing industry. Chicago's stockyards offered an ample supply of animal fat for the soap industry.

Chicago might have been the "soap capital of the world". William Wrigley, Jr. came to Chicago in 1891 at the age of 29 to sell the soap his father made in Philadelphia. He also sold baking powder and as a premium, Wrigley offered chewing gum made by the Zeno Manufacturing Company. The chewing gum sold very well and he promoted it with premiums: lamps, rugs, books, and even revolvers. By 1895, the old letterhead, which featured a young girl rising from the earth holding a bar of soap, was replaced by packages of Wrigley's Juicy Fruit and Pepsin Chewing Gum, and the words "Manufacturers of Chewing Gum".

In addition to Armour & Company, three other Chicago based soap companies—J. S. Kirk, N.K. Fairbank, and Swift & Company—produced large quantities of well-known soap brands.

Chicago Soap Companies

J. S. Kirk & Company (1859)

Across the Chicago River, and nearly opposite Old Fort Dearborn (built in 1803), stood the first house in Chicago, erected in 1795 by Jean Baptiste Pont Du Sable, who was popularly known as "The Father of Chicago". In 1804, it became the John Kinzie residence. James S. Kirk had boiled soap since 1839 in Utica, NY. He moved to Chicago in 1859 and built his plant on the site of Old Fort Dearborn. In 1867, he moved to a new plant on the historic site of the Kinzie residence.

This plant was destroyed in the 1871 Great Chicago Fire, but was rebuilt into the largest soap plant in America at the time. It was an imposing five-story factory, with a 182-foot chimney with the "Kirk" name on it. The factory walls had large advertising signs for Jap Rose, White Russian, Juvenile, and American Family Soaps. The plant was on the river, close to the Tribune Building and the Michigan Avenue Bridge. The volume of soap produced by Kirk was impressive even by today's standards. In 1886, the company sold 22 million pounds of White Russian Laundry Soap, claiming it to be the largest sale of any one brand of laundry soap on earth.

The 1893 Chicago World's Columbian Exposition was the 9th World Fair to be held. All previous fairs, including the first one in 1851, were in Europe except for the 1876 Fair, which was held in Philadelphia. Kirk had its own large exhibit at the Chicago Fair. In a special 1893 *Youth Companion Magazine, the World's Fair issue*, a Kirk advertisement indicates that they sold 47 million pounds of soap in 1892, a very impressive quantity even by today's standards. The Kirk plant was located on a prime location on Michigan Avenue in downtown Chicago. As the importance of Michigan Avenue grew, the local residents complained about the malodors coming from the plant's large chimney. The building was demolished and a new plant was built in 1916 at the North Avenue Bridge far from the downtown location. J.S. Kirk was sold to Procter & Gamble in 1930 for $10 million.

In 1990, after 60 years of producing all of the Kirk soap products, and later Procter and Gamble soaps, detergents, cooking oils, fatty acids, glycerine, and liquid cleansers, the plant was closed and demolished.

The best-known Kirk brands were: Kirk's Flake Soap, Kirk's Flake Chips, Kirk's Naphtha Soap, American Family Soap, American Family Flakes, Kirk Olive, Cocoa Hardwater Castile Soap, and Jap Rose Transparent Soap for the Toilet and Bath, which was the first advertised transparent soap in the United States.

American Family Free Premiums for Soap Wrappers.

Kirk's Soap-Bar or Chips Trolley Sign.

Jap Rose Soap Wooden Crate.

N. K. Fairbank Company (1865)

Nathaniel Kellogg Fairbank (1829–1903) born in Sodus, NY, went to Chicago in 1855. Seven years after his arrival, he invested in Smedley, Peck ,and Company, a lard and oil refinery. In 1865, he bought the firm and renamed it the N.K. Fairbank Company. The company grew to over 1000 employees and opened branches in many cities. Soap manufacturing began in 1882. Copco, Clarette, Chicago Family, Ivorette, Mascot, Santa Claus, Silver Dust, Sunny Monday Laundry Soap, Tom, Dick and Harry, Pummo Glycerine Pumice Soap, and other brands were produced

Two of the many products introduced in 1883 became very popular: Fairy Soap and the Gold Dust Washing Powder, also called Gold Dust Scouring Powder.

The "White, pure, floating" Fairy Soap package showed a drawing of a little girl sitting on an oval-shaped soap with the caption, "Have you a little '*Fairy*' in your home?" A drawing of the Gold Dust Twins, Goldie and Dusty, sitting in a washtub illustrated many washing products. Some people still remember the Gold Dust Twins, who became the symbol for the Fairbank Company. Gold Dust was a washing powder and the busy twins cleaned, "everything and anything from cellar to attic." Extensive magazine advertising was used with the slogan, "Let the Gold Dust Twins do your work." Today, Gold Dust items are very popular, and are also rather costly collectibles.

Fairy Soap Store Display Advertising Sign.

Another popular Fairbank product was Gold Dust Washing Powder.

Like most soap companies, Fairbank used a wide variety of advertising tools. The beautifully illustrated booklet "Fairy Tales" had poems, acrostics, and advertisements. An example is "The Fairy Acrostic."

F is for "Foremost," "Fairest" and "Fine";
A is for "Able" to do it each time;
I stands for "Ideal" in everything great;
R means the "Rarest" yet found up to date;
B is for "Better," "Brighter" and "Best"
A is for "Acme," that stands every test;
N means "no rival," and that is no jest;
K tells you "keep it" and you will be glad,
S is for "Standard" the best to be had.
F is for "Fairy" white, floating and pure;
A stands for "Always" the good kind, and sure;
I is for "Idol" of rich and of poor;
R is for "Real Merit," the sort that will stay,
Y is for "You," and you need it each day.
S is for "Soap," the boon of all health;
O is for "Our" kind, better than wealth;
A is an "Acrostic," you see it, we hope;
P is for "Perfect" and "Pure" FAIRY SOAP.

The following is another interesting Fairy Soap advertisement: "Sense Cents Scents" and, "People with Common Sense pay but five common cents for a soap with no Common Scents/ That's Fairy Soap." Lever Brothers purchased the Fairbank Company from Heckler & Company and phased out the Gold Dust Twins products in the late 1930s.

Swift & Company (1892)

Gustavus F. Swift moved his cattle dealing business from Cape Cod, MA, to Chicago. In 1892, he made his first soap product, Pride Washing Powder, with the help of two N.K. Fairbank Company employees who were hired. Pride Bar Soap and Cream Laundry Bar followed. In 1898, Swift bought the brand and the formula for Wool Soap from Raworth & Schotte. Wool Soap was heavily advertised using street banners, trade cards, postcards, booklets, magazine advertisements, and dealer premiums. Coupons were distributed in 1930 with the offer to, "Buy One Cake Wool Soap and Get One Wool Soap Free." Wool Soap trade cards showed two little girls, one dressed in a proper length nightgown and the other with a nightgown that shrank and showed her bare behind. The lucky one said, "My Mama used Wool Soap," and the other complained, "I wish mine had."

This is the story in a beautifully illustrated colored booklet entitled "Alphabet— Pretty pictures and truism about children's friend Wool Soap", "A true story told in verse and pretty pictures which we trust will interest both young and old."

A is for Alphabet read this one through and learn all the good that Wool Soap can do.
B is for Baby so pink and so white who is bathed with Wool Soap each morning and night.
C is for Children immersed in a tub now take some Wool Soap and give them a scrub.
D stand for Dip, which we take in the sea Wool Soap, comes in here, for you and me.
E for Early the time to arise and bathe with Wool Soap - that is if you are wise.
F stands for Faultless, as you surely will see that Wool Soap for the bath and toilet will be.

G is for Goose and of course he doesn't know Wool Soap is the best of all in this row.
H stands for Hurrah! We've found it last the famous Wool Soap, which can't be outclassed.
I is for Indian who sees with delight a really clean red man Wool Soap did it right.
J is for Judge he is healthy and stout, he uses Wool Soap—the secret is out.
K stands for Kisses the baby wants three bathe with Wool Soap she is sweet as can be.
L is for Laces the richest and best well washed with Wool Soap a critical test.
M stands for Model the word people use when quoting Wool Soap and stating their views.
N is for Nations progressive and great who use Wool Soap and are right up to date.
O stands for Object we have one in mind to talk for Wool Soap the best of its kind.
P is for Present the best time to try a bar of Wool Soap with quality high.
Q stands for Question which soap is quite pure? Why Wool Soap of course, in that rest secure.
R is for Ribbons as good as when new Wool Soap will do just the same thing for you.
S stands for Success which comes at our call if we use Wool Soap when soap's use at all
T is for Trial all soaps to compare Wool Soap win if the trial is fair.
U is stands for Uncle of American fame he uses Wool Soap let's all do the same.
V is for Victory Wool Soap has won, and yet its mission only begun.
W stands for Wool Soap remember its name keep singing its praises and spreading its name.
X is a cross, which we will all have to bear but using Wool Soap will lessen our care.
Y is the letter that still stands for You. It means use Wool Soap whatever you do.
Z stands for our Zeal of which we are proud when we talk Wool Soap we talk right out loud.

Swift offered many other soaps and cleansers: Maxine Elliott Complexion Soap, 1902; Sunbrite Cleanser, 1907; Pride Cleanser, 1909; Wool Soap Chips Borated (later renamed Arrow Borax Soap), 1912; Vanity Fair Beauty Soap, 1920; Quick Naphtha Chips, 1923 (name changed to Quick Arrow Chips in 1929, and Quick Arrow Flakes in 1931); Snow Boy Washing Powder, 1929; and Swift's Cleanser, 1945. The three best known and most widely advertised products were the following: Pride Soap for the laundry (Fig. 1.10); Pride Washing Powder for general cleaning purposes; and Wool Soap for toilet and bath, laces, fine fabrics, and woolens. In 1968, a new modern soap plant was built in Hammond, IN, for the production of both a generic floating soap and a new toilet soap named One Soap for the Whole Family. Swift could not establish a viable bar soap business against the major competitors, and the soap plant was closed in the mid-1970s.

(MY MUMMA USED WOOL SOAP)   (I WISH MINE HAD)

Wool Soap did not shrink woolens and was an "ideal bath soap."

Swift's Pride laundry soap.

Other Soap Companies from the Past and Their Brands
A partial list of the many soap companies from the past and some of their well-known brands is presented in Table 1.1.

Table 1.1. Other Soap Companies from the Past and Their Brands

Company	Brand
R.W. Bell & Co.	Soapona, Buffalo
Beach Soap Co.	White Lilly, White Pearl, Full Value
B.T. Babbitt's, Inc.	1776 Soap Powder, Best Soap
Comfort Soap Co.	Comfort Soap, Pearl White Naphtha, Tip Top
Cosmo Buttermilk Soap Co.	Buttermilk Toilet Soap
Cudahy Soap Works	Old Dutch Cleanser
David's Price Soap Co.	Goblin Soap, Old Dutch Cleanser
Enoch Morgan's Sons Co.	Sapolio, Hand Sapolio
Gowans & Strover's	Oak Leaf, Home Trade, Miners
The Grandpa Soap Co.	Grandpa Soap, Tar Soap
Hartford Chemical Co.	Lavine
Haskins Brothers Co.	Tribly Soap
Hecker Products Corp.	Sunny Monday Laundry Soap
James Pyle	Pearline Soap
C.L. Jones	Tulip Soap
Kendall Manufacturing Co.	Soapine, French Laundry Soap, Home
Kirkman & Sons, Inc.	Savonia, Kirkman's Floating Soap
Fairchild & Shelton	Ozone Soap
Larkin Soap Co.	Crème Oatmeal, Modjeska, Boraxine, Sweet Home
Lautz Bros. & Co.	Acme, Gloss, Marseilles White, Snow Boy
Los Angeles Soap Co.	White King, Cocoa Naphtha, Sierra Pine
Manhattan Soap Co.	Sweetheart Soap
Minnesota Soap Co.	Eureka, Top Notch, Peek A Boo, White Lily
G.E. Marsh & Co.	Good Will Soap
Oakite Products, Inc.	Oakite
Oberne, Hosick & Co.	Sweet Sixteen, German Mottled, White Prussian
The Packer Mfg. Co.	Packer's Tar Soap, Grandpa's Pine Tar Soap
Pacific Soap Co.	Citrus, Vogue
Potter Drug & Chemical Corp.	Cuticura
Resinol Chemical Co.	Resinol
The Rub-No-More Co.	Rub-No-More Washing Powder
Schultz & Co.	Star Soap, Gold
G.A. Shoudy & Son	Wonderful Soap, Telephone Soap, Tip Top Soap
W.M. Waltke & Co.	Lava, Oxydol
J.B. Williams Co.	Jersey Cream Soap, Shaving Soap
Allen B. Wrisley Co.	Olivilo, Carnation, Cucumber, Gardenia

The Big Soapers of Today
The Colgate Palmolive Company (1806)
William Colgate was an apprentice soap maker at the age of 15 in Baltimore; at 17, he went to New York to work for a soap maker. In 1806, at the age of 23, he rented a two-story brick building at 6 Dutch Street in lower Manhattan, NY, and converted it into a home, factory, and store. The first products were toilet and laundry soap, but he also sold starch and candles. When he started, most soap was homemade. They were crude, coarse, and harsh on the skin with unpleasant scents. Colgate offered a much improved quality perfumed soap to the urban crowd, and also provided a personal delivery service. Pale Soap was one of the first products.

Other soap companies and their products became part of the Colgate Palmolive Company. A timeline of the various mergers, company name changes, and products follows:

1806:   William Colgate opens a starch, soap, and candle shop on Dutch Street in New York City.
1807:   Francis Smith is made a partner in Smith and Colgate.
1857:   Colgate & Company is formed upon the death of William Colgate.
1864:   B.J. Johnson Soap Company opens in Milwaukee and later becomes Colgate & Company.
1872:   The three Peet Brothers (William, Robert, and James) start a soap company in Kansas City, KS.
1872:   Cashmere Bouquet soap is registered and patented.
1898:   Palmolive Soap is introduced by the B.J. Johnson Company.
1906:   On its 100th anniversary, there are 106 different kinds of toilet soap, and 625 varieties of perfumes.
1914:   The Peet Brothers build a soap plant in Berkeley, CA.
1914:   The Crystal Soap Company of Milwaukee is acquired.
1923:   The Palmolive Company office moves to Chicago, IL.
1926:   The Palmolive Company merges with Peet Brothers to form the Palmolive-Peet Company.
1928:   Colgate Company merges with the Palmolive-Peet Brothers Company.
1929:   The Kirkman & Son Company of Brooklyn, established in 1837, merges with the Colgate-Palmolive-Peet Company.
1953:   The present Colgate Palmolive Company Corporate name is adopted.
1970:   Irish Spring is launched in Germany under the name of Irische Frühling and in the rest of Europe as Nordic Spring.
1972:   Irish Spring is introduced in the United States
1987:   Colgate acquires the Softsoap liquid soap business from Minnetonka Corporation, creating Softsoft Enterprises.
1991:   Murphy Oil Soap is acquired
2006:   Colgate purchases Tom's of Maine
2007:   Kansas City Soap Plant sold to VVF Limited.

Colgate's sales volume reached $13.8 billion in 2008, with a net income of $1.7 billion.

The Palmolive Building
The Art Deco style Palmolive Building, a 37-story office building located on Chicago's Michigan Avenue (the Magnificent Mile), designed by the Holabird & Root architectural firm, opened in 1929 and has become an icon on the Chicago skyline. On its top was a revolving beacon, named in honor of Colonel Charles A. Lindbergh's flight over the Atlantic Ocean. President Herbert Hoover turned on the Lindbergh beacon with the push of a telegraph button in the White House on the night of August 28, 1930. Lindbergh refused the honor, and the beacon was renamed the Palmolive beacon. The beacon stopped operating in 1981. The building was home of the Gillette Safety Razor Company, *Esquire*, *Cosmopolitan*, and *Good Housekeeping* magazines, as well as other firms. From 1967 to 1987, it housed Playboy Enterprises and became known as the Playboy Building. In 2000, it was designated as a Chicago Landmark and in 2003 it was added to the National Register of Historic Places. It was decided in 2002 to transform this historic building into exclusive luxury condominiums and to rename it "The Palmolive Building".

The Old Water Tower and The Palmolive Building—Chicago.

Octagon Products
The octagon shape was the trademark of a light yellow Octagon Laundry Soap with rosin, first marketed in 1887. It was sold for general household purposes. Later, a white version containing silicate was introduced. Other Octagons brands followed: White Floating Soap, Naphtha White, Soap Chips, Soap Flakes, Soap Powder, Scouring Cleanser, and a toilet soap.

Octagon Coupons
From its early days, and lasting for many decades, each Octagon wrapper featured an octagon-shaped redeemable coupon. Beautifully designed "Octagon Soap Premium List" catalogs listed many premiums. A 32-page catalog from 1901 lists premiums redeemable for different quantities of coupons, from a very few to many: children's picture books, a collection of patriotic songs for 10 wrapper coupons, and for 1600 wrappers, a gentlemen's or a ladies' solid silver watch.

The Colgate Clock
For the 1906 centennial of the Colgate Company, an octagonal shaped 38 ft. diameter clock design was created, inspired by the Octagon laundry soap. A larger 50 ft. diameter clock replaced the old one, which was sent to Colgate's Clarksville, IN facility. The clock claimed to be the world's largest, and started marking time December 1, 1924 on top of the eight-story Colgate Building in Jersey City, NJ, overlooking the Hudson River. The clock was visible for up to 20 miles and became a landmark.

Next to the clock was "Colgate's Soaps Perfumes" signage which remained until 1985 when it was replaced with a large Colgate toothpaste tube sign.

Octagon-shaped Colgate clock became a landmark and was visible up to 20 miles away.

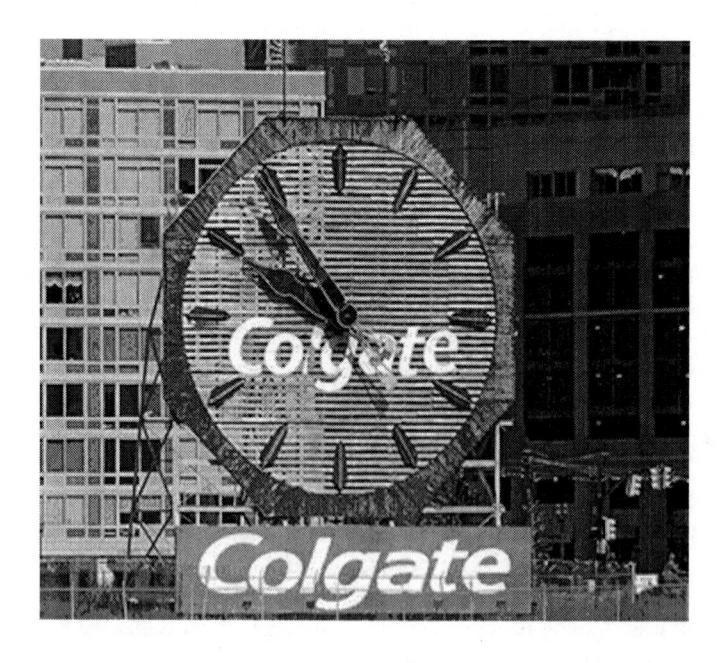

The Colgate clock in recent times.

In 1986, Colgate closed the Jersey City soap plant and razed all the buildings, saving the clock with the Colgate sign, and moving it to the Jersey City waterfront at Exchange Place to be part of a new project. The clock with the sign is still there, but as of 2008, the redevelopment project has not yet been realized.

The Procter & Gamble Company (1837)

The first major depression, known as "The Panic," took place in 1837, the year Martin Van Buren was elected President. In October of that same year, the Procter & Gamble Company was formed in Cincinnati, OH. William Procter, a candle maker, came from London, England, to Cincinnati. James Gamble, a soap boiler, emigrated from North Ireland. They were brothers-in-law, but it took many years before they decided to form Procter & Gamble with a total capital of $7,192.24. They made soaps and candles in a yard behind a small shop. James Gamble, 34 years old, ran the factory. William Procter, 36 years old, ran the office and store and also delivered the products in a wheelbarrow to customers. There were 18 soap and candle makers in Cincinnati at the time.

By 1840, Procter and Gamble had outgrown their simple place at 6th and Main Street and moved to their first factory, a group of small buildings adjacent to the Miami Erie Canal and close to the stockyards. Procter continued to attend only to sales and finance and seldom went to see the factory. Gamble never went to the downtown offices. They met on Saturdays. Their business grew; the factory had 80 employees and by 1859, sales exceeded $1 million, making it the largest manufacturing operation in Cincinnati. In the late 1850s, three of the five Procter boys joined the firm—William A., George H., and Harley T.—as well as three of the six Gamble boys—James N., David B., and William A. In 1878, the company was making 24 varieties of soap and the second generation was running the firm. William Cooper Procter, from the third generation, joined the family business in 1883.

The company grew rapidly, and to speed up the rate of growth, they bought established soap companies. In 1927, Procter purchased the William Waltke Company (founded in 1893), "Soap Makers and Chemists" from St. Louis, MO. Waltke had two major products: Oxydol and "Lava Chemical Resolvent Soap containing vegetable oils and pumice to quickly remove greasy, inky, and sticky substances from hands and face without injury to the skin." The dark gray-colored Lava was launched in 1928. Now it is a green-colored, fresh scent bar called The Hand Soap. The most remembered Lava advertisement was as follows: "World's worst bath soap, World's best hand soap." The Lava brand was sold to the Block Drug Company in 1995, and Block sold it to the WD-40 Corporation in 1999.

In 1930, Procter purchased the J.S. Kirk Company and its important American Family brand. Procter became a leader among 432 national soap manufacturers. Sales reached $10 million during the fiscal years of 1887-1890 with an average annual net profit of a half a million dollars. One hundred years later in 1987, U.S. sales reached $12.4 billion including all of the Procter products. Operating income was $1 billion.

In 2005, Procter & Gamble made a major acquisition by buying The Gillette Company for $57 billion. Procter & Gamble is the world's No.1 seller of household products. 23 brands reached over one billion US dollars in sales. Tide, Ariel, and Gain detergents, and the Olay product line are some of these billion dollar brands.

Total sales reached $83.5 billion in 2008, with a total net income of $12 billion.

Armour & Company (1867)

Armour & Company is remembered mainly for its food products, but few know that Armour was a major soap producer from the 1900s, long before Dial Soap was introduced in 1948. Philip Danforth Armour started a pork smoking, pickling, and rendering operation on Chicago's Archer Avenue and Halsted Street in 1867 when he purchased the Old Bell House for $160,000. Five years later, he moved to the Union Stockyards, which became the center of the meat packing industry in the United States.

In 1984, Armour purchased the Wahl Brothers Glue Works at 31st and Benson Streets. This plant made hide, bone glue, and fertilizer, and recovered grease. Grease was made into soap and around 1888, soap manufacturing began.

In 1896, a separate soap plant was built and Armour Soap Works began its operation. The first product, Armour's Family Soap, a laundry soap bar, was followed by other laundry soaps formulated for the heavy-duty jobs required for the households in those predetergent days: Armour's White Soap, Big Ben, Sail, Hammer, and White Flyer. Armour also became a leader in the manufacture of fine toilet soaps. As early as 1901, Fine Art, Armour's first toilet soap, was advertised in magazines. By 1927, Armour was producing ≈60 brands of toilet soaps: Sylvan, Milady and Flotilla (a white floating soap), La Satineuse, La Richesse, Florabelle Rose, Virgin Violet, Sultan Turkish Bath, and many others. A very special soap was Savon Mucha Sandalwood and Violet. The package was designed by Alphonse Mucha (1860-1939), the world famous Art Nouveau poster artist. After the 1930s, Armour slowly phased out the specialty soap business.

In 1964, when the world's largest, most modern soap making plant opened in Montgomery, IL, the company name was changed from Armour Grocery Products Co. to Armour-Dial, Inc. Since 1986, the company has been known as The Dial Corporation.

In 2004, Germany's Henkel GmbH purchased The Dial Corporation. In January of 2009, the Dial soap plant in Aurora, Illinois was sold to VVF Limited.

The Andrew Jergens Company (1880)

Andrew Jergens and his next door neighbor started their business in 1880. The Cincinnati plant was called Charles H. Geilfus & Company, proprietors of the Western Soap Company, Manufacturers of Fancy Toilet Soaps. In 1882, Andrew Jergens, Charles H. Geilfus, and W.T. Harworth formed a partnership. The newly named company, The Andrew Jergens Soap Company, at 180 Spring Grove Avenue, had 25 employees and a one soap kettle operation. The company was to manufacture, buy, sell, trade, and deal in soaps, oils, candles, flavoring extracts, perfumery, cosmetics, toilet articles, and glycerine.

Herman and Al Jergens joined later and on June 28, 1901, the company was incorporated under the laws of the State of New York as the Andrew Jergens Company, with an authorized capital stock of $1,250.00. Andrew Jergens was elected president, Herman F. Jergens became vice president, and Charles H. Geilfus assumed the duties of secretary and treasurer. In the early days, Jergens soap featured flower fragrances and beautiful package designs. Each brand had a distinct odor, true to the flower it represented. As early as 1911, the Andrew Jergens Company had 82 different brands of soap listed in its catalog; the majority were Jergens fragrance bars. During the 1920s, the Jergens soap line was reduced. The main lines that remained included the flower fragrance bars and Pure Castile, Hard Water, Pine Tar, Health Soap, and Baby Castile. Most of the soap wrappers had the notation, "made by the makers of Jergens Lotion". In 1988, the Japanese Kao Corporation acquired the Andrew Jergens Company.

Lever Brothers Company (1884)

The English soap industry was very large and well established in the early and mid-1800s. Many firms were established before Unilever, which became the world's largest, including Joseph Crosfield & Sons (1815), John Knight (1817), and R.S. Hudson (1837). In the 1851 Great Exhibition in London, a bewildering variety of soaps was shown by 103 manufacturers.

William Hesketh Lever, later Lord Levenhulme (1851-1925), entered his father's prosperous wholesale grocery business in Bolton, Lancashire, England, at the age of 16. His first job was to cut and wrap soap. Shopkeepers received disagreeably brown-colored, anonymous, long, 3 lb slabs, which they sliced into pieces and sold by weight. In 1884, Lever, at the age of 33 and already a wealthy man, felt that he had fully exploited the potential of the grocery business and contemplated retiring; instead, he decided to enter the soap business. He shrewdly anticipated the forthcoming great demand for soaps. The industrial revolution was underway; population, urban areas, and factories grew in number. The social and economic conditions were changing very quickly. A new middle class and a better paid working class demanded more soap as they became more educated about health and hygiene.

On February 2, 1884, Lever registered the name "Sunlight" in England and in all countries where the Trademark Act was in force. Once he had the name, he decided to break with tradition by wrapping a single bar in imitation parchment with the colorful, boldly printed Sunlight name (Fig. 1.23). At first, Sunlight was made for him by various manufacturers. As sales grew quickly in 1885, he leased Winter's Chemical Works in Warrington to make a better quality product with more vegetable oil and less tallow. The remarkably quick success of Sunlight demonstrated the potential for "branded" products, and helped to change the entire soap industry. Lever could not satisfy the increased demand. He decided to build a soap factory together with houses for the workers on the banks of the River Mersey. On March 3, 1888, Port Sunlight was born and by 1889, the first soap plant opened together with the first homes for employees. By mid-1890, 40,000 tons of Sunlight soap were sold in England alone.

Sunlight enamel advertising signs.

American ideas influenced Lever in the wrapping and "branding" of his soap, his novel promotions, and lively advertising. He signed contracts to place Sunlight plates in railway stations, positioned bright looking posters in grey looking streets, distributed puzzles, pamphlets, and helpful hints on health. Lever offered and gave a car and 11 bicycles to a prizewinner who saved 25,000 Sunlight wrapping papers. One promotion that lasted was the £1,000 reward offered to anyone who could prove that Sunlight "contained any harmful adulterant whatsoever". No one ever got the £1,000. But he did spend £2 million on advertising during his first 20 years of soap making. Famous illustrators, among them Harry Furniss, Tom Browne, and Phil May, were commissioned to design soap advertisements.

Lever founded the Lady Lever Art Gallery in Port Sunlight, which has one of the most important art collections in England. Lever started publishing the *Sunlight Year Book* and its companion, the *Sunlight Almanac,* in 1895. The name "Unilever" was coined in 1929 when Lever Brothers Limited and the Dutch Margarine Union merged.

In January 1999, Sunlight soap was discontinued in the UK, and the end of an era came in September 2001. Port Sunlight, once the world's largest soap manufacturing plant, closed after more than 100 years of operation. Its closing was due to the growth of shower gels and liquid soaps versus the traditional toilet bar soaps, as well as the old age of the plant.

Revenues in 2007 reached $58.6 billion, with a net income of $5.6 billion.

In 2008, the company's worldwide name was changed to Unilever.

Milestone Brands: Origins and Development of Bar Soap Categories

Soap Categories

The most important brands, starting with their introductory dates, are described under distinct "soap categories" based upon main functions, marketing claims, and positioning.

- Purity, Beauty, and Health Soaps (1872–1947)
- Deodorant and Skin Care Soaps (1948–1967)
- Freshness, Deodorant, and Skin Care Soaps (1968–1993)
- Antibacterial, Deodorant, and Moisturizing Soaps (1994–2000)
- Ingredient, Antibacterial, Deodorant, and Moisturizing Soaps (2000–2009)

In time, some soaps introduced under one category have evolved and changed into a different category—maintaining original claims while developing new claims.

Since some brands did not have many changes or line extensions, they are added in the original category for simplicity sake, while others with many changes and variants are under their respective categories with specific dates.

During the last decade several brands introduced many variants with emphasis on a large range of ingredients

Purity, Beauty and Health Soaps (1872–1947)

Cashmere Bouquet (1872)

Colgate's Cashmere Bouquet is the oldest U.S. made toilet soap. The name was registered in July, 1872, by Colgate & Company of John Street, New York. The soap was part of a line of toiletries that included perfume, talc, face powder, and lotion. Each soap bar was hand-wrapped and individually sealed with sealing wax as a sign of good taste and luxury. It sold for 25¢, a very high price for a soap product at that time. It became very popular with women who used it for its fragrance; it was also kept in drawers to scent linens, lingerie, and handkerchiefs. By 1883, it was claimed that more Cashmere Bouquet was sold than all of the imported toilet soaps from Europe.

The production costs of soaps were drastically reduced with the introduction of automatic soap presses in 1912, and wrapping machines in 1914. Cashmere Bouquet's shape was changed into a flat oval bar, which was wrapped without the hand-applied sealing wax. The price was dropped from 25 to 10¢.

Cashmere Bouquet—the oldest U.S. toilet soap—name registered July, 1872.

Cashmere Bouquet advertising was refined, stylish, and reflected luxury.

Sir Arthur Rackham (1867–1939), born in London, England, became very famous as an illustrator of children's books, including *Grimm's Fairy Tales, Peter Pan, Alice in Wonderland*, and others. He was compared to the American Howard Pyle, the illustrator of fantastic subjects. Colgate asked Sir Arthur to paint a series of aristocratic-style pictures because Cashmere Bouquet was known as the "Aristocrat of Toilet Soaps".

The four-color advertisements ran in the mid-1920s in *Ladies Home Journal Pictorial Review*, and other magazines. The original paintings were also exhibited in the Metropolitan Museum of Art in New York.

Beautiful 1922 magazine advertising.

In 1904, at the St. Louis World's Fair, a plodder and a new soap mill from France were exhibited. Both machines were purchased by Colgate, and Cashmere Bouquet became a milled soap. But it was not known until May 1926, when an advertisement for Cashmere Bouquet appeared in the *Ladies Home Journal*. The heading stated, "Now this 'hard milled' soap, used every day, keeps skin young and lovely." In the text, "hard milling" was explained in great detail: "It is 'hard milled' which means that it is put through special pressing and drying processes that give each cake an almost marble firmness. It is not the least bit squadgy. This special hardness is what makes it safe."

Roll mills are used to refine and homogenize soaps and they have nothing to do with "special pressing and drying". There is no "soft" or "hard milling". Consumers do not really know the meaning of "hard milled", "French milled", or "triple milled", but these claims connote quality and long lasting soap, so they are still used.

In 1991, a New, improved Cashmere Bouquet Mild Skin Care Bar, was introduced and in 1993, it was changed to Cashmere Bouquet Mild Beauty Bar.

Since 2000, only a Great Value Price 3-bar pack of Classic Fragrance Cashmere Bouquet Mild Beauty Bar has been available, and only in limited markets.

Ivory (1879)

The story is told that Ivory soap was invented by accident when one day in 1879, an operator left a soap crutcher (mixer) running during his lunch hour. The mix was lighter than normal due to the extra air whipped into it, and the soap bars floated. Instead of dumping it, the soap was sold to customers, who liked it and asked for more "White Soap", as it was called. Harley Procter wanted a catchier name, and he found it in church one Sunday in the same year, 1879, when this passage from Psalms 45:8 was read: "All thy garments smell of myrrh, and aloes, and cassia, out of the ivory palaces whereby they have made thee glad." The Psalm inspired Harley Procter, who proposed the next day to the members of the firm the name "Ivory". The name was approved. The official date of the first use of Ivory Soap as a trade name was July 18, 1879. The first cake of Ivory Soap was sold in October, 1879, a year that also saw the debut of the incandescent light bulb, the cash register, and the opening of Frank Woolworth's first 5 and 10¢ store in Utica, NY. In 2004, Mr. Ed Rider, the company's archivist, found an 1863 document in a notebook written by James N. Gamble, who wrote, "I made floating soap today. I think we'll make all or stock that way." So the old story of the accidental discovery of Ivory passed into history.

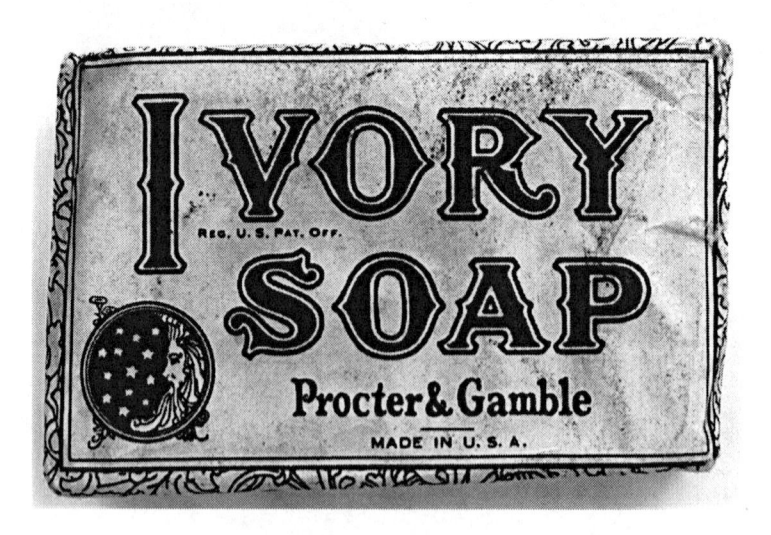

The basic, simple story about Ivory's quality, purity, and mildness has appeared in magazine advertising since the very first advertisement, which appeared in the *Independent Magazine* on December 21, 1882. The opening sentence read: "Ivory is a Laundry Soap with all of the fine qualities of a choice Toilet Soap, and is 99 44/100% pure."

The first advertisement started with a remarkable slogan and selling idea and has remained the same ever since. But Harley Procter had to fight for $11,000 to advertise Ivory because his partners were not convinced of the idea of advertising a single product directly to customers. Up to this time, small advertisements had been submitted to local papers by store owners, or by soap makers for the benefit of the store owners.

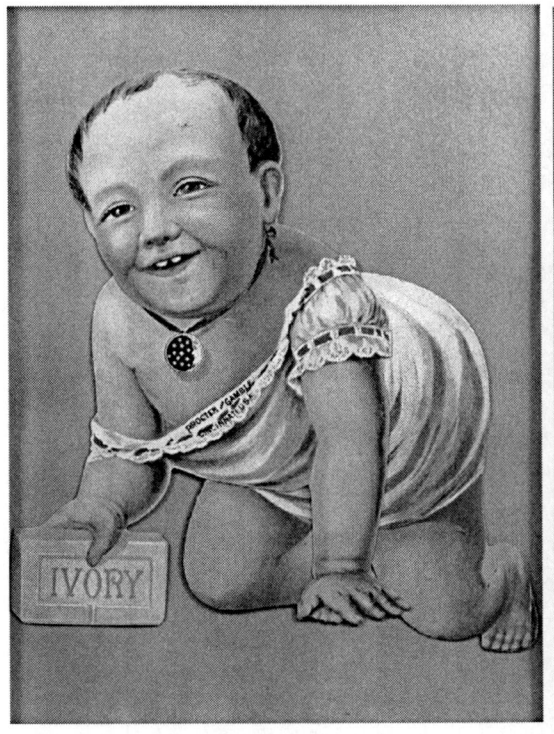

All Aboard! All Aboard!
Down the street we fly!
Faces clean, and fingers too,
See us flashing by!
See what makes us all so clean—
"Ivory Soap" is what we mean.

The first Ivory baby—1886.   Soap Box Derby—Soap packing crates were used.

As Ivory sales increased, new Ivory stories were needed. Famous illustrators drew Ivory babies and children. Competitions were held with large cash prizes for the best drawings—a new and revolutionary advertising ploy for its time. The "Ivory Baby" was used for more than 80 years before it was phased out. In October 2003, the "Ivory Baby" returned in Ivory advertising. It might well have been an Ivory advertisement that led to defining advertising as "salesmanship in print". In 1885, Ivory first advised buying "a dozen cakes at a time". Housewives were told to buy a dozen cakes at a time, remove the wrappers, and stand each cake on end in a dry place; for unlike many other soaps, the Ivory improves with age. Test this advice, and you will find the 12 cakes will last as long as 13 cakes bought singly." Old ideas are not really old. Today, most soaps are sold in multipacks, up to 20 bars per pack. New marketing ideas are not always new, as proven by the 1885 "buy a dozen Ivory soaps" promotion.

28 ● L. Spitz

The first full-colored soap advertising in magazines began in 1896. Quality color printing had to be done in Europe and the finished pages were sent back to the United States for insertion in the magazines. Later Procter & Gamble sent an employee to Europe to learn four-color printing. From the many Ivory babies, Maud Humphrey's (Humphrey Bogart's mother) *A Busy Day* painting from 1896, showing a little girl hanging up her doll's clothes, became a much sought after poster, which could be obtained for 10 Ivory Soap wrappers.

A Busy Day—1896 Advertisement by Maude Humphrey.

A few representative Ivory Soap magazine advertisements are listed:

> "Good health and pure soap the sample formula for beautiful skin; The beauty treatment of ten million babies; RX for your complexion; Ivory Soap kind to everything it touches; Approximately 99 44/100% pure; it floats; If you want a baby clean, baby smooth skin, use the Baby's Beauty Treatment—Ivory Soap; Ivory Soap now comes in new "purity sealed wrapper" dust and germs are sealed out; Keep your Beauty on Duty and give your skin Ivory Care Doctor's Advise; Sugar and Spice and a Skin so Nice; That Ivory look so clear so fresh so easily yours; Three generations prove: Young looking skin runs in an Ivory family; 1979—Thanks America we're celebrating our 100th birthday and you've made us your favorite soap."

What is the origin of the famous 99 44/100% purity claim? In 1883, Ivory Soap samples were sent to five colleges and independent laboratories for analysis and comparison with imported castile soaps, which were the standard of excellence then. At Cornell, they defined the purity of soap as 100% minus impurities. The chemical analysis indicated the following: free alkali, 0.11%; carbonates, 0.28%; mineral matter, 0.17%; total, 0.56%.

By subtracting the total from 100%, "99 44/100% pure" was born. Harley Procter and his associates combined "99 44/100% pure" with the phrase "It floats" and made it America's top slogan of all time.

After the great success of Ivory bars, other Ivory soap products followed. To help housewives who had been shaving Ivory bars into flakes or chips for laundry use, in 1919, Ivory Soap Flakes were first sold in grocery stores. After several years of experience in producing soap in beads or granular form, Ivory Snow, a new form of quick dissolving Ivory for dishwashing and fine washables, was marketed in 1930.

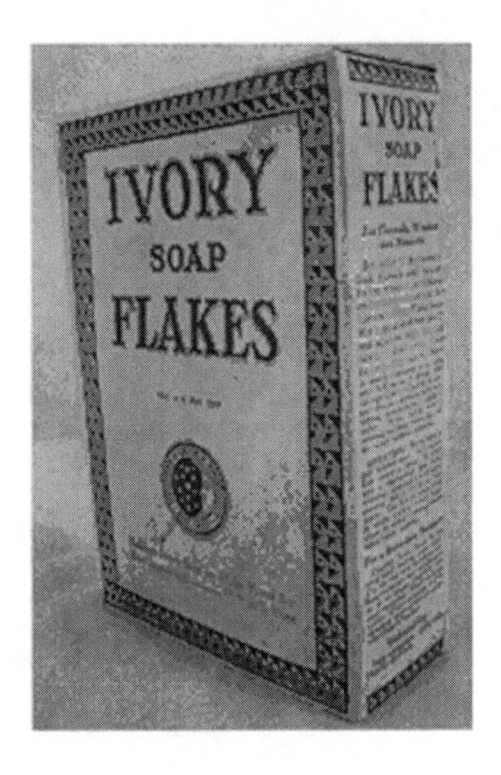

Ivory Soap Flakes—1919.

The "Sinking Ivory" Promotion

To celebrate the 120th anniversary of the birth of Ivory, a limited edition Ivory promotion with the original 1879 package design was offered in 2002. A total of 1501 non-floating bars were produced. The grand prize for finding the "sinking bar" was $100,000.

Jergens (1893)

The name and date of the first Andrew Jergens Company bar soap does not exist. The oldest reference is from 1893 listing: Uncle Sam Tar Soap, Carbolic, White Rose, and Castile. Some of the others that followed were: Dairy Milk Toilet Soap with Buttermik and Glycerine, Novelty Baby Soap (1895), Savon La Contessa, Pumice Chemical Soap (1904), Jergens Voilet Transparent, Rose of France (1908), Crab Apple and Turkish Bath Soap (1914). Many other soaps were introduced in later years.

Lifebuoy (1887)

In 1887, Lifebuoy was introduced in Great Britain. Because it contained carbolic acid and emphasized disinfectant properties, it became widely used in hospitals and on ships. At the time, disinfecting properties had become increasingly important because of the work of Joseph Lister, a British surgeon who began using dilute carbolic acid (phenol) as a germ-killing agent during his operations, Ignáz Semmelweis, a Hungarian physician who discovered in 1847 that "childbed fever" (puerperal fever) was contagious and simple hand washing could drastically reduce its occurrence. Doctors did not wash their hands regularly because the theory that germs carry diseases was not established. Lister spent over a decade further developing his ideas and trying to convince the medical community to accept them.

Lever brought Lifebuoy to America when he opened an office with a staff of 10 in New York in 1895. Three years later in 1898, he started manufacturing it when he purchased the Curtis Davis Company of Cambridge, MA, together with the rights to "Welcome Soap." In 1900, he also bought from Sydney and Henry Gross, Benjamin Brooks of Philadelphia and their "Crystal" and "Monkey Brand" soaps. Lifebuoy was first advertised as "Lifebuoy Toilet, Bath and Shampoo Soap," "The Friend of Health," "Skin Health," "A Life Saver," "A Sanitary Antiseptic, Disinfectant Soap Which Purifies While It Cleans." It was a heavily advertised product in all leading magazines. In 1916, the picture of the fisherman and the life preserver associated with Lifebuoy as a trademark was discontinued (Fig. 1.17). The antiseptic claim, which limited the sale for general use, was changed into a health claim. The name became "Lifebuoy Health Soap." The health appeal made the difference, especially during the 1918 influenza epidemic.

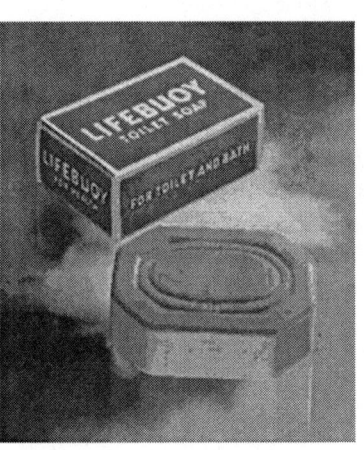

Lifebuoy Clean Hands Health Campaign

Beginning in the early 1920s, Lever Brothers Company in the U.S. started a very important handwashing campaign for children. To encourage children at an early age to be aware of the importance of personal cleanliness, in schools, every pupil received a Lifebuoy Wash-Up Chart and a free "School Size Lifebuoy Health Soap." A Health Pledge was printed on the top of each chart: "Cleanliness is the first law of health. I owe it to myself, my family, my school and my country to keep my body clean strong and healthy - free from dirt and germs. I'll try."

The chart had four tables, each marked with the days of the week. Each day, the children had to mark a square with an X when they washed their face or hands "Before Breakfast," "Before Dinner," "Before Supper," "After Toilet," and "Baths." Gold stars, merit badges, small Health Guards pins, and Clean Hands Campaign certificates were awarded to those who completed the chart. In addition, children learned about dirt, germs, and health using educational charts. This very successful campaign combined a fun game with important health-related issues.

The germ-fighting advertising copy was used until 1926, when something happened in a locker room. Mr. D.L. Countway, brother of Francis A. Countway, the President of Lever Brothers, on a hot May day after a game of golf, entered the locker room and greatly disliked the prevailing odors. Countway smelled something, indeed he did, and suddenly did something about it. The first "Perspiration Odors" advertisements were run, followed by "Body Odors" and then simply "B.O.," two letters that became part of everyday language. With B.O., Lifebuoy sales quadrupled from 1926 to 1930. In 1941, Zephyr Fresh Lifebuoy was introduced, and in 1948, the package and the shape were modernized.

In the 1950s there was a short-lived "New Pine Green Lifebuoy," which had all the ingredients, claims, and concepts that were present later in the United States and Europe using today's terminology, i.e., "freshness," "deodorancy," and "active people." In 1953, the medicinal odor was changed to a pleasant scent, the color became a soft coral and a new germicide TMTD (Puralin) was added.

In England, a number of changes took place. A white version was added in 1962; superfatting agents were incorporated in 1969 and were removed in 1980. In 1986, Lifebuoy "Fights More Germs—Lifebuoy Antibacterial Soap" was relaunched with a modern package design and softer perfume. A year later, the red variant was discontinued. Lifebuoy's largest market is India where in 2002, the bar was completely changed from a carbolic soap with cresylic perfume to a soap with a new "health perfume" and Active-B, an ingredient that protects against germs that can cause stomach infection, eye infection, and infectious cuts and bruises. This new Lifebuoy is advertised for today's discerning housewife and mother as "Family health protection for my family and me." There is a Lifebuoy Active Red, a Lifebuoy Active Orange, and two bars for the upper end of the market: Lifebuoy International Plus against germs and body odor, and Lifebuoy International Gold against germs that cause skin blemishes. Presently Lifebuoy is one of the world's largest selling bar soap.

Lifebuoy Health and Hygiene Education Program

Unilever began in 2002 "The Lifebuoy Swastyha Chetana" an ambitious rural Health and Hygiene education program in India aimed to reach 200 million people in rural villages about the importance of washing hands with soap and general hygiene. Health workers visiting villages with a "glo-germ" demonstration kit show how germs can stay on your hand unless you use soap and water. The program teaches that these "invisible germs" can cause infections.

Similar programs have started in many other Asian and African and Latin American Countries.

The World Health Organization estimates that each year 600,000 children under the age of 5 die in India and over 2.2 million worldwide die from diarrheal diseases and another 1.9 million from respiratory infections Studies have shown that handwashing with soap can cut deaths from diarrhea by almost 50 percent and from respiratory infections by 30 percent The internet provides many sites that offer extensive information. Three are listed.

- CDC Center for Disease Control and Prevention (www.cdc.gov/cleanhands)
- Clean Hands Coalition (www.cleanhands.org)
- The Handwashing Handbook: A Guide for Developing a Hygiene Promotion Program to Increase Handwashing with Soap (www.globalhandwashing.org)

To promote the use of soap and water life-saving habit, October 15, 2008, the first *Global Handwashing Day* was commemorated by millions of children in 52 countries on 5 continents.

Global Handwashing Day is supported by the Global Public-Private Partnership for Handwashing with Soap (PPPHW). Established in 2001, by the U.S. Agency for International Development (USAID), World Bank, Water and Sanitation Program, UNICEF, Unilever, Water Supply and Sanitation Collaborative Council, U.S. Centers for Disease Control and Prevention, Procter & Gamble, Colgate-Palmolive and the John Hopkins University School of Public Health.

Woodbury (1897)

John H. Woodbury founded the John H. Woodbury Dermatological Institute in the State of New York. He developed the John H. Woodbury Facial Soap, which he claimed to be the first soap that could be used safely on the face. In 1897, John H. Woodbury and Andrew Jergens entered into a contract that the Jergens Company would produce Woodbury Facial Soap. The concept of skin treatment with soap, facial cream, and powder was promoted in 1919 with "A Skin You Love to Touch" slogan considered a classic in American advertising history. *"The Dawn of a Great Discovery"* Woodbury Facial Soap with *"Filtered Sunshine"* was launched in 1936. Full-color advertising appeared in *Ladies Home Journal*, *Vogue*, *McCall's*, and other magazines and newspapers around the country. The ad showed a stylized photograph by Edward Steichen of a nude lady. This is claimed to be the first nude advertisement on record in the U.S.

Palmolive (1898)

Palmolive soap was a floating soap when first made by the B.J. Johnson Soap Company of Milwaukee in 1898. During a board meeting in 1911, Mr. Burdette Johnson and his sales manager, Charles S. Pierce, were discussing Galvanic Soap, then well-known laundry soap. Someone mentioned that they also had a green toilet soap made with palm and olive oils called Palmolive. Then someone else said that Cleopatra and other Roman beauties used palm and olive oils. That day Palmolive was born, or better reborn, as a different product. The first trial advertisement was supposed to run in Grand Rapids, MI, for $1000. Everyone considered this too much money for such an "uncertain venture" and for $700, the advertisement appeared in Benton Harbor, MI. By 1928, Palmolive was the largest selling toilet soap in the world.

Anthony and Cleopatra "washed" with a mixture of fragrant oils and fine white sand. This mildly abrasive mixture was rubbed on and cleaned the bodies. Cleopatra also had a milk bath but used no soap. Legend also tells us that Cleopatra's handmaidens bathed and massaged her from head to toe with gentle olive and palm oils. When Cleopatra is mentioned, we think of exotic beauty, velvety skin, mystery, and cosmetics. Palmolive Soap's advertising started with the Cleopatra theme from its very beginning in 1898. In 1984, the remnants of a nine-room perfume and cosmetics laboratory were unearthed on the Western Coast of the Dead Sea. Cleopatra built it.

Palmolive distinctive package with a band and seal.

Palmolive bar soap advertising is one of the finest in bar soap advertising history. The great variety, creative range, and excellence of execution of colored advertisements in such magazines as the Ladies Home Journal, Delineator, and Woman's Home Companion, are both a visual joy and a learning experience for all of us. Many well-known illustrators were hired to paint pictures for the Palmolive advertisements. After Cleopatra, a very famous phrase was created in 1924 by the company President, Mr. Charles Sumner Pierce: "Keep That Schoolgirl Complexion." The Palmolive Girl was very much part of the liberated 1920s (Fig. 1.18). She was active in sports, she traveled, and she always looked beautiful, delicate, and stylish. The Palmolive Girl appeared as a child, a young girl, a young woman, a mother, and a woman in the prime of life. By 1933, the "Keep That Schoolgirl Complexion" slogan became the most effective and sought after Palmolive billboard poster in Colgate history.

Palmolive Soap store display.

A sampling of Palmolive magazine advertising headings illustrates its extensive variety and dedication to beauty:

1904: Palmolive—The Dream of Past Generations.
1916: An Ancient Luxury Brought up to date—Palmolive.
1917: Cleopatra's vision Palmolive Soap.
1917: Four cakes a second 240 cakes a minute 14,400 cakes an hour for every working day. This is the enormous manufacturing volume required by the popularity of Palmolive Soap.
1918: Once a Queen's Secret—Now Your Favorite Soap.
1919: Palmolive—The Oldest of Toilet Requisites.
1920: The Cosmetics of Cleopatra.
1920: Wash your face every day.
1920: Tell me the truth about beautiful skin. Ancient women put the plea to tears. Their answer found carved in hieroglyphics dug up lately was "use palm oil and olive oil"; modern scientists give the same advice to women.
1927: Mother, I'll bet the Princess who looked just like you.
1930: More Palmolive Soap was sold in 1930 than in any year in Palmolive history.
1933: Invite romance by keeping that schoolgirl complexion.
1937: The Dionne Quintuplets use only Palmolive.
1943: Doctors prove 2 out of 3 women can get more beautiful skin in 14 days.
1950: You can have a lovelier complexion in 14 days with Palmolive Soap, Doctors prove!
1953: 100% Mild Palmolive Soap helps you guard that Schoolgirl Complexion look.
1957: New Palmolive Gives New Life to Your Complexion Safely . . . Gently!
1964: New Continental Palmolive Care can help you be younger looking, too.

Woodbury (1899)

The story of "For The Skin You Love To Touch" soap starts in 1876 when Mr. John H. Woodbury founded the John H. Woodbury Dermatological Institute in the State of New York. He developed the John H. Woodbury Facial Soap, which he claimed to be the first soap that could safely be used on the face. In 1897, John H. Woodbury and Andrew Jergens entered into a contract with The Andrew Jergens Company to produce Woodbury Facial Soap. The first published quarter-page advertising appeared on May 1911 in the Ladies Home Journal magazine and carried the slogan, *"For the Skin You Love To Touch."* This is considered, and rightly so, one of the classic slogans in American advertising and one of the most famous and best for selling soaps. The credit for this slogan goes to Helen Lansdowne Resor, secretary and later wife of Stanley Burnet Resor, president and chairman of J. Walter Thompson Company. This slogan also reflected a new marketing technique product segmentation targeting women instead of a general market. It is interesting to note that the Woodbury Facial Soap with its "neckless head" trademark on the package showing a man with a mustache is anything but feminine.

As sales increased, beautiful new advertisements appeared. The first full-page color advertisement in the Ladies Home Journal magazine from September 1915 was a fine illustration painted by F. Graham Coates, followed by an Alonso Kimball and Mary Green Blumenschien painting in 1916. A 1919 advertisement illustrates how early the concept of skin treatment with soap, facial cream, cold cream, and powder was promoted with the aid of a "Skin You Love To Touch" booklet. "The Dawn of a Great Discovery" Woodbury Facial Soap with "Filtered Sunshine" was launched in April 1936. Full-color advertising appeared in the Ladies Home Journal, Vogue, McCall's, and other magazines and newspapers around the country. The advertisement showed stylized photographs of a nude lady taken by Edward Steichen. This advertisement is claimed to be the first nude advertising on record. "Filtered Sunshine," vitamin D, was also a real first. Vitamin E was added to soaps in recent years.

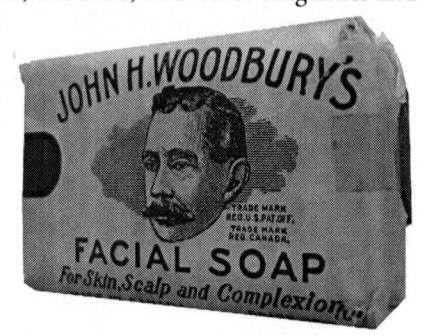

Lux (1925)

In 1899 Lux first appeared in the UK as Sunlight Flakes, but the name was changed in 1900. It was introduced in the USA in 1916 as a laundry for washing delicate fabrics by hand.

In 1924 and 1925, Lever Brothers tested "Lux Toilet Form," a white perfumed soap, and Olva, a green palm oil toilet soap. Olva was a round bar packaged as specialty soap. It was tissue wrapped, then placed into a carton that was overwrapped. The graphic design showed a beautiful lady drawn in an art deco style of the times. The claim "Super-creamed, made from vegetable and fruit oils" was ahead of its time. Lever looked at the market and thought that there might be a world potential for popularly priced, white, milled toilet soap. Cashmere Bouquet was a luxury item, Palmolive, the market leader, was green, inexpensive, and milled. At the time, no popularly priced, white, milled toilet soap existed. Lever spent two years searching for the right bar size and shape, the wrapper style, and the most preferred fragrance.

Consumers tested 40 different perfumes and eliminated 35 of them. Once Lux and Olva were launched, the preference for Lux was immediate and Olva was abandoned.

Lux, the Latin word meaning "light," was short, easy to remember and to pronounce in almost any language, a name that also had the advantage of sound association with "luxury". It is possible that a Liverpool trademark and preferred agent W.P. Thompson, who suggested the Sunlight name in 1884, might also have suggested Lux because of the association with light.

The first advertising for "Lux Toilet Form" appeared in newspapers in April 1925, and the copy read, "Now made just as France makes their finest Toilet Soaps/Ask for Lux Toilet Form today." In early 1927 the name was changed to Lux Toilet Soap and magazine advertising, combined with large scale sampling, began. The comparison to "French Soaps" was combined with an added economy note: "Yesterday . . . 50¢ for a French Toilet Soap/Today . . . the same luxury for . . . 10¢."

In 1928, the "talkies" (talking motion pictures) were being developed and the use of Hollywood movie stars and directors as endorsers was introduced, one of the most successful soap marketing ideas ever used.

The first testimonial advertisements referred to "Exquisite smooth skin, women's most compelling charm" says the 25 leading motion picture directors".

More personal testimonials followed endorsed by practically all the Hollywood stars who signed up to testify or testified to the virtues of Lux Soap, Lever never paid any star for their endorsements.

A few of the over 400 stars who endorsed Lux were: Shirley Temple, Ginger Rogers, Elizabeth Taylor, Joan Crawford, Marlene Dietrich, Lana Turner, Ava Gardner, Rita Hayward, Esther Williams Debbie Reynolds, Raquel Welch, Kim Novak, Marilyn Monroe, Ursula Andress, Brigitte Bardot, Sophia Loren.

Since 1929 when Lux was introduced in India over 50 leading Indian stars endorsed Lux soap. The first star was actress Leela Chitnis.

Now the most famous "Bollywood" stars endorse Lux. Actress Shriya is the current Lux brand ambassador. Other famous "Lux" stars are Priyanka Chopra, Kareena Kappor and Aishwarya Rai.

During the 1930s and 1940s, many black and white Lux beauty-related and in the 1950's onwards color advertisements were used in many magazines.

Some of the slogans were: "9 out of 10 Screen Stars are Lux Girls!", "Be Lux Lovely", "You're beautiful! You're adorable! You're lovely ! You're irresistible!", "To him you're just as lovely as a movie star".

In 1957, in addition to the white Lux in gold foil, four new colors were introduced: pink, green, yellow, and blue. Few people remember the unique looking Lux advertised in 1962 as "Like no other soap in the world New Lux with three deep beauty bands. Now three beauty bands go deep into new Lux. Its moisturizing creamy lather says: forget your dry skin worries." There were two gold-colored bands on the face of a white bar and one gold band in the middle on the back. This bar predated all of the multicolored products that followed years later. In the United States, Lux has become an economy bar. In 1987 Lux was repackaged and was called the "The Pure Beauty Soap." By the mid-90s, Lux was removed from the U.S. market. It was redesigned and relaunched in recent years with traditional and new positioning as well as new packaging.

New Lux bars are being introduced mainly in Brazil and India.

Camay (1926)

Camay's birth was not easy. The name "Camay" was adopted from the French word "cameo." To a lady, a cameo suggests fine treasured things. There were many pro and con arguments for a new product that would have to compete not only against Lux, Palmolive, and Cashmere Bouquet, but against Procter's own Ivory Soap. It would have to be advertised as a perfumed soap while Ivory was advertising against "heavenly smelling soaps." Cooper Procter, as chief executive, decided to test market Camay in 1923: he affirmed an important Procter & Gamble principle, competing with itself. By 1926, Camay was a national brand, but initially, it did not do very well.

Neil Mosley McElroy, who joined the company in 1925 fresh out of Harvard, worked as an advertising department mail clerk. It was his idea that the marketing of each brand should be the full responsibility of a specialized manager. This idea evolved into the basic principle of brand management, i.e., that each brand would be operated as a separate business. Camay was not selling well because Camay executives were not allowing it to compete freely against Ivory. McElroy suggested the "one man one brand idea" in connection with advertising and on May 13, 1931, wrote a historic memorandum in which he suggested that the "brand manager" should devote single-minded attention to all aspects of marketing a brand with the help of a support team. Richard R. Deupree, then president, approved the concept and with it Procter & Gamble's marketing philosophies and practices changed forever.

The concept of competing against similar company brands as vigorously as against competitor's products in similar price categories was a new concept in American industry. Cars from the same maker did compete but they were in very different price categories. Thus, Camay not only became a world renowned beauty bar but was responsible for a new consumer marketing concept of great future importance, one that has been copied by nearly every packaged goods company.

Camay was called "The Soap of Beautiful Women" Many advertisements were signed by Helen Chase, the famous beauty consultant. In the mid 1940's "Go on the Mild Soap Diet Tonight" for 30 days . . . let no other soap touch your skin" was used later followed by "Just One cake of Camay and your skin will feel softer, smoother!" and by other variations such as "Your first case of Camay brings you a lovelier complexion." Cold Cream was added to Camay in 1954. "Your skin will love Camay's Caressing Care!" "There is fine Cold Cream in Camay."

In 1983 "Camay with Creamy Coconut Soap" claimed to be enriched with one-half creamy coconut soap to provide for more rich and luxurious lather. A year later for the first time for a Procter and Gamble soap brand, three "Personalized Skin Care Camay" bars, one for each skin type, were introduced: Normal Skin Formula, Oily Skin Formula and Dry Skin Formula.

In 1987, Camay was reformulated and redesigned into a "Moisture Cleansing Bar" in scented and unscented versions. In Europe, "Camay Classic, Chic and Light" were launched in a new upscale package. Camay was updated in France into an attractive pink pearlized type scented luxury bar.

A year after the 1989 European launch, the new redesigned cameo and a fuchsia, ivory and black package with three fragrances were test marketed and nationally distributed in the United States in 1990. P&G abandoned the skin care approach and switched to fragrances. "More than 3 New Fragrances, 3 New Feelings" "Camay Petal-Soft Scented all Over" was the tag line for: Camay Classic/ Softly Romantic Classic/Pink; Camay Natural/Fresh, Clean, Natural/White; Camay Flair/ Exciting Sex/Flair/Peach.

In 1992, the three bars became "The Camay Fragrance Collection" bars all with more moisturizers—fragrance drops of natural moisturizers "Uncover a New You and Be Softer Too" The Camay Classic and Flair kept their name but the Natural was changed to "Camay Innocent." In 1993, Camay with a low market share and no longer the beauty bar of the past, was repositioned to the mid-price soap category; 2-bar bundles were changed to 3-bar bundles without a change in price. The bar was also reformulated to give more lather and a less slick rinse feel.

The many changes did not help to maintain Camay as an important beauty soap brand. In 2008 two Camay bars are left in limited distribution in the USA, Camay Classic with softly scented natural moisturizer with a floral romantic scent and Camay Flair with softly scented natural moisturizer and a sexy exotic scent. In Canada there is only one type; Camay Natural

Jergens (1947–2005)

Jergens Mild Soap was first introduced in 1947 when Jergens Lotion was popular in the market. In 1951, Jergens Lotion Mild Soap was introduced as an economy mild soap, and later in 1976, it was reintroduced with a new fragrance and new shape as an "Economy Model–Mildness at a very mild price" bar. The mild antibacterial, deodorant version appeared 47 years later in 1994.

In 1984, Jergens, ahead of the later trends for ingredient bars, introduced an Aloe & Lanolin and a Vitamin E & Lanolin Skin Conditioning bar. In 1995, three new Jergens Natural Ingredient bars were targeted for "Skin Care for the Whole Family." The three bars were Jergens Naturals Skin Care with Baking Soda, with Aloe and Lanolin, and with Vitamin E and Chamomile. In 2005, three more Jergens bar soaps were offered: Jergens "Trust the mildness" White All Family, Jergens Natural Skin Care Aloe, and Lanolin with Vitamin E and Chamomile.

Deodorant and Skin Care Soaps (1948–1967)

Dial (1948–2008)

In 1946, Armour & Company wanted to offer a completely new soap product. An employee suggested a soap containing a deodorant. A research chemist, Mr. Robert E. Casely, remembered that three years earlier, Dr. William Gump from Givaduan-Delawaaña, Inc. had left a sample of a new germicide. This new chemical, hexachlorophene (G-11), was introduced in 1943. It was offered to all of the major soap companies but none felt that there was a market for a germicidal soap. Armour tested the product in house and with outside laboratories, and it was confirmed that when combined with soap, hexachlorophene was nonirritating to the skin and reduced bacteria on the skin which, in turn, reduced perspiration odor. Dial was born on July 1, 1948.

After considering over 700 potential names including Revoke, Secure, and No No, Dial was chosen. The Dial name suggested 24-hour protection, and the first slogan, "Keep Fresh Around the Clock," later changed to "Round the Clock Protection" was coined. The first advertisement's heading stated "Stops Odor Before it Starts." To promote a soap that can effectively control perspiration odor, rather than a medicated soap for antiseptic purposes was a new concept, a novel idea. On July 1, 1948, Dial was introduced in Oklahoma City and Omaha, favorite test market cities. It sold for 25¢ a bar in drug and department stores, making it twice as expensive as all of the other soaps. In August the first "fragranced" advertising appeared in the Chicago Tribune. It is said that in August 1948, four Armour employees snuck into the Chicago Tribune's press room and poured Dial perfume into the ink supply of the presses. The next day as people read the newspaper in the buses, trolleys, subways, and trains, the first public transportation air freshener with the help of Dial, was achieved. The success of this novel approach was immediate. Another "first" used in the Dial introduction was to show a young lady taking her bath in the window of a drugstore.

By 1949, Dial was available in grocery stores throughout the country. Three years after its introduction, Dial passed Lifebuoy, which dropped its familiar medicinal odor. In tonnage sales, it overtook Palmolive to take fourth place behind Lux, Ivory and Camay.

Dial's commitment to advertising and marketing support shows how expensive it is to launch a new soap and that profits are usually not made for some time even for a highly successful new product such as Dial. The advertising budget for the 1.5 years (1948 and 1949) was <$1 million. In 1950, it was $2 million while the selling price was cut from 25 to 19¢. In the first two years, Armour lost $3 million, and in 1950 $800,000. In 1951, a profit of $200,000 was achieved. In 1952 and 1953 the combined earnings reached $4 million, canceling all of the previous combined losses. In 1954, the profits reached $4 million before taxes. These profits can be best appreciated in comparison to the same $3 million of Armour's pretax earnings on its entire $2 billion business in the same year.

In 1953, the still current famous "Aren't you glad you use Dial?" theme started. It was created by Fairfax M. Cone, Chairman of Foote, Cone and Belding. The full slogan now is "Aren't you glad you use

Dial, don't you wish everybody did?" This slogan is currently used in a series of creative, humorous TV commercials. The much imitated "wet head" full page magazine advertisements, showing close-ups of men and women enjoying a shower followed. The same year in 1953, Dial had become the nation's number one selling soap in dollar volume. A unique distinction for Dial was being chosen to be the first "space age soap." In the Smithsonian Institute's National Air and Space Museum in Washington, D.C., a bar of Dial soap is displayed in the Astronaut's Tools and Equipment exhibit. Alan Shepard, the Mercury 7 astronaut, carried a Dial soap on the historic first U.S. manned space flight above the earth on May 6, 1961.

Since 1986 various Dial bar soaps were introduced, some of which are still on the market. These include: Mountain Fresh Dial (1988), blue marbleized "New Mountain Fresh Dial" refreshing deodorant soap, "Invigorating as a mountain breeze and as clean and refreshing as a mountain stream"; Ultra Moisture Dial Antibacterial with Vitamin E (1991) (discontinued); Dial Plus, a moisturizing antibacterial body soap in versions for normal-to-dry skin and sensitive skin. This was a pearlized bar introduced in 2000 (discontinued); Skin Conditioning Dial Ultra Skin Care # 1 Antibacterial Soap (1998), the first bath soap in the United States sold in supermarkets next to the mass market commodity bars. Each bar is packaged individually in a clear stretch film and labeled like specialty soaps, with two bars packed into a window-style printed carton; Dial with Vitamins (2002); Dial with Vitamins E, A, and B5. It is an amber-colored translucent bar, which replaced the green translucent Dial Ultra Skin Bar.

In 2005, Dial Daily Care bars for "smooth soft skin" were offered in Green Tree & Vitamin E, Lavender & Oatmeal, and Aloe versions.

In 2007 Dial for Men "The ultimate Clean Full Force and Blue Grit with Microscrubbers" entered the men's products market.

The 2008 Dial variants are the ever present Gold "Round the Clock" Odor Protection soap, and the Clean & Refresh type bars; White, Spring Water, Tropical Escape, Mountain Fresh, and Aloe. All these bars are antibacterial deodorant soaps.

The Clean & Soft are the 2008 White Tea & Vitamin E, Spa Minerals & Exfoliating Beads, and the 2009 Cranberry Antioxidant Glycerine Soap bars.

In 2009 Dial for Men 3-D Odor Defense deodorant soap with three claims; "1. Destroys odor causing germs, 2. Deep cleans dirt and odor, 3. Defends against odor all day" was added to the men's line.

Vel (1948–2008)

Colgate's Vel Beauty Bar introduced in 1948 was the first synthetic detergent (syndet) soap-free complexion bar in the United States. Consumer Reports magazine evaluated 76 toilet soaps in October 1948 and found Vel less alkaline than normal soaps. Vel lathered easily in hard water, rinsed off the skin easily and did not leave a deposit (ring - soap scum) in the bathtub.

There is a 2004 dermatologist tested non-drying creamy formula "Vel Mild Skin Care Bar" for soft, smooth skin sold only in limited markets.

Vel never reached national distribution and always lacked promotional backing in spite of being the first "syndet bar.

Jergens (1951–2008)

In 1951, Jergens Lotion Mild Soap was introduced as an economy mild soap. In 1976, New Fragrance Jergens Lotion Mild Soap was reintroduced in a new shape as the "Economy Model—Mildness at a very mild price." In 1985, the bar was renamed "Jergens Mild." Jergens Gentle Touch Soap with Baby Oil: "Discover the feeling of baby soft skin" was a marbleized soap. Later, a 1984 magazine advertising explained "You have almost 3000 square inches of skin/let Gentle Touch with Baby Oil Baby all of them." Gentle Touch was discontinued in 1998.

Jergens Skin Conditioning Bars

In 1984, Jergens introduced the Aloe & Lanolin Skin Conditioning Bar, and its companion, the Vitamin E & Lanolin Skin Conditioning Bar. Aloe, vitamin E, and lanolin are the most preferred bar soap ingredients other than moisturizing/cleansing cream. Cocoa butter, baby oil, glycerine, and bath oil follow in order of importance. The well-known Jergens Skin Conditioning Lotion was offered in combination with the Skin Conditioning Soap.

In 1993, the Skin Conditioning bars of 1984 were restaged into an improved Jergens Aloe & Lanolin Soap Skin Care bar, and a Vitamin E & Lanolin Skin Care bar; both were packaged in a new carton. The bars were advertised together with the Jergens Skin Conditioning Lotion for "Natural Skin Conditioning". Jergens indicated to potential customers that "Aloe Works Wonders on Wet Skin (showing the soap) and Miracles on Dry Skin." The bar soap and the lotion were offered as a pair of complementary products with two distinct functions: washing followed by moisturizing, by the same ingredients and by the same brand. Jergens was the first company to offer the soap and lotion combination in the United States. Today, it is common practice to offer bar soaps with other moisturizing skin care products. Many times the bars are offered as a free promotion.

Jergens Mild Antibacterial Deodorant Soap appeared in 1994, 47 years after Jergens Mild Soap.

Jergens Naturals

In 1995, a new line of three Jergens Natural Ingredient bars was targeted for, "Skin Care for the whole family. . . Naturally!" They included Jergens Naturals Skin Care Bar with Baking Soda—Deodorant; Jergens Naturals Skin Care Bar with Aloe and Lanolin—for Normal to Dry Skin; and Jergens Naturals Skin Care Bar with Vitamin E and Chamomile—Unscented, Hypoallergenic for Sensitive Skin.

In 1997, two bars, the Jergens Moisturizing Body Bar—Refreshing for Touchable Soft Clean Skin, and The Jergens Moisturizing Body Bar—Extra Moisturizing for Touchable Soft Clean Skin, were introduced together with two Jergens Moisturizing Body Shampoos. A year later, more bars were added; Jergens Moisturizing Sensitive Skin Body Bar/Clean Rinsing/Hypoallergenic and Jergens Extra Moisturizing Body Bar/Soft Clean Skin/Extra Softness That Lasts.

In 2004, only three bar soaps were offered: Jergens/Trust the mildness/White All Family Bar, and the two 1995 Jergens Naturals Skin Care Bar with Aloe and Lanolin, and with Vitamin E and Chamomile.

In 2008, only the Jergens Mild bar remains on the US market.

Zest (1952–2008)

Procter & Gamble's combination soap-synthetic (combo) Zest Beauty Bar was introduced in 1952. A *Consumer Reports* April 1958 review of 102 brands of toilet soaps reported that Zest, like Vel in the 1948 report, was less alkaline, lathered well even in salt water, and left no ring in the bath tub. In 1956, Zest became a Deodorant Beauty Bar advertised as, "More than just a soap. Zest gives you both glorious new cleansing action and new deodorant action! Feel Really Clean. Get that Zest Glow From Head to Toe!" The Zest tub test TV commercial showed, "an awful looking residue on the bathtub with soap", and another one showed eyeglasses rinsed with Zest and with soap. "Give up sticky soap film and feel cleaner with Zest", as well as, "Zest, the deodorant bar that leaves no sticky film."

"Rinses cleaner than soap" and "Zestfully Clean" slogans have been used for decades. The latest slogan is "Energizing Refreshment".

Many new Zest variants have appeared since 2001:

2001: Zest Energy Rush - Zestfully Clean/Refreshingly Fun

2003: Ultimate Clean /Anti-bacterial, Deodorant Protection
Aqua Pure Deodorant Bar
Whitewater Fresh Deodorant Bar/Smooth
Spring Burst
Cool Xtreme with Refreshing Mint
Citrus

2008: Aloe Splash with Aloe
Tangerine Mango Twist
Hydrating Effects Aqua Pure with Vitamin E
Stimulating Effects Ocean Energy with Ultra Beads
Mint Explosion
Marathon with Scent Caps System
Ocean Energy with Scent Caps System
Hair & Body 2 in 1 Cleansing Formula

Some of the Zest bars sold in the United States are made in P&G's Mexican soap plant.

Zest, the "Zestfully Clean" bar.

Dove (1955)

In 1955, Dove was introduced as a new toilet bar that "looks like a soap, it's used like a soap, but it is not a soap." Lever explained that *"one quarter* of the content of each bar was rich emollient *cleansing cream*; it was completely neutral, nondrying, and kind to tender skins." In addition, it left no soap scum or ring in the bathtub. In 1979, the *cleansing cream* content claim was replaced by *moisturizing cream*. An Unscented White Dove appeared in 1989, and a Sensitive Skin Formula bar in 1995.

The main advertising claim is still based on the original concept: "Dove is one quarter moisturizing cream", and, "Dove won't dry your skin like soap." Advertising campaigns were based on testimonial letters from housewives, teachers, and even doctors who praised Dove's superior quality and performance.

In 1989, the US made Dove bar was test-marketed in Italy with very good results. A global roll out started in 1991, and by 1994, Dove bars were sold in 55 countries. It was the first time ever that a soap bar (actually a synthetic bar) became a popular skin care bar worldwide.

The success of the Dove bar established the basis for what has become Unilever's Dove "mega beauty brand" range of personal care products.

Dove is claimed to be the world's largest cleansing brand sold in more than 80 countries.

Safeguard (1963–2008)

Procter & Gamble moved slowly into the growing deodorant soap segment. In 1963, Safeguard New Deodorant and Anti-Bacterial Soap with RD 50 "for complexion and bath" was introduced.

The higher coconut oil content "Richer, Livelier Lather Safeguard - The Perfect Family Soap", appeared in 1970, and in 1979, a "Fresh Scent" version appeared.

From 1988 to 1991, Safeguard DS Deodorant Soap for Dry Skin Protection, treating both body odor and dry skin, in white and beige colors, was on the market. Safeguard became the All Family Germ Fighter Antibacterial Deodorant Soap - Mild enough for the Whole Family in 1992.

The original pillow shape bar was changed in 1999 to a dog bone, "Hugs your hand", shape for easier holding and to "get a better grip on fighting germs."

For better shelf appearance, new packaging graphics were introduced in 2001, and the dog bone bar shape was changed back to the classic pillow shape.

Over the years, Safeguard's deodorant and antibacterial claims were switched several times.

Safeguard Antibacterial white with Aloe and Safeguard Antibacterial beige are the two versions offered since 2006.

Safeguard, the "All Family Germ Fighter Antibacterial Deodorant Soap—Mild Enough for the Whole Family".

Tone (1968–2009)

Armour's Tone "The Moisturizing Soap" with Cocoa Butter, has emphasized skin care since its introduction in 1968. Today it is called The Skin Care Bar. In 1979, the "Beauty Begins with Tone" TV campaign used a hit song from the 1960s, "Pretty Woman". "Bye, Bye Dry" magazine advertisements followed in 1985. The "new formula with 50% more Cocoa Butter" in its original yellow and recent cream color has a new health-oriented theme: "After you tone up your body, tone up your skin. Tone gives your healthy looking body that healthy looking skin. Tone up with Tone."

The Tone bars currently on the market are: Original (2000), Island Mist (2000—discontinued in 2009), Mango Splash (2005), and Sugar Glow (2007—discontinued in 2008), Almond Milk (2009) and Antioxidant with Blueberry Extract and Vitamins (2009).

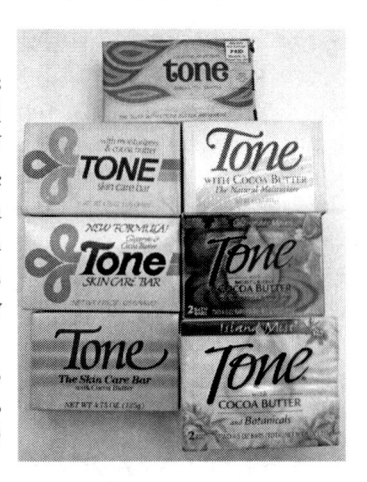

Freshness, Deodorant and Skin Care Soaps (1968–1993)

Fa (1968)

In 1954, die seife Fa (The Fa Soap) produced by Henkel, appeared on the German market. 1968, a brand new die frische Fa (The Fresh Fa), a green and white marbleized, natural, freshness soap with the "Wild Freshness of Limes" was launched, and created a brand new category of "freshness multicolored" soaps. Fa's natural, fresh, youthful theme, emphasizing the use of soap all over the body from head to toe, became very successful and led the way to a new generation of fresh, natural, self-indulgent soaps with marbleized effects.

In a Fa Brand Manual published in 1997, an interesting and unique "Fa Mission of Freshness" summarizes "The Unique Dimensions of Fa Freshness" on different levels: "(i) physical level: Fa sensations—rejuvenation, naturalness, cleanliness, fluidity, lightness, airiness, vitalizing, stimulating; (ii) social level: Fa behavior/attitudes—confidence, energy, vitality, new drive, dynamism, harmony, inner equilibrium with others; (iii) psychological level: Fa sentiments—surprise, discovery, new feeling, revival, replenishment, joy, lightness, inspiring liberating; and (iv) metaphysical level—life, birth, spring, youthfulness, radiance, vitality, mythical, space and time, transcendence, a short moment in eternity, freedom, spontaneity, untamed, unlimited."

From 1968 until 1984, Fa had minor changes, except when, as part of a new line of toiletry products, Henkel introduced new concepts for product differentiation and market positioning using English words. Fa Fresh, Fa Soft, and Fa Beauty bars were introduced.
This novel approach became popular and has been widely used for soap and many other products worldwide.

"Die frische Fa" (The Fresh FA) introduced in 1968 the "freshness soap category".

Atlantic and Pacific (1969)

In 1969, Lever introduced Atlantic on the German market. It was a multicolored blue bar with seaweed extract, shaped like a sea shell, and packaged in a carton that showed a photograph of the bar. Later, Pacific, a companion bar, was launched. The Atlantic advertising campaign used large, risqué, very stylish, outdoor poster and magazine print advertising. These interesting, attractive bars were discontinued in the late 1970s.

Irish Spring (1972–2008)

Colgate's Irish Spring was the first American "freshness" category bar. Introduced in 1972 as "A Manly Deodorant Soap" with "The Freshness of Irish Morning," Irish Spring was a green and white marbleized product in a black carton. Irish Spring bars with different names were launched in many countries.

In 1986, the product was changed into a, "deodorant soap with skin conditioners", with an invigorating scent, "to keep you feeling fresh and clean". The package became green.

Since 1994, several variants were added:

1994:   Irish Spring Original Deodorant Soap
1996:   Irish Spring Sport, a deodorant soap with antibacterial protection for active lifestyle people – (discontinued)
1999:   Irish Spring Aloe, Deodorant Soap
2000:   Irish Spring Fresh, a variation of the original with a revitalizing scent (formerly the 1994 Irish Spring Waterfall Clean) (discontinued)
2002:   Irish Spring Vitamins, Deodorant soap with a provitamin E formula. – (discontinued)
2003:   Irish Spring Icy Blast, Deodorant Soap
2005:   Microclean with MicroBeads Deodorant Soap
2006:   Moisture Blast with HydroBeads Deodorant Soap
2008:   Reviving with Mint Deodorant Soap

Caress (1972–2009)

The original white "Caress with medicated cream" of 1968 did not succeed on the market. In 1972, a peach colored, special shaped "Caress Body Bar with Bath Oil with 101 drops of bath oil" was introduced using the "Before you dress/Before you dress" slogan.

In 1988, the bar was repackaged and changed to "Caress with more bath oil". The slogan was, "Skin feels best when it's Caressed". "Light Caress Body Bar with Bath Oil containing light fragrance" was added in 1992 and "Fresh Deodorant Caress" in 1997.

Other Caress Beauty Bar variants followed: Waterfresh Breeze (2000),Berry Fusion (2002), Shimmering Body Bar (2003), Evening Silkening with jasmine rich moisturizers (2005), Glowing Touch with shea cream and gentle skin brighteners (2005), Berry Indulging Silkening with Vitamin E (2006), Daily Silk with silk protein and natural moisturizers (2006), Exotic Oil Infusions Morrocan Ultra Rich Cream Oil with cassis cream and starflower oil (2007), Tahitian Renewal, Exfoliating with beads, pomegranate seed oil and Tahitian palm milk extracts (2008), Evenly Gorgeous with burnt brown sugar and karite butter (2009).

Coast (1974–2009)

In 1974, Procter & Gamble launched the blue and white marbleized Refreshing Deodorant Coast, "The Eye Opener". There was also for a short time a yellow marbleized Sun Spray version. The Dial Corporation purchased the Coast brand from Procter & Gamble in 2000. Dial continues to offer the original Coast bar. Coast Max, with a blend of botanical extracts (aloe vera and papaya), was added in 2003 followed by Arctic Surf and Pacific Force in 2005. Arctic Surf was replaced by Arctic Boost in 2009.

Pure & Natural (1985–2008)

Dial introduced Mild & Gentle Pure and Natural Soap in 1985 as the purest and mildest soap on the market. It was the first time that a non-floating soap was positioned directly against Ivory. In 1990, a new version packaged in a carton became A Mild and Gentle Body Soap with Great Lather with the tag line, "The Natural Clean Your Family Deserves."

In 2008, offering an economy, "pure & natural hypoallergenic", dermatologist tested bath bar, Dial launched three special hypoallergenic all vegetable pure and natural bars in three variations: Renewing Grapefruit and Pomegranate, Cleansing Rosemary and Mint, and Almond Oil and Cherry Blossom. These 98% natural origin bars are packaged in 100% biodegradable, recyclable cardboard cartons. On the package it is printed: "Plant the carton in soil, water, and watch your plants grow." Dial Basics— "Hypoallergenic – Dermatologist Tested" all vegetable base soap replaced the original Pure & Natural bar in 2008.

Lever 2000 (1987)

Lever 2000, a soap/synthetic combination (combo) bar is the first product with the company name. It was test marketed and went into full distribution only two years later. Lever 2000 was formulated to provide a combination of superior mildness with superior combined antibacterial deodorant protection. It was advertised as mild enough for use by the whole family, children, teens and adults: "Lever 2000 The Deodorant Soap That's Better for Your Skin". "Lever has special skin care ingredients, so it's good for all your 2000 body parts" (the advertising showed mama parts, papa parts, baby parts). Lever 2000's advertising was creative and different. The soap became Lever's most successful new bar soap since the introduction of Dove in 1955.

Lever 2000 "The Deodorant Soap That's Better for Your Skin".

"Antibacterial, Deodorant, and Moisturizing Soaps" (1994–2000) and "Ingredient, Antibacterial, Deodorant, and, Moisturizing Soaps" (2000–2009)

Antibacterial soaps contain germ-fighting ingredients that inhibit odor-causing germs for hours. Since the early 1990s, antibacterial properties were emphasized over deodorancy, with many soaps using the combination "Antibacterial/Deodorant" protection claims. In addition, moisturizing (skin care) became very popular for most cosmetic products and, as a consequence, many moisturizing bar soaps were introduced.

Most bar soaps since 2000 are the "ingredient" type. Soaps with added ingredients existed in the past, but since 2000, customers welcomed products with a larger range of ingredients. The soap companies complied, launching many variants of older brands, repackaged and newly formulated. Many of these products still maintain the well established "Antibacterial, Deodorant, and Moisturizing" claims, but the main emphasis now is on the "ingredients". Brand new bars have also been introduced.

Oil of Olay (1994–2008)

Richardson-Vick's Oil of Olay Beauty Bar appeared in 1983 as a line extension of their internationally successful Oil of Olay Beauty Cream, Beauty Cleanser, and Night of Olay products. A magazine advertisement used the slogan, "Give Yourself the Bubble Facial and Discover a Secret of Beautiful Skin."

In 1985, Procter & Gamble purchased Oil of Olay from Richardson-Vick. The original soap-based bar was reformulated by Procter into a synthetic (syndet) product in 1994. The new bars were Oil of Olay Bath Bar (White and Pink) and Oil of Olay Bath Bar for Sensitive Skin–Unscented/Hypo-Allergenic.

In 2000, the name was simplified to Olay and the bars were renamed: Olay Scented Skin Care Bar and Olay Unscented, Hypo-Allergenic Skin Care Sensitive Skin Bar. The tagline "Olay—love the skin you're in" was introduced.

There were three short lived Ohm by Olay beauty bars from 2002 to 2005: Sandalwood & Chamomile, Jasmine & Rose, and Citrus & Ginger as part of an Ohm by Olay "Holistic Beauty from head to soul" line consisting of Body Wash, Body Mist, Body Scrub, and the three beauty bars.

A new customized collection of four Olay Bars introduced in 2003 were Normal Skin, Sensitive Skin, Dry Skin with Shea Butter, and Unscented for Sensitive Skin.

A limited edition seasonal Olay Winter Retreat for extra dry skin vanilla essence bar has been offered each winter since 2005.

The 2003 Dry Skin with Shea Butter bar was restaged in 2005 as Extra Dry with Shea Butter. The Normal Skin was re-named Normal Skin with Silkening Moisturizers and the Unscented for Sensitive Skin remained the same.

An Olay Dry bar with oatmeal was added in 2006.

The 2008 collection of Olay Body bars are Daily Purifying with microbeads, Ultra Moisture with shea butter, Ultra Moisture vanilla indulgence, Quench, Age Defying, Fresh Reviving, and Calm Release.

The Oil of Olay cosmetic line consists of a continually expanding variety of personal care products. Olay has become one of the world's largest cosmetic brands.

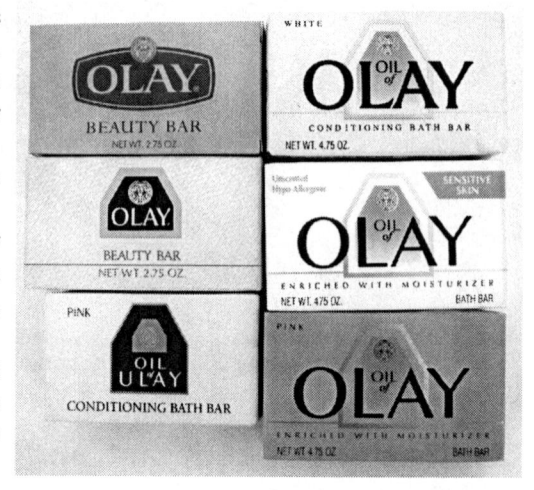

Palmolive Bars (1999–2008)
Several new Palmolive Bar Soap categories have been introduced worldwide since 1999.

Palmolive Botanicals
The most important innovation started with the introduction in 1999 in Mexico of the Translucent Palmolive Botanicals with glycerine and essential oils. Bars wrapped with a clear BOPP (biaxially oriented polypropylene) film and coated with humidity-resistant acrylic on both sides were introduced. The four bars are listed in Spanish and English versions: (i) Palmolive Botanicals: Energizante—Girasol y Acacia en Agua de Manantial/Energizing—Sunflower and Acacia; (ii) Revitalizante—Romero e Ylang Ylang en Agua de Manantial/Revital-izing—Rosemary & Ylang Ylang; (iii) Relajante—Botones de Rosa y Malva en Agua de Manantial/Relaxing—Rose and Mallow; (iv) Acariciante—Manzanilla y Calendula en Agua de Manantial/Soothing—Chamomile & Marigold.

Palmolive Fruit Essentials
Colgate's Softsoap liquid products are the large selling personal care liquid products in the United States. Two Softsoap Translucent Bars made in Mexico and distributed in Canada have been added to the Softsoap line of products: Softsoap Fruit Essentials/Essentiels aux fruits—Juicy Melon/ Melon juteux, and Softsoap Fruit Essentials/Essentiels aux fruits—Fresh Picked Raspberry/ Framboises fraiches.

In some markets, Softsoap Fruit Essentials are called Palmolive Fruit Essentials.

Palmolive Aromatherapy
(i) Palmolive Aroma Therapy—Energy with Pure essential oils, Mandarin & Ginger, Green Tree Extract and (ii) Palmolive Aroma Therapy—Anti- Stress with Pure essential oils, Lavender, Ylang Ylang & Patchouli.

The Botanicals, Fruit Essentials, and Aromatherapy types are specially formulated, extruded translucent soaps. Worldwide, several companies now offer this type of translucent soap, all wrapped in a clear film.

Palmolive Naturals
The opaque Palmolive Naturals are offered with many ingredients: (i) Silkening Care Palmolive Naturals with Milk and Honey Extracts, (ii) Revitalizing Care with Grapeseed and Orchid Extracts, (iii) Balanced Care with Jasmine and Rose Extracts, and (iv) Extra Moisture Care with Aloe and Olive Extracts.

Palmolive Marseille Soaps
In Europe, there has been a revival of the age-old Marseille. Colgate introduced the translucent Palmolive Marseille bars: Palmolive Marseille—Milk & Lemon; Peach & Sweet Almond; and Chevrefeuille & Olive. The French Palmolive Les Bien Faits de Marseille translucent bars were the Energisant, Revitalizant and Apaisant types. The English translation of the name and the product claims are as follows: Palmolive Marseille Energizing—Enriched with Essential Oils—Sunflower, Acacia, & Glycerine; Revitalizing—Enriched with Essential Oils—Rosemary, Ylang Ylang, & Glycerine; Relaxing—Enriched with Essential Oils—Chamomile, Calendula, & Glycerine.

There are other cube and standard shaped Marseille type bar soaps, soap flakes, and even powder and liquid detergents on the European markets: Le Petit Marseillais and Le Chat in France, Sole Marseille Soap Flakes, Omino Bianco Detersivo Lavatrice Marsiglia, and Ace Detersivo Marsiglia Detergents in Italy—regardless of the product type, all packages are illustrated with a Marseille Bar Soap.

In the United States, only Palmolive Classic Scent Mild All Family Soap, and Palmolive Gold Deodorant Soap "Great Value Price" bars are offered.

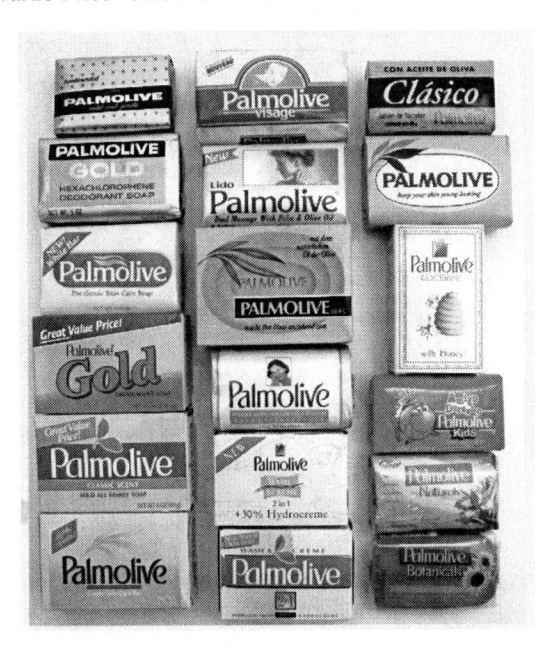

Ivory Moisture Care (1997–2000)

In 1997, 118 years after the birth of the 99 44/100% floating Ivory bar, Procter & Gamble introduced Ivory Moisture Care Bath Bar and Body Wash, two companion products, "for ultra mild cleaning and moisturizing benefits for the entire family." Ivory Moisture Care Bath Bar is a nonfloating synthetic detergent (syndet) bar. It was offered in Light, Fresh Scent, and Unscented versions. These syndet bars were discontinued in 2000.

Safeguard (1999–2008)

The easy to hold, hand-hugging, "dog-bone" shaped bar of 1969 reappeared in 1999. The tag line associated with the easy to hold shape was, "get a better grip on fighting germs". In 2001, the bar shape was changed back to the original classic pillow shape. Since then, minor formulation and package graphic changes have been made. Over the years, Safeguard switched deodorant and antibacterial claims several times. Safeguard today is an antibacterial, deodorant soap, offered in white and beige colors.

Safeguard Deodorant and Antibacterial Soap with RD SO "for complexion and bath" was introduced by Procter & Gamble in 1963. "Richer Livelier Lather Safeguard –The Perfect Family Soap" bar, with a higher coconut oil content, followed in 1970, and a "fresh scent" version appeared in 1979.

Safeguard became the All Family Germ Fighter Antibacterial Deodorant Soap–Mild Enough for the Whole Family in 1992.

Safeguard Antibacterial white with Aloe, and Safeguard Antibacterial beige are the two versions offered since 2006.

Safeguard, the "All Family Germ Fighter Antibacterial Deodorant Soap—Mild Enough for the Whole Family".

Lux (2000–2009)

For over a decade, Unilever concentrated on the worldwide expansion of the Dove bar, simultaneously expanding and strengthening Lux, except in the United States where Lux has been discontinued.

Several Lux soaps types were introduced in India, Latin America, and elsewhere.

Lux Ingredient Bars

The Lux ingredient bars include three Lux Beauty Soaps: Good Day Sunshine—with Orange Extracts & Apricot Kernel Oil; Milk and Honey—with Honey & Almond Milk; Morgentau—with Lemon Grass & Herbal Essence; and two Lux Beauty Moments: Milk & Honey—with Honey and Almond Milk; and Garden of Japan—With Lotus Blossoms and Ginkgo Extracts.

In 2000, there were four Lux Milk Cream ingredient type bars in India with the tag line, "New Lux brings out the star in me"; they are Lux Rose Extract & Milk Cream, Lux Almond Oil & Milk Cream, Lux Fruit Extracts & Milk Cream, and Lux Sandal Saffron & Milk Cream.

In 2003, these bars were re-launched as Skin Care bars with a marbleized effect: Lux Skin Care with Rose Extracts, Lux Skin Care with Almond Oil, Lux Skin Care with Fruit Extracts, and Lux Skin Care with Sandal Saffron. In India, the high end bar soaps are sold as toilet soaps, and the more economical ones as bathing bars. Lux is sold in over 100 countries, and it is now the world's largest selling bar soap brand.

Lux Tanslucent Bars

In 2003, translucent Lux bars wrapped in a clear BOPP film, with a very attractive Lux logo and the face of a woman, went on the market for the first time in Brazil, and later in other countries. It was also the first time that a mass-marketed product's appearance, i.e., transparency, was mentioned. The amber and green colored translucent bars with Portuguese and Spanish text were Lux Glicerina - Transparencia Irresistible. The attractive BOPP clear film was also used for the new opaque Lux bars: Lux Perfeiçaó Cremosa—Perfeccion Cremosa; Massagem Marinho—Masaje Marino; Nutriçaó Radiante/Nutrición Radiante; Rosa Aveludada / Aterciopelado; Toque de Suavidade /Toque de Suavidad, and Lux Beleza Negra /Belleza Negra.

In Australia in 2003, the new translucent Lux Skin Sense Translucent Body Bars, packaged in a carton with a rectangular clear window to show the product, were launched: Lux Skin Sense Awaken Translucent Body Bar—Grapefruit & Lemongrass; Lux Skin Sense Calm Translucent Body Bar—Chamomile & Geranium; Lux Skin Sense Refresh Translucent Body Bar—Green Tea & Lime; Lux Skin Sense Embrace Translucent Body Bar—Neroli & Ylang Ylang.

Special Lux Bars

Special, new types of Lux soaps have been introduced in Brazil, India, and in other countries.

In India in 2002, the International Lux Skin Care Sunscreen Formula with sunscreen lotion claimed that a clinically proven patented triple sunscreen system forms an invisible layer of "UV Guard" against the sun's UV rays, protecting the skin and preventing it from darkening.

In 2003, the Brazilian Lux Skin Care - Morena e Negra bar specially targeted dark-skinned women.

The 2005, Lux Firmassage in Brazil, contained vegetable gelatin and grape seed oil. The bar shape is designed for a massaging effect.

Lux Chocolate Seduction of 2006 is a chocolate colored fragrance bar. It is rich with cocoa cream and strawberry vitamins, and is made in India as well as in Brazil.

Introduced in India in 2007, Lux White Glow is a two-tone (white and gold) bar with a 3 step beauty system consisting of a "Fruitscrub" (to remove dead cells), "Turmeric Extract" (to enhance skin complexion), and "Sunscreen" (to protect from sunlight).

54 • L. Spitz

The Lux Luxo Provocateur of 2008 from Brazil is a black bar containing extracts of black roses and violets for a hypnotizing skin. The bar was also introduced in India.

Introduced in India in 2008, Lux Crystal Shine with "Shine of Crystals and Drops of Moisturizer" is promoted by Priyanka Chopra Miss World for a "Luminous Sensation and Envied Skin". Two new "Chunky" Lux fruit extract variants, with small visible strawberry colored "chunks" in the Lux Strawberry & Cream bar, and peach colored chunks in the Lux Peach & Cream bar, were launched. Both bars contain a blend of "succulent fruits & luscious Chantilly cream that melts down into your skin making it soft and smooth". The famous Bollywood star Shriya was chosen to be the brand ambassador.

Lever 2000 (2000–2009)

Lever 2000 was reformulated in 2000 into five Lever Moisture Response Bars: Anti-Bacterial, Perfectly Fresh with Vitamin E, Pure Rain with Vitamin E, Fresh Aloe with Aloe Vera, and Sensitive Skin Unscented Deodorant soap with Vitamin E.

Very creative magazine print advertising follows the original Lever 2000 approach. Different advertisements show two people, or mother, father, and child touching, embracing different body parts, which are numbered. The advertisement indicates, for example, "Part 64 Meets Part 1820", with the tag line, "I like every part of you." Other taglines: "You are the best part of my life", and, "You touch me."

In 2005, new packaging and modified names were given to fresh scented variants: Hydrate with Fresh Aloe; Energize with Vitamins A, B_5, E and Ginseng; Refresh with Pure Rain scent; and the Original Perfectly Fresh.

Early 2007, four Lever 2000 with Vaseline Intensive Care bars were introduced: Original with nutrients & minerals; Refresh with minerals and Vitamin E; Fresh Aloe with aloe vera and cucumber extract; and Energize with Vitamins A, B5, E, and Ginseng.

Lever 2000 with Vaseline Intensive Care is the first bar soap which includes a very familiar ingredient, namely Vaseline, as an additive.

The 2008 reformulated Lever 2000 with Vaseline, and Lever 2000 with Vaseline for Men were introduced in 2008.

The newest Lever 2000 bars are the three reformulated, repositioned bars from 2009, called Lever 2000 Generous Skin Hydrators bars: "Original refreshes and replenishes with minerals and nutrients", "Pure Rain cools and replenishes with minerals and vitamin e", and "Fresh Aloe with crisp cucumber and allow extract."

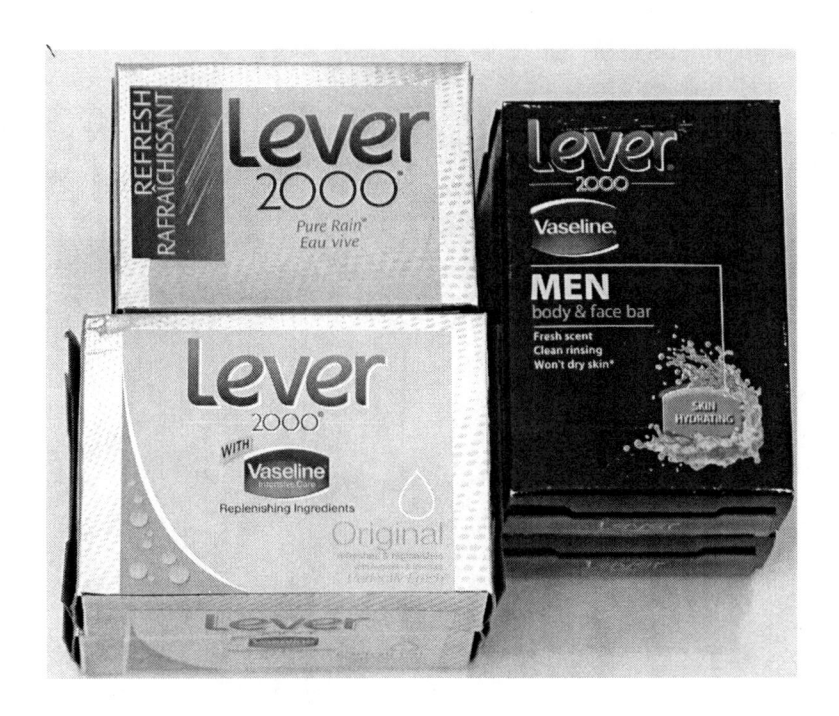

Dove (2001–2009)

Many Dove Beauty Bars have been added since 2001.

Dove Nutrium Nourishing Dual-Formula Skin Conditioning Bar with Vitamin E was introduced in 2001, and was the first new Lever soap since the 1986 Lever 2000 bar. Dove Nutrium is a striped, multicolored, syndet bar composed of white stripes which contain Dove's gentle moisturizing cleanser, pink stripes of nutrient-enriched lotion with Vitamin E. Advertising emphasized that Dove Nutrium goes, "beyond cleansing and moisturizing—it replenishes skin's essential nutrients" and it, "looks different because it is different."

- Sensitive Skin Unscented (2001)
- Gentle Exfoliating with ultra fine exfoliating beads (2003)
- go fresh Cool Moisture with cucumber, green tea extracts and fresh hydrating lotion (2005)
- go fresh Energizing with grapefruit and lemongrass scent (2005)
- Unscented (2005)
- Energy Glow (2006)
- Calming Night with soothing honey and vitamin A (2006)
- The original Nutrium was transitioned into Nutrium Cream Oil (2007)
- Sensitive Skin Unscented (2007)
- Cream Oil Ultra Rich Velvet with ¼ moisturizing cream and rosewood & cocoa butter scent (2007)
- Pro-Age with ¼ moisturizing cream enriched with olive oil (2007)
- Winter Care (2008)
- go fresh Burst with nectarine and white ginger scent (2009)

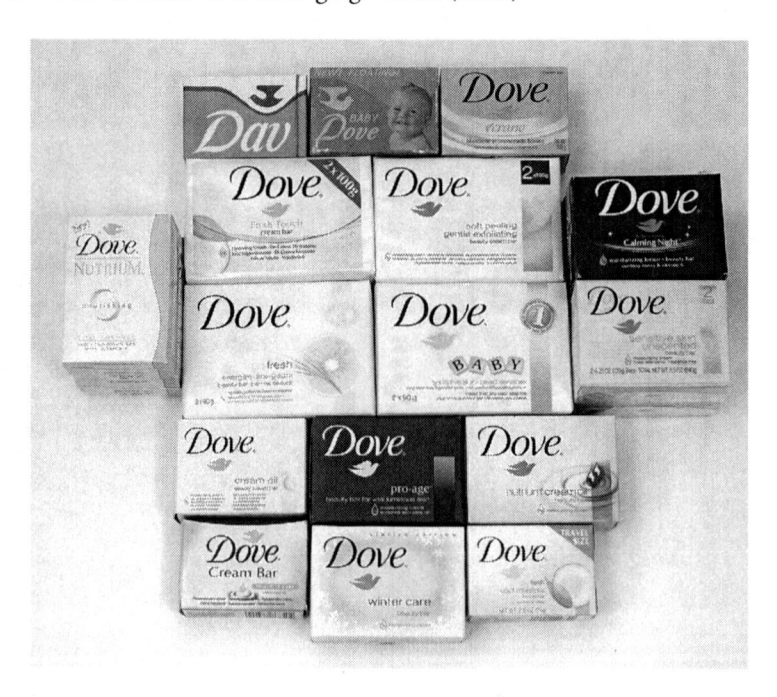

Fa (2002–2008)

Henkel's Fa soaps are produced and sold in many countries, but not in the United States. Many Fa bars were launched since 2002.

The "Wellness System" Moisturizing and Revitalizing with Sea Minerals translucent bars launched on the European market in 2002, and were packaged in a carton with a window were the Orange Blossom SPA, Bamboo Essential SPA, Sea Extract SPA, and Green Tree SPA. Like most mass marketed translucent soaps, they did not succeed and were discontinued in 2004.

There have been many Fa line extensions. Some of them include the "Wild Freshness" Body Care bars: Caribbean Lemon, Exotic Fruits, Papaya Grape, and Aqua.

Many other Fa ingredients bars followed: Fa Caring Bar Soap - Palm Milk; Fa Energizing - Ginkgo Extract; Fa Moisturizing - Grape Extract; Fa Refreshing - Caribbean Lime; Fa Sensitive - Aloe Vera Milk; and Fa Vitalizing - Water Plant Extract.

Fa Asia Spa Jasmine Blossom, Lotus Blossom and Fa Yoghurt Aloe Vera, Yoghurt Vanilla Honey, Yoghurt Coconut, and Yoghurt Sensitive all with Yoghurt Protein are the latest Fa bars.

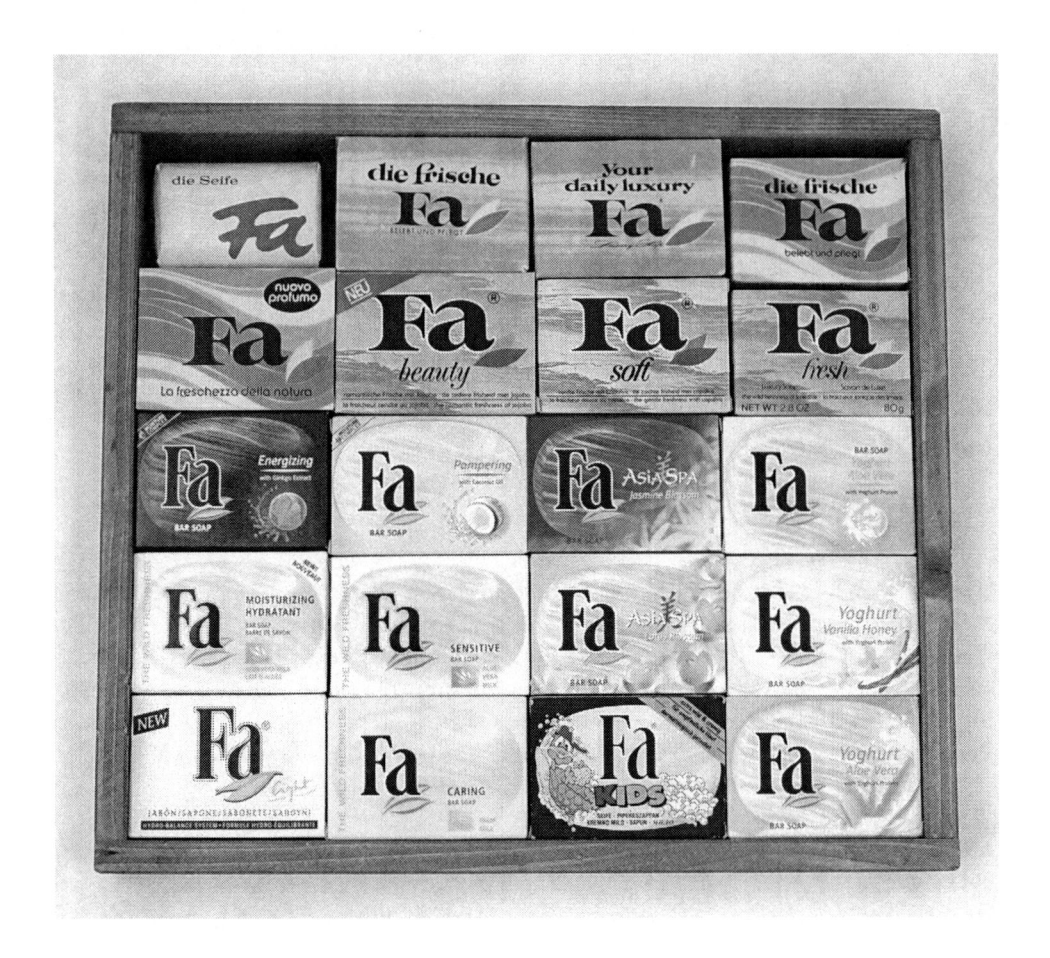

Lifebuoy (2003–2008)

Lifebuoy is no longer manufactured in the United States. Lifebuoy is Unilever's largest selling bar soap brand in India where it has been repositioned since 2003 as a premium deodorant soap.

The current bars are: Lifebuoy Active Red—Active Protection for Complete Family Health; Lifebuoy Active Orange—Active Protection for Complete Family Health; Lifebuoy Active Gold—Active Protection for Complete Family Health—with soothing Milk Cream; Lifebuoy Active Green—For Complete Family Health—with nature's Tulsi and Neem. The Active Green bar is the first Lifebuoy bar with natural ingredients, unlike the traditional health related bars.

Pears (2004–2008)

A new line extension was added to the Classic Pears Transparent Soap: a green colored Pears Transparent Oil-Clear Bathing Bar. The Classic version is recommended for dry skin, and because it is "Pure and Gentle", it is fine for babies. The new special Oil-Clear formula, "helps to gently clean excess oil on the surface while retaining the essential oils and moisture on the skin." Presently all Pears soaps are produced only in India. Pears is offered now in these variants: the traditional amber colored Classic, a green Oil Control with lemon flew extracts, and a blue Germ Shield with mint Extract .

Ivory (2004–2008)

The best proof of the longevity of this classic brand is that the latest Ivory package has the original story: "The Soap That Floats—99-44/100% Pure—Simple Naturally Clean."

The first major change to the original white "99-44/100% Pure–It Floats" Ivory in 125 years was in the 2004 introduction of a floating, green colored Ivory Aloe Rediscover the Mildness bar. A purple Ivory Lavender Gentle Clean Soothing Scent Bar for Baby Smooth Skin was launched in 2006. In 2007, the bars were repackaged and renamed as Simply Ivory Lavender and Simply Ivory Aloe, and a Simply Ivory Classic was added. All the three types are white.

The History of Soaps and Detergents ● 59

Discontinued Soaps

Many bar soaps have not passed beyond the original test markets, and others were discontinued after their market shares descended below sustainable levels. A listing of these soaps by the five major soap companies follows.

Colgate

Nature's Chlorophyll Green Palmolive (1952); Aromatic Balsam Essence (1958); Spree (1960); Choice, Beauty Bars for Normal, Dry and Oily Skin (1960); Cleopatra Beauty Soap with five fragrant oils (1963); Spree Deodorant Bar (1965); Petal Deodorant Beauty Bar (1966); Skin Mist Complexion Bar (1969); Cadum (1971); Dédoril Deodorant Soap (1971); DP-300 Antibacterial Beauty Soap (1971); Irish Spring Sunshine Yellow (1980); Softsoap Bar (1980); Experience Beauty Bars, Creamy Milk Bath Essence, Rosewater, & Glycerine Essence (1980); Dermassage Moisture Bar for Dry Skin with Protein (1980); Cleopatra Beauty Soap with five Fragrance Oils (1984), Hypo-Allergenic Palmolive (1989); Palmolive Essential—Hydrating Cleanser—Sensitive Skin (1995).

Dial

Glad—Soap/Synthetic (Combo) Bar (1958); Princess Dial—Superfatted Bar (1958); Soaprize—The floating fun soap with the prize inside—Alvin Alligator, Sylvester Sub, and Willie Whale (1965); Nutrelle Face and Body Bar with Vitamin E (1986); Spirit—Refreshing Deodorant Bar—Blue and White Striped Bar—Liquid scent burst with extra freshness (1990); Spirit—Three Soaps in One—Cleanses, moisturizes, deodorant protection—Solid blue color (1991).

Jergens

Nature Scents—Wild Flower, Herbal & Lavender (1975); Gentle Touch—Bath Bar with Baby Oil—Marbleized (1977); Duo-Care—Deodorant Soap with Moisturizers (1980); Fiesta Refreshing Deodorant Soap—Marbleized (1982); Jergens Clear Complexion Bar—Transparent Medicated Bar for Problem Skin (1972).

Lever

Olva Supercreamed Soap (1925); Swan—Pure White Baby Soap (1941); Praise 3 soaps in 1—Anti-Blemish, Deodorant ,and Cleansing Cream (1959); Pine Green Lifebuoy Deodorant Soap with Puralin Plus (1961); Lux with Beauty Bands—a unique bar with longitudinal bands across the two sides of the soap, two on the front and one centered across the back. The bands we were a different color than the rest of the soap (1962); Phase III—Deodorant Beauty Bar with Cream (1966).

Procter & Gamble

Dawn—Synthetic Floating Deodorant Beauty Bar (1958); Blossom—Facial Soap—Pampers your Skin—Will not leave an unsightly bathtub ring (1961); Velvet Skin (1944–1947 and test marketed again in 1963); Monchel with Moisturizing Glycerelle (1982–1988); Zest Free (1987–1988); Safeguard DS Dry Skin Protection (1988–1990). Old Spice High Endurance Deodorant Soaps (2002-2005)

Handcrafted Artisan and Specialty Soaps

During the last two decades, handcrafted artisan soap making has grown from a hobby into small businesses. This soap category grew, and "The Handcrafted Soap Makers Guild, Inc" (HSMG) was formed in 1998. The Guild is an international non-profit professional trade association with the mission, "to promote the handcrafted soap industry; to act as a center of communication among soap makers;

and to circulate information beneficial to soap makers". The membership, which now includes soap makers and vendors from the United States, and 14 other countries, reached 800 in 2008. The Guild's website, http://www.soapguild.com, provides extensive resource material and details on the HSMG Annual Conferences.

Specialty gift soaps, luxury fragranced soaps, personal care soaps, and medicated soaps are offered by Bath and Body Works, Body Shop, Caswell Massey, Crabtree & Evelyn, Estee Lauder, Johnson and Johnson, L'Occitane, and other firms.

Laundry Washing Products

Laundry soaps were used in all households until the introduction of synthetic detergents. There were key factors that led to the replacement of soaps by detergents. The inability of soap to clean cloth adequately over a range of temperatures, and the formation of lime soap curd in hard water were two important drawbacks. There was a shortage of fats and oils for soap production before and after World War II. Synthetic detergents made laundering much easier and simpler, and less time consuming. New automatic washing machines started replacing the old wringer-type washers and washboards. In the early 1950s, 3% of homes had automatic washers; by 1960 it was greater than 30%, and in 2000, it reached 77%.

Laundry Bar Soaps

Laundry soaps in bar form were used since the turn of the 19th century in the United States and Europe, and they are still widely used in many developing countries as the sole washing agent or in combination with powdered detergents. There are various types of laundry bar soaps, e.g., soaps made of fats and oils with and without fillers, combination soap/synthetic, and all synthetic types. Chapter 6 covers in detail laundry products in bar form. Some of the best known laundry bars were mentioned earlier in the company histories. Table 1.2 summarizes those already mentioned together with many other popular brands of the past. The date of introduction for many laundry bars could not be found.

Laundry Soap Powders

As washing machines were introduced in developing countries, soap powders in flake, chip, granule, and bead form were introduced. Table 1.3 lists the main brands, their manufacturer, and the product's introductory date.

Powder Laundry Detergent

The early history of the search for synthetic detergents goes back to 1831. Edmond Frémy, a French chemist, reacted sulfuric acid with olive oil. When the resulting thick brown liquid was diluted in water and neutralized with caustic soda, it had a soapy appearance: it foamed and removed some grease from greasy objects. Frémy had actually produced a crude form of soap. It was many decades later before further work was done.

Persil (1907)

In 1907, Henkel & Cie of Düsseldorf, Germany introduced Persil, the world's first "self-acting" washing powder. The product consisted of soap with perborate as the bleaching agent, and silicate (waterglass). The Persil name is derived from perborate and silicate. In 1909, the English firm J. Crosfield & Son, Ltd., a competitor of Lever Brothers, purchased the Persil name. In 1919, Lever acquired Crosfield and together they promoted the sale of Persil in the UK.

Today, Persil is a major brand for Henkel worldwide, except in the UK, France, and the Netherlands where Lever Brothers Limited (now Unilever) acquired the rights to sell Persil. Henkel's Persil advertising presented the public with beautiful images of young ladies, and young girls washing clothes. The most famous is the *Lady in White* painted in 1922 by the German artist Kurt Heiligenstaedt. Later, there were other "ladies in white," all projecting the Persil image of cleanliness. Persil continues to be a widely used powder detergent in Europe and elsewhere, but not in the United States.

Fewa (1932)

Henkel's Fewa, introduced in 1932, claimed to be the world's first synthetic detergent for fine fabrics. The inventor, Bruno Wolf, believed that it was a completely new product, but nobody else did. The original package indicates that, "it is alkali free." Fewa's advertising used a figure called "The Wool Man", which appeared on the first package, and later was replaced by the "The Cleaning Fewa-Johanna" figurine used until the 1950s.

Fewa, introduced in 1932, claimed to be the world's first synthetic detergent for fine fabrics.

Dreft (1933)

Dreft, originally called "Drift", America's first synthetic detergent, was introduced two years after Robert Duncan, a Procter & Gamble researcher, visited two firms in Germany. At the I.G. Farben Research Laboratories they showed him a "wetting agent" sold to the textile industry that let dye solutions penetrate fibers uniformly. After Farben chemists learned that a small textile firm used a product made from cattle bile instead of soap, they reproduced it synthetically and sold it in a product called Igepon. At Deutsche Hydrierwerke (purchased by Henkel in 1932), Duncan was shown a product under development, which was to compete against Igepon. He had 100 kg of this paste-type product, called a "surface-active agent" (surfactant), sent to Cincinnati for immediate testing. After quickly obtaining licensing agreements in 1932, and a great deal of work, Dreft was test marketed in the United States in 1933. Dreft did eliminate the hard water soap curd problem, but it worked well only for lightly soiled clothes. Since it became a detergent in 1958, Dreft had been targeted for laundering baby clothes; its positioning in the market has remained the same (Fig. 1.47).

Dreft, the first U.S. powder detergent, was introduced in 1933.

Tide (1946)

It took more than 10 years after Dreft to find the right "builder" to help a surfactant to do a "heavy-duty job" on all fabrics. In the early 1940s, complex phosphates were discovered; among them was sodium tripolyphosphate, a key chemical for formulating an all-purpose powder detergent. Procter & Gamble applied for a patent in 1944, but restrictions on raw materials delayed the introduction of Tide until 1946, when it was introduced as the "New Washing Miracle Tide—Oceans of Suds—Whiter Clothes—Sparkling Dishes" (Fig. 1.48). The "Tide's In, Dirt's Out" slogan was used on radio and TV against P&G's own Chipso, Duz, and Oxydol, and competitive soap powders and flakes.

Tide is America's best selling powder detergent.

The success of Tide was immediate; by 1949, it became America's best selling laundry detergent, and it has been number one ever since. In 1996, for its 50th anniversary, P&G estimated that of ≈35 billion wash loads, 100 million tons of clothes, done each year in the United States, one third are washed with Tide.

There have been many innovations and changes for Tide powder and liquid detergents. In 2008, over 30 different varieties were offered. A timeline of the most important ones follows:

- Tide Powder Detergent (1946) - The first heavy-duty powder detergent with phosphate builder introduced. Full distribution achieved in 1949
- Tide XK (1968) - First U.S. powder detergents formulated with XK enzyme that removes protein stains. (XK meant Xtra Kleaning)
- Liquid Tide (1984) - Non-Concentrated version
- Unscented Tide (1986) – Contains masking fragrance
- Tide Powder with Bleach (1988) - First with color-safe bleach
- Ultra Powder (1990) - Compact powder detergent
- Liquid Tide with Bleach Alternative (1991) - Non-Concentrated version
- Ultra Liquid (1992) - More concentrated than original
- Tide Powder (1992) - Phosphates replaced by zeolites
- Tide Free Liquid (1992) - Free of dyes and perfumes, replaced Unscented Tide
- Ultra 2 Powder (1996)
- Tide HE (High Efficiency) (1997)
- Liquid Tide Clean Rinse (1998) - Clean Rinse technology with a polymer that "captures dirt in the rinse cycle."
- Tide Rapid Action Tablets (2000) - First Tide in tablet form; discontinued in 2004.
- Tide Clean Breeze (2002) – New scent added to powder and liquid
- Tide with a Touch of Downy Liquid and Powder (2004)
- Tide Coldwater (2005)
- Tide with Febreze (2005)
- 2X Ultra Tide Coldwater Liquid (2007)
- 2X Ultra Tide Liquid and Powder (2007)
- 2X Ultra Tide with Febreze Liquid and Freshness Powder (2007)
- Tide Total Care Liquid (2008)
- 2X Ultra Tide with Dawn Stain Scrubbers (2008)
- 2X Ultra Concentrated Tide with Bleach Alternative (2008)
- Tide Liquid Total Effects (2008)
- 2X Ultra Tide Pure Essentials Liquid with Baking Soda in White Lilac Scent (2008)
- 2X Ultra Tide Pure Essentials Liquid with Citrus Extract and Lemo Verbena Scent (2008)

Detergent powders did not change much until the revolutionary, first, ultracompact, powder detergent Attack was introduced in 1987 in Japan by the Kao Corporation.

In 1989, new compact powders appeared in Europe and the United States. These new products were welcomed and by 1995 in the U.S. powder detergents had 55% of the market of which 90% were compacts. Powders have been losing market share against liquids and are projected to be only 10% of the total detergent market by 2010.

Table 1.2. Laundry Bar Soaps

Brand	Company	Year
Sunlight	Lever Brothers Co.	1884
White Star Soap	N.K. Fairbank & Co.	1884
Mascot	N.K. Fairbank & Co.	
Sunny Monday	N.K. Fairbank & Co.	
Chicago Family Soap	N.K. Fairbank & Co.	
Santa Claus	N.K. Fairbank & Co.	
Lenox Soap	Procter & Gamble Co.	1884
P and G White Naptha Soap	Procter & Gamble Co.	1905
OK Laundry Bar	Procter & Gamble Co	1930
P and G White Laundry Soap	Procter & Gamble Co.	
Luna	Procter & Gamble Co.	
Blue Barrel	Procter & Gamble Co.	
Blue Ribbon	Procter & Gamble Co.	
Star Soap	Procter & Gamble Co.	
Armour Family Soap	Armour & Co.	1888
Lighthouse White Naptha	Armour & Co.	
Big Ben Laundry Soap	Armour & Co.	
Hammer	Armour & Co.	
American Family	J.S. Kirk & Co.	1890
White Russian	J.S. Kirk & Co.	1892
Kirk's Flake White Laundry Soap	J.S. Kirk & Co.	
Fels Naptha	Fels & Co.	1894
Swift's Pride Soap	Swift & Co.	1895
Cream Laundry Bar Soap	Swift & Co.	1900
Classic White Laundry Soap	Swift & Co.	1910
Quick Naptha Soap	Swift & Co.	1914
T.N.T. Laundry Soap	Swift & Co.	1925
Octagon	Colgate-Palmolive Peet	1896
Crystal White Family Soap	Colgate-Palmolive Peet	1897
White Eagle Family Soap	Colgate-Palmolive Peet	
Galvanic	The Palmolive Company	
Good Will Soap	E. Morgan & Sons	
Big Jack Laundry Soap	Fitzpatrick Bros. Inc.	
Big Master Soap	Lautz Bros. & Co.	
Grandma's White Laundry	The Globe Soap Co.	
Tag The Rich Soap	Werk Soap Co.	
Omaha Family	Haskins Bros. & Co.	
Blue Barrel White Laundry Soap	Purex Corporation	
White King Laundry Soap	Los Angeles Soap Co.	

Table 1.3. Soap Powders

Brand	Company	Year
Oxydol	Waltke & Co.	1896
Octagon	Colgate	1898
American Family	J.S. Kirk	1898
Lux Flakes	Lever	1905
Persil	Henkel	1907
Persil	Lever	1918
Ivory Flakes	P&G	1919
Rinso Granulated Soap	Lever	1919
Fab	Colgate	1921
Chipso	P&G	1921
Fels Naptha Golden Soap Chips and Soap Granules	Fels & Co.	1922
Silver Dust	Lever	1925
Oxydol	P&G	1927
Super Suds	Colgate	1927
Duz	P&G	1929
Ivory Snow	P&G	1930
Palmolive Beads	Colgate	1930
American Family	P&G	1930
Concentrated SuperSuds	Colgate	1939
Chiffon Soap Flakes	Armour & Co.	1942
Perk Granulated Soap	Armour & Co.	1945

Table 1.4. U.S. Powder Detergents

Brand	Company	Year	Features and Claims
Dreft	P&G	1933	First synthetic detergent in the U.S.
Vel	Colgate	1939	
Breeze	Lever	1947	
Sterox (later called All)	Monsanto	1947	First low-sudsing detergent for front-loading machines
Fab	Colgate	1948	Name is derived from Fabric
Surf	Lever	1949	With Solium the Sunlight ingredient
Rinso	Lever	1950	
Surf	Lever	1952	
Cheer	P&G	1952	Blueing
Oxydol	P&G	1952	Changed from soap to detergent
American Family	P&G	1952	Changed from soap to detergent
Felso	Fels & Co.	1952	
Super Suds	Colgate	1953	With White and Blue Lighting Granules
Omo	Lever	1954	
Dash	P&G	1954	Concentrated low suds
Duz	P&G	1956	Changed from soap to detergent
American Family	P&G	1956	Speckled
All	Lever	1958	
Ajax	Colgate	1963	
Cold Power	Colgate	1966	For cold water laundering
Bold	P&G	1965	
Gain	P&G	1966	
Cold Power	Colgate	1968	Heavy Duty with Enzymes
Drive	Lever	1969	With Enzymes
Arm & Hammer	Church & Dwight	1970	First U.S. Phosphate-Free Detergent
Purex	Purex	1971	Phosphate-Free Detergent
Trend	Purex	1975	
Era	P&G	1978	
Fresh Start	Colgate	1978	Concentrated Detergent in a Plastic Bottle
Surf	Lever	1983	Detergent for removing dirt and odors
Wisk	Lever	1986	Discontinued January 2004
Wisk Powerscoop	Lever	1989	First U.S. Superconcentrate
Bold, Cheer, Dash, Gain Dreft Oxydol converted to Ultra versions	P&G	1990-1991	
Ultra Purex	Dial	1991	
Fab and Ajax Ultra	Colgate	1991	
All, Surf, and Wisk	Lever	1991	Concentrated Ultra versions
Surf Micro	Lever	1992	
Persil Micro	Lever	1992	
Ultra Purex	Dial	1993	
Ultra Ivory Snow	P&G	1993	Changed from soap to detergent
Ultra Tide	P&G	1994	
Bounce	P&G	1998	
Fab Sensitive Skin	Colgate	1997	
Cheer	P&G	1999	With Fabric Protecting Liquifiber
Purex Advanced	Dial	2000	Double Action
Arm & Hammer	Church & Dwight	2007	Plus the Power of OxiClean
Purex Ultra Concentrate	Dial	2007	Natural Elements
Purex with Rezunit	Dial	2007	
Pures with Color Safe Bleach	Dial	2007	
2X Ultra All	Unilever	2007	

Liquid Laundry Detergents

The first liquid detergents contained only a limited amount of phosphates; therefore, they lacked the cleaning power of powdered detergents and also cost more. But customers liked the convenience of using liquids. Beginning in the 1970s, liquids were phosphate-free and environmentally friendly. In time, the major manufacturers improved the performance of their products.

Lever's Wisk, introduced in 1956, was the first liquid detergent in the U.S. and remained the market leader for three decades. P&G's first entry into liquids was in 1973 with Era, but its big success came with Liquid Tide in 1984. It is documented that 400,000 hours of research and development, and $30 million were spent on Liquid Tide.

Colgate entered the market with Dynamo in 1974.

Concentrated liquid products followed concentrated powders. The first one appeared in Europe in 1991, and a year later in the United States.

In 1995, liquids had 45% in the U.S., of which 75% consisted of the concentrated versions. Liquids kept gaining market share at the expense of powders.

By 2000, liquids reached a 50% market share, and in 2008, over 85%.

A Wal-Mart and Unilever partnership started the development of concentrated liquid detergents. Unilever, upon the request of Wal-Mart, introduced in February 2006 the three-times concentrated 3X All Small and Mighty in a 32 oz bottle which can wash the same number of loads than the standard All in a 100 oz bottle.

In September 2007, Wal-Mart announced that by May 2008, all Wal-Mart Stores will only sell concentrated liquids detergents, and started working with other suppliers asking them to offer their own concentrated products. In a short time many new products came onto the market. Wal-Mart sells about 25% of all the liquid detergents sold in the U.S.

The benefits for the consumer and the environmental impact of the new 2X and 3X products are very significant: less water use for the new formulas; smaller bottles require less plastic resin; are easier to handle, store, and transport; and require less packaging materials. Fuel costs and landfill waste are reduced.

Unit Dosing Laundry Products

In the mid- and late 1980s, a number of "unit dose" products were introduced, including Fab One Shot, Tide Multi-Action sheets, Cheer, Clorox, and others. Cheer and Clorox offered unit doses in PVA film. By the early 1990s, all of these products were withdrawn from the market.

Tablets comprise the other unit dose type of product. The first detergent in tablet form was Salvo by P&G followed by Lever's Vim, both introduced in the 1960s. Salvo was withdrawn from the market in the 1970s. Tablets were successful in Europe and they were tried again in the United States. Salvo came back in 2000 and others followed: P&G's Tide Rapid Action Tablet (2000), Tide with Bleach (2001), Quick Dissolve (2002), Lever's Wisk and Surf Tablets (2001).

In spite of their convenient use, U.S. households have not accepted tablets, which reached only a 2.5% market share in 2002; for this reason, all those listed were discontinued in January 2004. Tablets are best suited for the European front-loading washing machines. In the United States, most of the washing machines are top-loading, and users like to set the detergent and water levels to suit the load size. Due to the current emphasis on the green trend (sustainability) in the cleaning market, detergent tablets might return again

Table 1.5. U.S. Liquid Detergents

Brand	Company	Year	Features and claims
Wisk	Lever	1956	First Heavy-Duty All-Purpose Liquid Detergent
Biz	P&G	1956	Heavy-Duty Liquid Detergent
All	Lever	1960	First Low-Sudsing Liquid Detergent
News	Purex	1960	
Cold Water All	Lever	1963	
Dawn	P&G	1972	
Era	P&G	1973	
Dynamo	Colgate	1974	Concentrated Heavy-Dut
Purex	Purex	1975	Concentrated Phosphate Free
Trend	Purex	1977	
Solo	P&G	1979	
Tide	P&G	1984	Non-concentrated version
Bold	P&G	1985	
Surf	Lever	1985	
Cheer	P&G	1986	
Dreft	P&G	1989	
Ivory Snow	P&G	1989	
Purex	Dial	1990	All Temperature
All, Surf and Wisk	Lever	1991	Double Power versions
Tide with Bleach	P&G	1991	Non-concentrated version
Tide Ultra	P&G	1992	More concentrated than the original
Tide Free	P&G	1992	Free of dyes and perfumes
Gain	P&G	1993	
Ultra Power Fab, Ajax, and Dynamo	Colgate	1993	
Fab Sensitive Skin	Colgate	1997	Dye and perfume free
Tide Clean Rinse	P&G	1998	Clean Rinse technology to suspend soil in the wash water
Downy Care	P&G	1998	
All	Lever	1999	The Stain Lifter
Surf	Unilever	1999	With Active Oxygen
Purex Advanced	Dial	2000	Double Action
Wisk Sport	Lever	2002	Tackles Tough Grass Stains and Dirt
Wisk 3X Ultra Concentrate	Lever	2006	
Wisk Bleach Alternative	Unilever	2007	
3X All Small and Mighty	Unilever	2007	
Arm & Hammer Essentials	Church & Dwight	2007	With Baking Soda
Arm & Hammer	Church & Dwight	2007	Arm & Hammer Plus the Power of Oxiclean
Purex Ultraconcentrate	Dial	2008	Natural Elements
Purex Baby	Dial	2008	
Ultra Gain with Baking Soda	P&G	2008	
Ultra Cheer with Bounce	P&G	2008	

Selling Soap

Sixty years ago a bar of soap sold to the trade for ≈4.5¢. Raw materials cost was 2¢, labor cost was 0.25¢, selling cost was 1.25¢, and overhead and profit were 1¢. Compare today's costs and profits with those six decade old figures. Soaps and detergents were always highly competitive products to sell, and today they still are. In the past, many sales methods and tools were used, some of which are still used in our electronic/internet age.

In addition to extensive magazine advertising, soaps were promoted beginning in the 1880s with a variety of selling tools. Billboards, enamel plaques, thermometers, store displays, trading cards (hand bills), booklets, streetcar (trolley) cards, school posters, coins, a large variety of premiums, and colorful advertising posters redeemable for soap wrappers were used extensively.

In 1897, the Peet Brothers used a new method to introduce the white laundry bar. Not having money to advertise it, they added one white bar of to each wooden shipping soap box containing 100 yellow laundry soap bars. As customers did not object, they put two in each box and as demands for the white bar grew, they increased the quantity in each crate and started to market it under the soap under the name of Crystal White—"The perfect Family Soap"—which was made from vegetable oils.

Piggly Wiggly, the first self-service grocery store, opened September 11, 1916, in downtown Memphis, TN. The founder, Clarence Saunders, had the idea to have one clerk offer quick service at low prices. By 1922, there were 1241 Piggly Wiggly stores in the country. A full-scale replica of the first store, with many of the soaps sold in the 1920s, is in the Pink Palace Museum of Arts and Industry in Memphis. Most soaps in the United States are sold in multipacks, up to 20 bars per pack. In some promotions two or more bars of the entire pack are offered free.

Magazine Advertising

Advertising records and illustrates the lifestyles, social customs, interests, dress styles, and even the problems of any given period in history. The sole purpose of an advertisement is to persuade the consumer to buy the product. But to persuade anyone to buy anything takes persuasive effort, a fact examined humorously in "Hints to Intending Advertisers", by Thomas Smith, London, 1885.

> *The first time a man looks at an advertisement, he does not see it.*
> *The second time he does not notice it.*
> *The third time he is conscious of its existence.*
> *The fourth time he faintly remembers having seen it before.*
> *The fifth time he reads it.*
> *The sixth time he turns up his nose at it.*
> *The seventh time he reads it through and says, "Oh brother!"*
> *The eighth time he says, "Here's that confounded thing again!"*
> *The ninth time he wonders if it amounts to anything.*
> *The tenth time he will ask his neighbor if he has tried it.*
> *The eleventh time he wonders how the advertiser makes it pay.*
> *The twelfth time he thinks it must be a good thing.*
> *The thirteenth time he thinks perhaps it might be worth something.*
> *The fourteenth time he remembers that he has wanted such a thing for a long time*
> *The fifteenth time he is tantalized because he cannot afford to buy it.*
> *The sixteenth time he thinks he will buy it some day.*
> *The seventeenth time he makes a memorandum of it.*
> *The eighteenth time he swears at his poverty.*
> *The nineteenth time he counts his money carefully.*
> *The twentieth time he sees it, he buys the article or instructs his wife to do so.*

Great advances were made in printing and publishing during the last two decades of the 19th century. Soap advertising was both beautiful and creative from its very beginning. Soap advertising in magazines began in the 1880s and was widely used from the start of the 20th century. Famous illustrators were commissioned to produce original work for soap advertisements in magazines. *Delineator, Woman's Home Companion, Ladies' Home Journal, Harper's, Saturday Evening Post,* and *Good Housekeeping* carried many soap advertisements, including the work of prominent illustrators: Maxfield Parrish, *The Dutch Boy* for *Colgate's Cashmere Bouquet*; Sir Arthur Rackham, a series of Cashmere Bouquet advertisements; Maud Humphrey (Humphrey Bogart's Mother), *The Busy Day* (Ivory Girl), P&G's first color print advertisement; Saida (Henrietta W. LeMair), FAB illustrations for Colgate.

Howard Pyle ("the father of American illustrators"), Jessie Wilcox Smith, Joseph C. Leyendecker, Andrew Loomis, Willie Pogany, Coles Phillips, Palmer Cox, N.C. Wyeth, Steven Dohanos, Sewell Collins, and Elizabeth S. Green all illustrated many soap advertisements—some even signed their names.

Artist-illustrated soap advertising was used until the early 1940s when advertising became more "photographic" rather than "graphic", and illustrations were phased out.

Soap advertising practically disappeared from the magazines in the late 1970s, but returned in the 1980s and had been growing to the mid 2000s, but has been diminished considerably since.

An example of a rather unusual full page advertisement was published in the December 11, 1886 issue of the *Youth Companion Magazine*:

Try The Frank Siddalls Soap.

An Eminent Divine says: "The Advancement of the World and the Spread of Civilization and Christianity depend on interchange of thought among people, and their willingness to learn, and that the man or woman who opposes the introduction of new improvements, the trial of new ways and the use of new things, should be condemned as not being good and useful members of society. Every word in this advertisement is the truth. Don't Be A Clam. Clams are not a proper model for human beings to copy after for they open their shells to take in their accustomed food, but they shut up very tight when anything new comes along for they are clams. Although it seems strange to use the same soap that is recommended for kitchen use, for toilet, shaving, etc. still, sensible people know that the world moves, and will be glad to try The Frank Siddalls Soap."

Another example is from an old, small soap company in California which had to be different to be competitive. They advertised Strykers soap as, "An Honest Confession from G. Stryker Suddsfaster, the Old Soapmaster: Strykers does not contain phoolium, hooeyum, hotairium, baloney um or any other mysterious ingredients you can't understand. It's just good soap."

The Soap Operas: Radio and TV

Radio Advertising and Soap Operas

In 1920, Station KDKA, in Pittsburgh, was the first American station to begin broadcasting. Its first broadcast on November 2, 1920, was the election of Warren G. Harding as President of the United States. In 1921, there were 393 radio stations, and two years later, there were 573. In 1924, radio station WEAF New York, WGY Schenectady, and KDKA Pittsburgh hooked up for the first commercial network radio programs in history. The first radio commercial was in 1922 on station WEAF in New York for the Hawthorne Court apartment complex in Jackson Heights, New York. The *Palmolive Show* for Colgate's Palmolive Soap started in 1927. It was a mid-week musical comedy program of classical and popular music. On August 19, 1929, over the stations of the NBC's Blue Network, the *Amos 'n' Andy* serial started for Lever's Pepsodent Tooth Paste and later for Lever soap products. The program was an instant success and became one of the longest running and popular comedy radio programs of all time. *Amos 'n' Andy* was brought to television in 1951 by Freeman Gosden and Charles Correll, the two men who created and started the radio show. The television series with an all-black cast lasted only two years on CBS because civil right groups complained about the show's racial stereotyping.

Amos 'n' Andy sponsored by Lifebuoy.

In 1930, the first Ivory radio show, *Mrs. Reilly*, was broadcast over 62 radio stations. Ivory advertisements were appearing regularly in 38 leading magazines. Ivory's success as an advertising pioneer was being emulated by other company brands. In 1930, Emily Post's *Etiquette Chats* and Helen Chase's *Beauty Forums* talked about Camay. There was also *George, the Lava Soap Man*.

In 1930, Lever's first radio show, *Peggy Winthrop*, was introduced and on October 14, 1934, the *Lux Radio Theater* took to the air. Market research found that housewives liked to be entertained by radio, and not instructed by it. The "Soap Opera" was born. The Colgate-Pamolive-Peet Company was the first firm to sponsor a serial for a soap product. On January 27, 1931, the first broadcast for Super Suds "Fast Dissolving" Soap Beads nightly program featuring three gossiping housewives Clara (Clara Roach), Lu (Lu Casey), 'n' Em (Emma Kueger) aired on NBC's Blue Network. December 4, 1933 is considered a historic day in network broadcasting. On NBC's Red Network, the first episode of Procter & Gamble's *Oxydol's Own Ma Perkins* program was presented. Ma Perkins was a self-reliant widow with business problems. Oxydol was a very popular granulated soap because it made laundering easier and faster.

In 1933 NBC's Red Network presented the first episode of Procter & Gamble's Oxydol's Own Ma Perkins.

Procter & Gamble began to dominate the radio and television serials, and because many of the serials were sponsored by soap products, somebody unknown coined the slang name "soap opera". In 1938, Procter & Gamble had 21 radio shows, using five hours per day of radio time, and spending $6 million, the largest amount for radio advertisement.

In 1934, Procter & Gamble's American Family Soap and American Family Flakes sponsored an NBC Red Network Chicago-based serial called *The Songs of the City*. During the commercial the announcer said, "It's cheaper to buy a new soap than to buy new clothes." This statement meant a lot to people trying to recover from the Great Depression. Each American Family wrapper had a coupon that could be redeemed at American Family outlet stores for various items. The American Family coupons became very popular and are still remembered by many people today. They were last redeemed in June 1971.

Lever Brothers spent $5 million in 1939 on radio for two daytime and four evening programs. This included the $1.4 million *Lux Radio Theater Shows*, with movie stars who were paid up to $5,000 for each appearance.

Radio serials were sponsored by Ivory Soap, such as, *The Road to Life*. In 1937, *The Guiding Light* was sponsored by Duz, a granulated soap, which used the selling slogan: "Duz does everything."

Television Advertising and Soap Operas

On April 30, 1939, President Franklin D. Roosevelt opened the New York World's Fair. NBC provided 3.5 hours of live coverage, thus starting the first regular television programming in the United States. Transatlantic Clippers opened regularly scheduled air service to Europe in 1939. Announcer Red Barber, between innings of a Brooklyn Dodgers Cincinnati Reds baseball game, "sold" an Ivory contest over a brand new experimental TV to the few New York "lookers". Ivory was pioneering again with a completely new form of visual advertising.

Procter & Gamble made a TV pilot of *Ma Perkins* in the late 1940s, but it was not well received. It was in 1951 when TV viewers finally liked a 15-min serial called *Search For Tomorrow*. In 1952, *Guiding Light* became the first radio serial to succeed on TV, and it has become the longest running program of any type or any medium. Many other "soap operas" became famous and some are still running, including *As The World Turns* and *Another World*. The *Colgate Theatre* debuted on network TV on January 3, 1949.

It was a half-hour show featuring short story adaptations from the *Ladies' Home Journal* and *Collier's* magazines.

Other Advertising Media

Trade Cards and Trading Cards

Trade cards were colorful advertising cards from the Victorian Era. They were chromolithographs that were given away free to promote a product, a business, or a service. The trade cards evolved from the so-called "handbills" or "shopkeeper bills" and "flyers". The merchants used the backs of handbills to tally the buyer's account. They were introduced in 1876 at the Philadelphia Centennial Exhibition and faded away completely by the 1904 St. Louis World's Fair. In spite of being short-lived, a large number and variety of soap-related trade cards were offered by soap companies. Many of these survived and are now sought after collectables. In the United States, trade card collector publications and clubs exist. England, Italy, and other countries also produce trade cards.

It is interesting to note another type of card that was very popular but was not used for soap promotions. Trading cards, different from the trade cards, were called "gum cards" before World War II. One or two cards were packed inside each package of bubble gum. They featured sports and other subjects. Baseball-related trading card collecting is still very popular in the United States.

Trolley Car Advertising Signs

The first horse drawn streetcar was introduced in New York in the year 1832. It was a half century later in 1873 that the horses were replaced by cable cars in San Francisco and in Chicago.

In 1888, the first electric streetcar was installed in Richmond, Virginia, and from then on many cities converted to the electric-powered streetcars—later renamed trolley cars.

Above the windows inside each trolley car, there was a metallic holder system designed for the 21 × 11 inch size cardboard advertising signs. Most of the soap manufacturers used trolley advertising signs for different soap and other products until the 1930s. Few of the colorful soap trolley signs survive and are now collectible items.

Trolly car sign advertising Ivory Cake or Flakes.

Soap Posters

The first posters created for advertisements appeared in France in the mid-19th century. Jules Chéret, called "the Father of the Poster", and many other famous artists created posters and elevated their status to an art form. Many soap posters were produced.

A few soap posters by well known artists are listed: the most famous soap poster is *Bubbles* by Sir John Millais (1886). *Chiozzi e Turchi Fabbrica di Saponi* by Adolph Hohenstein (1899); *Cosmydor Savon* by Jules Chéret (1891); *Savon le Chat* by J. Calao (1910); *Cadum Baby* by Arsene-Marie Le Feuvre (1912); *The Persil White Lady* by Kurt Heiligenstaedt (1922); *Cadum* by Georges Villa (1925*)*; *Palmolive* by Emilio Villa (1926); *Savon La Tour* by L. Capiello (1930); *Savon Fer a Cheval* by H. Le Monnier (1931); *Monsavon* by Charles Loupot and Jean Carlu (1940); *Monsavon au Lait* by Raymond Savignac (1949/1950).

Premiums

All of the important companies published very attractive, colorful premium catalogs for their main products starting in the early 1900s. Some catalogs included practical information on how to use the products. The premiums were toys, school supplies, sporting goods, cutlery, kitchen utensils, china, glassware, towels, linens, purses, toilet articles, leather goods, jewelry, silverware, watches, clocks, books, furniture, and many more items. The number of coupons to be sent for the premiums varied from 10 to 2000, depending on the value of the item.

Product Booklets

Beautiful, colored, freely distributed product booklets were widely used as promotional tools. Poems or fairy tales, such as "Little Red Riding Hood" or "Mother Goose", and topics such as American History, were the subject matter. "A Handbook for Mothers, How to Bring Up a Baby", a 40-page booklet, published by Proctor & Gamble in 1906 and written by Elizabeth Robinson Scovil, a nurse and author of several mother and child books, was given to expectant mothers. This nicely written, illustrated, and printed guide naturally included tips on how to use Ivory Soap Flakes and bars. It was reprinted until the mid-1920s. The foreword reads, "The little book is sent to you with the compliments of the manufacturers of Ivory Soap." Likely, their doctors gave the books to new mothers. A similar Procter booklet called "Bride" was distributed to brides.

Soap Sculpting

In 1924, Proctor & Gamble sponsored the first National Ivory Soap Sculpture Competition held at the Art Center in New York. There were 300 entries and cash prizes were awarded. The competition became an annual event for many years. The National Soap Sculpture Committee, with headquarters at 160 Fifth Avenue in New York City, was responsible for organizing an annual nationwide competition for small sculptures made of Ivory Soap. Monetary prizes were awarded in three categories: Advanced, for over 18 years old (only nonprofessionals were eligible to participate); Senior, from 14 to 18 years old; and Junior, for under 14 years of age. The winning entries were sent around the country to be shown in schools, libraries, art centers, and even museums. Soap sculpting became a fun educational event and an everyman's art activity. Due to rationing of soap during World War II, the annual event stopped from 1942 to 1947. It was restarted in 1949 and finally ended in 1961.

Ivory Stamp Club

Doug Storer died in 1986 at the age of 86. He created, operated, and sold the largest stamp promotion business in U.S. history. In 1933, he learned that Proctor & Gamble had $10,000 left over from an unspent advertising budget of the year before. He was given a free hand to use it. One day a stamp collector, Captain Tim Healy, an Irish World War I hero, came into Storer's office. Healy was a great storyteller. Storer was impressed and thought that kids would like the stories. He booked Healy on small radio stations in Hartford, CT and Worcester, MA. Healy went on the airwaves for 15 minutes twice a week. His radio show opened with, "Hello boys and girls, mothers and fathers, grandpas and grandmas, we are going to hear some fascinating stories today about stamps." He seized the interest of children and asked them to send for his stamp album—10¢ and two Ivory Soap wrappers. "Keep clean with Ivory Soap, collect stamps, and learn about history, geography and people," gained the support of parents and teachers and helped to sell more Ivory Soap. The kids asked friends and neighbors to buy Ivory Soap so they could get extra wrappers and, with them, buy more stamps.

Coupons

The first grocery coupon worth one cent was used for Grape Nuts Cereal by the C.W. Post Company in 1895. Since the 1930s, due to the Depression, households, to save money, started using coupons. "Couponing" grew and become a favorite pastime and hobby. By the 1960s, half of the American households used coupons. By the 1990s, besides the packaged food industries, others began using coupons. The Sunday edition of most newspapers have many inserts which include a large number of coupons for various products. In 2002, over 3.6 million coupons were redeemed with an estimated saving value of $3 billion.

Kirk's American Family laundry bar wrapper premium coupons started after the Chicago fire in 1871. The American Family Soap coupons, the most popular and still remembered by many, were last redeemed in June of 1971, 100 years after they were first offered. Kirk was also the first to use colorful trading cards as advertising and public relations tools. Baseball cards appeared later.

Coupons advertising American Family Flakes.

Slogans and Jingles

Some of the slogans, jingles, and product positioning statements were described before, and they are summarized along with others in the following list:

Ajax Laundry Detergent:	The White Knight—Stronger Than Dirt
Camay:	The soap of beautiful Women
Caress:	Before you dress, Caress
	Skin feels best when it's Caressed
Coast:	The Eye Opener
Dial Soap:	Aren't you glad you use Dial? Don't you wish everybody did?
	People who like people like Dial
Dove:	is one quarter moisturizing cream
Duz Granulated Soap:	Duz does everything
Jergens Body Bar:	For touchable soft, clean skin
Irish Spring Soap:	Feel Fresh and clean as a whistle.
Ivory Soap:	99 and 44/100% pure
Lever 2000:	Deodorizing and Moisturizing for all your 2000 parts
Lux:	The Soap of Hollywood Stars
Oil of Olay:	A lifetime of beautiful skin
Palmolive Soap:	Keep that schoolgirl complexion
Pears Soap:	Matchless for the complexion
Tide Detergent:	Tide in dirt out
Wisk Laundry Detergent:	Ring around the collar
Woodbury Soap:	A skin you love to touch

Zest:	Zestfully Clean
Rinso:	Happy Little Washday Song
	Rinso White or Rinso Blue? Soap or Detergent it's up to you
	Both wash whiter and brighter than new.
	The choice, lady, is up to you.

During the last decade, more new soap and detergent products and line extensions have been introduced worldwide than ever before. As customers have more choices, they demand better and also more economical products. Many challenges and opportunities lie ahead for everybody in the very active and very competitive soaps and detergents industry.

Albert Einstein did not agree that we need many types of products, for him one type of soap was enough. When asked why he used only one soap for bathing and shaving, he replied, "Two soaps? That is too complicated."

This journey down the soap and detergent memory lane ends with a beautiful picture entitled *Soap Bubbles* painted in 1733 by the French painter Jean-Siméon Chardin and with the lyrics of "I'm Forever Blowing Bubbles", a song composed and written in 1919 by Jaan Kenbrovin and John William Kellette.

Soap Bubbles by the French painter Jean Siméon Chardin—1734.

I am dreaming dreams/I am scheming schemes
I am building castles high/They're born anew/Their days are few
Just like a sweet butterfly/And as the daylight is dawning/They come again in the morning
I'm forever blowing bubbles/Pretty bubbles in the air
  They fly so high nearly reach the sky/ Then like my dreams they fade and die
Fortune's always hiding/I've looked everywhere
I'm forever blowing bubbles/Pretty bubbles in the air.

I'm Forever Blowing Bubbles, a very popular song from 1919.

Acknowledgments

Mr. Ed Rider, chief archivist of the Procter & Gamble Company helped me with hard-to-find historical and current information on soap and detergent-related subjects. I extend a very special note of gratitude for his assistance that extends over three decades.

I appreciate the information provided by The Colgate Palmolive Company, The Dial Corporation, Henkel GmbH, The Andrew Jergens Company, KAO Corporation, Unilever USA, Lever UK, and Lever Faberge UK.

Suggested Readings and References

100th Anniversary 1867–1967, Armour Magazine, Chicago, February 1967.

150 Years of Excellence Through Commitment and Innovation, Moonbeams, Procter & Gamble Co., August 1987.

150 Years of Procter & Gamble, Advertising Age, Chicago, August 20, 1987.

20th Century Advertising and the Economy of Abundance, Advertising Age, Chicago, April 30, 1980.

Bergwein, K., The Early History of Soap, Dragoco Report, Vol. 6 (1968).

Boys, C.V., Soap Bubbles, Their Colors and Forces Which Mold Them, Dover Publications Ltd., New York (1959).

Cashmere Bouquet, Soap & Chem. Specialties, May 1955.

Celebrating 100 Years—Ivory Soap, Moonbeams, Procter & Gamble Co., 1979.

Colgate Palmolive Company, Fortune, April 1936.

Colgate Palmolive Company, Historical Review, 1980.

Dunn, S.W., and A.M. Barban, Advertising, Its Role in Modern Marketing, Dryden Press, Hinsdale, Illinois, 1978.

Dyer, D, Dalzell F, Olegario, R, Rising Tide, Lessons from 165 years of Brand Building at Procter and Gamble, Harvard Business School Press, Boston, Mass (2004)

Encyclopedia of Advertising, edited by J. McDonough and K. Egolf, K, Taylor & Francis Group, London (2003).

Encyclopedia of Consumer Brands, edited by J. Jorgensen, Vol. 2, St. James Press, Detroit, (1994).

Fa Brand Manual, Fa a Mission of Freshness, Schwarzkopt & Henkel Cosmetics, 1997.

Gathmann, H.; American Soaps; Henry Gathmann: Chicago, (1893).

Goodrum, C., and H. Dalrymple, Advertising in America—The First 200 Years, H.N. Abrams, Inc., New York (1990).

Green, V.W., Cleanliness and the Health Revolution, Soap & Detergent Association, New York (1984).

Handley, J. Art Deco Landmark Getting New Life as Pricey Condos, Chicago Tribune, February 2, 2003.

James, B., P. Taylor, and M.A. Murray Pearce, The House of Yardley, Tillotsons Ltd., London, 1970.

Johnson, B., Dial: How a Number One Product Stays on Top, Product Marketing, February 1977.

Kiefer, D. M., The Tide Turns for Soap, Today's Chemists at Work, October 1996.

Klaw, S., How Armour Cleaned up with Dial, Fortune, May 1955.

Leblanc, R., Le Savon: De la Prehistoire au XXIeme Siecle, Editions Pierann (2001).

Lever Brothers Company, Fortune, April 1939.

Levey, M, The Oldest Soap in History, Fortune, December 1957.

Linard, J. E., Laundry Products: A Glance at the Past, a Look to the Future, Book of Papers International Conference and Exhibition, American Association of Textile Chemists and Colorists, 1997, pp. 133–140.

Luckman, Charles, Twice in a Lifetime, W.W. Norton & Co., New York (1988).

McCoy, M., Soaps and Detergents, Chem. Eng. News 82:23–28 (2004).

Memorable Years in History, Procter & Gamble Publication, 1982.

Moonbeam Special Edition, Procter & Gamble Co., 1954.

Opie, R., Packaging Source Book, Chartwell Book, Secaucus, NJ (1989).

Osteroth, D., Soap Through the Ages, Dragoco Report, April 1981.

Our First 150 Years, Procter & Gamble Publication, 1987.

Procter & Gamble, Fortune, December 1931, April 1939, March 1956.

Qureshi O, ed, Lux Inspiring Beauty, Bennet Coleman & Co. Ltd. Mumbai, India (2007)

Reader, W.J., Fifty Years of Unilever, William Heinemann Ltd., London, 1980.

Riggs, T., ed., Encyclopedia of Marketing, The Gale Group, Farmington, MI (2000).

Schisgall, O., Eyes on Tomorrow, J. G. Ferguson Publishing Co., Chicago, 1981.

Sivulka, Juliann, Stronger Than Dirt, Humanity Books, Amherst, New York (2001)

Soap A Monthly Magazine for Soapmakers; The Dorland Company, 1925, 1(1).

Soap History, Ciba Geigy Review, Basel, Switzerland, April 1947.

Soap Through the Ages, Unilever Educational Booklet, Unilever House, London, 1952.

Spitz, L., ed., Soap Technology for the 1990s, AOCS Press, Champaign, Illinois, 1990.

Strategies for the 21st Century, in Proceedings of the 4th World Conference on Detergents, edited by A. Cahn, AOCS Press, Champaign, Illinois, 1999.

Swasy, L., Soap Opera, Time Books, New York, 1993

The Armour Magazine; Chicago, March 1912, July 1914, July 1915.

The Moon and Stars, Procter & Gamble Co., 1963.

The Palmoliver; Colgate & Company Magazine, 1928, 10.

The Pulse; Colgate & Company Magazine, 1930, 2; 1931, 3.

The Shape of Things to Come the Next 20 Years in Advertising and Marketing, Advertising Age, Chicago, November 13, 1980.

Then, Now and Tomorrow, Inside the Archives, Moonbeams, Procter & Gamble Co., April 1989.

Toilet Soap, Consumer Reports, October 1942, September 1944, October 1948, April 1953, September 1957, March 1978, March 1981, January 1985.

Tom Branna – Spin Cycle, Happi, Vol. 46 No.1 (2009)

Tom Branna, A Rising Tide, Happi, Vol. 45 No. 1(2008)

Tom Branna, Clean Well, Happi, Vol. 44 No. 1 (2007)

Trumann, B.C.; The World's Fair - Columbian Exposition; Mammoth Publishing Co.: Chicago, (1893).

Weill, A., The Poster, G.K. Hall & Company, Boston, 1985.

Weinroff, L.A., The Sweet, Sweet Scent of Soap, Chicago Historical Society Magazine, Vol. VI, No. 1, Spring 1977.

What Soap, Consumer Union Reports, May 1936.

Williams, E., The Story of Sunlight, Unilever PLC, London, 1988.

Wilson, Charles, The History of Unilever, Vol. 1, Cassel & Company Ltd., London (1970).

Your Servant Soap, Association of American Soap & Glycerine Producers, New York, October 1941.

Implications of Soap Structure for Formulation and User Properties

Norman Hall
Continúa Consultancy Service, United Kingdom

1. Objectives

Sections 2–4 discuss how soaps of various chain lengths from an oil blend interact to create lather and some other user properties.

Processing influences the arrangement of the various soaps within the bar, and Sections 5–11 discuss how changes in formulation and processing can modify the bar's structure, and how this structure influences user properties, such as bar hardness, mush, and cracking.

2. Basic Chemistry

The fundamental principles of soap composition and performance are generic and apply to all soaps. The chemistry of soapmaking is very simple. Soaps are the salts of (mainly) saturated and unsaturated fatty acids having carbon number C10 to C18. The source of the fatty acids is always a blend of natural triglyceride oils. However, relatively few manufacturers make soap by neutralising a blend of fatty acids. Most create soap directly from the blend of oils.

To make soap directly from oil, a blend of glyceride oils is reacted with a strong sodium hydroxide solution to give the soap, plus glycerine—and a lot of heat. Separating the soap from the glycerine byproduct is not easy, and may not even be necessary.

Fats/ Oils   +   Caustic Soda   ▶   Soaps   +   Glycerine

$$R_1COOCH_2 \qquad\qquad\qquad R_1COONa \qquad CH_2OH$$
$$R_2COOCH \; + \; 3\,NaOH \;\; \blacktriangleright \;\; R_2COONa \; + \; CHOH$$
$$R_3COOCH_2 \qquad\qquad\qquad R_3COONa \qquad CH_2OH$$

The alternative is to split the triglyceride oil into fatty acids and glycerine using high temperatures and high pressures. In this case the fatty acids and glycerine can easily be separated. The separated fatty acids are normally distilled, blended, and then neutralised with NaOH solution to form soap.

Fats/ Oils   + Water   ▶   Fatty Acids   +   Glycerine

R_1COOCH_2
|
R_2COOCH + 3 H_2O ▶ R_1COOH CH_2OH
| R_2COOH + $CHOH$
R_3COOCH_2 R_3COOH CH_2OH

Distillation

Fatty Acids + Caustic Soda   ▶   Soaps   +   Water

$R_{1,2,3}COOH$ + NaOH   ▶ $R_{1,2,3}COONa$ + H_2O

Whether it is better to make soap by classical saponification or by a fatty acids route involves economics, supply chain issues, raw material, and finished product qualities. The availability of equipment and of operators skilled in the relevant processes can be important issues to consider.

3. Glycerides

The alkyl chains R_1, R_2, and R_3 on the triglyceride molecule of the oil or fat include both saturated and unsaturated fatty acid types, of different carbon atom chain lengths (carbon numbers), for example as follows:

CH_2-O-CO-(CH_2)x-CH_3 (saturated)
|
CH-O-CO-(CH_2)y-CH=CH-$(CH2)$z-CH3   (unsaturated)
|
CH_2-O-CO-(CH_2)x-CH_3 (saturated)

This example shows a triglyceride with two saturated fatty acid chains (say, palmitic acid, C16) and one unsaturated fatty acid chain (say, oleic acid, C18:1). Several different combinations of saturation/unsaturation and chain length are possible on a given glycerine backbone, giving oils differing widely in their characteristics—especially their melting points. These differences are very important for edible applications, where the oils are used as triglycerides, but are much less important for soap applications, because in soapmaking all the fatty acid chains are separated from the glycerine backbone.

For soaps the only important factors are the relative proportions of saturated to unsaturated fatty acids (measured by the *iodine value*—the grams of iodine reacting with the unsaturated component in 100 g of oil or fat), and the lengths of the fatty acid chains (carbon number).

The proportions of saturated and unsaturated fatty acids and the chain lengths of those acids are a characteristic of the oil or oil blend used for making soap.

3.1 Why Use a Mixture of Oils?

A blend of oils is almost always used to make toilet soaps or laundry soaps. The most common oils are coconut oil (CNO) or palm kernel oil (PKO), which, understandably, are generally called nut

Implications of Soap Structure for Formulation and User Properties • 85

oils, and tallows (AT or T) or Palm Oils (PO), which are generally called non-nut oils. To get the best performance from soaps, you need both types of oils, that is, the non-nut oils and the nut oils.

3.1.1 The Non-nut Oils (Tallows or Palm Oils)
These oils provide long-chain-length saturated fatty acids (C16/C18—palmitic and stearic acids), which give soaps that are almost insoluble at normal user temperatures and so do not lather. To put this insolubility into context, consider that calcite is more soluble in water at 25°C than is sodium stearate. However, these almost insoluble soaps add to lather stability and add hardness, which makes the soap solid.

Lather stability is to a large extent governed by the rate at which a liquid film drains under gravity from between the bubbles. When the liquid film becomes so thin at any point that the film sides touch, then the bubbles burst. Any mechanism that slows the rate of liquid drainage will increase lather stability.

When the liquid film contains insoluble particles, these can sometimes "bridge" the film sides and cause premature lather instability. This can happen with very large particles of filler materials. However, the particles of insoluble soaps are very small, and they are asymmetric. There are two consequences of this asymmetry.

First, in the same way that sticks of wood flowing in a fast-moving river will align with their longest axis parallel to the river flow, so the insoluble, asymmetric particles of sodium stearate and sodium palmitate will align so that their long axis is parallel to the flow of liquid draining from between the bubbles of lather. The narrow axis of the stearate/palmitate particles is very small, so it is unlikely that the particles will bridge the film sides until most of the liquid has drained.

Second, and most important, at the junctions between the air bubbles there is a region of much slower liquid flow. This is the Gibbs-Plateau border. In this region, motion of the particles of sodium stearate/palmitate become much more random, like wooden sticks floating randomly in a slow-moving pond of water.

Just as sticks floating in a slow pond can cause a logjam that inhibits the water flow, so the particles of sodium stearate/palmitate can collect in the the Gibbs-Plateau border to an extent that they will block the flow of liquid draining from the lather. This increases the time it takes for the bubble film sides to touch and for the bubbles to burst—and so increases lather stability

Tallow or palm oils also provide long-chain-length unsaturated fatty acids (mainly C18:1), and the chain length gives soaps with reasonable solubility but only moderate lather stability and rather poor lather volume. However, the moderate lather stability of sodium oleate from tallow or palm oils is improved significantly by the lather-stabilising effect of the insoluble sodium stearate/palmitate soaps provided by the same oils. This means tallow soap or palm oil soap alone can give a reasonable amount of lather and have good lather stability, provided the use temperature is high enough (over 25–30°C) to dissolve the sodium oleate.

3.1.2 The Nut Oils
These oils provide short-chain-length fatty acids (especially C12, lauric acid) that give soaps of moderate solubility, but with high lather when they are dissolved. Solubility can be increased by using the soap at a higher temperature, but sodium laurate does not dissolve significantly until the temperature is over 40°C. However, another important mechanism helps sodium laurate to dissolve at a much lower temperature.

4. The Importance of the Oleate:Laurate Eutectic Mixture

It is a common but mistaken idea that nut oils make soap lather because the C12 soap has a high lather. It is true that C12 soap has a higher lather than C18 soap (which effectively does not lather at all) and a higher lather than C18:1 soap. However, the lather of a soap containing both tallow/palm and nut oil is much higher than that of C12 soap alone. The reason for this is fundamental to many aspects of the performance of soap tablets.

When a system contains both C12 and C18:1 fatty acid soaps together, the mixture has a solubility much higher than the solubilities of either of the individual components. For example, a 1:1 mixture of C12 sodium soap and C18:1 sodium soap is very soluble and has a very high lather. This mixture will also make the soap softer. Consider the Krafft temperatures (T_k) of

$$\text{Sodium C12 soap (sodium laurate)} = 42 \text{ °C}$$
$$\text{Sodium C18:1 soap (sodium oleate)} = 28 \text{ °C}$$

The *Krafft temperature* is the temperature, or more precisely the narrow temperature range, above which the solubility of a surfactant increases rapidly. At this temperature the solubility becomes equal to the critical micelle concentration. When micelles form in the solution, the detergent can dissolve in the micelles more easily than in the water, and the solubility increases rapidly.

However, the Krafft temperature of the sodium oleate/sodium laurate 1:1 mixture is below 0 °C, which means it is very soluble even in cold wash water, and so lathers quickly and gives a good amount of lather.

Solubility of the Eutectic Mixture from C12 and C18:1 Soaps

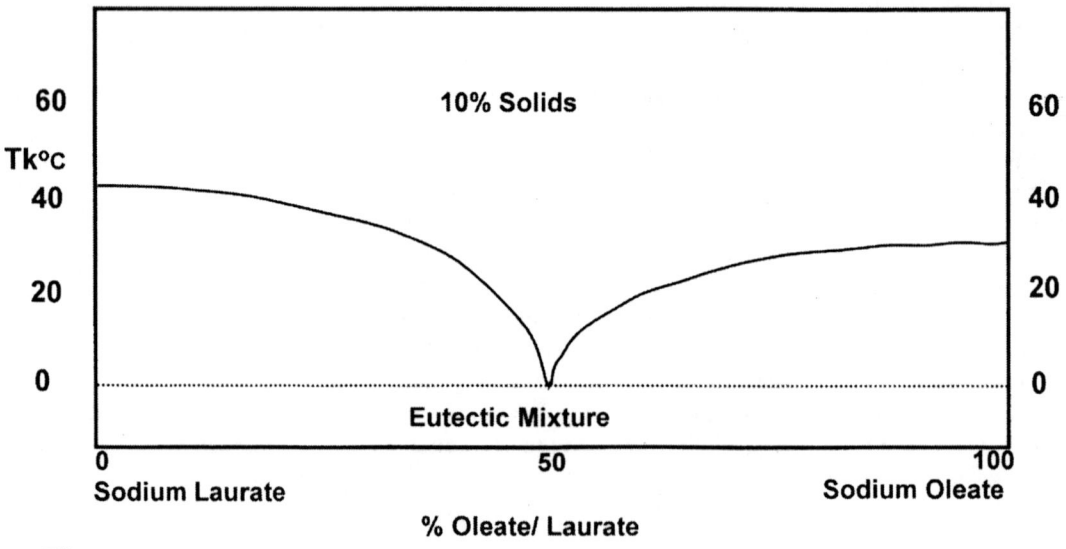

Tk = Krafft Temperature

A note on eutectics:

> *A eutectic property is a material property that exhibits very differently in a mixture than it does in any of the components of the mixture.*
>
> *Eutectic mixtures usually arise when the molecules of two (or more) of the materials have totally different shapes or sizes. Then the molecules of the mixture cannot pack together neatly. In the example mixture considered here, C12 saturated soaps have a straight chain of carbon molecules, whereas C18:1 unsaturated soaps usually have a "U" shaped (cis) carbon chain. Each*

molecule "sabotages" the other so that the mixture cannot crystallise normally, and the abnormal crystals have higher solubility.

In (almost) all soap tablets there will be more soap derived from C18:1 fatty acids than from C12 fatty acids, that is, most soap tablets contain more sodium oleate than sodium laurate. It therefore follows that as the C12 content is increased—as the soap contains more nut oil—so the amount of 1:1 oleate:laurate mixture will increase, and the lather from that eutectic mixture will increase.

There is also the potential for soap softness to increase with increasing nut oil level, which, however, is not observed. The explanation of why this could happen, but does not to any significant extent, can be explained on the basis of the first of two models, discussed next, relating soap structure and properties.

5. Soap Structure–Performance Models

Soap structure models are helpful because they provide ways to more easily visualise the links between formulation, processing characteristics, and user properties. There are two useful models of soap structure.

Model 1 is a very simple macro model based on solid-soap-to-liquid-soap phase ratios. This model is used to explain many aspects of processing characteristics.

Model 2 is a molecular model that considers the crystallisation changes occurring during soap drying and subsequent processing, and can be used to explain aspects of the user properties of soaps.

5.1 Model 1—The Macro Model

Soap bar hardness at constant moisture content and electrolyte content is a function of the balance between the following components:

a. Solid (insoluble) soaps—mainly Na C16 and C18 saturated fatty acid soaps (sodium palmitate/stearate), but with some content of Na C12 saturated fatty acid soap (sodium laurate) when this has not had an opportunity to form the 1:1 eutectic mixture with Na C18:1 unsaturated fatty acid soaps (sodium oleate).

b. Soluble soaps dissolved in the free-water content of the soap bar. These are mainly the very soluble 1:1 oleate:laurate eutectic soaps, plus some free Na C18:1 soap (sodium oleate), plus electrolytes, glycerine, and so on.

The very simple macro model
Soap hardness at constant moisture content is
a function of the balance between :

Solid / insoluble soaps	:	Softer/ soluble soaps
Na C16/ 18 and any undissolved C12		C12 / C18:1 eutectic + excess C18:1/18:2 in the water of the bar

| solid phase | : | liquid crystal phase in the bar water |

more solid/less liquid = harder
more liquid/ less solid = softer

The simple "bar chart" model of soap structure and hardness can be very useful to explain many formulation and processing effects, and will be referred to several times in what follows. This model simply shows a schematic balance between the solid soaps and the dissolved soaps. In real life the "liquid phase" will be an up to about 25% solution of dissolved soaps, and such a high-concentration soap solution will have a solution liquid crystal structure. This means that although the soaps are dissolved, the soap molecules are arranged into large domains of micelles, and these have a regular packing structure. In most soaps, that structure will be an hexagonal close-packed arrangement of cylindrical or rod micelle domains.

Although it has a regular structure, the liquid phase remains mobile and fluid. Therefore, the more liquid phase there is in soap, relative to solid phase, the softer the soap will be.

5.1.1 Predictions of Effects from Formulation/Process Changes
Some of these effects will be intuitively obvious, but try to keep in mind how the model *explains* them.

a. Higher temperature
At a higher temperature, more soap will dissolve from the solid phase into the liquid phase. The liquid phase volume will increase in size relative to the solid phase, and the soap should be softer.

b. Higher water content in the soap
There are two effects of this change. First, the extra water will directly increase the liquid phase volume. Second, the liquid phase will (potentially) be able to dissolve more soap, giving a further increase in liquid phase volume. Both actions will lead to a softer soap, so a relatively small change in soap water content (say 1–2%) can have a big influence on product softness.

c. More electrolyte in the soap
By common ion and ionic strength effects, electrolytes (usually sodium salts of stronger acids) will salt-out soaps from solution in the liquid phase. The result will be less liquid phase volume and more solid phase, and a harder soap.

The hardening effect of electrolyte can be large. For example, an increase from 0.5% NaCl in soap to 1% NaCl will almost double the soap hardness from typically 2×10^5 N/m^2 to 3.8×10^5 N/m^2. The effect is large because all of that electrolyte will be dissolved in only the free water of the soap, which is typically 10% water. The overall 0.5% electrolyte is therefore present as a $0.5 \times 100/10 = 5\%$ solution, and the 1% electrolyte as a 10% solution. That is a big difference.

Always think of soluble soap component concentration effects in this way.

d. More nut oil in the soap oil blend
Clearly, if a higher percentage of nut oil in a soap gives more lather because it allows the formation of more 1:1 oleate:laurate eutectic, then this same mechanism should mean that a higher nut oil content will give a softer soap.

If all else is equal, then this is true. However, all else is not normally equal. During manufacture, soap is washed with a strong solution of NaCl to remove the glycerine byproduct. As the soap's nut oil content increases, a higher concentration of NaCl solution is needed to separate soap from the glycerine lye, and the residual soap will always contain more of the NaCl. This will increase the soap's hardness.

The softening effect of the eutectic and the hardening effect of the electrolyte just about balance out, and in practice increasing nut oil content up to about 40% in a blend with 60% tallow or palm oil does not give a softer or harder soap. Above 40% nut oil, the tallow or palm oil no longer provides enough C18:1 to form more of the 1:1 eutectic. The additional C12 therefore stays as solid phase, and the soap will quickly get very hard as the nut oil content increases beyond 40%.

e. Free fatty acids

To improve their performance, some toilet soaps can be made to contain significant amounts, typically more than 5%, of free fatty acids (superfatting). Such free fatty acids can be part of the solid phase or solubilised in the liquid phase. In the liquid phase they will pack within the micelles, and, because they do not have a strongly polar head group, they will dilute the charge density at the micelle surfaces. The result is that the usual cylindrical or rod micelles can form a different, more stable structure—a lamellar structure. A lamellar liquid phase structure can inherently "hold" more dissolved soap and is much less viscous than a hexagonal liquid phase structure. The result is a larger liquid phase volume, and certainly a lower-viscosity liquid phase—and a softer soap.

f. Glycerine

Glycerine is very soluble in water and so will always locate in the liquid phase. The first few percent of glycerine added to, or left in, soap will simply increase the size of the liquid phase (slightly) and will give a slightly softer soap. Glycerine levels of up to about 1% in a toilet soap, and perhaps 2% in a laundry soap, can have this effect.

At higher levels of glycerine, say 6%, such as would be present if you did not remove any of the residual glycerine from soapmaking, the effective concentration of glycerine in the liquid phase becomes very high. It can be close to 40% in toilet soap:

$$6\% \text{ glycerine in } 10\% \text{ free water} = (6/(10+6)) \times 100 = 37.5\%$$

Soap will not dissolve in strong solutions of glycerine, and if glycerine is added or increased to levels over about 2%, some dissolved soaps will be displaced from the liquid phase and become solid. In effect, a high level of glycerine acts like an increased level of electrolyte.

g. Minerals added to soaps

At moderate levels (say 5%, perhaps higher), any minerals will simply be dispersed evenly between both the solid and liquid phases and have little influence on soap hardness.

At higher levels (say over 10%) minerals can disrupt the packing of the soap crystal domains which are the solid phase. Such disruption can lead to the domains being able to "slide over each other" more easily, giving the impression of a softer soap. However, the softness is due to weakness in the solid phase rather than to any change in the solid-to-liquid ratio.

h. Perfume

Perfume acts like free fatty acids—it quickly softens soap, probably because some of the polar ketone and aldehyde components quickly migrate into the liquid crystal phase. As more perfume components migrate into the liquid phase in the hours/days after soap manufacture, the liquid crystal phase can probably also change from the viscous hexagonal structure to the more fluid lamellar structure. This is an area in need of research.

5.2 Model 2—The Molecular Model

This model considers the crystallisation changes that occur during soap drying and subsequent processing and can be used to explain aspects of the user properties of soaps.

The mixture of oils from which soaps are made have a wide range of saturated and unsaturated fatty acid chain lengths, from below C8 to greater than C20. For the purposes of this model, only the three major chain length groups will be considered.

- Sodium palmitate/stearate (NaP/St), C16/C18
- Sodium oleate (NaOL), C18:1
- Sodium laurate (NaL), C12

Soapmaking by normal alkali saponification, washing, and fitting produces a liquid neat soap phase with a 70% solution of soap at 100 °C minimum. This neat soap has no formal crystal structure, or at least no solid crystal structure. To make toilet soap noodles, the neat soap is preheated to typically 135–145 °C and is sprayed into a vacuum chamber, where it loses water to become 85% soap with around 12% water (and some residual electrolyte, glycerine, and so on). During the vacuum-spraying process the soap cools and solidifies very, very quickly (in 0.5 sec) from, say, 140°C to below 50°C. This sort of very rapid cooling and solidification does not give enough time for optimum crystallisation.

Instead, the soaps of all the different chain lengths crystallise into a metastable solid phase. Only some of the most soluble chain length soaps will remain in the residual water of the soap—sodium oleate (NaOL) and, in much lesser amounts, linoleate, and short-chain saturates such as C6 and C8.

Although it does not have the most stable arrangement/packing of the mixed-chain-length soaps, the solid phase produced at the drier does always have a well-defined way in which all the long carbon chains of the fatty acid parts of the molecules pack together. The details of this structural arrangement can be seen using X-ray diffraction techniques.

The structure of the solid phase formed immediately after drying is called the *kappa phase* (or the *omega phase* by some authors). Remember that the kappa phase will contain all the soap chain lengths:

Kappa phase = Na P/St, NaL, and even some NaOL.

Remember from the preceding that the best lather is obtained from Na laurate (NaL), but only when it is dissolved by forming the 1:1 eutectic mixture with NaOL. In the solid kappa phase resulting directly from the drying stage, almost all the NaL is "locked up," and so soap directly from the drier has relatively poor lather.

The question is how to free the NaL from the solid phase so that it can form the eutectic mixture with the NaOL dissolved in the liquid phase? The answer is to provide "activation energy" in the form of mechanical work energy. Applied work will allow the metastable kappa phase to release some NaL into the liquid phase, where it will form the thermodynamically more stable eutectic with NaOL. As soon as some eutectic forms in the liquid phase, the liquid phase contains more soap and is a better oleophilic solvent, and so it will more easily dissolve even more NaL from the solid phase. This process is self-perpetuating so long as the work energy continues to be applied.

In the limit all the NaL could move out of the solid kappa phase into the liquid phase. As this movement of NaL progresses, it leaves "holes" in the kappa phase structure and makes it weak. As more and more NaL moves from the kappa phase to the liquid phase, eventually the kappa phase structure starts to collapse and then recrystallise.

Soap is not only mixed/ milled and refined in order to evenly distribute perfumes, colour etc.

It is mixed/ milled and refined in order to change the crystal phase structure and to give soap with better user properties.

5.2.1 Implications of the Molecular Model
The nature of the phase to which the kappa phase will recrystallise depends on the type of soap. In particular, it depends on the solvent power of the liquid phase, that is, the water content and whether it contains any solubilised free fatty acids.

If the water content of a 90/10 to 70/30 type soap is at or above 15% (preferably 18%), then with enough supplied work energy almost all of the sodium laurate will move from the solid kappa phase into the liquid phase. If enough NaL moves from the solid to the liquid phase to cause restructuring of the solid phase, then the solid phase will crystallise to what is called the *zeta phase* (or the *beta phase* by some authors).

Remember that the zeta phase will now contain only NaP/St soaps because all or most of the NaL has moved into the liquid phase. A characteristic of the zeta phase is that the crystal size is very small. This means there is less scattering (diffraction/reflection) of incident light, and so soaps with high zeta phase content are often translucent.

Scattering is also minimised when the refractive index (RI) between two surfaces is at a minimum, here at the boundary between the solid phase and the liquid phase. Glycerine has a high RI (1.5), close to that of soap, and is very soluble in the water of the liquid phase. A high glycerine content will therefore increase the RI of the liquid phase to be closer to that of the solid phase, and will always enhance the optical effect of soap translucency.

It is relatively easy to get toilet soaps to form some zeta phase solid by applying mechanical work energy. A higher water content (at least 16%, rather than the usual 13%) will help the movement of NaL from the solid to liquid phases and, with enough work energy, can result in almost complete conversion from the kappa phase to the zeta phase.

High zeta phase soaps will always have better lather properties, because the liquid phase will contain a much higher proportion of the high-lathering NaOL:NaL eutectic mixture.

5.3 Formulation Limitations for the Molecular Model

5.3.1 Nut Oil Content

If a soap has less than 10% nut oil (equivalent to about 5% sodium laurate), then even when all the NaL moves from the solid phase to the liquid phase, this movement will not distort the kappa phase enough to cause any major recrystallisation to the zeta phase.

If the soap contains more than 30–35% nut oil, then even when the liquid phase contains as much NaL as is possible, there will still be enough NaL remaining in the kappa phase to keep that phase stable.

5.3.2 Water Content

Most toilet soap has about 20% nut oil and about 13% water content, and only trace amounts of free fatty acids. Even if the maximum realistic amount of work energy is applied to such a soap formulation, only part of the NaL will move from the solid kappa phase into the liquid phase. There is simply not enough solvent power in 13% water to move enough of the NaL to cause the kappa phase to fully recrystallise to the zeta phase.

Normal soaps therefore contain only a limited amount of zeta phase and are not translucent.

5.3.3 Fatty Acids

Soap with free fatty acid (FFA) content at more than 5% of the total fatty acid (TFA) content are often called *superfatted* soaps. The presence of such FFA can significantly modify the soap structure and user properties. Because the presence of FFA imposes significant processing constraints, superfatted soaps are less common today than they were in the period from 1970–1990. Most toilet soaps now contain zero or only fractional amounts of FFA.

If the soap being worked contains >5% free fatty acids, then both the solid phase and the liquid phase structures will change. The liquid phase structure will contain a lot of solubilised free fatty acids, and, as explained earlier, this will cause the liquid phase structure to become lamellar rather than hexagonal.

Under the influence of work energy (mixing or milling), the movement of NaL from the solid phase to the liquid phase is exactly the same as for non-superfatted soap, and so the kappa phase becomes more and more NaP/St-only. However, when this structure collapses, it forms a phase with crystals much larger than those in the kappa phase (rather than much smaller, as is the case with the zeta phase). These larger crystals are long and ribbon-like and highly intertwined. This structure is called the delta phase.

Delta phase can only be formed in superfatted soap if all work energy is input at temperatures ideally below 38°C, and certainly below 40°C. If these ribbon-like delta phase crystals can be formed in the soap due to work energy input, then there can be significant advantages for the final soap performance. In particular, the mush can be reduced.

All soap mush is delta phase, whether it forms from a soap in which the solid phase is in kappa or zeta form, or is already in delta form. During soap mush formation, water penetrates into the soap bar. If the bar solid is kappa phase (NaP/St+NaL), then during the relatively long soap/water contact time needed to form mush, the NaL will dissolve out into the surrounding water, leaving the residual NaP/St to recrystallise into the delta phase in the presence of a lot of water. That water becomes trapped amongst the long, ribbon-like crystals of the delta phase and gives a highly mushing product.

The net result is that kappa phase soaps inherently have high mush.

If the solid phase is zeta phase, then there is no NaL present to dissolve out into the surrounding water, but eventually, by Oswald ripening, the small zeta phase crystals will grow into the long, ribbon-like crystals of the delta phase, and will still trap the penetrated water. This growth probably takes a longer time to create the delta phase than does the NaL-loss-induced recrystallisation by which it forms from the kappa phase. Therefore, all else being equal, a zeta phase soap should take longer to give the impression that it is a highly mushing product. But all is normally not equal. To encourage the initial formation of the zeta phase, a soap will normally have a higher water content, and, if it is to be a translucent soap, probably also a high glycerol content. Higher water or glycerol content both result in much greater water penetration, and this more than compensates for the slightly longer mushing time expected from the zeta-to-delta phase transformation.

The net result is that highly worked soaps can form mush more slowly, but usually other formulation characteristics that promote significantly increased mush are also present.

However, if the solid phase is already delta phase because the soap contained free fatty acids and was worked at temperatures below 38°C, then there will be less mush than from soaps with a kappa or zeta solid phase. The preformed delta phase does not trap the penetrated water in the same way that it traps water when the delta phase forms due to mushing.

6. *Soap/Water Interactions*

So far emphasis has been on how formulation and processing influences soap structure. Now it is appropriate to consider how soap structure influences bar/water interactions, and how these in turn affect bar properties.

There are numerous examples of how aspects of the formulation of a soap bar influence its user performance properties. For example, higher water content usually gives a softer soap that will absorb more water and generate more mush as it stands, wet, on the wash basin. That same soap then usually shows more cracking during use. Similarly, a soap bar formulated with a high percentage of nut oil in the oil blend will, through formation of more of the oleate:laurate eutectic mixture, generally lather more easily and create more lather during use. Whether that extra lather is appreciated or desired is another issue that depends on wash habits and the hardness and temperature of the available water.

Earlier sections have shown how the processing of the soap bar influences its user properties. For example, again considering soap with higher water content, supplying a significant amount of extra work energy (through intensive mixing or milling) will make the product very hard and impart significant translucency, by causing a kappa-to-zeta phase change.

A general relationship can be shown diagrammatically as follows:

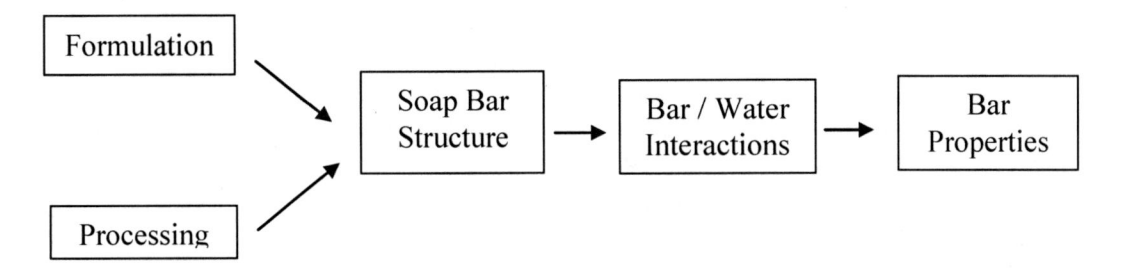

```
Formulation ──┐
              ├──▶ Soap Bar Structure ──▶ Bar / Water Interactions ──▶ Bar Properties
Processing ───┘
```

That is, there is a direct link between formulation/processing and soap properties, with soap structure and bar/water interactions as intermediate steps. A key element of these relationships is how soap/water interactions influence the bar's user properties.

7. Mush and Cracking

Soap mush and cracking are two of the most visible negative properties that can result from soap/water interaction. Soap mush and cracking effects are related.

The relationships between the two effects can be seen by making some assumptions about soap cracking:

- The fundamental cause of cracking is stress applied to inherent weaknesses in the structure of plodded soap bars.
- This weakness in the structure arises when processing makes solid soap crystals align in specific ways.
- The types of soap crystals formed in a soap bar can be influenced by the formulation.
- The stress on the soap structure that leads to cracking occurs when soap swells due to water uptake by the liquid phase during mushing, and shrinks when the mush dries.
- The amount of mush, and therefore the resultant stress, are influenced by both formulation and processing.

These propositions imply that to understand the causes of cracking requires an understanding of

- how soap structure is influenced by processing and formulation, and
- how water interacts with soap to form mush.

8. The Source of Structure Weakness

8.1 The Fundamentals of Structure

The fundamentals of soap structure were described earlier in Section 5.2, but repeating the key points.

Rapid cooling of soap in a drier gives a product with most of the sodium laurate and even some sodium oleate locked into a relatively insoluble solid kappa phase, along with sodium palmitate and sodium stearate. Work energy during subsequent processing will provide the activation energy needed to cause most of the laurate soap (C12) and some of the oleate soap (C18:1) trapped in the solid kappa phase to move into the liquid phase. In the liquid phase, which already contains some of the more soluble sodium oleate, the oleate and laurate will form a 1:1 eutectic solution with greatly increased solubility. The increased solubility and amount of soap present will be sufficient to allow the soluble soaps in the free water to form into micelles. The micelles will pack into either hexagonal close-packed or lamellar liquid crystal structures.

Soap is therefore a mixture of two components:

1. Asymmetric solid-phase crystals, predominantly of insoluble sodium palmitate and sodium stearate but probably including some residual sodium laurate. These crystallise into relatively large asymmetric crystals or domains of crystals.
2. A liquid phase consisting of water containing soluble soaps, electrolyte, glycerol, free fatty acids, perfume, and so on.

In most soap tablets, the liquid phase will mainly be a liquid crystal, rather than an isotropic solution. The liquid crystal forms when the solution contains enough soap to form large micelles composed of very soluble soaps, such as the oleate:laurate eutectic mixture, and of very soluble, very short-chain-length soaps, such as C6 to C10 types. The liquid crystal phase may also contain free fatty acids, perfume, and other compounds, solubilised in the micelles.

Most 80/20 non-nut oil/nut oil type soap tablets will contain about 80 parts solid phase and 20 parts overall liquid phase, with a hexagonally structured liquid phase that occupies twice the free water content. The liquid phase of soap containing high levels of free fatty acids can have lamellar structure and occupy about 3 times the free water content.

8.2 Pressure Effects during Plodding

Neither the solid phase nor the liquid phase is continuous throughout a final soap bar, but individual, microscopic areas of the bar (domains) may have continuous solid or liquid phase. Importantly, if you "squeeze" soap hard enough, some of the liquid phase will separate from the solid phase. The pressures experienced during soap plodding are typically 60 kg/cm^2 and are high enough to give some separation of this type. This is especially true if the liquid phase contains some low-viscosity isotropic solution as well as the usual high-viscosity hexagonal liquid crystal phase. The liquid phase that separates will normally also contain some glycerol and electrolyte residual from soapmaking.

The separating liquid phase has two effects:

1. It coats the solid-phase crystals, making it much more difficult for the plodder to make them coalesce into a homogeneous bar. Thus, too high a plodding pressure can increase soap cracking.
2. Solid soap crystals are relatively hydrophobic, and water will not penetrate through them. So when soap is immersed in water, the water will first enter the soap via any physical channels between crystals. These physical channels are regions between coalesced soap noodles. If these regions contain liquid phase material that has been squeezed away from the solid phase during plodding, then the liquid phase will swell, because it absorbs water to try to dissolve. The swelling will further reduce cohesion between the soap noodles.

A corollary is that if you add water to soap noodles/pellets after drying, then it may dissolve only a small amount of soluble soap and remain as isotropic solution mixed with the solid phase and the liquid crystal phase. This isotropic solution has low viscosity relative to the much more concentrated and structured liquid crystal phase, and therefore is more likely to be squeezed out by plodder pressure and is more likely to increase cracking. Thus, adding water at the mixer can increase cracking.

8.3 Crystal Orientation during Plodding

Recall that if you float sticks of wood in a river, the sticks will turn to have their long axes parallel to the direction of the water flow. The domains of solid-phase crystals in a soap bar are asymmetric, like the sticks, and as the soap mass moves during plodding many of the asymmetric soap crystals will similarly align themselves parallel to the direction of soap movement.

Maximum alignment occurs when there is shear applied to the moving soap mass. In particular, this will occur when part of the moving soap mass is next to a relatively stationary surface, for example:

- a part of the plodder, such as the barrel wall,
- the surface of the plodder screw, which often moves at a different rate from the soap mass, and
- other areas of the soap mass that are not moving at the same time or at the same rate.

During extrusion with a screw-action plodder there are many opportunities to generate such shear and to cause large-scale crystal alignment.

Consider the first case above, shear between the soap mass and the plodder barrel. Although there already may be some alignment of crystals in the pellets that are fed to the plodder (because of shear during screw extrusion from the drier), this will largely be offset by the random mixing of pellets as they feed into the plodder. Once the pellets are compressed, they will shear against the barrel wall and also against the screw surface.

This shear will introduce an order to the crystals in the soap mass such that, at the point of entering the plodder cone area from the screw, the soap crystals are typically perpendicular to the main longitudinal axis of the plodder. The appendix shows details of why this is so.

Now remember that at any point in time soap flows into the cone from only a small section at the very end of the plodder screw. Soap does not flow uniformly from the screw across the whole input face to the cone. Once the cone of a plodder has become packed with soap, most of the soap will be moving forward at a much slower rate than that entering the cone (under pressure) from that small section at the end of the screw. Therefore, in the cone there will be significant shear between faster-moving and slower-moving areas of soap. That shear causes reorientation of the soap crystals. The crystals that entered the cone oriented perpendicular to the longitudinal axis of the plodder (and therefore perpendicular to the soap flow) will reorient themselves to become parallel to the longitudinal axis of the plodder and to the soap flow. This reorientation is likely responsible for much of what is often called "work softening" of soap in a plodder and significantly influences the direction and extent of subsequent water penetration and cracking during product use.

It is sometimes easier to visualise this reorientation and subsequent flow in the plodder cone by considering a plodder fitted with a multihole pressure plate between the screw and the cone areas. Generally such a pressure plate is not recommended because it will cause a temperature increase. However, imagining this configuration sometimes makes it easier to visualise how some candles of soap entering the cone will move faster than others and how the shear between them will encourage crystal orientation to be parallel to the direction of soap flow.

The rods or candles of soap formed by a pressure plate will deform as they travel through the cone. They will retain their individual candle structure, but they will become thinner. They become thinner because soap accelerates as it moves through the cone. Soap entering a 300 mm diameter cone at 1 cm/s (2.5 tons/h) must leave at 25 cm/s through a typically 60 mm final orifice plate aperture. As circular candles of plastic soap are forced together by pressure in the cone, they adopt a hexagonal packing.

When the soap flow in the cone reaches the orifice plate there will be some distortion of the soap crystal patterns as soap builds up behind the plate. Then, as the soap flows around the edges of the orifice, there will be more shear and another reorientation of soap crystals such that the compressed candles of soap in the extruded bar will again have crystals oriented parallel to the direction of soap flow.

8.4 Visualising Soap Structures in the Plodder

An alcohol immersion procedure allows visualisation of crystal orientation during soap flow through a plodder cone, in the extruded bar, cut slug, and final stamped tablet.

The steps of the procedure are as follows:

1. Cut a smooth surface on the soap section (ideally with a microtome or carpenter's plane).
2. Immerse the cut surface in deionised water at 20 °C for an hour. Water penetrates into the soap via the liquid phase surrounding any solid phase crystals and causes the liquid phase to swell.
3. The wet soap surface is then soaked in a solution of at minimum 95% alcohol for an hour, during which the penetrating water and associated liquid phase is rapidly removed from the soap.
4. Finally the soap surface is allowed to dry naturally at ambient temperature.

Typically, after about 1 hour of drying, it is possible to see a pattern that corresponds to the areas of solid phase and liquid phase in the original soap.

The alcohol immersion procedure can be applied to the surfaces of sections of soap cut from the soap block removed from a plodder cone. This shows that the individual candle structure is retained through the entire cone.

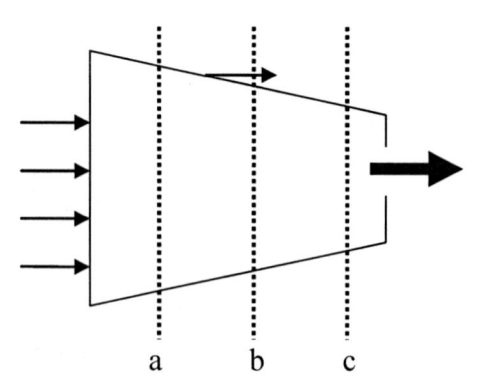

a   b   c

If you take slices cut from the plodder cone at the points a, b, and c, and then perform the alcohol immersion procedure on one side of each slice, the pattern of the original candles is seen to survive more or less unchanged through the cone. All that happens is that the candles become slightly smaller in diameter and more hexagonal in profile as they come closer to the orifice plate and the cone pressure increases.

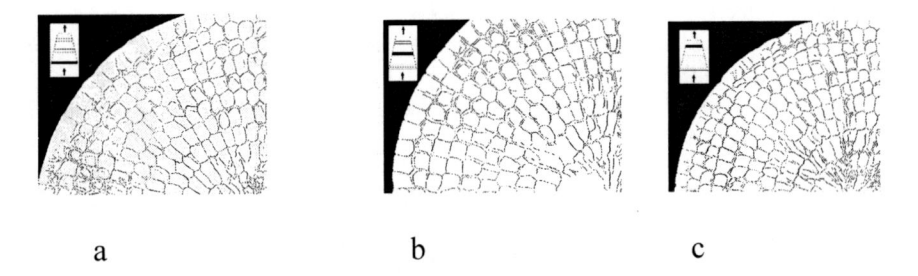

Implications of Soap Structure for Formulation and User Properties ● 97

a b c

If you take a block of soap from just behind the orifice plate of a plodder cone, and cut it in another direction, parallel to the direction of extrusion, then the alcohol immersion procedure shows a representation of the general orientation of solid-phase crystal domains parallel to the direction of soap flow, and a much more pronounced orientation in the same direction being introduced as the soap encounters significant shear at the orifice plate.

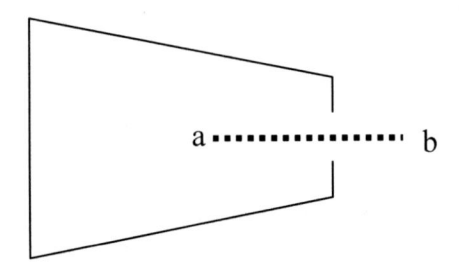

a ················· b

A section through a…b will show the flow line pattern made by orientation of the solid phase crystal domains.

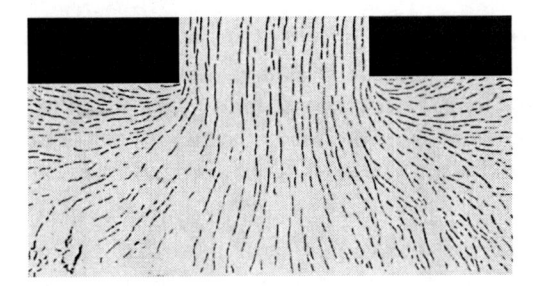

The greater parallel orientation in the final slug is quite obvious, as is the orienting effect of shear at the corner of the orifice plate.

8.5 The Overall Effects

In the cone and in the bar or slug that emerges from the orifice plate, there are compressed candles of soap with surfaces made of soap crystals oriented with the long axis parallel to the direction of extrusion. Surfaces with such parallel crystal orientation do not easily adhere together.

Consider an analogy with the mineral talc. Talc is smooth and slippery because the talc crystals are flat plates, and these become oriented in the same direction as the bulk mineral experiences shear when rubbed between layers of skin.

Because of pressure effects in the cone, these candles may have surfaces coated with relatively low-viscosity liquid-phase material containing soaps, electrolyte, and glycerol. This will significantly reduce the adhesion between soap candles.

Indeed, an electrolyte+glycerol solution is sometimes used as a lubricant to stop soap from sticking to soap dies during stamping.

Therefore, plodded soaps inherently have planes of structural weakness. The lines of weakness are parallel to the direction of extrusion, and will form whether or not you use a plodder pressure plate.

Another simple way to see these lines of weakness is to take a freshly plodded length of soap, grip each end, and "twist" the bar along its length with alternate clockwise and anticlockwise hand movements. The weaknesses in the structure will cause the bar to delaminate quite quickly into the remnants of the original candles.

9. Consequences of Structural Weakness

Soap does not crack until it is used in water and allowed to dry (unless you have done something very, very wrong in the formulation or process). It is not the water that makes soap crack—it is the behaviour of mush as it is formed when water penetrates into the soap bar, and especially as it dries. Therefore, to understand cracking we must understand mush.

9.1 What Is Mush?

Mush is the soft, paste-like layer that forms on soap when it is in contact with water. Typically it forms when soap is left on the side of the wash basin after use.

Mush consists of very long, twisted, insoluble soap crystals intermixed with a liquid, which may be an isotropic solution or in the liquid crystal phase, or both. Very close to the soap surface the liquid has a high soap content and is a liquid crystal, whereas further from the surface it is isotropic solution because there is not enough dissolved soap to have a defined structure. From the soap surface to the true solution, mush therefore has a variable composition and a somewhat variable structure, and it is dynamic—the water penetrates and the soap dissolves during all the time that they are in contact.

9.1.1 The Solid Phase of Mush

Most of the insoluble solid phase of mush is sodium palmitate and sodium stearate (C16 and C18 soaps). These soaps have the long, twisted crystal form (delta phase) because during the long water contact time of mushing, any residual, more soluble soaps (sodium laurate or even sodium oleate) in the bar's solid kappa phase will dissolve. The solid phase in most soap bars ,for example, the common 80/20 type, will be either:

1. Kappa phase (relatively large crystals containing all the soap chain lengths). This sort of solid phase is found only if soap has not been worked very hard by mixing, milling, and other processing steps.

2. A mixture of kappa phase and zeta phase. This phase mixture is found in most soaps. The zeta phase component consists of very small crystals of sodium palmitate and stearate, formed by recrystallisation of the kappa phase when work energy input encouraged the more soluble sodium laurate to move to the liquid phase.

During prolonged contact with water:

1. any kappa phase solid will lose laurate to the water, the phase structure becomes weakened, and the kappa phase will recrystallise to the large crystals of the delta phase, and

2. any zeta phase crystals may grow (via Oswald ripening), because this will decrease the area of the boundary between the solid and liquid phases: a few big crystals have a smaller surface area than

many small crystals. As they grow, the structure becomes better considered as in the delta phase rather than the zeta phase.

Remember, zeta phase and delta phase are chemically the same: primarily sodium stearate and sodium palmitate soaps. They differ only in molecular arrangement and crystal size.

To summarize: the solid phase of mush is large crystals of delta phase, and because the crystals are large they are always opaque, even when they come from energetically worked translucent soaps!

9.1.2 The Liquid Phase of Mush

Water penetrates into a bar of soap via the liquid phase. The liquid phase absorbs water by osmotic pressure, which tries to dilute the liquid phase. This causes the phase volume to increase and is observed as swelling of the wetted soap surface.

There are now two effects:

1. The water penetrating the soap dilutes the liquid phase. In the limit, it can dilute the liquid phase enough that an existing liquid crystal component no longer has a high enough concentration to remain as a liquid crystal, and reverts to a simple isotropic solution.

2. During the long soap/water contact time of mushing, some sodium laurate and sodium oleate will dissolve from the solid phase into the water. These dissolving species can now increase the concentration of dissolved soap and make it more likely that the liquid phase is a liquid crystal.

Which effect predominates depends on where you are in the mush layer.

Clearly, it is most likely near the soap surface that the penetrating water will contain a lot of dispersed solid phase and have a liquid crystal character. Further away from the surface, the less structure the liquid will have and the less dispersed solid phase it will contain, until eventually it becomes isotropic solution only, with no solid phase.

A pictorial model of the mushing process can be given as follows:

Initial state

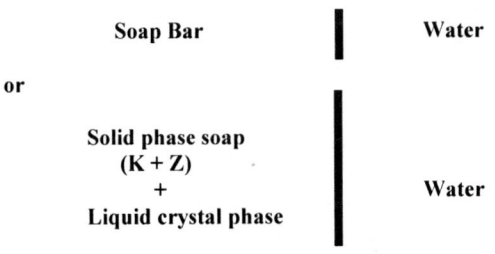

Soap Bar | Water

or

Solid phase soap
(K + Z)
+
Liquid crystal phase | Water

Initially there is a sharp boundary that separates the bar's solid and liquid crystal phases from the water with which the bar is in contact.

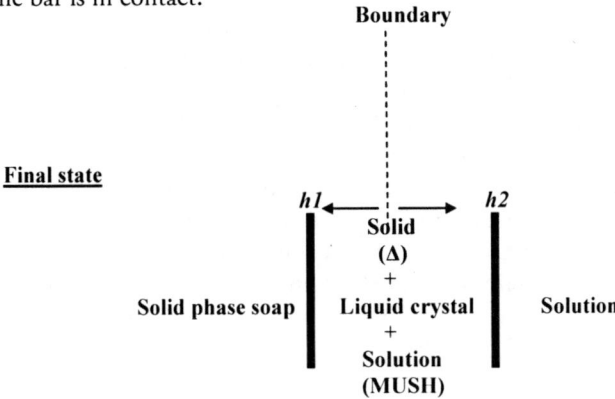

Boundary

Final state

$h1 \leftarrow\ \ \ \rightarrow h2$

Solid phase soap | Solid (Δ) + Liquid crystal + Solution (MUSH) | Solution

After a period of time water will have penetrated into the bar to a depth of $h1$. The water cannot penetrate via the solid soap phases. The water penetrates via the liquid crystal phase, so that the liquid phase now contains more water and has a bigger phase volume, that is, it will swell. The swelling action then pushes the soap-to-solution boundary forward to point $h2$. Some of the soap will have diffused completely out of the soap into the water to form true solution, so the outward movement to $h2$ is not as great as is theoretically possible.

The total thickness h of the mush layer is the sum of water penetration + liquid phase swelling, that is, $h = h1 + h2$.

A number of processes contribute to bar/water interactions. Water uptake by diffusion through the liquid phase. This can be impeded by the solid phase crystals co-mixed with the liquid phase and adding to the tortuosity of the diffusion route.

Swelling of the liquid crystalline phase

This depends on the structure of the liquid crystal phase (hexagonal or lamellar) and will be less if the soap in the liquid crystal phase is more soluble, in particular if work energy applied to the soap has enabled formation of more of the 1:1 oleate:laurate eutectic mixture.

Dissolution of the liquid crystalline phase

This also depends on the solubility of the soaps in the liquid crystal. Diffusion of surfactant micelles through the swollen layers and out into the water to form isotropic solution.

The water absorption and swelling of the liquid phase will eventually continue until all the soluble soap has dissolved and diffused away from the bar surface into true solution. However, that process is hindered by the presence of solid soap phase, which in general is in the form of large, ribbon-like delta phase crystals that can significantly impede diffusion.

Recrystallisation of the solid phase

This can change the morphology of the solid phase and therefore the structure of the mush layer, or it may release soluble soaps, which then move into the liquid phase to further influence dissolution and swelling of that phase.

9.2 Mush Is Delta Phase

That is, mush from any soap is *always* the long, ribbon-like crystals of sodium palmitate and sodium stearate. If delta phase is formed because of mushing, that is, by water penetration, then close to the bar surface the delta phase structure traps much of the penetrated water. The result is perceived to be a soap bar with high mush.

If the soap bar contains preformed delta phase before contact with water, then that implies that work energy has already persuaded a lot of the laurate trapped in the kappa phase at the drier to migrate into the liquid phase. In normal 80/20 soaps, this work-energy-induced migration causes the kappa phase to recrystallise to the small-crystal zeta phase. When the bar is subsequently in contact with water for a long time, the zeta phase *slowly* changes to delta phase through Oswald ripening. However, for some types of soaps (superfatted soaps = higher nut oil soaps with >5% free fatty acids) the recrystallisation induced by work energy input is from kappa phase to delta phase, rather than to zeta phase. Made under the correct process conditions (high shear at temperature <37°C), these soaps have a low perceived mush because the delta phase is preformed in the bar and no longer traps the penetrating water during mushing.

9.3 Stress

When a soap forms mush, it also swells because of the water uptake into the liquid phase. However, that water uptake alone is not enough to deform the soap structure and cause cracking. Cracking occurs when the mush dries. When a mushed soap dries, the swollen soap shrinks again. A contracting soap film adhering to a surface will put stress on that surface. When the stress is across a plane of weakness in the soap structure formed by the alignment of crystals in the soap mass during plodding, the result is cracking.

Initially the crack may not be big enough to be visible—call that a "micro-crack." However, the next time the soap is wet:

- Water will penetrate more deeply into that micro-crack.
- The mush formed in the micro-crack will take longer to dry out than mush on the normal surface of the bar.
- There will be an extra stress when the mush does dry out.

So the micro-crack eventually becomes a visible crack.

9.4 The Fundamentals of Water Penetration

Osmotic pressure difference is the fundamental cause for water penetration into soap. Obviously, water does not penetrate into soap via the solid phase crystal domains. These are solid and hydrophobic.

Water penetrates via the liquid phase. The liquid phase is a relatively concentrated solution of electrolyte, of glycerol, and especially of soluble soaps. The driving force for water to penetrate will remain so long as there is a concentration difference between the liquid phase and the surrounding water.

In theory, there will always be a concentration difference, because the surrounding water can be regarded as infinite in extent and containing no dissolved material. In practice, soaps normally form mush when they are left in contact with a relatively small amount of water in a soap dish or on the side of a wash basin. Eventually this surrounding water contains so much dissolved electrolyte, soap, and so on, that the rate of further water penetration is very low. In effect, all of that small amount of water has become part of the mush.

9.5 Formulation and Water Penetration

The types of soaps in the liquid phase will influence the water penetration. If the soaps are very soluble—for example, are present as the oleate:laurate eutectic mixture + very-short-chain types such as C6, C8 soaps—then much less water needs to be absorbed into the liquid phase to get those soaps into a low-viscosity isotropic phase where they can diffuse away from the bar.

Conversely, If the liquid phase contains relatively low levels of oleate:laurate but significantly more of the much less soluble free oleate, then much more water will need to be absorbed to form an isotropic solution.

No matter how much soap is present in mush, consumers perceive mush with more water as "high mush" and mush with less water as "low mush." However, that is not relevant to a story about cracking.

What is relevant is that, in both high and low mush cases, the mush contains the same amount of soap, just different amounts of water. Remember that the only mush seen by the consumer is mush on the soap bar, which is made up of both soap and water. Another way to think about this is that low mush soaps have more soluble mush that "disappears" into solution, and so it is not seen as mush.

Note: Strictly, the above statement applies to mush determined by the immersion test. In that test, the soap, cut to a defined size, is in contact with a nominally 'infinite' volume of water, so that there is always a concentration gradient and water always penetrates into the soap bar. The same statement will hold for soaps in contact with a limited amount of water, but the numerical data may be quite different.

Other major elements of the formulation also influence water penetration:

- Glycerol – At the levels normally found in soap, glycerol generally increases the osmotic driver for water penetration and so increases mush. However, note that at high levels of glycerol (>5%) there is an opposing effect: glycerol limits the solubility of soap in the liquid phase, effectively decreasing the overall phase volume. Apart from also giving a harder soap, the lower phase volume decreases the water penetration.
- Electrolyte – This has a very similar effect to glycerol, but at much lower levels, for example 0.5%, rather than 5%.

Remember that these solubility-limiting effects from glycerol and electrolyte are in addition to those from the effects of pressure in the plodder cone "squeezing" liquid phase from the soap and causing it to inhibit adhesion at the candle interfaces. The net effect is still the same—more glycerol or electrolyte in the liquid phase means poorer candle adhesion.

9.6 Interesting Effects of Mush

The following effects have no practical importance and are included here only because they can help understand soap structure.

9.6.1 Annealing

If you heat a bar of soap for 16 hours at 75 °C, sealed so that there is no water loss (annealing), then the mush will halve as measured by the immersion test. A typical 80/20 soap will have mush reduced from 7.5 g/50 cm^2 of originally immersed surface area to about 3g/50 cm^2. That is a very significant change, and one that would easily be noticed by the consumer. Although I have no record of it ever having been examined, I believe the soap cracking will also decrease significantly.

A mass balance experiment shows that the reduced mush of annealed soap is due to much less water being present in the mush removed from the soap. The soap contents of the mush from soap are the same with and without annealing.

The hypothesis is that during the high-temperature annealing process, some of the normally insoluble soaps (sodium palmitate and sodium stearate) also dissolve in the liquid phase, but then crystallise out again when the soap is cooled. However, no shear or stress is applied to the soap during the relatively long time of natural cooling, and this gives a very compact and well-ordered soap structure. It is much more difficult for water to penetrate into such a compact structure—there are fewer channels of structural weakness for water penetration (equivalent to a significant increase in tortuosity), less mush formation, and less crack-producing stress when the mush dries.

9.6.2 Mush and Bar Moisture Content

Water penetration and mush formation increase with the water content of the soap *at the time of processing* (mixing, milling, and especially plodding). Post-processing water loss will not markedly change the subsequent water penetration and mush. It is as if the soap has a "memory" of the water content at which it was originally made!

This effect occurs because during natural drying of a soap bar, some soluble components from the liquid phase will mainly precipitate as solid-phase oleate:laurate eutectic mixture (sometimes called eta phase), but as very, very small crystals dispersed as an intimate mixture with the liquid phase.

When this intimate mixture is rehydrated during water penetration and mush formation, the small crystals will dissolve very rapidly—so rapidly that the rehydration step will not influence the rate of water penetration.

10. The Implications of Structure and Mush for Cracking

10.1 Control the Formulation

We should control, as far as possible, the formation of soap structure weaknesses that run parallel to the direction of extrusion. This is not easy because the formation of such weaknesses is inherent in the way extruded soap is made. However, the following can reduce weaknesses:

- A formulation low in electrolyte and glycerol to improve candle adhesion
- A formulation with appropriate plasticity

Unfortunately, nobody has ever been able to define or quantify what is an "appropriate" amount of plasticity, other than to say it is an amount that gives minimum cracking with a given equipment configuration.

10.2 Processing to Give Phase Optimisation Can Help to Minimise Cracking

There are two phase optimisation goals:

1. Maximising the recrystallisation from kappa to zeta phase in normal soaps (those without significant levels of free fatty acids). This also means moving laurate from the kappa phase into the liquid phase, thus giving a liquid phase with more of the oleate:laurate eutectic mixture—and hence less water penetration, less liquid phase swelling, and less mush.

2. Generally, this means processing normal soaps at 45–47°C to improve solubility of laurate in the liquid phase. It may be necessary to adjust the oil blend to allow such temperatures to be used and to also get a good bar finish and line efficiency.

Similarly, maximising the recrystallisation from kappa to delta phase in soaps with significant levels of free fatty acids will also give a liquid phase with more oleate:laurate, less water penetration, and less mush.

However, this type of soap containing significant amounts of free fatty acids must be processed at 35–37°C, and certainly below 40°C. At any higher temperature the free fatty acids solubilised in the liquid phase act as an excellent lipophilic solvent and will also dissolve some sodium palmitate and sodium stearate from the solid phase. When such soap is subsequently cooled, the sodium laurate from the liquid phase co-crystallises with the sodium palmitate and sodium stearate to reform kappa phase. This leaves the liquid phase containing an even greater excess of sodium oleate as soap. Sodium oleate is much less soluble than the oleate:laurate eutectic, and much more water must be absorbed before it will dissolve. This means more swelling with water penetration, more mush, and more stress on the soap when it dries, that is, potentially more cracking.

It is very difficult to process soap at below 37°C through all of the working stages of a finishing line.

10.3 Processing Can Help with Soap Flow and Orientation Effects

10.3.1 Soap Flow

A shaped bar (tablet) of soap is normally formed from a rectangular block or slug produced from the plodder by compressing it, quickly, between two shaped surfaces (stamping dies).

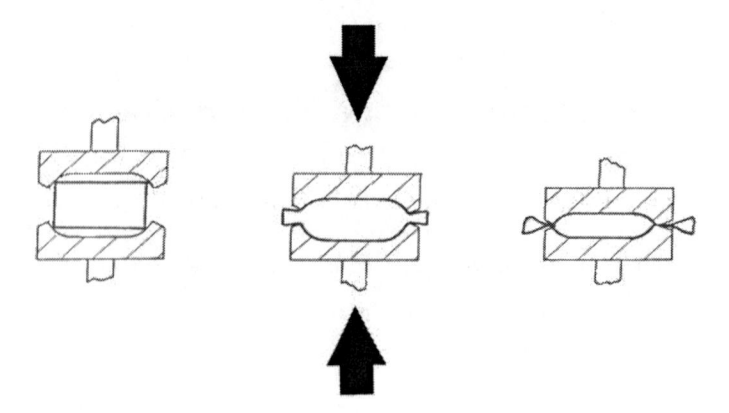

The soap flows as it changes shape in the die. Pressure and soap flow in the die can help remove some of the structural weaknesses introduced by the plodder, but unless both the slug shape and the final bar shape are properly designed, they can introduce new potential weaknesses, which may lead to cracking. In the simplest case, and especially for bandless bars, this means maximising the flow of soap in the die, so a square or round slug is best to produce a generally rectangular bar.

Often it is noted that cracking occurs more frequently on one side of a finished tablet face than on the other. The following sections explain how such an effect can occur and how to minimise the problem.

10.3.2 Orientation Effects at Extrusion
Soap cracking generally, and especially cracking that is predominant on one soap face, can be significantly reduced by appropriately selecting the extrusion orifice plate and the orientation of the cut soap slug at the stamper.

Earlier, in Section 7.5, it was described how structure lines are produced in a slug as it is extruded through a single, rectangular orifice plate (die plate/eye plate).

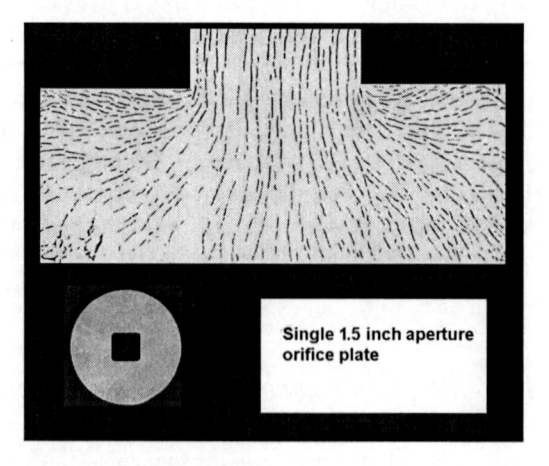

The pattern of flow lines (lines of potential structural weakness visualised using the alcohol immersion procedure) are reasonably even on both sides of the slug.

Implications of Soap Structure for Formulation and User Properties • 105

Now consider the situation where two slugs are produced side-by-side and close together from twin rectangular apertures in the same orifice plate. This is a common manufacturing method.

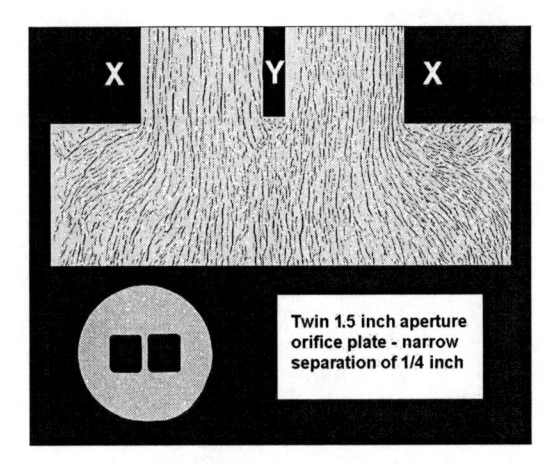

Twin 1.5 inch aperture orifice plate - narrow separation of 1/4 inch

The pattern of flow lines is now uneven. There are more flow lines and more regular flow lines on the outside faces of the slug (X) than on the inside faces (Y). This implies some significant structure difference between the outside and the inside slug faces.

Now consider the situation where the rectangular apertures in the orifice plate are much further apart, ideally a minimum of 45 mm at commercial scale.

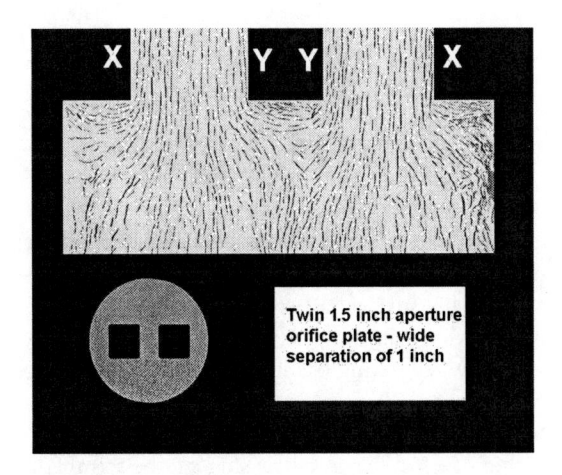

Twin 1.5 inch aperture orifice plate - wide separation of 1 inch

Now the pattern of flow lines is more even on both the inside faces (Y) and the outside faces (X) of each slug. Although the patterns at X and Y are not exactly the same, separating the apertures on the orifice plate has made each slug have a structure much more like that of a slug produced from a single aperture.

The implications for soap cracking of uneven flow lines on the inner and outer faces of a slug are best explained by again using the alcohol immersion procedure to show the flow line patterns in cross section through slugs and stamped bars, perpendicular to the direction of extrusion.

A square slug from a single-aperture orifice plate has a uniform pattern of stress lines. When stamped across any opposing faces, it will give a bar with a uniform pattern of stress lines, in which the outer flow lines follow the profiles of the slug edge and, in particular, the tablet edge.

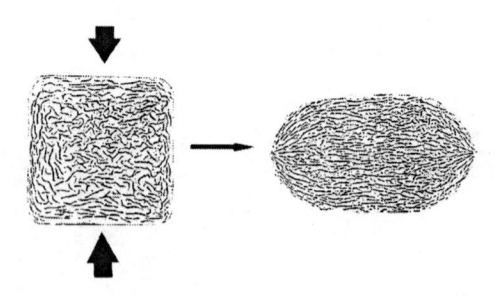

Remember that the slugs produced through two apertures that are close together had more flow line orientation on the outside faces of the slugs than on the inside faces. This shows well in the cross-sectional pictures from rectangular or round slugs.

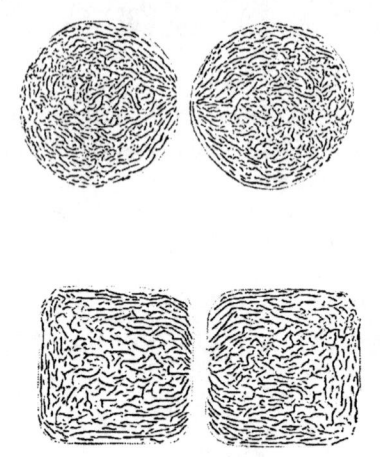

Note that between the slugs, where the previous view (horizontally through the cone end and slug) showed fewer and more irregular flow lines, there is now a clear indication that the flow lines are significantly perpendicular to the slug surface rather than parallel.

The effect is very similar to what will occur if a larger slug is simply cut vertically at the centre, which is, in effect, what the narrow separating bar in the orifice plate has done.

10.3.3 Orientation Effects during Stamping

There are now two ways in which such an asymmetric slug can be stamped:

Inner face of the slug on the bar face:

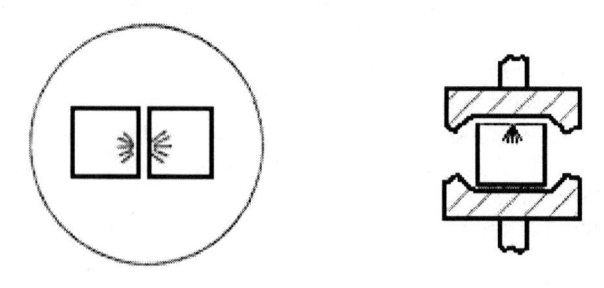

Inner face of the slug on the bar edge:

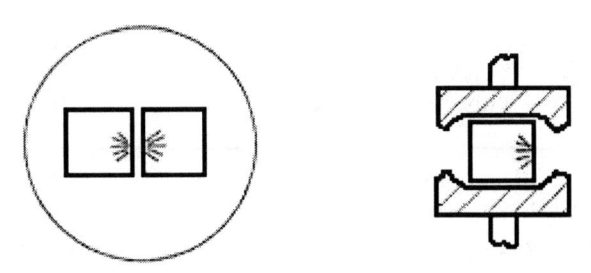

View these together with the final bar structures:

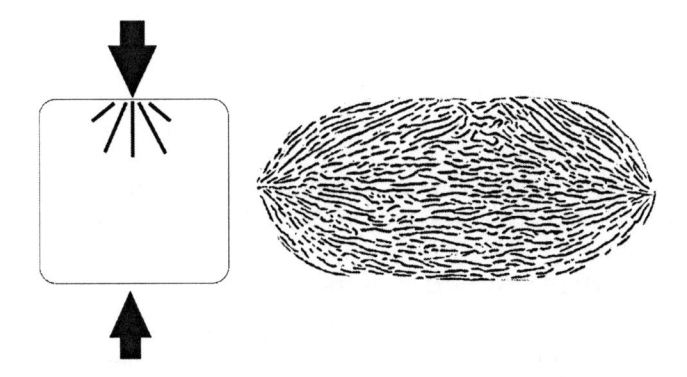

Clearly, when the final face of the tablet originates from the inside face of the slug it also retains the poorer structure of the slug. The poorer tablet structure allows easier water penetration, more mush, and eventually more cracking. Importantly, the cracking will predominate on one face of the tablet.

Typically, 15% of such bars will crack in a home use placement test. The majority of the cracking will be on just one face of the bar.

When the soap slug is stamped with the inner face presented to what will eventually be the peripheral edge of the tablet, the flow line pattern is much more even at the tablet faces, and the irregularities from the inner face of the slug are largely removed with the soap flash from the stamping die.

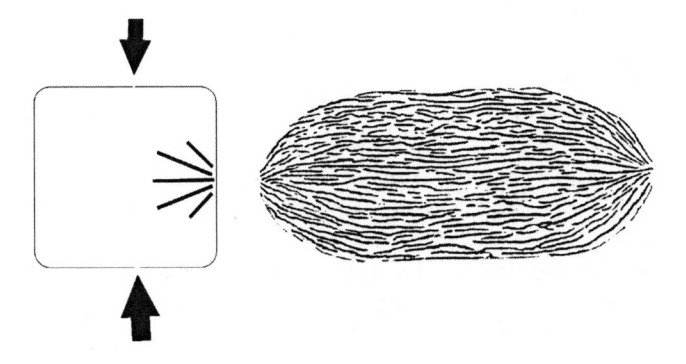

Typically only 4% of such bars will crack in a home use placement test. Any face cracking is likely to be equal on both bar faces.

End or "flash line" cracking may increase slightly, but consumers are not so deterred by such cracking.

The best practical solutions are therefore the following:

- Ideally, only extrude soap using a single-aperture orifice plate with a nearly square aperture.
- If double-aperture orifice plates must be used, then use maximum separation between the apertures. At least 45 mm is recommended.
- If double-aperture orifice plates must be used, then make sure that when the slugs are stamped, they are oriented so that the inside faces of the slugs are on the bar edges.

11. Effects of Formulation on Cracking

11.1 General Notes

The precursors to cracking are:

- water penetration,
- mush formation and swelling, and
- mush drying and shrinkage.

These factors will lead to stress on any structure with inherent weaknesses, for example from crystal alignment. Any factor that influences these precursors will therefore also influence cracking.

11.2 Special Note on Materials Added at the Mixer

If the soap structure is not homogeneous because it is not well mixed, then this will lead to different rates of water penetration into different parts of the bar, and to correspondingly different levels of mush, mush drying, shrinkage, and stress. Such differential effects will certainly lead to increased cracking.

In particular, a soap may be inhomogeneous because the process route seeks to mix relatively high levels of soluble or fatty components after the basic soap noodle has been dried. In the simplest case, it is never advised to add more than 1% of water to soap noodles in a ribbon mixer; also, it is easy to observe that adding high levels of perfume (>1.5%) or fatty acids will make soap noodles very slippery and difficult to mix into a properly homogeneous mass. Homogeneous mixing is possible, but is difficult and time consuming.

Very high levels of hydrophilic additions can dilute the liquid crystal phase so much that it will revert to a isotropic solution with very low viscosity, which is much more likely to be forced onto the surfaces of candles in the plodder, giving rise to increased cracking.

11.3 Temperature Effects

Temperature is a physical condition, rather than a formulation characteristic, whose main effect is to modify the relative compositions ("formulations") of the solid and liquid phases.

Generally, for most soaps increased temperature during processing gives lower cracking. The exception is any soap for which increasing the temperature also increases mush. This is the case for superfatted soaps containing >5% free fatty acids.

Higher temperature gives less cracking because at higher temperatures more laurate dissolves from the solid phase into the liquid phase, where it forms very soluble oleate:laurate. This leads to a significant increase in the liquid phase volume. More liquid phase and less solid phase in turn means a softer, more plastic soap (Section 5.1.1a). The softer, more plastic soap candles adhere better during plodding without the use of excessive pressures that could force liquid phase onto the candle surfaces.

In addition, if the liquid phase is rich in oleate:laurate, then it will contain more ordered, hexagonal, liquid crystal. The hexagonal liquid crystal is viscous and so is more resistant to being forced out of the solid phase and onto the surfaces of the soap candles in the plodder cone.

Less liquid phase on the candles gives better candle-to-candle adhesion and less cracking. Because there is less liquid on the candles, less extra water is absorbed between the candles during mushing (there is less electrolyte, glycerol, and soap to dilute), so that there is less inter-candle mushing and less stress when the mush dries.

11.4 Soap-to-Water Ratio at Manufacture

In the context of soap properties involving water penetration, it is always better to consider the soap-to-water ratio, rather than the % water content or % TFM. This is particularly important in comparisons with soaps containing fillers or high levels of nonsoap components. For simplicity we will consider products containing only soap and water as major components. Higher water content therefore means a lower % TFM.

There are competing effects, but in general, for modest increases in water content (up to 3% more water), higher water content means less cracking:

- Higher water = more liquid phase = softer soap = better candle adhesion = less cracking

However, the following are also true:

- More liquid phase = more water penetration = more swelling = more mush = more cracking
- More water = a more dilute liquid phase, possibly with some conversion of liquid crystal to isotropic solution = more opportunity for liquid to wet the candles in the plodder = more cracking

The consensus is that for soap with water content up to ~15%, the dominant factor is the increased softness from the increase in liquid phase volume, leading to less cracking. However, and importantly, the extra water must be included in the soap at the drying stage. If an extra 3% of water is added at a subsequent mixing stage, then it will very likely simply dilute the liquid crystal and form isotropic solution. This results in poorer mixing, producing inhomogeneity in the soap, and can lead to significantly increased cracking.

At water contents of 15–20%, there is enough water present for almost all of the laurate to move from the solid phase to the liquid phase. The solid phase will then recrystallise from kappa phase to become substantially zeta phase—small crystals with good packing. The liquid phase then contains a lot of oleate:laurate and has a viscous liquid crystal structure. The result is that soaps subjected to work energy at >15% water content will become hard and tough (and also translucent).

The harder soap will significantly increase pressure in the plodder, forcing liquid phase between the candles, resulting in poorer candle-to-candle adhesion and potentially more cracking. The greater amount of liquid phase forced to the candle surfaces attracts more water penetration to those regions during mushing, giving more swelling and more stress on the structure when the mush dries, again encouraging more cracking.

Note that although mush increases with water content at any moisture level above normal (12%), at much higher moisture contents, say 20% or slightly higher, the mush will decrease if the soap receives sufficient work energy input. This can explain why many laundry soaps have less in-use mush than toilet soaps.

As moisture content increases beyond 20%, water penetration will continue to increase, but mush will decrease. There is no contradiction in this statement. The decrease in mush as moisture content increases over 20% occurs because the kappa and zeta phases are progressively replaced by delta phase. If the long, ribbon-like crystals of delta phase are preformed in a soap, then they will not trap water as it penetrates during mushing. This means less soap swelling relative to soaps containing only kappa and/or zeta phase. Less swelling can outweigh the continued increase in penetration/dissolution loss during mushing, giving an overall decrease in mush and decrease in cracking.

11.5 Changing the Titer (Iodine Value) of the Non-lauric Soapmaking Oils

Increased titer (approximately melting point) means decreased iodine value, that is, less unsaturation in the oils.

Again there are competing effects, but in general, higher IV/lower titer corresponds to less cracking. Increased IV means more sodium oleate and (usually) more sodium linoleate in the oil blend. These soaps are soluble, especially linoleate, and some will be dissolved in water as the liquid phase. An increase in the IV results in an increase in liquid phase volume, a softer soap, better candle adhesion, and hence less cracking.

At constant levels of nut oil (constant sodium laurate), there is a smaller opposite effect, because the greater amount of unsaturated soaps, especially oleate, distributed between the solid phase and the liquid phase has lower solubility than oleate:laurate. This means more water will need to be absorbed to dissolve the extra oleate, and that potentially means more swelling, more mush, more stress when the mush dries, and so more cracking. However, this is a much smaller effect than that from the increase liquid phase volume.

11.6 Increased Nut Oil in the Soap Blend

There are competing effects, but in general, increased nut oil means less cracking. More nut oil (up to about 40% in a blend) means there is more opportunity for laurate to move into the liquid phase, resulting in more oleate:laurate formation, greater liquid phase volume, softer soap, better candle adhesion, and less cracking. In addition, more of the very soluble oleate:laurate in the soap means much less water absorption, less swelling, less mush, less stress when mush dries, and less cracking.

On the other hand, more nut oil potentially leads to more residual electrolyte in the soap, hence more in the solution phase, which may migrate to the candle surfaces in the plodder, resulting in poorer candle adhesion and more cracking. In practice this is a minor effect unless electrolyte levels are significantly above specification.

11.7 Increased Glycerol

In general, increased glycerol leads to increased mush and increased cracking. However, the effect is not as big as expected. The same amount of extra water will give a bigger increase in mush.

At high levels of glycerol, this is probably because glycerol located in the liquid phase displaces some soap, resulting in a smaller overall liquid phase volume. In contact with water and relative to the soap it has displaced, the glycerol dissolves easily, thus there is less water uptake, less swelling, less mush, and less cracking.

Also, less soap in the liquid phase means a smaller liquid phase, so glycerol at higher levels has a hardening effect on soap. A harder soap means good work energy input, and so the laurate will move to the liquid phase, where it forms soluble oleate:laurate, again resulting in less water uptake, less swelling, less mush, and less cracking.

Once again, the effect is not as big as expected, but it is still present—high glycerol does give increased mush and cracking relative to a soap with lower levels of glycerol.

11.8 Perfume Effects

Again, there are competing effects, but in general, increased perfume means less cracking. Perfume mainly resides in the liquid crystal phase, increasing the phase volume and so the soap's plasticity, which leads to better candle adhesion and less cracking.

Importantly, although the effects of different perfumes vary, higher levels of perfume almost always change the structure of the liquid crystal phase from the viscous hexagonal structure to the much less viscous lamellar phase structure.

In theory, the lower viscosity makes the liquid phase more likely to migrate to the candle surfaces under plodder pressure, giving poorer adhesion and more cracking. This probably occurs in practice, but the softening effect of the hexagonal-to-lamellar change predominates, so that, overall, more perfume means less cracking.

11.9 Free Fatty Acids (Superfatting)

Superfatting involves a minimum of 5% free fatty acids (some say 7.5%) and preferably at least 30% nut oil in the overall oil blend. These levels affect various performance attributes of such soaps, notably lather. For the cracking issue, we will consider any soap with free fatty acids >5%.

There are competing effects, but in general, superfatting decreases soap cracking. The liquid crystal phase of soap is highly oleophilic, and free fatty acids will locate in that phase. This will increase the liquid phase volume, giving softer soap, better candle adhesion, and less cracking. Also, free fatty acids will change the structure of the liquid crystal phase from the viscous hexagonal form to the less viscous lamellar form. This will further soften the soap and reduce cracking. All else being optimal, these effects will predominate over any effects from greater mobility of the lower-viscosity lamellar phase leading to reduced candle-to-candle adhesion.

On the other hand, if superfatted soap is processed at too high a temperature, say>35–37°C, then the liquid phase will contain only a limited amount of laurate soaps. The laurate will be trapped in a solid kappa phase. When the soap is in contact with water, a lot of water is absorbed to try to dissolve this laurate, and when it does dissolve, the kappa phase changes to delta Phase, trapping the water onto the bar surface. Thus, there is more mush and more cracking. However, the soap-softening effects will probably predominate.

11.10 Electrolyte

More electrolyte gives more cracking. The major effect is that electrolyte will "salt out" soaps from the liquid phase by ionic strength and common ion effects, resulting in less liquid phase volume, harder and less plastic soap, poorer candle adhesion, and more cracking.

In addition, there will be more electrolyte in the liquid phase, and when it migrates to the candle surfaces under the influence of plodder pressure, it will be more effective at reducing candle-to-candle adhesion, also resulting in more cracking.

11.11 Overall Composition Effects

There are always competing effects, even for an individual composition change. One composition change may introduce another, for example, increased nut oil will almost always increase electrolyte.

In very broad terms, the influences on cracking are as follows:

- Major effects from electrolyte and especially from temperature if free fatty acids are present
- Moderate effects from nut oil level, from low levels of free fatty acids, and from higher levels of glycerol
- Smaller effects from perfume, oil blend IV, and low levels of glycerol

The following schematic graph is a starting point for discussion of the magnitude of effects from different formulation changes. The scales quoted represent the author's experience, but much more work is needed to complete the picture.

Influences on Cracking

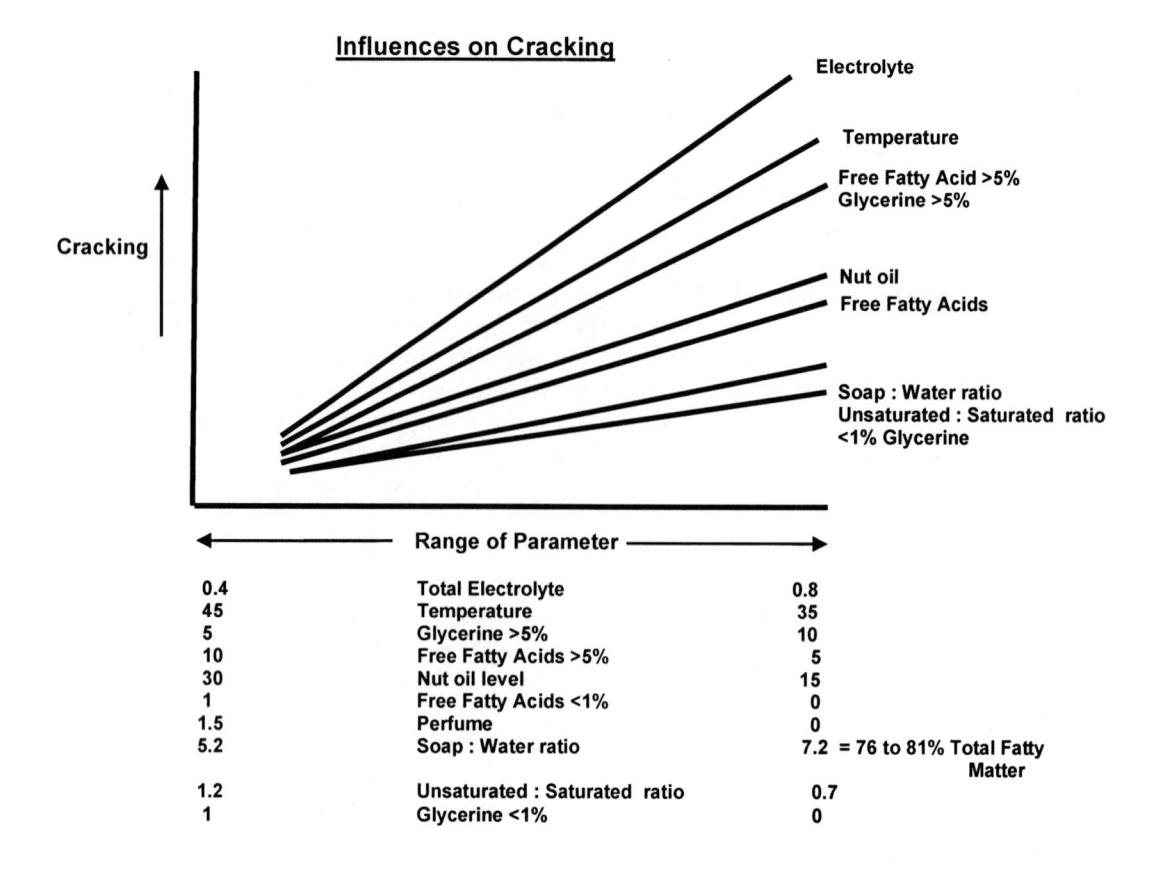

Cracking ↑ vs Range of Parameter →

Lines (from steepest to shallowest):
- Electrolyte
- Temperature
- Free Fatty Acid >5%
- Glycerine >5%
- Nut oil
- Free Fatty Acids
- Soap : Water ratio
- Unsaturated : Saturated ratio
- <1% Glycerine

0.4	Total Electrolyte	0.8
45	Temperature	35
5	Glycerine >5%	10
10	Free Fatty Acids >5%	5
30	Nut oil level	15
1	Free Fatty Acids <1%	0
1.5	Perfume	0
5.2	Soap : Water ratio	7.2 = 76 to 81% Total Fatty Matter
1.2	Unsaturated : Saturated ratio	0.7
1	Glycerine <1%	0

Appendix

One technique to help visualise soap flow and crystal orientation in a plodder is to remove a soap-filled screw from a plodder and draw lines on the soap surface showing where it was in contact with the plodder barrel wall. The lines should represent where the soap will shear against the barrel wall. On a soap-filled screw they will be perpendicular to the longitudinal axis of the screw and oriented in the direction shown below (if the screw rotates counterclockwise when viewed from the cone end).

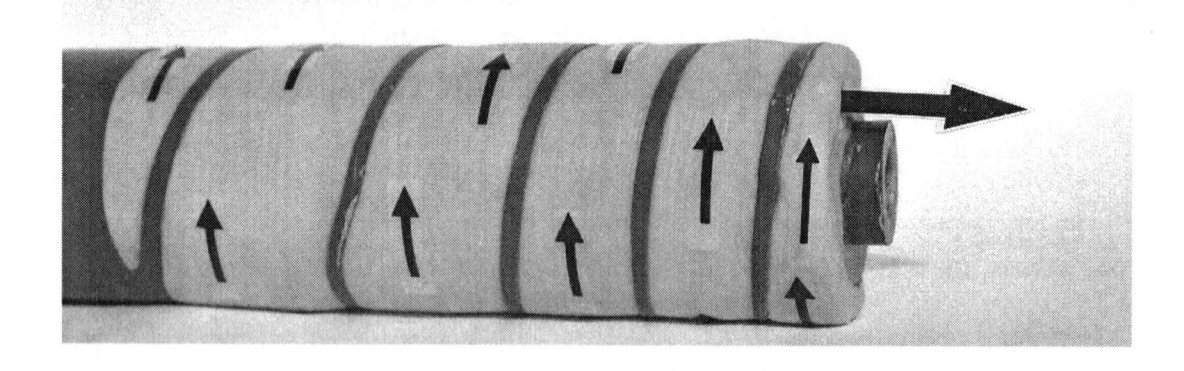

Implications of Soap Structure for Formulation and User Properties ● 113

A second picture shows a closer view of the shear direction at the point where the soap leaves the screw and enters the cone, and emphasises a sharp change in the direction of soap flow

While the soap is still warm and plastic, remove it from the screw and open out the piece of soap into one long strip. The lines originally drawn perpendicular to the longitudinal axis of the screw will now be seen as running parallel to the length of the total strip of the soap removed from the screw.

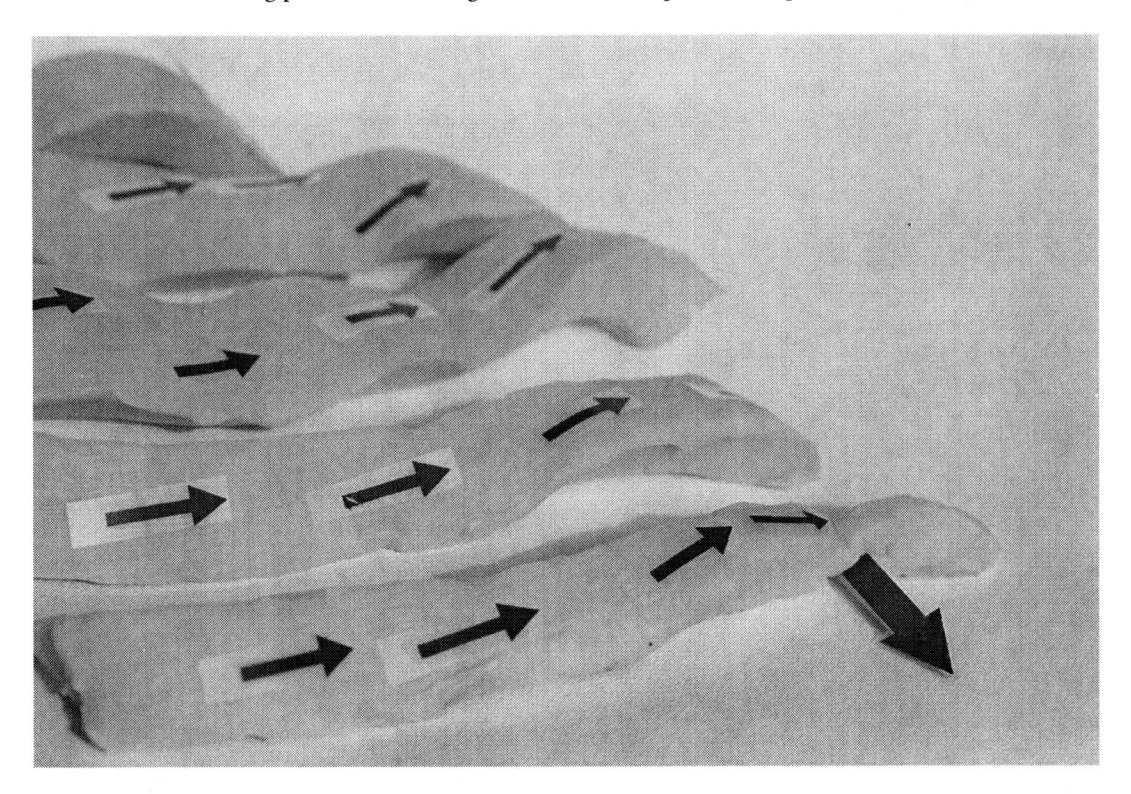

As described in the text, the asymmetric soap crystals will align parallel to the direction of shear indicated by the arrow marks in the photograph. At the point just before the soap leaves the screw flight, the soap will have been subjected to considerable shear.

However, at the point where soap enters the cone, there is a very sharp change in soap flow direction, which implies that soap crystals entering the cone will have a longitudinal axis perpendicular to the direction of soap flow. In the cone, the flow of soap against soap will realign the crystals so that they are again parallel to the direction of soap flow. The realignment process requires energy and is probably associated with the phenomenon known as "work softening."

Soap Structure and Phase Behavior

Michael Hill[1] and Teanoosh Moaddel[2]
[1]*Columbia University, New York, New York, USA;* [2]*Unilever HPC, Trumbull, Connecticut, USA*

Introduction

While soap has been used since antiquity, soap production has historically been more of an art than a science. For example, the soap-boiling process, widely used for centuries, manipulated a soap mass around and through various phases with such cryptic names as nigre, middle soap, neat soap, kettle wax, and curd (Vold, et al., 1941).

In the twentieth century, however, soap scientists developed a coherent understanding of the structure and phase behavior of soap (Vold, et al., 1941; McBain & Lee, 1943; McBain & de Bretteville, 1942; Buerger, 1942; Buerger, et al., 1942; Buerger, et al., 1945; Ferguson et al., 1942; Ferguson, 1944; Palmqvist, 1983; Zajic, et al., 1968; Dumbleton, & Lomer, 1965; Yang et al., 1987; Vold et al., 1952; Lewis et al., 1969; Tandon et al., 2000; Tandon et al., 2000; Mantsch et al., 1994; Laughlin, 1994; Small, 1986). As a result, it is now appreciated that the complex behavior of soap systems can be fully explained in terms of the molecular phenomena common to surfactant systems. This chapter will attempt to elucidate the principles of soap behavior in terms of its structure and phases. In addition, the way these soap phases organize relative to one another to form the final bar structure will be discussed in terms of both processing effects and user properties.

Soap Molecular Structure

Soap, commonly defined as the salt of a fatty acid, is the reaction product of aqueous caustic soda with fats and oils from natural sources. As a surfactant molecule, soap contains a hydrophilic head (the carboxylate group) and a hydrophobic tail (the aliphatic chain). This dual character gives soap its ability to dissolve both aqueous and organic phases, its ability to form monolayers at the air–liquid interface (as in foam generation and stability), and its ability to cleanse. The extent to which a particular soap has these properties is determined both by the counter-ion(s) and the aliphatic chain(s) that are present (Piso & Winder, 1990; Murahata et al., 1997; Rosen, 1978).

Depending on the source of the fat or oil used, the distribution of the aliphatic chains can vary as shown in Table 3.1, including chain lengths from C8 to C22 as well as a range of unsaturation, including oleic (C18:1), linoleic (C18:2), and linolenic (C18:3) chains.

Soap Phase Structure

Solid Soap

As with all pure materials, pure single-chain sodium soap (anhydrous sodium salt of a single fatty acid) will form a solid crystal structure when sufficiently cool.

This structure generally consists of packed bilayers of soap molecules, arranged head-to-head and tail-to-tail. If water is also present, a hydrated crystal structure will form, consisting of packed bilayers of soap molecules with the water of hydration in the region between the packed carboxylate heads.

Solid soap crystals have been probed by x-ray diffraction. The observed diffraction patterns can be divided into two groups: the long spacings that correspond to the perpendicular separation

Table 3.1. Typical Composition of Natural Oils and Fat

Common name	Chemical name	Chemical formula	Symbol	Tallow	Lard	Coconut	Palm kernel	Soybean
			Saturated fatty acids					
Caprylic	Octanoic	$C_8H_{16}O_2$	C8			7	3	
Capric	Decanoic	$C_{10}H_{20}O_2$	C10			6	3	
Lauric	Dodecanoic	$C_{12}H_{24}O_2$	C12			50	50	0.5
Myristic	Tetradecanoic	$C_{14}H_{28}O_2$	C14	3	1.5	18	18	0.5
Palmitic	Hexadecanoic	$C_{16}H_{32}O_2$	C16	24	27	8.5	8	12
Margaric	Heptadecanoic	$C_{17}H_{34}O_2$	C17	1.5	0.5			
Stearic	Octadecanoic	$C_{18}H_{36}O_2$	C18	20	13.5	3	2	4
			Unsaturated fatty acids					
Myristoleic	Tetradecenoic	$C_{18}H_{26}O_2$	C14:1	1				
Palmitoleic	Hexadecenoic	$C_{18}H_{30}O_2$	C16:1	2.5	3			
Oleic	Octadecenoic	$C_{18}H_{34}O_2$	C18:1	43	43.5	6	14	25
Linoleic	Octadecadienic	$C_{18}H_{32}O_2$	C18:2	4	4	1	2	52
Linolenic	Octadecatrienic	$C_{18}H_{30}O_2$	C18:3	0.5	0.5	0.5		6

Source: Bartolo & Lynch, 1997.

between carboxylate heads in the bilayers (longrange order), and the short spacings that correspond to the lateral separation between parallel aliphatic chains (short-range order), as shown in Fig. 3.1. As expected, crystals of pure sodium soaps (anhydrous or hydrated) have a long spacing that correlates with aliphatic chain length. One set of researchers noted four different short spacing patterns in different soap samples. This was taken as evidence of four different molecular arrangements, with four distinct solid soap phases. These solid soap phases were named alpha (α), beta (β), delta (δ), and omega (ω) (Ferguson et al., 1942; Ferguson, 1944). However, another set of researchers concluded that these four were not single phases but were mixtures. They also saw evidence of many more solid crystal phases, including gamma (γ), epsilon (ε), eta (η), zeta (ζ), kappa (κ), mu (μ), and sigma (σ) (Buerger, et al., 1942; Vold et al., 1952), leaving soap technologists with a Greek "alphabet soup" of soap phases. While many of these inconsistencies can be attributed to different sample preparation methods (McBain & de Bretteville, 1942; Buerger, 1942; Buerger, et al., 1942; Buerger, et al., 1945), this resulted in two sets of nomenclature to describe the more common soap phases. A summary of this nomenclature is contained in Table 3.2.

The various types of solid soap phases differ in their degree of hydration, although there is still debate as to whether hydration follows stoichiometric rules or results from a solid solution of soap and water (Buerger, 1942; McBain & Lee, 1943; Buerger, 1942; Perron & Madelmont, 1973; Perron, 1976; Madelmont & Perron, 1976; De Mul et al., 2000). While the existence of multiple solid phases resembles polymorphism, this is not true polymorphic behavior, since these solid soap phases differ in composition.

When soap containing a range of chain lengths is crystallized, solid crystals containing a mixture of chain lengths will form. This is a true solid solution since the molecular composition present in the solid crystals may vary continuously. In addition, multiple solid phases may form and coexist. These solid crystals have been characterized as the same solid crystal phases that form from the pure chain length soaps and water.

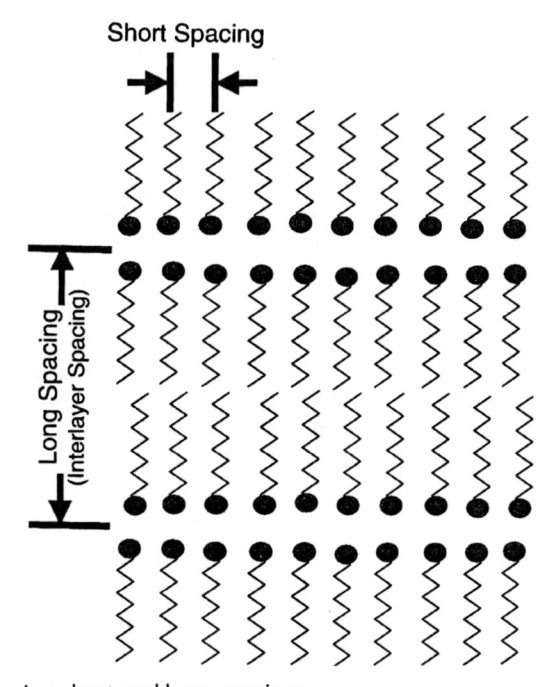

Fig. 3.1. Solid soap crystal showing short and long spacings.

Table 3.2. Nomenclature for Solid Soap Phases.

Buerger Nomenclature (6) (Unilever)	Ferguson Nomenclature (7,8) (Procter & Gamble)
Eta (η)	Omega (ω)
Kappa (κ)	Omega (ω)
Zeta (ζ)	Beta (β)
Delta (δ)	Delta (δ)

Sources: Buerger, et al., 1945; Ferguson et al., 1942; Ferguson, 1944.

It is well known that morphology and hydration affect crystal dissolution behavior (McCrone, 1965; Bernstein, 2002; Grant & Higuchi, 1990). Therefore it would not be surprising if the various types of solid soap crystals differed from one another in properties related to dissolution, such as lathering or mush properties. However, while such properties have been reported for bars containing various types of solid crystals (Ferguson et al., 1942; Ferguson, 1944; Palmqvist, 1983; Zajic, et al., 1968; Bartolo, & Lynch, 1997), the dissolution properties for the single crystals in isolation have never been reported. This is especially significant since commercial soap bars contain multiple components that are divided among multiple phases (as will be discussed in the section on Soap Colloidal Structure), so that the recrystallization of one dispersed solid crystal phase into another solid crystal phase must be commensurate with corresponding changes in the composition and/or phase structure of the continuous phase. While it is tempting to attribute differences in soap bar behavior solely to the type of solid soap crystal present, the effect of changes in the continuous phase must not be ignored.

Liquid Crystalline Soap

A liquid crystal is defined as a class of material that has both liquid- and solid-like properties. Liquid crystalline soap phases can form either when anhydrous soap is heated or mixed with water. These phases are classified as thermotropic and lyotropic liquid crystals, respectively. In either case, the tail portions of the soap molecules become more fluid resulting in a loss of short-range order, as in a liquid, while still maintaining their long-range order as in a solid. The various types of soap liquid crystals are discussed in the following sections.

Thermotropic Phases. When anhydrous soap is heated, it passes through numerous phases prior to melting. These phases, called subwaxy, waxy, superwaxy, subneat, and neat II, are thermotropic liquid crystals, since their formation is primarily determined by temperature (as opposed to composition), as shown in Fig. 3.2. These soap phases have long been recognized by soap boilers, who named all these Waxy phases. However, as these phases do not occur in commercial soap bars at temperatures typically encountered, they are of primarily academic interest to soap technologists.

The term "curd phase" that is commonly encountered in soap literature has not been well defined. This term is sometimes used to refer to an anhydrous soap prior to its transition to the subwaxy anhydrous polymorph (Vold, et al., 1941; Skoulios & Luzzati, 1961), as shown in Fig. 3.2. In other instances it has also been used to refer to crystal hydrate (Laughlin, 1994).

Lyotropic Phases

A variety of lyotropic liquid crystal phases in a well-defined sequence can form when a surfactant molecule is mixed with water. The generic sequence of these liquid crystal phases with increasing water is shown in Fig. 3.3. In soap–water binary systems, however, only two of these phases are apparent. When anhydrous soap and water are mixed and allowed to equilibrate at the appropriate temperature (dependant on chain length and degree of unsaturation), two lyotropic liquid crystalline phases can

Soap Structure and Phase Behavior ● 119

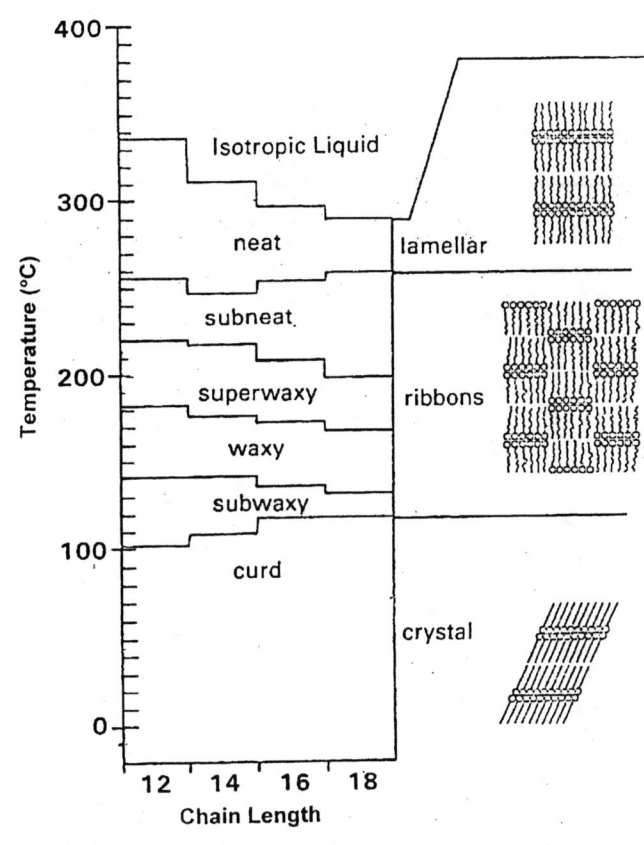

Fig. 3.2. Thermotropic phase behavior of sodium soaps. Skoulios & Luzzati, 1961.

● *Water Rich Composition*

I₁ - Normal Cubic Liquid Crystal

H₁ - Normal Hexagonal Liquid Crystal

V₁ - Normal Bicontinuous Cubic Liquid Crystal

L_α - Lamellar Liquid Crystal

V₂ - Inverse Bicontinuous Cubic Liquid Crystal

H₂ - Inverse Hexagonal Liquid Crystal

I₂ - Inverse Cubic Liquid Crystal

● *Surfactant Rich Composition*

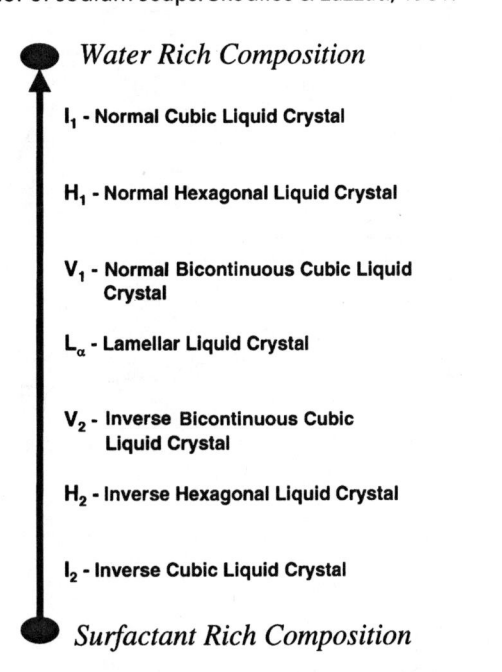

Fig. 3.3. Generic sequence of liquid crystal phases.

form: a lamellar liquid crystal phase at lower moisture and a hexagonal liquid crystal phase at higher moisture (Fig. 3.4).

The structural transition between the lamellar and hexagonal phase will invariably pass through what is commonly referred to as an intermediate phase. The structure of these intermediate phases remain uncertain and can range from distorted rod-shaped aggregates (Seddon, 1990; Kekicheff, 1989) to pierced lamellar planes (Holmes & Charvolin, 1984; Luzzati et al., 1968; Funari et al., 1992). The intermediate, Waxy, and various solid crystal phases are not shown in Fig. 3.4 for sake of simplicity.

As depicted in Fig. 3.5, the lamellar phase is ordered along one dimension. It can easily be seen that the structure of a lamellar liquid crystal phase is essentially the same as that for a solid soap crystal, except that the hydrocarbon tails in the liquid crystal are in a "fluid" rather than "rigid" state. It is this order in only one dimension that causes the lamellar phase to be the liquid crystal phase with the lowest viscosity.

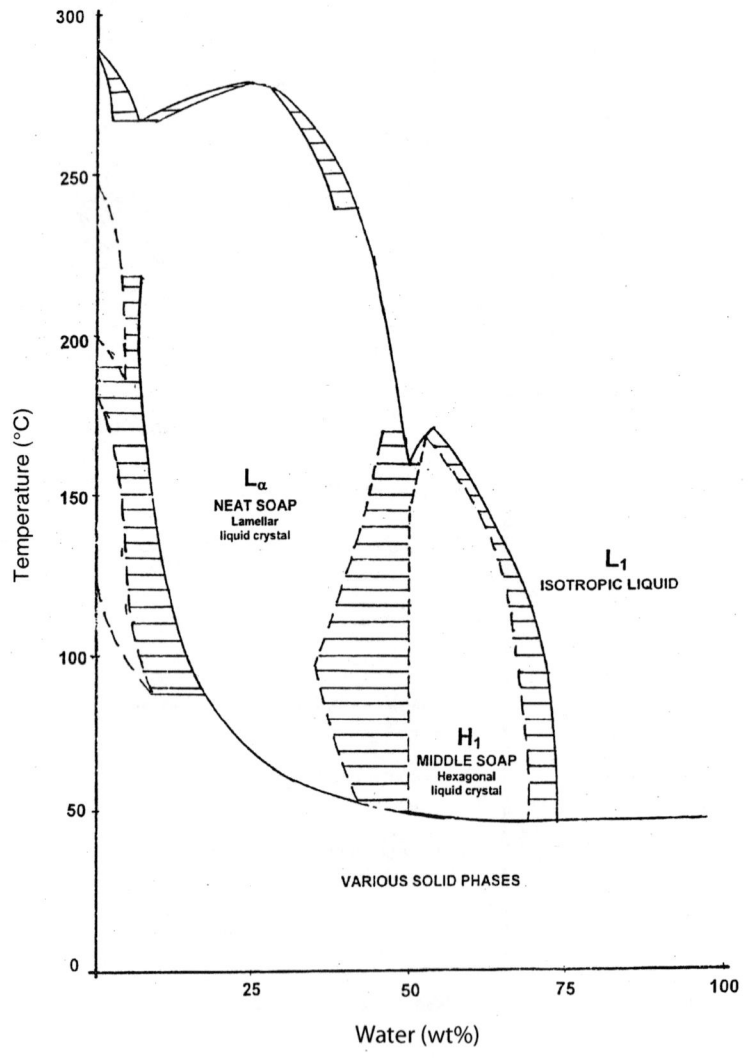

Fig. 3.4. Generic binary soap–water phase diagram. Abbreviations: Lamellar liquid crystal, L_α; Normal hexagonal liquid crystal, H_1; Micellar phase, L_1. Palmqvist, 1983

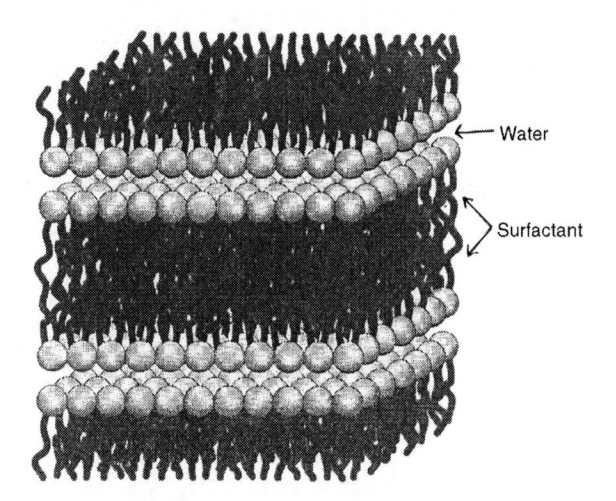

Fig. 3.5. Lamellar liquid crystal. Fennell & Wennerstrom, 1999.

The soap lamellar phase was regularly observed by the old soap boilers. When boiled soap was allowed to settle, it separated into two layers: an upper layer of a lamellar phase and a lower layer of an isotropic soap solution. Since most of the impurities settled into the lower layer, the lamellar phase was relatively clean, and hence was named "neat soap" (Thomssen & McCutcheon, 1949).

While the lamellar phase can be identified by its low angle x-ray diffraction pattern, it is most easily identified by examination with an optical microscope fitted with cross polarizers. The characteristic optical pattern associated with the lamellar phase is shown in Fig. 3.6.

The lyotropic liquid crystalline hexagonal phase consists of close packing of long cylindrical micelles with the soap molecules aligned so that the hydrophilic heads are on the cylinder surface and the hydrophobic tails point toward the center. These structures are ordered along two dimensions (Fig. 3.7). If these cylindrical micelles were to be viewed end on, it would be apparent that they are arranged in a hexagonal pattern with each cylinder surrounded by six others.

On a macroscopic scale these long cylinders form an entangled network, causing the hexagonal phase to have a very high viscosity. Soap boilers knew this phase as "middle soap" because of its location between neat soap and isotropic soap solution in the binary soap–water phase diagram (McBain & Elford, 1926). Well aware of its high viscosity, soap boilers carefully avoided forming middle soap, knowing that once it was formed it would be difficult to process further.

It is well known that soap boilers would keep a ladle of salt on hand to add to boiling neat soap (lamellar phase) to stop the conversion into middle soap (hexagonal phase) (McBain & Lee, 1943). The impact of salt on the transition from lamellar phase, a structure with zero mean aggregate curvature (Fig. 3.8) to hexagonal phase, a structure with positive mean aggregate curvature (Fig. 3.9) can be explained in terms of its effect on the packing of the soap molecules.

Aggregate curvature results from the balance of attractive and repulsive forces across a surfactant film. These forces generally consist of head group interactions, interactions at the polar/nonpolar interface, and alkyl chain interactions (Seddon, 1990). More specifically, these include dispersive (Van der Waals'), steric, and electrostatic forces (Van Oss, 1991). Depending on the system conditions and composition, it is the interplay between these noncovalent interactions that controls aggregate curvature (Seddon, 1990; Holmberg et al., 2003). The negative charge on the carboxylate group of a soap molecule leads to electrostatic repulsion between adjacent head groups, tending to push them apart and increase the curvature of the surfactant film, favoring hexagonal phase. However, this effect can be mitigated through electrostatic shielding by the addition of an electrolyte, allowing closer packing of adjacent soap

122 M. Hill and T. Moaddel

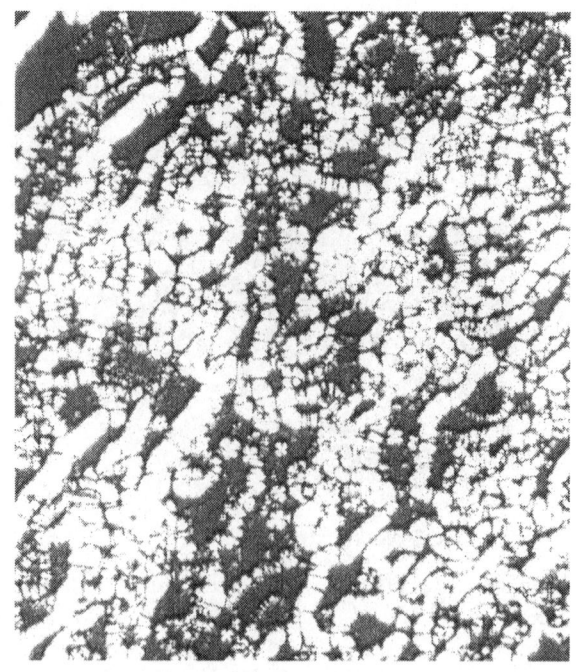

Fig. 3.6. Characteristic optical pattern of a lamellar liquid crystal viewed through a microscope fitted with cross polarizers.

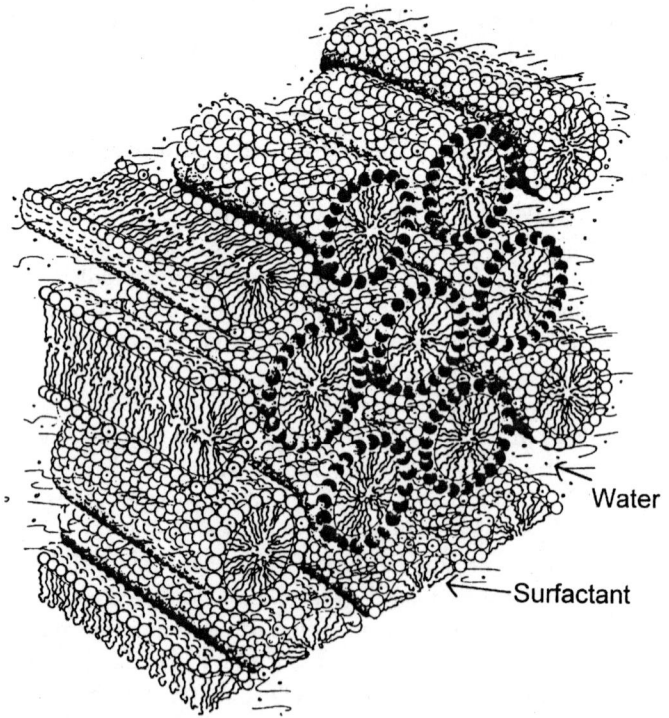

Fig. 2.7. Hexagonal liquid crystal. *Source:* Rosevear (43).

Fig. 3.7. Hexagonal liquid crystal. Rosevear, 1968.

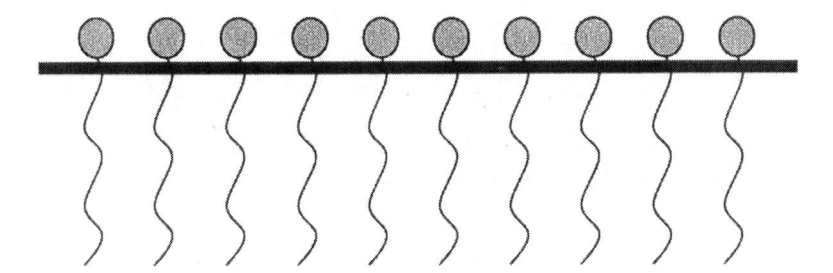

Fig. 3.8. Surfactant film with zero mean curvature. Seddon, 1990.

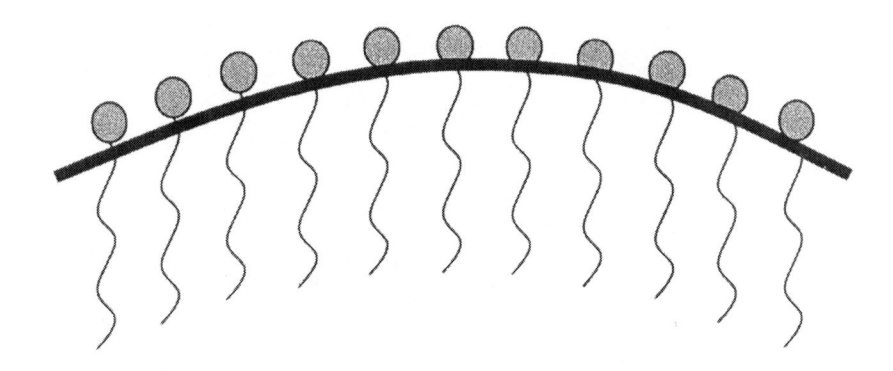

Fig. 3.9. Sufactant film with positve mean curvature. Seddon, 1990.

molecules, favoring lamellar phase. Similarly, the addition of uncharged surface active species (e.g., fatty acid or fatty alcohol) can serve to separate the charged species within the packed structures, allowing closer packing of adjacent molecules and thereby favoring lamellar phase. In addition, if the added material is a fatty acid or fatty alcohol, an attractive hydrogen bonding force with the carboxylate head group of soap (Lynch et al., 2001) will also allow for closer packing of molecules.

As with the lamellar phase, the hexagonal phase is most easily identified by examination with an optical microscope fitted with cross polarizers. The characteristic optical pattern associated with hexagonal phase is shown in Fig. 3.10.

Isotropic Soap Solution

As with all surfactants, at very low concentrations soap molecules exist in water as monomers. However, as the concentration of soap increases, a point will be reached when the monomers start to form spherical aggregates known as micelles, with the carboxylate groups on the surface of these superstructures and the hydrophobic tails all pointed toward the center (Fig. 3.11). The concentration where this phenomenon occurs is known as the critical micelle concentration (CMC). As the soap concentration increases beyond the CMC, more and more micelles will form. Eventually the micelles will start to distend, and will develop a rod-like shape; at even higher concentrations, they develop a worm-like shape (Fig. 3.12). All of these solutions are collectively referred to as isotropic soap solution. As the soap concentration is increased further, the worm-like micelles will begin to form a positionally ordered two dimensional lattice structure having long-range order—the hexagonal liquid crystal phase discussed in the previous section.

124 M. Hill and T. Moaddel

Fig. 3.10. Characteristic optical pattern of a hexagonal liquid crystal viewed through a microscope fitted with cross polarizers.

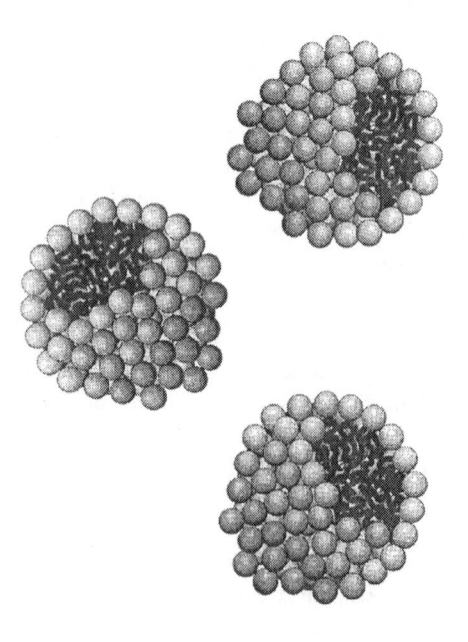

Fig. 3.11. Spherical micelle. Fennell & Wennerstrom, 1999.

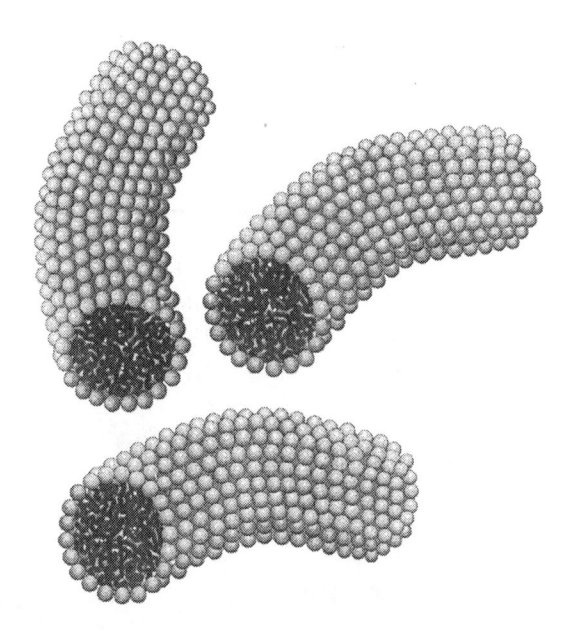

Fig. 3.11. Worm-like micelle. Fennell & Wennerstrom, 1999.

The solubility of soap in water is strongly temperature dependent. Solid soap crystals will dissolve to form either monomers or micelles in solution depending upon whether the temperature results in soap solubility that is below or above the CMC, respectively. The temperature at which solubility equals the CMC is known as the Krafft temperature, and corresponds with a rapid increase in soap solubility (Fig. 3.13). In soap literature, reference is sometimes made to a Krafft Boundary temperature. This value will typically lie 15–20°C above the Krafft temperature for a particular soap (Laughlin, 1994). In the case of soap it typically corresponds to the temperature at which solid soap crystal melts to form hexagonal liquid crystal.

The Krafft temperature of a particular soap is determined both by the counter-ion(s) and the aliphatic chain(s) that are present (Holmberg et al., 2003). For example, for a given aliphatic chain, Krafft temperature decreases as the counter-ion is changed from sodium to potassium to triethanollammonium (Murray & Hartley, 1935). Similarly, for a given counter-ion, Krafft temperature decreases as the aliphatic chain decreases in length and/or increases in degree of unsaturation (Krafft & Wiglow, 1895).

When an isotropic soap solution is viewed through a polarized microscope, the solution will appear completely black, demonstrating the absence of aggregates larger than micelles.

Isotropic solution phase was regularly observed by the old soap boilers. As noted previously, by the end of the soap-boiling process all colored impurities would settle into the lower layer of isotropic soap solution. As a result, this lower layer was darkly colored and hence was termed "nigre" (Thomssen & McCutcheon, 1949).

What is the state of matter of a bar of soap? A soap bar certainly appears solid, yet it can be deformed under sufficient pressure, as demonstrated by the manufacturing processes for commercial soap bars (e.g., extrusion). In reality, a bar of soap may typically contain multiple phases, including various types of solid crystals and liquid crystals, as described previously.

The number of phases that may exist in any system at equilibrium is determined by Gibbs' Phase Rule (Atkins, 1986). For example, when a soap bar contains the salt of a single fatty acid and water (two components) at room temperature and atmospheric pressure there can be at most two phases at equilibrium. But as soap is generally made from natural fats and oils, it will typically contain the salts of several fatty acids. This allows the simultaneous existence of multiple phases at equilibrium in a soap bar.

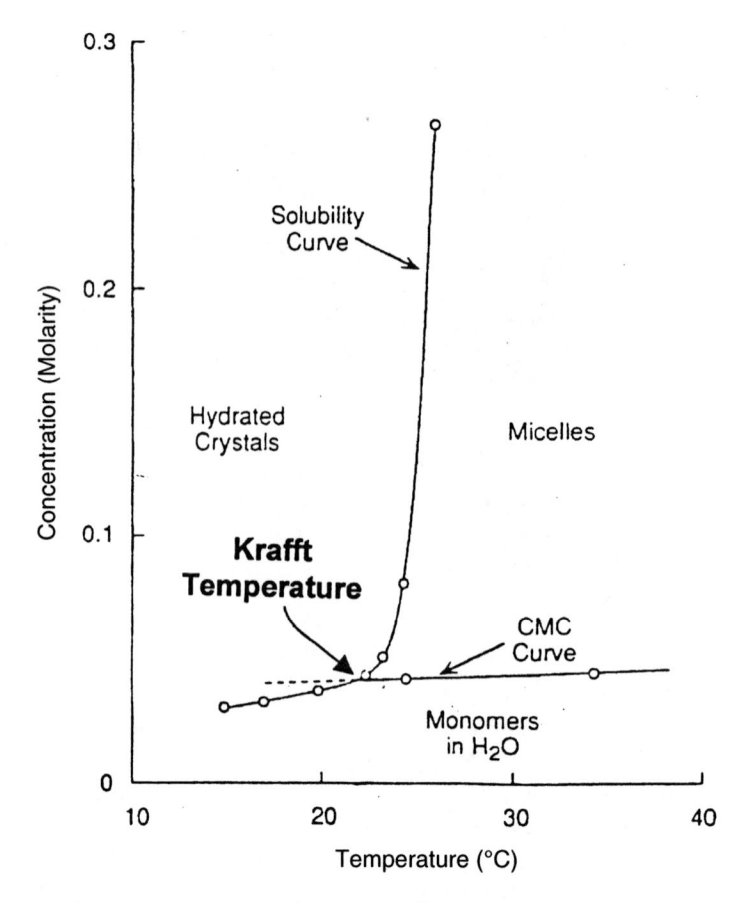

Fig. 3.13. Soap concentration vs. temperature, showing Krafft temperature. Abbreviation: Critical micelle concentration (CMC). Shinoda, 1963.

Commercial soap bars contain multiple components divided among multiple phases. This is further complicated by the fact that observed phases may be metastable. For example, the rapid removal of water from neat soap (lamellar phase) during vacuum spray drying precipitates out solid crystals of mixed chain lengths in the form of kappa phase (omega phase with the Ferguson et al. nomenclature [1942; 1944]). The fractions of short-chain saturated soap (C8–C14), long-chain saturated soap (C16–C24), and unsaturated soap (C18:1; C18:2; C18:3) making up this kappa phase will depend on the fat source. However, the fat source will also impact the final equilibrium bar structure. In conventional toilet soap, a mixture of two separate crystal types will form at thermodynamic equilibrium: one crystal type composed of the less soluble saturated long-chain soaps (e.g., C16 and C18), referred to as the delta phase, that is dispersed in a continuum of another crystal type composed of the more soluble saturated short-chain soaps and unsaturated soaps (e.g., C12 and C18:1), referred to as the eta phase. This configuration of less soluble soaps dispersed in a continuum of more soluble soaps can be compared to "bricks and mortar."

Phase structure notwithstanding, many of the desired properties of soap bars require the rapid dissolution of the more soluble soap components. Significantly, water can more easily interact with the continuous phase in a soap bar than with the dispersed phase. Hence it is desirable for the more soluble saturated short-chain soaps and unsaturated soaps to be in the continuous phase (i.e., in the equilibrium state) not the metastable state.

Mechanical work (shear and extensional flow) during soap finishing (intensive mixing, milling, and plodding) provides the surface renewal to facilitate the migration of molecules to their equilibrium phase through recrystallization. Thus, variations in soap-processing conditions will lead to macroscopic soap property variations, either through changes in the extent to which the actual phase structure is able to migrate to the equilibrium state (determined by the quantity of mechanical work), or through changes in the equilibrium state itself (determined by the process temperature and water concentration).

This transformation toward the equilibrium state can be monitored by changes in the low- and wide-angle x-ray diffraction patterns of the solid crystal phases, or by differential scanning calorimetry. Taking kappa phase soap through a soap-finishing line redistributes the composition of these solid crystal phases, as is evident from the peak broadening and upward shift in the interlayer spacing in the low-angle x-ray pattern (Fig. 3.14). Since the long spacing for all solid soap crystals of typical commercial soap blends exists within the range of 40–45 Å, the broadened primary peak for the finished soap in Fig. 3.14 reflects the presence of a second solid soap crystal at a slightly higher long spacing. This change corresponds to a transition from the metastable kappa phase to the equilibrium mixture of eta and delta phases.

Corresponding changes in the wide-angle pattern and in the Differential Scanning Calorimetry trace are shown in Figures 3.15 and 3.16, respectively. In each case, a finished soap produces sharper peaks than an unfinished soap because the mixture of eta and delta phases is more ordered than the mixed chain length kappa phase.

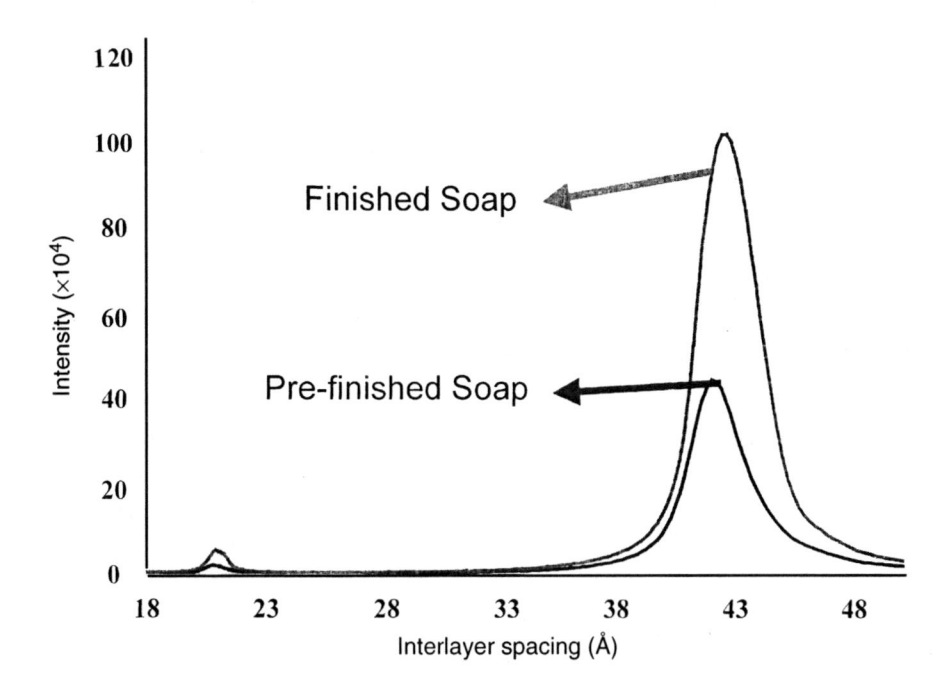

Fig. 3.14. Low-angle x-ray pattern of soap, pre- and post-finishing.

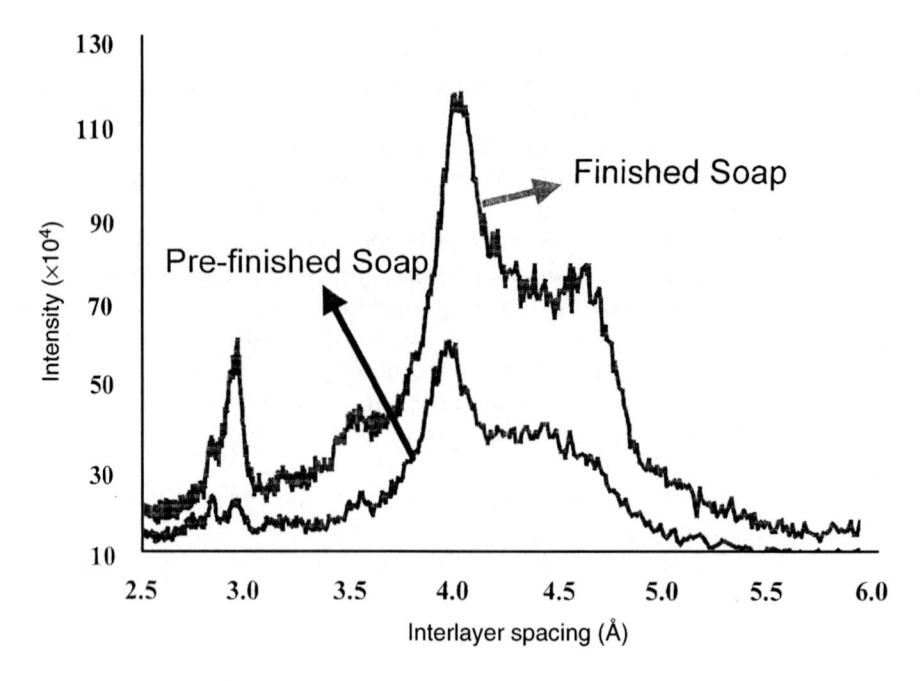

Fig. 3.15. Wide-angle x-ray pattern of soap, pre- and post-finishing.

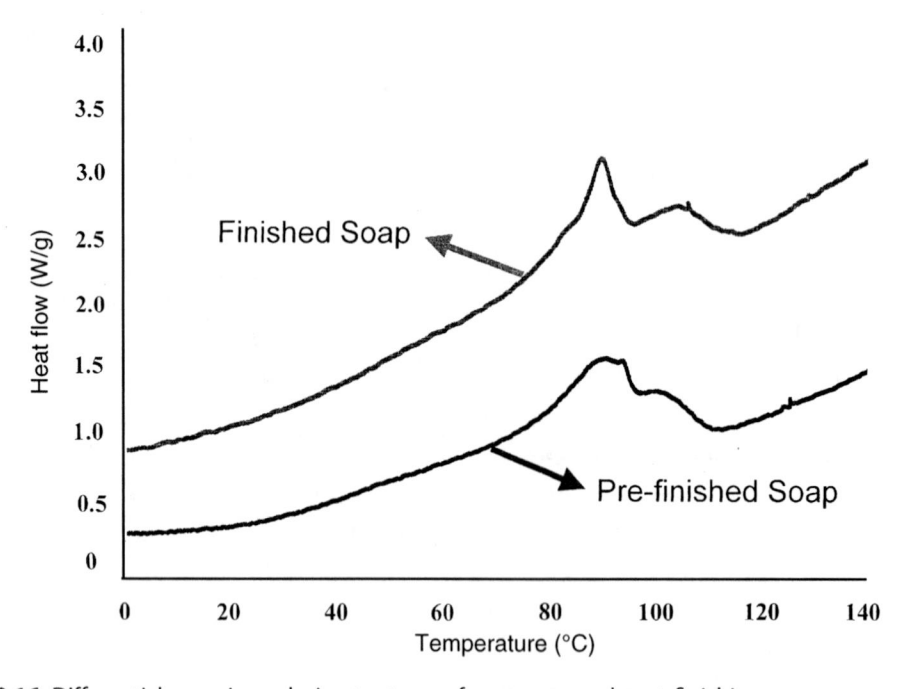

Fig. 3.16. Differential scanning calorimetry trace of soap, pre- and post-finishing.

Superfatted Soap

Commercial soaps may occasionally be formulated to contain excess or free fatty acids, and are then commonly referred to as "superfatted" soaps. While free fatty acid was originally added to soap to ensure the absence of unneutralized caustic soda, properly formulated superfatted soap bars have been observed to produce a high volume of rich dense lather (Piso & Winder, 1990; Bartolo, & Lynch, 1997). This may ultimately be traced to the impact of fatty acid on soap bar structure and phases.

The strong hydrogen-bonding interaction between the carboxylic acid group of the fatty acid and the carboxylate head group of the soap (Lynch et al., 2001; Lynch, 1997) gives rise to acid–soap complexes that may exist as solid crystals. While these complexes appear to form in definite stoichiometric ratios (Lynch et al., 2001; Lynch, 1997; Lynch et al., 1996), the question arises as to whether these complexes are thermodynamically stable states or are "kinetically" frozen states that arise as a consequence of the preparation method. Constructing an equilibrium binary phase diagram of an acid–soap is problematic due to the inability of the mixture to equilibrate rapidly in the solid state (Lynch et al., 2001; Lynch, 1997; Lynch et al., 1996; Kung & Goddard, 1968). Attempts to resolve some of these issues by careful sample manipulation resulted in a slight revision to the binary phase diagram of Na Palmitate–Palmitic Acid first reported by McBain (McBain & Field, 1933), which is shown in Fig. 3.17 (Lynch et al., 1996).

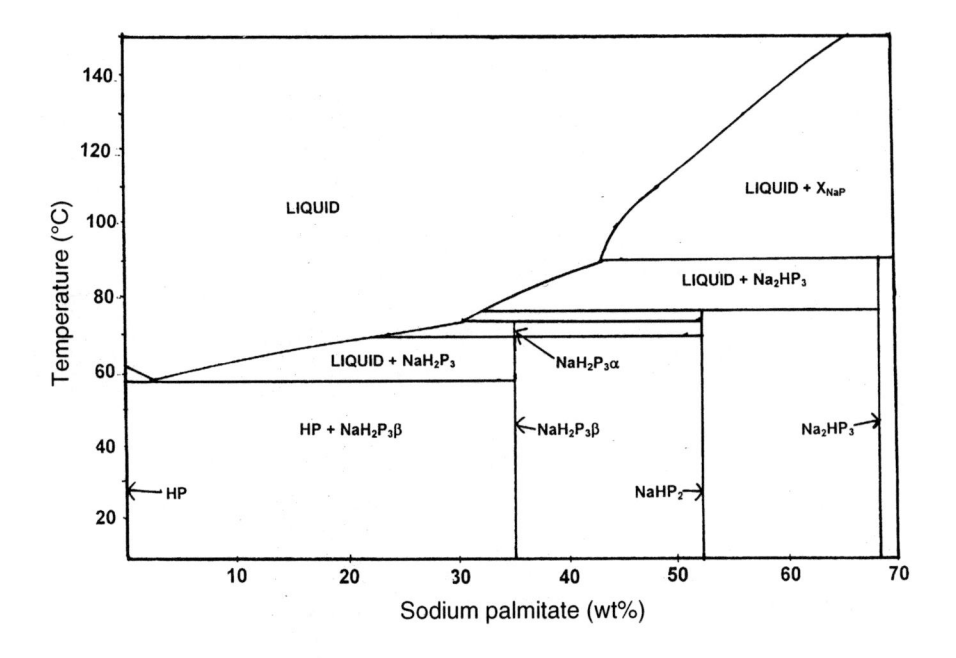

Fig. 3.17. Binary acid–soap phase diagram for palmitic acid-Na palmitate. Abbreviations: Palmitic acid, HP; Molecular complex of one molecule sodium palmitate and two molecules of palmitic acid (α and β are polymorphs), NaH_2P_3; Molecular complex of one molecule sodium palmitate and one molecule of palmitic acid, $NaHP_2$; Molecular complex of two molecules sodium palmitate and one molecule of palmitic acid, Na_2HP_3; Sodium palmitate crystal, X_{NaP}, Lynch et al., 1996.

130 M. Hill and T. Moaddel

When water is added to a fatty acid–soap system, a rich variety of liquid and liquid crystalline structures will form (Cistola & Small, 1986; Cistola et al., 1988). The temperature at which the different types of structures can form will depend on the nature of the hydrocarbon chain and degree of unsaturation of the fatty acid and the soap, the ratio of fatty acid to soap, and the amount of water. Fig. 3.18 is an illustration of types and expected location of the various phases in a fatty acid–soap–water ternary phase diagram. It should be noted that this ternary phase diagram describes the general behavior of all fatty acid–soap systems consisting of a single chain length and single degree of unsaturation, provided that system is above the melting temperature of the fatty acid.

What accounts for the superior lathering properties of superfatted soaps? While some have attributed this to the presence of solid crystals of acid–soap within the bar (Bartolo, & Lynch, 1997), the lathering properties can also be impacted by the bar's other phases. For example, a typical superfatted bar will contain roughly 81% soap, 7% fatty acid, and 12% water, plus minor ingredients. Since this corresponds to a fatty acid to soap ratio of 9%, a single fatty acid–soap system at this composition would be in a multiphase region, where one of the phases would be lamellar liquid crystal (Fig. 3.18). A lamellar liquid crystal is similarly present as one of the phases of superfatted soap systems made from blends of fatty acids. The facile interaction between the lamellar liquid crystal and water gives rise to rapid dissolution and hence higher lather volume.

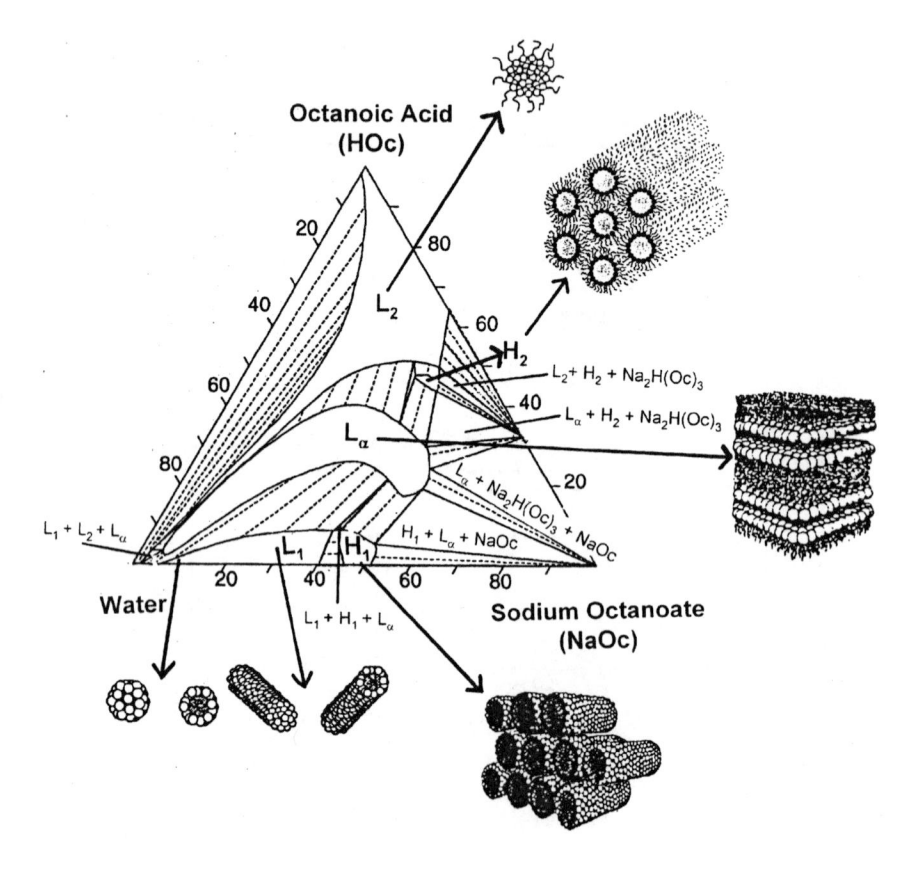

Fig. 3.18. Ternary acid–soap–water phase diagram for octanoic acid–Na octanoate water (phase diagram, pictures). Abbreviations: Octanoic acid, HOc; Sodium octanoate, NaOc; Molecular complex of two molecules of sodium octanoate and one molecule of octanoic acid, Na$_2$H(Oc)$_3$; Inverse hexagonal liquid crystal, H2; Inverse micellar phase, L2. See Fig. 3.2. Ekwall, 1975; 59. Davis, 1994.

Transparent and Translucent Soap

A soap bar is generally opaque because incident light is scattered by the heterogeneous domains within the bar. This light scattering can be significantly reduced either by matching refractive indices of the various domains, or by sufficiently reducing the domain size of the dispersed phase. Commercial transparent and translucent soap bars capitalize on these phenomena.

Cast Transparent Bars. It has long been known that if certain soap compositions are dissolved in hot ethyl alcohol, they may be cast into molds of the desired shape and allowed to solidify and age for alcohol evaporation, yielding a transparent bar (Cristiani, 1881). It is also possible to make similarly transparent bars using triethanolamine rather than ethyl alcohol, eliminating the need for aging or solvent evaporation to achieve full transparency (Fromont, 1958). These cast transparent bars are sometimes called poured or molded bars.

Cast transparent bars are typically made as a blend of 50% soap and 50% solvent. Depending on the desired bar properties regarding firmness and latherability, the soap blend employed can range from mixtures rich in long-chain saturated soaps to mixtures rich in short-chain saturated and unsaturated soaps (Instone, 1991). The solvent may contain ethyl alcohol, glycerine (or other polyols), sugar, and/or rosin (Murahata et al., 1997; Instone, 1991). The hot soap and solvent solution must itself appear transparent, showing the absence of any solid or liquid crystalline soap phases, or else the mixture will not give rise to transparent bars when cooled (Instone, 1991).

Little has been published on the structure and phase behavior of these systems. Since x-ray evidence indicates the presence of solid soap crystals, it was inferred that these crystals must be smaller than the wavelength of visible light, thereby permitting transparency (McBain & Ross, 1944). More recently it has been noted that the solvents both reduce the quantity of solid soap crystal and limit crystal size (Murahata et al., 1997; Jungermann, 1990). Nevertheless, this does not address the full colloidal structure of these bars.

The manufacture of cast transparent bars is a labor-intensive process since they must be de-molded when cool. Hence these bars have been mass produced only at a premium cost.

Extruded Translucent Bars

It was long noted that ordinary milled soap at high (e.g., 20–30%) water levels appears translucent if the opaque whitening agent is omitted (Ferguson & Rosevear, 1954; Kelly & Hamilton, 1957). While bars with this water level are soft and deform on aging, it was found that replacing some of the water with glycerine both enhances translucency and helps maintain hardness. These compositions achieve translucency both through matching refractive indices of the various domains and by sufficiently reducing the domain size of the dispersed phase.

The difference in refractive indices of the various domains present in a soap bar ultimately stem from the difference in refractive index between soap ($n = 1.5$) and the solvent, water ($n = 1$). Addition of co-solvents, such as polyols (glycerine, sorbitol, or propylene glycol), triethanolamine, or their mixtures, can raise the refractive index of the solvent to approach that of soap and improve translucency.

The domain size of the dispersed phase can be reduced by recrystallization. As discussed previously, mechanical work provides the surface renewal to facilitate the migration of molecules to their equilibrium phase. Within the right window of process temperature and water concentration, the equilibrium state for the dispersed phase will consist of small crystallites of zeta phase (beta phase in the Ferguson et al. nomenclature [1942; 1944]) (Ferguson & Rosevear, 1954). Hence sufficient mechanical work under the right process conditions will also improve translucency. This mechanical work may be delivered to the formulated soap mixture through the use of intensive mixers, roll mills, and/or refiner/plodders. Although they are somewhat less transparent than cast transparent bars, translucent extruded bars can be manufactured on conventional soap-making equipment and hence mass-produced at costs similar to that for opaque soap.

Conclusion

As this chapter has shown, a coherent understanding of the structure and phase behavior of soap has emerged over the last century, moving soap production from an art to a science. The complex behavior of soap systems can now be fully explained in terms of the molecular phenomena common to surfactant systems. Nevertheless, some questions involving soap systems persist. Some of these questions are of academic interest, such as those involving the behavior of solid soap at very low water levels. Other questions are of industrial interest, such as those involving the influence of a particular set of process conditions on soap bar properties. No doubt continued study will shed further light on the answers to both types of questions.

References

Atkins, P.W. Physical Chemistry, 3rd ed., W.H. Freeman and Co., New York, 1986, pp. 192–211.

Bartolo, R.G.; and M.L Lynch. Soap, in Kirk-Othmer Encyclopedia of Chemical Technology, edited by Jacqueline I. Kroschwitz and Mary Howe-Grant, John Wiley & Sons, 1997, Vol. 22, pp. 297–326.

Bernstein, J. Polymorphism in Molecular Crystals, Oxford University Press, New York, 2002, p. 243.

Buerger, M.J. The Characteristics of Soap Hemihydrate Crystals, *Proc. Nat. Acad. Sci.* **1942,** *28,* 529–535.

Buerger, M.J. The Characteristics of Soap Hemihydrate Crystals, *Proc. Nat. Acad. Sci.* **1942,** *28,* 529–535.

Buerger, M.J.; L.B. Smith; A. de Bretteville, Jr.; and F.V. Ryer. The Lower Hydrates of Soap, *Proc. Nat. Acad. Sci.* **1942,** *28,* 526–529.

Buerger, M.J.; L.B. Smith; F.V. Ryer; and J.E. Spike, Jr. The Crystalline Phases of Soap, *Proc. Nat. Acad. Sci.* **1945,** *31,* 226–233.

Cistola, D.P.; D. Atkinson; J.A. Hamilton; and D.M. Small. Phase Behavior and Bilayer properties of Fatty Acids: Hydrated 1:1 Acid–Soaps. *Biochem.* **1986,** *25,* 2804–2812.

Cistola, D.P.; J.A. Hamilton; D. Jackson; and D.M. Small. Ionization and Phase Behavior of Fatty Acids in Water: Application of the Gibbs Phase Rule. *Biochem.* **1988,** *27,* 1881–1888.

Cristiani, R.S. A Technical Treatise on Soap and Candles, Henry Carey Baird and Co., Philadelphia, 1881, p. 423.

Davis, H.T. Factors Determining Emulsion Type: Hydrophile–Lipophile Balance and Beyond, Coll. Surf. A: Physicochem. *Eng. Aspects 91,* **1994,** 9–24.

De Mul, M.N.G.; H.T. Davis; D. Fennell Evans; A.V. Bhave; and J.R. Wagner. Solution Phase Behavior and Solid Phase Structure of Long Chain Sodium Soap Mixtures, *Langmuir* **2000,** *16,* 8276–8284.

Dumbleton, J.H. and T.R. Lomer. The Crystal Structure of Potassium Palmitate (Form B), *Acta Cryst.* **1965,** *19,* 301.

Ekwall, P. Advances in Liquid Crystal, edited by G.H. Brown, Academic Press, New York, 1975, Vol. 1, pp. 1–142.

Fennell, E.D. and H. Wennerstrom. The Colloidal Domain, Where Physics, Chemistry, Biology, and Technology Meet, 2nd ed., Wiley-VCH, New York, 1999, pp. 16–17.

Ferguson, R.H. and F.B. Rosevear. U.S. Patent 2,686,761 (1954).

Ferguson, R.H. The Four Known Crystalline Forms of Soap, *Oil Soap* **1944,** *21,* 6–9.

Ferguson, R.H.; F.B. Rosevear; and R.C. Stillman. Solid Soap Phases, *Indust. Eng. Chem.* **1942,** *35,* 1005–1012.

Fromont, L.E.G.H. U.S. Patent 2,820,768 (1958).

Funari, S.S.; M.C. Holmes; and G.J.T. Tiddy. Microscopy, X-ray Diffraction, NMR Studies of Lyotropic Liquid Crystal Phases in the C22EO6/water system. A New Intermediate Phase. *J. Phys. Chem.* **1992,** *96,* 11029.

Grant, D.J.W. and T. Higuchi. Solubility Behavior of Organic Compounds, in Techniques of Chemistry, Vol. 21, edited by W.H. Saunders and A. Weissberger, Wiley Interscience, New York, 1990, pp. 32, 33.

Holmberg, K.; B. Jonsson; B. Kronberg; and B. Lindman. Surfactants and Polymers in Aqueous Solution, 2nd ed., John Wiley & Sons, 2003, pp. 89–93.

Holmes, M.C.; and J. Charvolin. Smectic–Nematic Transition in a Lyotropic Liquid Crystal. *J. Phys. Chem.* **1984,** *88,*

810.

Instone, T. U.S. Patent 5,041,234 A (1991).

Jungermann, E. Specialty Soaps: Formulations and Processing, in Soap Technology for the 1990s, edited by L. Spitz, AOCS Press, Champaign, Illinois, 1990, pp. 230–243.

Kekicheff, P. Phase Diagram of Sodium Dodecyl Sulfate–Water System, *J. Colloid Interface Sci.* **1989**, *131*, 133.

Kelly, W.A. and H.D. Hamilton. U.S. Patent 2,970,116 (1957).

Krafft, F. and H. Wiglow. Ueber das Verhalten der fettsauren Alkalien und der Seifen in Gegenwart Von Wasser. III Die Seifen als krystalloide. *Ber.* **1895**, *28*, 2566.

Kung, H.C. and E.D. Goddard. Molecular Association in Fatty Acid Potassium Soap Systems. II, *J. Coll. Inter. Sci.* **1968**, *29*, 242–249.

Laughlin, R.G. The Aqueous Phase Behavior of Surfactants, edited by R.H. Ottewill and R.L. Rowell, Academic Press, New York, 1994, pp. 106–136, 380.

Lewis, E.L.V. and T.R. Lomer. The Refinement of the Crystal Structure of Potassium Caprate (Form A), *Acta Cryst.* **1969**, B*25*, 702–710.

Luzzati, V.; A. Tardieu; and T. Gulik-Krzywicki. Polymorphism of Lipids. *Nature* **1968**, *217*, 1028.

Lynch, M.L. Acid–Soaps. *Curr. Opin. Coll. Inter. Sci.* **1997**, *2*, 495–500.

Lynch, M.L.; F. Wireko; M. Tarek; and M. Klein. Intermolecular Interactions and The Structure of Fatty Acid–Soap Crystals. *J. Phys. Chem. B.* **2001**, *105*, 552–561.

Lynch, M.L.; Y. Pan; and R.G. Laughlin. Spectroscopic and Thermal Characterization of 1:2 Sodium Soap/Fatty Acid–Soap Crystals. *J. Phys. Chem.* **1996**, *100*, 357–361.

Madelmont, C.; and C. Perron. Study of the Influence of the Chain Length on Some Aspects of Soap/Water Diagrams. *Coll. Polym. Sci.* **1976**, *254*, 581–595.

Mantsch, H.H.; S.F. Weng; P.W. Yang; and H.H. Eysel. Structure and Thermotropic Phase Behavior of Sodium and Potassium Carboxylate Ionomers. *J. Mol. Struct.* **1994**, *324*, 133–141.

McBain, J.W. and A. de Bretteville. X-Ray Evidence for a Third Polymorphic Form of Sodium Stearate, *Science* **1942**, *96*, 470.

McBain, J.W. and M.C. Field. Phase-Rule Equilibria of Acid Soaps. Part II. Anhydrous Acid Sodium Palmitate. *J. Chem. Soc.* **1933**, 920–932.

McBain, J.W. and S. Ross. The Structure of Transparent Soap. *Oil Soap* **1944**, *21*, 97–98.

McBain, J.W. and W.J. Elford. The Equilibria Underlying the Soap-Boiling Processes. The System Potassium Oleate-Potassium Chloride-Water. *J. Chem Soc.* **1926**, 421–438.

McBain, J.W. and W.W. Lee. Application of the Phase Rule to Soap Boiling. *Ind. Eng. Chem.* **1943**, *35*, 917–921.

McBain, J.W. and W.W. Lee. Vapor Pressure Data and Phase Diagrams for Some Concentrated Soap Water Systems Above Room Temperature. *J. Am. Oil Chem. Soc.* **1943**, *20*, 17–25.

McBain, J.W.; and W.W. Lee, Sorption of Water Vapor by Soap Curd. *Indust. Eng. Chem.* **1943**, *35*, 784–787.

McCrone, W.C. Polymorphism, in Physics and Chemistry of the Organic Solid State, edited by D. Fox, M.M. Labes, and A. Weissberger, Wiley Interscience, New York, 1965, Vol. 2, p. 726.

Murahata, R.I.; M.P. Aronson; P.T. Sharko; and A.P. Greene. Cleansing Bars for Face and Body: In Search of Mildness, in Surfactants in Cosmetics, edited by M.M. Rieger, and L.D. Rhein, Marcel Dekker, Inc., New York, 1997, pp. 307–330.

Murray, R.C. and G.S. Hartley. Equilibrium Between Micelles and Simple Ions, with Particular Reference to the Solubility of Long Chain Salts. *Trans. Faraday. Soc.* **1935**, *31*, 183.

Palmqvist, F.T.E. Soap Technology—Basic and Physical Chemistry, Alfa-Laval, Stockholm, Sweden, 1983, pp. 1–28.

Perron, R. Le Point sur les Diagrammes de Phase des Savons de Sodium. *Revue française des Corps Gras.* **1976**, *23*, 473–482.

Perron, R.; and C. Madelmont. Donnees Recentes en Physicochimie des Savons Alcalins, Revue francaise des Corps Gras. **1973,** *20,* 261–268.

Piso, Z.; and C.A. Winder. Soap, Syndet, and Soap/Syndet Bar Formulations, in Soap Technology for the 1990s, edited by L. Spitz, AOCS Press, Champaign, Illinois, 1990, pp. 209–229.

Rosen, M.J. Surfactants and Interfacial Phenomena, John Wiley & Sons, New York, 1978, pp. 83–122.

Rosevear, F.B. Liquid Crystals: The Mesomorphic Phases of Surfactant Compositions. *J. Soc. Cosm. Chemist.* **1968,** *19,* 581–594.

Seddon, J.M. Structure of the Inverted Hexagonal (HII) Phase, and Non-lamellar Phase Transitions of Lipids. *Biochim. Biophys. Acta.* **1990,** *1031,* 1–69.

Shinoda, K. The Formation of Micelles, in Colloidal Surfactants, edited by E. Hutchinson and P. Van Rysselberghe, Academic Press, New York, 1963, Vol. 28, p. 7.

Skoulios, A.E. and V. Luzzati. La Structure des Colloides d'Association. III. Description des Phases Mesomorphiques des Savons de Sodium Purs, Rencontrees audessus de 100°C. *Acta Crystallogr.* **1961,** *14,* 278.

Small, D.M. The Physical Chemistry of Lipids, edited by Donald J. Hanahan, New York and London, Plenum Press, 1986, pp. 285–340.

Tandon, P.; R. Neubert; and S. Wartewig. Phase Transitions in Oleic Acid as Studied by X-ray Diffraction and FT-Raman Spectroscopy, *J. Mol. Struct.* **2000,** *524,* 201–215.

Tandon, P.; R. Neubert; and S. Wartewig. Thermotropic Phase Behavior of Sodium Oleate as Studied by FT-Raman Spectroscopy and X-ray Diffraction. *J. Mol. Struct.* **2000,** *526,* 49–57.

Thomssen, E.G. and J.W. McCutcheon. Soaps and Detergents, MacNair-Dorland Co., New York, 1949, pp. 115–218.

Thomssen, E.G. and J.W. McCutcheon. Soaps and Detergents, MacNair-Dorland Co., New York, 1949, pp. 134–139.

Van Oss, C.J. Interaction Forces Between Biological and Other Polar Entities in Water: How Many Different Primary Forces Are There? *J. Dispersion Sci. Tech.* **1991,** *12,* 201–219.

Vold, M.J.; M. Macomber; and R.D. Vold. Stable Phases Occurring between True Crystals and True Liquid Crystals for Single Pure Anhydrous Soaps, *J. Am. Chem. Soc.* **1941,** *63,* 168.

Vold, R.D.; J.D. Grandine; and H. Schot. Characteristic X-Ray Spectrometer Patterns of the Saturated Sodium Soaps, *J. Phys. Chem.* **1952,** *56,* 128.

Yang, P.W.; H.L. Casal; and H.H. Mantsch. ATR Infrared Spectra of Oriented Sodium Laurate Multilayers, *Appl. Spectr.* **1987,** *41,* 320–323.

Zajic, J.; M. Malenicky; M. Brotankova; and M. Bares. Influence of Technological Process on Changes of the Internal Structure of Soap, Scientific Papers of the Institute of Chemical Technology, *Prague Food,* **1968,** E *22,* 103–110.

Formulation of Traditional Soap Cleansing Systems

Edmund D. George and David J. Raymond
Bradford Soap Works Inc., West Warwick, Rhode Island 02893, USA

Introduction

This chapter discusses soap base compositions, including a brief historical perspective on traditional soap and matrix effects which may affect the formulation process. Typically, any saponifiable oil can be made into soap, and the characteristics of the end result are dependent on the type of oil used. Traditional soap consists of sodium or potassium salts of triglycerides and fatty acids, notably from beef tallow, coconut oil, and palm kernel oil, and to a lesser extent from such oils as grape seed, sweet almond oil, rice bran oil, and others. Varieties of soap include transparent, opaque, and translucent soaps, and specialty bases such as shaving, nonmarring opaque, cream/paste, and powdered soaps. Since the production of these bases is well documented in the literature, they will not be discussed further in this chapter (Chambers et al., 1990; Chambers & Instone, 1990; Davidsohn et al., 1953; Dawson & Ridley, 1989; Deweever & Carrol, 1975; George & Serdakowski, 1987; Joshi, 1985; Jungennan et al., 1988; Nagashima et al., 1981; O'Neill et al., 1974; Swern, 1979; Thomsenn & Kemp, 1937; Verite & Caudet, 1981; Wood-Rethwill et al., 1989; Woolatt, 1985).

The formulation of soap bars has become more complex over the years due to an ever-increasing number of soap bases that incorporate more and more additives. The "green" and "natural" market segments have led to soap products with new materials. Also, consumers have become more accustomed to multifunctional products offered by the cosmetic industry, including conditioning shampoos, antiperspirants, sunscreens, lotions, and creams. Traditional soaps were designed for cleaning skin and clothes, but as time passed soaps came to be used as a delivery system for perfumes and superfatting agents. Today, the cleansing aspect seems almost secondary to the effects of the various additives that are delivered through the soap system. The 2006 *International Cosmetic Dictionary* published by the Cosmetic, Toiletry, and Fragrance Association, now called the Personal Care Products Council, lists over 13,000 monographs of INCI names as well as 3,000 suppliers and 57,000 trade names reportedly used in cosmetic applications (Gottschalck & McEwen, 2006).

As with any drug or cosmetic product, matrix effects must be considered when developing soap formulas. Among these are additive-base interactions, pH effects, additive-additive interactions, fragrance effects, and processing effects. Any combination of these effects may influence the physical and aesthetic characteristics of the final product. Sometimes it is difficult to predict the consequences of matrix effects, and only with time and experience does the formulator begin to understand these interactions.

Additive-base Interactions

These can occur when acidic compounds are added that may interact with the soap base by changing its physical or chemical characteristics. With enough acidic material, alkaline soap bases may break down into fatty acids, which renders the soap ineffective. This may not be immediately noticeable, since the soap base does not have sufficient water to behave like a solution, but it can occur over time after processing and storage.

pH Effects

Most stability problems arising from pH effects occur with additives in traditional soaps, as opposed to cosmetic and personal care products, which have an acidic pH. Certain compounds, including some quaternary compounds and fragrance ingredients, are unstable under the alkaline pH conditions found in traditional soaps. Also some OTC active ingredients, such as salicylic acid and benzoyl peroxide, are most stable in combo systems, which have a neutral to acidic pH.

Additive-additive Interactions

Additive-additive interactions are similar to the additive-base interactions mentioned earlier, and the two are handled in the same manner.

Fragrance Effects

Fragrance effects develop from fragrance compounds, such as aliphatic and aromatic acids, esters, ketones, and glycols. These compounds can profoundly affect the processing characteristics by increasing the softness and tackiness of the soap or, in the case of translucent or transparent soaps, altering the optical clarity. Fragrance diluents or solvents appear to soften and/or cloud transparent soaps, making an already difficult base even more difficult to work with. These include certain glycols, diethyl phthalate (DEP), and dipropylene glycol (DPG). Also, vanillin is known to cause severe browning in soaps due to chemical reactions in the alkaline pH range. Some newer ingredients on the market can be added to the formulation or fragrance to retard this effect. Formulators should therefore work closely with suppliers to optimize fragrance selections prior to finalizing the formulation. In order to ensure proper delivery and stability of the fragrance, fragrance suppliers must be briefed on the type of base that their fragrances will be incorporated into.

Processing Considerations

Processing must be considered when formulating a product. Process parameters that must be monitored or controlled include temperature, shear, scrap recycle, viscosity, vacuum, die refrigeration and shape, type of equipment, milling vs. refining, and plodder speed. Although this subject is beyond the scope of this chapter, it is an essential aspect of a successful product and must be given due consideration.

Properties of Soap Bases

Chemical Properties of Bar Soaps

Variations in the primary materials of the base formula influence the chemical properties of the soap. For instance, transparent bases can be made from detergents and fats and oils using combinations of sodium hydroxide, potassium hydroxide, and alkanlolamines such as triethanolamine. Synthetic systems can be plasticized with saturated fatty acids, fatty alcohols, or a combination. The ratio of fats and oils (e.g., 80% tallow/20% coconut oil vs. 70% tallow/30% coconut oil) and the choice of manufacturing process (e.g., continuous, full-boiled vs. semiboiled) also affects the chemical properties. However, the choice of the preservation system is critical to the long-term chemical stability of the cleansing system.

Antioxidants are useful as fat, oil and fatty acid preservatives. However, when these materials are converted to soap, we have found that chelators provide better protection than antioxidants such as BHT. We have found that as a preservative in traditional soap systems at a pH of 10, chelators provide better protection than antioxidants such as BHT. Additionally, BHT can cause severe yellowing of soap products when stored under certain wrapping and warehouse conditions. It is hypothesized that certain quinones form when BHT reacts with nitrogen-based exhaust products produced by warehouse motor trucks and lifts, which lead to the undesirable yellowing.

Virtually all of the preservatives listed in Table 4.1 are chelators. This is because pro-oxidant metals, such as iron, copper, zinc, and magnesium, have extremely negative effects on soap chemical stability, and therefore need to be deactivated. Pentasodium pentetate and tetrasodium etidronate are particularly effective preservatives for color and odor stability in these systems, and are often needed at levels below 0.10% for each. Chelators may be most functional when used in combination, depending on the type of metals (i.e., speciation) needing to be chelated. A single chelator may only be effective against certain metals and not at all effective against others. The formulator must be familiar with these properties in order to develop additive packages that are stable and functional.

Physical Characteristics of Bar Soaps

The physical characteristics of the soap bar also influence the amount and types of additives that are incorporated into the final formula. With each formulation, the following important bar characteristics and their parameters should be established in order to generate complete product profiles: wear rate, crack resistance and sloughing, washdown, lathering, color, and odor. Additives tend to influence some or all of these aspects of the soap bar, and the potential for negative impacts on them must be determined prior to production.

Table 4.1. Typical Soap Preservatives

INCI Name	Abbreviations
Chelators	
Diphosphoric Acid	HEDP
Tetrasodium Etidronate (Na$_4$HEDP)	Sodium HEDP
Etidronic Acid	EHDP
Ethylenediamine Disuccinic Acid	EDDS
Tetrasodium Etidronate (Na$_4$HEDP)	Sodium HEDP
Pentasodium Pentetate	Na$_5$DPTA (DPTA)
Tetrasodium EDTA	Na$_4$EDTA
Trisodium EDTA	Na$_3$EDTA
–	NaLED$_3$A (surfactant-chelator compound)
Tetrasodium Etidronate	Sodium HEDP (Na$_4$HEDP)
Tetrasodium Etidronate and Pentasodium Pentetate	Sodium HEDP (Na$_4$HEDP) & Na$_5$DPTA (DPTA)
Tetradibutyl Pentaerythrityl Hydroxyhydrocinnamate	
Trisodium Ethylenediamine Disuccinate	Na$_3$EDDS
Tetrasodium Glutamate Diacetate	GL
Citric acid and salts	
Gluconic acid and salts	
Antioxidants	
BHT	BHT

Wear Rate

Wear rate describes the lasting power of the soap bar under use conditions. It is influenced by the solubility of the base, which is determined by the titer of the fats and oils, the type of alkali used, and the amount of water.

For instance, transparent soaps have relatively high wear rates due to the use of high levels of solvents, such as glycols, water, and alcohols, which are needed to maintain clarity. These soaps may also contain a surfactant system that aids in maintaining clarity but has high solubility. This combination tends to let the bar "melt away."

Crack Resistance and Sloughing

Crack resistance relates to the tendency of soap bars to crack and/or disintegrate when subjected to repeated wet/dry cycles. It is measured in the laboratory by submersing one half of the bar in ambient water for 4 to 24 hours, then air-drying it until completely dry. Cracks will appear if the system is prone to cracking. Sloughing is often described as the dissipation, crumbling or shedding away of soap during use. Related to wear rate, the amount of sloughing is similarly determined by submersing, drying, and calculating the weight loss.

Industry experience indicates that translucent soaps traditionally have poor crack resistance. One theory suggests that translucent systems lack the "grain," or internal crystalline structure, that traditional opaque soaps have. This grain allows the soap to hydrate and dehydrate uniformly, whereas translucent soap lacks sufficient structure to provide this stability. Sucrose and certain solvents in fragrances can also promote cracking.

Washdown

The feel of a bar during use can be determined by a washdown test. This is usually performed at a relatively low temperature, such as 85–90°F, in order to determine if there is any grit, drag, or sandiness in the bar. Synthetic and combo systems are prone to this problem, as well as formulas containing sodium cocoyl isethionate. Causes include improper processing when the base and bar are made, or hard particles in the surfactant system. When a bar is allowed to get too cold during processing or shipping, this can create a potentially unpleasant feel for the user, although grit may disappear at temperatures encountered during normal use. Sandiness may occur in traditional opaque soap bases when excessively dry particles form during the vacuum-drying process. This problem is less prominent in soaps with a relatively high water range, for example 12–14%.

Lathering

Although lathering and detergency are not necessarily related, and foam may actually be just a visual aid allowing the user to see where the product has been applied, consumers associate quick, copious foam with quality and cleaning. Foaming characteristics can be influenced by many factors, including the types and ratios of fats and oils or, in the case of synthetics, the types of surfactants and plasticizers. Many additives that are oily in nature tend to act as defoamers if incorporated at high levels, such as in superfatted bars. Traditional soaps will lather poorly in hard water and seawater, whereas synthetics, if properly formulated, will foam well. Standard foam height tests should be performed when determining the product profile.

Color

Soap bases tend to yellow, so that the color of the final bar formulations will also change. This, coupled with additive and fragrance instability, can produce color variations over a short period of time. Accelerated stability testing in oven, sunlight, and/or fluorescent light can help predict the stability of the system.

A reflectance colorimeter is used to record the color of a sample. This instrument mathematically calculates the color as the human eye sees it. Measurements can be stored in a computer database, enabling the color to be recalled in the future as a reference standard. This method can be used for all types of soaps, including translucent and transparent soaps, as well as for determining the yellowness of soap bases. It is a particularly reliable tool for color characterization. Other, more sophisticated color-measuring devices can also be employed that not only measure color, but also formulate and correct soap batches, thus allowing more precise and efficient color matching.

Odor

Olfactory evaluation of soap bases and finished soap products is as important as any other measurements of physical characteristics. Consumers tend to view fragrance perceptions as being at least as important as any other product characteristic. Therefore it is important that fragrances be formulated for soap to ensure as much stability as possible. Odor stability can be evaluated under similar conditions as color stability. Trained technicians and odor evaluation panels usually review the results of olfactory tests.

Colorants

The use of colorants very much depends on the type of product that is being produced and government regulations governing the product. Colorants can be divided into three categories:

- color additives subject to certification (certified additives)
- color additives not subject to certification (noncertifiable additives)
- color additives not certified (noncertified additives)

In the United States, these distinctions are used in applying the federal Food, Drug, and Cosmetic (FD&C) Act (US Code of Federal Regulations, Title 21, part 1–99) and the Fair Packaging and Labeling (FP&L) Act, which set forth the definitions and labeling requirements for drugs and cosmetics. Drugs must use certified colors, whereas cosmetics can use certified and noncertifiable colors. Soap is exempt from these requirements, and a product can use any combination of colors, as long as it meets the definition of soap and is designated as such. However, any soap that makes drug and/or cosmetic claims must be labeled accordingly, with the appropriate colorants included in the ingredient list. In countries other than the United States, resources such as the CTFA dictionary can be consulted to determine acceptable and approved colorants.

Tables 4.2 and 4.3 list some of the more common certified and noncertifiable colorants. Certified colors are common for drugs and cosmetics, whereas noncertifiable colors are primarily used in cosmetic soaps.

Certified Colorants

Certified colorants (see Table 4.4) may be water-soluble, oil-soluble, or oil-dispersible. They also include the corresponding metal lakes. The solutions or dispersions are typically made in the lab at the 1–2% level, but higher loads of 30–50% may be obtained from vendors who have specialized equipment for grinding and dispersion.

Solubility and dispersion tables should be consulted to determine the optimum concentrations for even, complete dispersal. To keep dispersant and additive levels low, it is recommended that colorants be used at the lowest but most effective concentration possible. Certified colorants tend to have the lowest stability of the three categories, and they are affected by a number of factors, including pH, sun and fluorescent light, heat, and additive interactions. For instance, FD&C Green 3 renders a green color below pH 7 but a royal blue above pH 7. Stability stations should be used to determine the overall color stability in the surfactant system.

Table 4.2. Color Additives Subject to Certification (FDA Summary of Colors-March, 2007)

Color Additives	Color Index Number	Use in Cosmetics
FD&C Blue 1	42090	
FD&C Green 3	42053	Except in eye area
FD&C Red 4	14700	Externally except in eye area
FD&C Red 40	16035	
FD&C Yellow 5	19140	
FD&C Yellow 6	15985	Except in eye area
D&C Blue 4	42090	Externally except in eye area
D&C Brown 1	20170	Externally except in eye area
D&C Green 5	61570	
D&C Green 6	61565	Externally except in eye area
D&C Green 8	59040	Externally except in eye area (0.01% max)
D&C Orange 4	15510	Externally except in eye area
D&C Orange 5	45370	Externally except in eye area Lip products (5% max) Mouthwashes, Dentifrices (GMP)
D&C Orange 10	45425	Externally except in eye area
D&C Orange 11	45425	Externally except in eye area
D&C Red 6	15850	Except in eye area
D&C Red 7	15850	Except in eye area
D&C Red 17	26100	Externally except in eye area
D&C Red 21	45380	Except in eye area
D&C Red 22	45380	Except in eye area
D&C Red 27	45410	Except in eye area
D&C Red 28	45410	Except in eye area
D&C Red 30	73360	Except in eye area
D&C Red 31	15800	Externally except in eye area
D&C Red 33	17200	Externally except in eye area Lip products (3% max) Mouthwashes, Dentifrices (GMP)
D&C Red 36	12085	Externally except in eye area Lip products (3% max)
D&C Violet 2	60725	Externally except in eye area
D&C Yellow 7	45350	Externally except in eye area
D&C Yellow 8	45350	Except in eye area
D&C Yellow 10	47005	Externally except in eye area
D&C Yellow 11	47000	Externally except in eye area
D&C Black 2	77266	
D&C Red 34	15880	
Ext. D&C Violet 2	60730	
Ext. D&C Yellow 7	10316	

Formulation of Traditional Soap Cleansing Systems • 141

Table 4.3. Color Additives Exempt from Certification (FDA Summary of Colors-March, 2007)

Color Additives	Color Index Number	Use in Cosmetics
Aluminum powder	77000	Externally including the eye area
Annatto	75120	No restrictions
Beta-carotene	75130/40800	No restrictions
Bismuth citrate	–	Scalp hair dye only
Bismuth oxychloride	77163	No restrictions
Bronze powder	77400	No restrictions
Caramel	Caramel	No restrictions
Carmine	75470	No restrictions
Chlorophyllin-copper complex	75810	Cosmetic dentifrices (0.1% max)
Chromium hydroxide green	77289	Externally including the eye area
Chromium oxide greens	77288	Externally including the eye area
Copper powder	77400	No restrictions
Dihydroxyacetone	–	Externally including the eye area
Disodium EDTA-copper	–	Cosmetic shampoo only
Ferric ammonium ferrocyanide	77510	Externally including the eye area
Ferric ferrocyanide	77510	Externally including the eye area
Guaiazulene	–	Externally except the eye area
Guanine	75170	No restrictions
Henna	–	Scalp hair dye only
Iron Oxides	77489 77491 77492 77499	No restrictions
Lead acetate	(Prohibited)	Scalp hair dye only (0.6% Pb w/v max)
Luminescent Zinc Sulfide	–	
Manganese violet	77742	No restrictions
Mica	77019	No restrictions
Pyrophyllite	–	Externally including the eye area
Silver	77820	Nail polish only (1% max)
Titanium dioxide	77891	No restrictions
Ultramarines (blue, green, pink, violet)	77007	Externally including the eye area
Zinc oxide	77497	No restrictions

Noncertifiable Colorants

Noncertifiable colorants (see Table 4.4) tend to be very stable compounds and are used extensively in cosmetics, eye shadows, mascaras, and facial makeup. In some products they can be used as primary and secondary colorants and can stabilize color drifting previously caused by certified colorants. These compounds should be dispersed or wet out before use to maximize their color value. A dispersion aid, such as a 30–40% potassium cocoate solution, helps disperse these colorants well. The formulator can develop a color by approximating the amount of dry pigment and dispersing the powder in the cocoate solution. "Blooming" can occur, where color migrates slowly into the soap system because of large pigment particles, if the colorant is not dispersed properly or dries out before use. Unfortunately this can happen over hours or days of production and cause a blotched color in the finished product. However, there are newer materials on the market that contain predispersed colorants for ease of use.

Opacifiers, such as titanium dioxide, zinc oxide, and bismuth oxychloride, also are used to give uniformity to the color system. Without them the soap may appear to have various light and dark colored areas caused by compression from processing and pressing operations. Titanium dioxide is offered in both rutile and anatase forms, with the water-dispersible anatase USP grade used most often. Traditionally, rutile types typically do not give the brightness or the hiding power that the anatase gives. However, there are newer rutile versions whose ability to provide opacity, stability, and brightness appears to be comparable to, or better than, the anatase types.

Table 4.4. Common Soap Colorants

Certified Colorants	Noncertifiable Colorants
Green 3 (CI 42053)	Caramel
Red 4 (CI 14700)	Chromium hydroxide green
Red 40 (CI 16035)	Chromium oxide greens
Yellow 5 (CI 19140)	Iron oxides
Green 5 (CI 61570)	Mica
Green 6 (CI 61565)	Titanium dioxide
Green 8 (CI 59040)	Ultramarines (blue, green, pink, red, violet)
Orange 4 (CI 15510)	Zinc oxide
Red 17 (CI 26100)	
Red 33 (CI 17200)	
Violet 2 (CI 60725)	
Yellow 10 (CI 47005)	

Noncertified Colorants

In the United States, noncertified pigments are often used in bar soaps that conform to the definition and labeling requirements of soap. These colors tend to be very stable and less sensitive to pH and to provide a wide range of brightness and hues. They often are supplied as dispersions with concentrations of 25–50%. Also, only small amounts of these pigments are required to achieve dark colors in most soap bases.

Table 4.5 highlights some of the more common noncertified colorants that are used in soaps.

Table 4.5. Common Noncertified Colorants in Soap

Pigment Name	Pigment Type	Color Index Number
Phthalo Blue RS	Blue 15	74160
Phthalo Blue	Blue 15:1	74250
Phthalo Blue GS	Blue 15:3	74160
Phthalo Green	Green 7	74260
Quinacridone Magenta	Red 122	73915
Napthol Red ITR	Red 5	12490
Carbazole Violet	Violet 23	51319
Hancock Yellow G	Yellow 1	11680

Natural Colorants

The so-called biocolorants, which reportedly have bioactive properties, are another interesting group of materials. These materials are not true colorants, as defined by the preceding categories, but are typically plant extracts that impart a variety of bright colors to the product. They come in the form of water-soluble liquids and powders, as well as oil-soluble waxes and powders.

Fragrances

Virtually any scent can be created to fit a product profile and marketing concept. All fragrances should be developed to ensure stability and potency for use in a particular soap matrix. Use levels will vary depending on composition, but ranges typically start at 0.25–0.50% for masking purposes and can go as high as 3–4% for prestige fragrance bars.

At high use levels it is important to use vacuum in the extrusion process to minimize surface bubbling or blistering caused by the fragrance components. With the vacuum system, a very small amount of fragrance will be sacrificed in order to obtain a uniform overall extrusion.

As mentioned previously, fragrances may affect the processing of soaps to the extent that they will not extrude or press (stamp) at the desired rates. This is seen in synthetic and combo systems and, more dramatically, in translucent systems where clarity and firmness are critical. As indicated earlier, two particular fragrance additives used as solvents cause problems in translucent soaps. Dipropylene glycol (DPG) and diethylphthalate (DEP), tend to make translucent soaps sticky to the point where the soap does not extrude or press at all. When these solvents are removed from fragrance formulations, a noticeable increase in productivity can be observed. Certain resinoid compounds used in fragrances may also cloud translucent systems due to solubility and particulate effects. Other components may produce similar effects, which should be investigated through experimentation.

Additives

Most soap products include compounds added to achieve a desired functional and/or marketing position. This has become the prime focus of soap formulators. The *CTFA Cosmetic Handbook* (16) and *McCutcheon's Functional Materials* (Manufacturing Confections Publishing Co., 2006) contain information on many of these functional materials. We discuss several important categories:

- Emollients
- Humectants
- Moisturizers
- Occlusive agents
- Dermabrasive agents
- Drug components
- Anti-irritants
- Foam boosters
- Miscellaneous compounds

Evaluation of additive benefits in soap bases is a difficult task for the formulator. Most often, expert panels evaluate performance and aesthetic perceptions. More sophisticated and highly technical techniques can also be employed, such as measurements of transepidermal water loss (TEWL) and skin elasticity, tracer studies for residual ingredients after rinsing, spectroscopic and fluorescence studies, and human skin models. Kajs and Garstein review the use of these methods in evaluating cleansing products (Kajs & Gartsein, 1991).

Emollients

Emollients are compounds that are used to impart and maintain skin softness and pliability and to generally improve the skin's overall appearance. Fatty esters, alkoxylated ethers, and alkoxylated alcohols comprise the bulk of the ingredients in this category (Table 4.6). Typical use levels are 1–3%. These compounds are stable under normal bar soap conditions; however, care should be taken to review the physical properties of each of these materials in order to determine the best way to incorporate the ingredients into the base, as well as to identify any stability considerations, such as pH and temperature.

Humectants/Moisturizers

Humectants and moisturizers (Table 4.7) are skin-conditioning agents that increase the moisture levels and moisture-retaining abilities of the skin. Their effectiveness depends on the humidity in the environment. Typical use levels vary widely, from 0.1% to 10%. Dahlgren and coworkers describe the results of instrumental methods of evaluating glycerin in various surfactant systems, as well as perceived skin benefits (Dahlgren et al., 1987). The results indicate that high levels of glycerin provide improved skin feel and softness. Monosaccharides or simple sugars, such as fructose and glucose, darken under alkaline conditions, and this process is accelerated by the small amount of free alkali that is present in traditional soap bases. The neutralization of the free alkali by the addition of a fatty acid, such as coconut or stearic acid, helps to reduce the darkening effect. The discoloration will vary depending on the type and the amount of the sugar. Disaccharides, such as sucrose, are relatively stable under mild alkaline conditions. Under humid conditions, water droplet formation or sweating may occur with high levels, 10% or higher, of glycols and alcohols, such as glycerin, sorbitol, and propylene glycol. This is a persistent problem with transparent soaps that employ high levels of multiple humectants that also serve as solvents and solubilizers. This hygroscopic effect appears to be less of a problem with other types of soap systems that employ much lower levels, under 5%, of these materials.

Occlusive Agents

Occlusive agents (Table 4.8) include ingredients that are designed to prevent moisture evaporation from the skin, thereby helping to maintain soft and smooth skin. They typically are lipid in nature and are added to achieve the desired effect in so-called dry-skin products. Use levels are in the 1–10% range. The higher levels may cause extrusion and pressing problems by making the soap excessively soft and sticky, but this can be reduced by incorporating the ingredients in the base-making stage or in the premilling/refining stage prior to extrusion.

Dermabrasive/Exfoliating Agents

Dermabrasive agents work in conjunction with cleansers to remove the outermost layer of stratum cornea by scrubbing, thus producing a smoother feel to the skin. Loden and Anders describe a method to measure the effects of dermabrasive cleansers and have determined that they are perceived to improve skin smoothness (Loden & Anders, 1990).

Natural components, such as bran, loofah, oatmeal, jojoba beads, and various seeds, are popular, since they can be blended to achieve a desired look in the cleanser and a desired feel on the skin, and yield a "natural" product. If a natural component is not needed, polyethylene beads of various sizes and colors can also be used effectively. These agents can be added to the amalgamator process under normal conditions, with some precautions. When using refiners, the screen sizes should be large enough to allow the scrub agents to pass through. Otherwise they will be filtered out. Likewise, when using mills, the milling action should not break down the component into fine particles, which will reduce the scrubbing effect. This can be accomplished through proper gapping of the roller mills and the

Formulation of Traditional Soap Cleansing Systems • 145

Table 4.6. General Emollient Groupings and Common Ingredients

Esters	Ethers	Alcohols	Oils and Oil Type	Polyols	Lanolin Derivatives	Silicone Derivatives
Butyl Myristate	Methyl gluceth-1	Cetearyl Alcohol	Castor oil	Glycerin	Acetylated lanolin	Dimethicone Copolyol
Cetyl palmitate	Methyl gluceth-20	Cetyl Alcohol	Jojoba oil	Diglycerin	Lanolin	Dimethicone
Glycerol stearate	PPG-10 methyl glucose ether	Stearyl Alcohol	Mineral oil		Lanolin esters	Silicone fluids
Isopropyl myristate	PPG-20 cetyl ether	Propylene Glycol	Mink oil		Lanolin alcohols	Silicone Quaterium-18
Isopropyl palmitate	PPG-20 lanolin alcohol ether		PEG-glycerides		Lanolin fatty acids	Behenyl Dimethicone
Octyl palmitate	PPG-20 methyl glucose ether		Petrolatum		PEG-lanolin	
Tridecyl neopentanoate	PPG-20 oleyl ether		Shea Butter		Ethoxylated lanolin alcohols	
Avocado-Oil IPeg-8 Esters	PPG-20 myristyl ether		Cocoa butter			
Glycereth-7 Benzoate			Sweet almond oil, wheat germ oil, olive oil, grapeseed oil			
Isostearyl lactate						

Table 4.7. Typical Humectants and Moisturizers

Saccharides	Ethers	Alcohols	Polyols	Miscellaneous Compounds	Botanicals
Fructose	Methyl gluceth-10	Mannitol	Glycerin	Acetamide MEA	Aloe Barbadensis Leaf Extract
Glucose	Methyl gluceth-20	Sorbitol	Diglycerin	Hydrogenated starch hydrosylate	Hypnea Musciformis Extract
Honey	PPG-10 propylene glycol	Propylene Glycol	Sorbeth-20, 30, 40	PCA	Gellidiela Acerosa Extract
Lactose		Octyl Decanol		Sodium PCA	Sargassum Fillpendula Extract
Sucrose		Tetradecyleicosanol		Urea	Avena Sativa (Oat) Kernel Flour
Xylitol		Polyglycerin compounds		Silicone Quaterium-18	Fruit extracts
Galactoarabinan					Essential Oils and Oil Extracts
β-Glucan					Nonfat Dry Milk Cranberry Seed Oil

use of low-shear mills. If this is not possible, then a scrub agent can be added in a postmilling step or possibly in the finish extruder with no screens. This problem occurs mostly with larger-sized natural components, such as loofah and oatmeal, and to a much lesser extent with the polyethylene beads. Finally, mild exfoliation can be achieved using spherical exfoliants, such as polyethylene and jojoba beads, since they tend to roll over the skin. Irregularly shaped material, such as crushed seeds, leaves, and shells, tend to provide a more aggressive effect.

Drug Components

The drug categories in which bar soaps are most commonly included are the antimicrobial and anti-acne products. In the United States, each drug category is governed by a monograph published by the federal Food and Drug Administration, the FDA. Each monograph sets the conditions under which a product in a particular category may be sold. Organizations such as the CTFA and SDA, and others, have been working diligently with the FDA over the last several years to finalize the antimicrobial monograph to include personal-care antimicrobial cleansers as a separate product class within this category.

The formulator should consult the relevant monograph to obtain specific information for a category prior to beginning development work on a product containing a drug component, and such products should be manufactured or controlled according to good manufacturing procedures (GMP), since the drugs are subject to FDA auditor inspections.

Over-the-Counter (OTC) Active Ingredients in Soap

Typical soap categories containing OTC active ingredients include antimicrobial and topical anti-acne products. Again, OTC/GMP practices need to be enforced in accordance with current United States and European Union monograph requirements.

Table 4.8. Typical Occlusive Agents

Beeswax
Castor Oil
Coconut Oil
Dimethicone
Hydrogenated Oils
Jojoba Oil
Jojoba Wax
Mineral Oil
Natural and Synthetic Waxes
Olive Oil
Petrolatum
Shea Butter
Theobroma Grandiflorum Seed Butter
Cocoa Butter
Astrocaryum Murumuru Butter
Sweet Almond Oil
Tallow
Wheat Germ Oil
Hydrogenated Soybean Oil (and) Hydrogenated Soybean Polyglycerides (and) C15-23 Alkane
Hydrogenated Soybean Polyglycerides (and) C15-23 Alkane
Dilinoleic Acid

Antimicrobial Compounds

Of the antimicrobials, triclosan and triclocarbon are the two compounds most widely used in bar soaps. Typical use levels are 0.3–1% maximum for triclosan and 1–1.5% maximum for triclocarbon. They are incorporated at the amalgamator stage and may be predispersed or dissolved in a suitable solvent, such as a fragrance, prior to addition. As with all drug products, the soap line should be validated to ensure that the finished product is homogenous and that the proper level of antimicrobial is in the soap.

Anti-acne Compounds

The most common approved anti-acne ingredient used in traditional bar soap is sulfur, at 3–10%. Salicylic acid at 0.5–2% and benzoyl peroxide in the 5–10% range are also approved ingredients, but are used mostly in synthetic or combo bases due to stability concerns. Sulfur is stable in both alkaline and acid bases and is normally added at the amalgamator stage and processed under normal conditions.

Anti-irritants

There are several ingredients on the market for which anti-irritant properties are claimed. Among these are sucrose esters, alpha-bisabolol, lactylates, and ethoxylated vegetable oils. Also, surfactant mixtures incorporating materials such as amphoterics, sarcosinates, ethoxylated surfactants, glucosides, and isethionates can reduce the overall irritancy of the product. Most of these compounds have been evaluated in shampoo systems, in which skin and eye irritation levels are reportedly reduced, but they can also be effective in bar soaps.

The overall mechanisms of action are not fully understood, but are believed to involve several factors, including binding to or formation of complexes with irritants, blocking sites that are prone to irritation, and prophylaxis that covers the skin and thereby reduces or prevents irritant contact. Before adding an anti-irritant to a formulation, a baseline test should determine the irritation level of the test material without the anti-irritant ingredient to serve as the negative control. This can be achieved by several different methods that are currently on the market. The so-called soap chamber test, developed by Frosch and Kligman, and the zein test, as described by Gotte, are in vitro methods that measure the ability of surfactants to solubilize the vegetable protein zein (Frosch & Kligman, 1979; Gotte, 1966). Newer in vitro methods continue to be developed that utilize human skin factors in benchtop assays and will benefit the formulator in screening various additive packages quickly. There does not appear to be one single test that is an adequate predictor of irritancy. Rather, several testing protocols may be needed to evaluate a soap product or surfactant.

Secondary Surfactants

Secondary surfactants are often used to increase the performance of the bar, resulting in improved skin feel, reduction of irritation caused by the primary surfactant, improved solubility, or improved quality and quantity of the foam. Typically they are added at low levels, under 5%, as an adjunct to the primary surfactant.

Theoretically, most surfactants found in shampoos and liquid soaps may be used in bar soap systems as long as they are compatible with the system and are stable at the bar's given pH. However, most of them are pastes and liquids, which tend to make soap systems tacky, resulting in increased processing problems. This is due not only to their physical state, but also to their tendency to lower the viscosity of soap systems, thus causing softening of the system. Additives and fragrances may complicate the situation, as previously discussed.

Table 4.9 lists several of the more common surfactants that are used in the industry, including acyl isethionates, amphoterics, sarcosinates, sulfosuccinates, and sulfoacetates. These ingredients may be added to the amalgamator and processed under normal conditions. In particular, sodium cocoglyceryl sulfonate has good foam boosting, after-feel, and processing characteristics. When working with powdered surfactants, care should be taken to safely handle dust produced. An alternate method for introducing these additional surfactants is to include them in the base-making process so that they become an integral part of the base. For synthetic and combo bases, these surfactants may be added during the normal compounding process. Additionally, Rieger has developed an excellent list and overview of surfactant types available to the formulator (Rieger, 1993).

Table 4.9. Secondary Surfactants

Type	Example	Chemical Type	Form
Acyl isethionates	Sodium cocoyl isethionate	Anionic	Solid
Acyl isethionates	Sodium cocoyl methyl isethionate	Anionic	Solid
Sulfonic Acid	Sodium cocoyl taurate	Anionic	Solid
Sulfonic Acid	Sodium methyl cocoyl taurate	Anionic	Solid
Amphoterics	Disodium cocoamphodiproprionate	Amphoteric	Paste
Sarcosinates	Sodium cocoyl sarcosinate	Anionic	Liquid-solid
Sulfosuccinates	Disodium lauryl sulfosuccinate	Anionic	Liquid-solid
Sulfoacetates	Sodium lauryl sulfoacetatess	Anionic	Solid
Alkyl glyceryl sulfonate	Sodium cocoglyceryl ether sulfonate	Anionic	Paste
Alkyl glucoside	Lauryl Glucoside	Nonionic	Liquid
Amine Oxides	Cocamidopropylamine Oxide	Nonionic	Liquid-Solid
Amine Oxides	Lauramine Oxide	Nonionic	Liquid-Solid
Amino Acid	Sodium Lauroyl Hydrolyzed Silk	Anionic	Liquid
Fatty Acid Salts	Potassium Cocoate	Anionic	Liquid
Fatty Acid Salts	TEA Oleate	Anionic	Paste
Alkylamido Alkylamines	Disodium Cocoamphodiproprionate	Amphoteric	Liquid-Paste
Alkylamido Alkylamines	Disodium Lauroamphodiacetate	Amphoteric	Liquid-Paste

Miscellaneous Materials

Several miscellaneous compounds are commonly used in soap formulations. Optical brighteners are sometimes used to shift the appearance of a product from a yellow tone to a bluer tone. For products labeled as soap, one ingredient, disodium distyrlbiphenyl disulfonate, is particularly effective for this purpose at approximately 0.3% in traditional soap base. Currently this compound is a noncertified colorant in the United States and is not acceptable in products labeled as cosmetics or drugs. However, it is approved in the European Union for use in cosmetics. Color stabilizers have recently been introduced that help reduce color fading caused by photolytic effects.

Sodium benzotriazolyl sulfonate, sodium benzotriazoyl butylphenol sulfonate and buteth-3, tributyl citrate, and bumetrizole are a few examples of such compounds. They can be added at the amalgamator.

Encapsulated products can be useful to the formulator. The challenge in using an encapsulated product in soap products is that the capsule must survive during processing, yet have a small enough particle size to minimize the drag or grit feel after washing. If this is accomplished, ease of processing and the delivery of sensitive materials during washing results in enhanced effectiveness and stability. Recently, United States Patent 6,403,543 (George, 2002) demonstrated the suspension of microencapsulated beads in clear glycerine soap. The beads can contain many of the functional materials that have been discussed. This system not only provides an interesting visual effect, but also isolates the functional material from the surrounding soap matrix, thus offering additional stability.

Herbal extracts have gained increased popularity recently due to heightened consumer awareness of the environment and the subsequent offering of "natural" themes in many product categories, including aromatherapy, spa, and fragrance-free products. Extracts of rosemary, chamomile, sage, and aloe, among others, are often used as masking agents to replace fragrances while imparting whatever skin benefits are claimed for them. The current CTFA dictionary has an extensive cross-index of botanicals in the form of powders, oils, and extracts. As with any naturally derived material, specifications should be developed to ensure acceptable purity and consistency with every shipment. Many botanicals are offered in a standardized version, which gives the formulator a degree of confidence as to the strength of the material. Since some of the components in a natural product are unknown, adequate stability studies need to be performed on all formulations to determine whether there are any color or fragrance issues. For instance, certain phenols and tannins may unacceptably discolor the soap. A recent publication (Naturals Encyclopedia, *Cosmetics and Toiletries*) outlines many of the natural and botanical ingredients available today.

Other Traditional Cleansing Systems

Cream soaps and cleansing grains are other types of non-bar cleansing systems that have been developed. The cream products can be viewed as modified combo bars that have a paste-like consistency. These formulations include a combination of traditional soap with synthetic detergents, plus structuring agents and beneficial additives. Colorants and fragrances can be added, as well as special ingredients, such as clays and botanicals. The resulting product is paste-like and generates excellent foaming, cleansing, and after-feel. They can be dispensed in a variety of containers, and when pressurized provide an excellent mousse product. Cleansing grains are formulated with a powdered soap and/or a detergent cleanser, together with beneficial additives, colorants and fragrances, and flow agents. Liquid additives should be kept at a minimum to ensure that the product flows and dispenses properly. Encapsulating the liquid material in a solid matrix, whereby the liquid is delivered when the product is used, allows a greater amount of liquids. These products offer additional opportunities for the consumer to take advantage of advances in personal-care skin-cleansing additives.

Alternative soap bases can be made from nontraditional oils. These include, but are not limited to, grape seed, olive, sweet almond, and rice bran oils. When saponified, these oils yield a finished soap that can be extruded and pressed. The resulting soap bars are slightly softer than traditional soap made from tallow or palm oil, but are extremely smooth when used, providing excellent lather and a nice after-feel. The all-vegetable oil components may be attractive for market sectors in spa, aromatherapy, and natural product lines. Standard soap additives, as well as botanicals, can be incorporated into these bases.

In conclusion, the formulator is urged to take a balanced approach when developing surfactant systems. Consideration must be given to base type and function, which dictates acceptable additives for the product category. In turn, process considerations must be addressed in order to efficiently manufacture a quality product. Otherwise, surfactant systems can be overwhelmed as one tries to achieve higher and higher end-product functionality in a world of ever-changing consumer cultures and regulatory policies.

References

Chambers, J. G.;. Instone, T.; Joy, B. S.; Salmon, T.M.F. Detergent Bar. European Patent 385,796, September 5, 1990.

Chambers, J. G.;. Instone, T. Detergent Bar. European Patent, 0350,306, January 10, 1990.

Dahlgren, R.M.; M.F. Lukacovic; S.E. Michaels; M.O. Visscher. Effects of Bar Soap Constituents on Product Mildness. In *Second World Congress on Detergents*, Baldwin, A.R., Ed.; Amercian Oil Chemists' Society: Champaign, IL, 1987.

Davidsohn, J.; E.J. Better; A. Davidsohn. *Soap Manufacture*; Interscience: New York, 1953; vol. I.

Dawson, G.G.; Ridley;G. Toilet Composition. European Patent 311,343, April 12, 1989.

Deweever, E. M.; Carroll, T.E. Cleansing bar. U.S. Patent 3,903,008, September 2, 1975.

Frosch, P.J.; A.M. Kligman. The Soap Chamber Test. *J. Am. Acad. Dermatol.* **1979**, *1*, 35–41.

George, E.; J. Serdakowski. Computer Modeling in the Full Boil Soap Making Process. *HAPPI* **1987**, *24* (1), 34–47.

George, E. D. Soap with suspended articles. U.S. Patent 6,403,543, June 11, 2002.

Gotte E. *Asthet Medzin* **1966**, *15*, 313.

Gottschalck, T. E. ; G.N. McEwen. *CTFA International Cosmetic Dictionary*, 11th ed.; The Cosmetic, Toiletry, and Fragrance Association: Washington, D.C., 2006.

Joshi, D. Translucent soaps and processes for manufacture thereof. U.S. Patent 4,493,786, January 15, 1985.

Jungermann, E.; Hassapis, T.; Scott, R.A.; Wortzman, M. Compositions and processes for the continuous production of transparent soap. U.S. Patent 4,758,370, July 19, 1988.

Kajs, T .M.; V. Gartsein. Review of the Instrumental Assessment of Skin: Effects of Cleansing Products. *J. Soc. Cosmet. Chem.* **1991**, *42* (4), 249–279.

Loden, M.; A. Anders. Mechanical Removal of the Superficial Portion of the Stratum Corneum by a Scrub Cream: Methods for the Objective Assessment of the Effects. *J. Soc. Cosmet. Chem.* **1990**, *41*, 111–121.

Manufacturing Confections Publishing Co. *McCutcheon's Functional Materials*, North American ed.; Glen Rock, NJ, 2006; vol. 2.

Naturals and Botanicals Encyclopedia, *Cosmetic and Toiletries* 118(11) 73-91.

O'Neill, J.; Komor, J.; Babcock, T.; Edmundson, R.; Shay, E. Transparent soap composition. U.S. Patent 3,793,214, February 19, 1974.

Rieger, M.M. *Surfactant Encyclopedia*; Allured Publishing Co.: Wheaton, IL, 1993.

Swern, D., Ed. *Bailey's Industrial Oil and Fat Products*, 4th ed.; J. Wiley and Sons: New York, 1979; vol. I.

Thomsenn, E.G.; C.R. Kemp. *Modern Soap Making*; MacNair-Dorland: New York, 1937.

United States. Code of Federal Regulations, Title 21, part 1–99. April 1, 2008.

Verite, C; Caudet, A. Transparent Soap Containing an Alkanediol - 1,2. European Patent 336,803 .September 30, 1992.

Wood-Rethwill, J.C.;. Jaworski, R.J.; Myers, E.G.; Marshall, M.L. Process for making translucent soap bars. U.S. Patent 4,879,063, November 7, 1989.

Woolatt, E. *The Manufacture of Soaps, Other Detergents and Glycerine*; Halstead Press: New York, 1985.

Nagashima, T.; Usuba, Y.; Ogawa, T.; Takahisa, O.; Takehara, M. Transparent detergent bar. U.S. Patent 4,273,684, June 16, 1981.

Chemistry, Formulation, and Performance of Syndet and Combo Bars

Marcel Friedman
Consultant, 1 Yair Stern Street, 45100 Hod Hasharon, Israel

Soap is undoubtedly the oldest surfactant and skin cleanser. For thousands of years, soap has been obtained from saponification of oils and fats by alkali—the oldest recipe being boiling animal fats and wood ashes. Soap is chemically defined as the alkali salt of fatty acids. In general parlance, the term "soap" has taken on a more functional definition, by which any cleansing agent, regardless of its chemistry, is considered a soap. This sometimes misleading definition will be further considered as this chapter deals with the chemistry of the synthetic detergents and the soapless soap revolution. Conventional soaps that are alkaline salts of fatty acids are characterized by multiple attributes, such as very good emulsification, detergent, and usage properties, as well as ease of manufacture and low cost.

Conventional soaps suffer from two main disadvantages. Soaps hydrolyze in water and release caustic alkali. This subsequently increases the pH of the alkali retained in soap during manufacturing. The high pH of natural soaps affects the skin's natural acidic protective mantle by changing the skin's pH (originally between 5 to 6.5). Even if the pH is restored relatively quickly after washing, this is probably the main cause of the well-known negative soap effect. Another major disadvantage of natural soap, first perceived during the second World War, is its behavior in hard or salt water. When water is very hard or contains a high level of electrolytes (such as seawater), the foaming performance and cleansing efficiency of soap are seriously inhibited, if not eliminated, as natural soap forms insoluble and inactive salts in the presence of magnesium and calcium contained in hard water:

$$2\ RCOONa + Mg^{++} \rightarrow (RCOO)_2Mg + 2Na^+$$
$$2\ RCOONa + Ca^{++} \rightarrow (RCOO)_2Ca + 2Na^+$$

In addition, the insoluble salts precipitate on the surfaces of sinks and bathtubs as a gray fatty mass that is unsightly and difficult to remove. Unfortunately, the identical phenomenon happens during skin cleansing, on which calcium and magnesium salts precipitate. The dermatological drawback of this is obvious.

The synthetic surface-active agents (surfactants) do not have these important disadvantages, and it is for this essential reason that they have been used in the manufacture of dermatologically recommended syndet bars, also known as soap-less soaps or alkali-free cleansing bars. Surfactants are also used in combination with soap for the formulation of "combo" (also called "combi") bars. Initially, the surfactants were included as minor ingredients to disperse lime soap, and then as co-surfactants at higher percentages.

Soap pH

Conventional soaps neutralized with caustic alkali (sodium hydroxide) have an alkaline pH between 10 to 11 (Lux, Palmolive, Camay, or Ivory) or between 8.5 to 10, when they are partially neutralized with milder alkali amines, such as triethanolamine (Neutrogena). According to their formulation the popular combo bars have a pH between 8.5 to 10.0 when the major cleansing agent is soap (such as Lever 2000, Zest, Nivea Milk Bar, or Satina) or even as low as pH 7.5, when only a small amount of soap is incorporated (Dove, Caress, or Olay).

Pure soap-free syndets have a pH between 5.5 to 7.0, matching the pH of the skin (i.e., Neca 7, Vel, Eubos, Sebamed, pH 5 Eucerin, or Basis pH). For this reason it is believed that the syndets' impact on the natural physiological skin equilibrium is extremely limited, making it the product of choice for cosmetic and baby cleansing bars. In particular applications, primarily medicated/dermatological, the pH can be even lower, between 3.5 to 5.0, for example in iodine cleansing bars (such as Neca Polydine) in which the iodine complex is shelf stable only in acidic media. A syndet base has the particular advantage of being the most pH suitable media for such applications. Fig. 5.1 shows the pH ranges of the various bars.

Chemistry of Synthetic Surfactants

A surface-active agent is defined as a chemical substance that, even at low concentration, absorbs at the surface, reducing the free surface energy at the interface of any two-phase system, such as gas–liquid, liquid–liquid, or liquid–solid. To achieve this, the surfactants must dissolve in each of the two phases. This is accomplished by the presence of two distinct groups in their molecular structure. In a water-oil system, one group will be easily soluble in water (hydrophilic); the other will be insoluble in water (hydrophobic) but soluble in oil (lipophilic). The balance between hydrophobic and hydrophilic features governs the application of the surfactant as a detergent, wetter, or emulsifier.

A known scale, characterizing the surfactant according to its hydrophilic-lipophilic balance (HLB), was presented by Griffin. According to this system, highly hydrophobic surfactants have low HLB values, starting at 1, whereas highly hydrophilic molecules are given high HLB values, up to 40. Detergents, for instance, have values in the range of 13 to 15, compared to only 4 to 6 for waterin-oil emulsifiers.

There are four main types of surfactants, classified by the nature of their hydrophilic head: anionic, cationic, amphoteric, and non-ionic. The first three are charged molecules. Anionic surfactants possess a negative charge that has to be neutralized with an alkaline or basic material before full detergency is developed. Cationic surfactants are positively charged and therefore have to be neutralized by a strong acid before they can develop surface properties. Amphoterics include both acidic (negative) and basic (positive) groups in their molecules and are positively or negatively charged, according to the pH of the solution. The non-ionic surfactants contain no ionic constituents, and have no electric charge (Fig. 5.2).

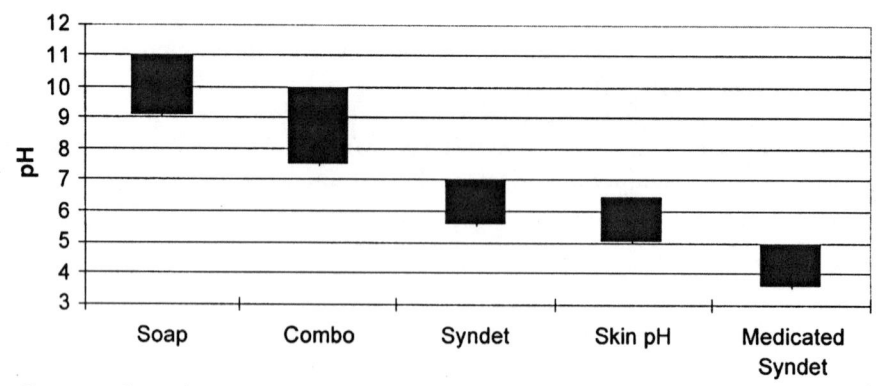

Fig. 5.1. Cleansing formulations and pH range.

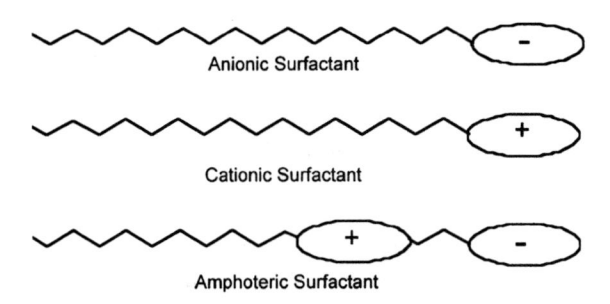

Fig. 5.2. Diagram of surfactant types and charge distribution.

Soap is the simplest anionic surfactant. Since the saponification reaction is a simple hydrolysis of natural materials, soap is often considered a "natural" surfactant. All other surfactants, obtained by many simple or highly sophisticated reactions, are considered synthetic surfactants. Like soap, most surfactants used in personal cleansing bars are anionic. A list of the anionic surfactants (including soap) that are used as active ingredients in cleansing bars is given in Table 5.1.

Table 5.1. Anionic Surfactants Used as Active Ingredients in Cleansing Bars

Surfactant Name	Chemical Formula[a]
Sodium carboxylate (soap)	$RCOONa, RCOOMg$
Alkyl sulfate	$ROSO_3Na, ROSO_3K$
Alkyl sulfosuccinate	$ROCOCH(SO_3Na)CH_2COONa$
Amido sulfosuccinate	$RCONHCH_2CH_2OCOCH(SO_3Na)CH_2COONa$
Acyl isethionate	$RCOOCH_2CH_2SO_3Na$
Alkyl glyceryl ether sulfonate[b]	$ROCH_2CHOHCH_2SO_3Na$
Monoglyceride sulfate	$RCOOCH_2CHOHCH_2OSO_3Na$
Linear alkyl benzene sulfonate	$RC_6H_5SO_3Na$
α-Sulfo fatty acid esters	$RCH(SO_3Na)COOCH_3$
Acyl taurate	$RCON(CH_3)CH_2CH_2SO3Na$
Alkyl sulfoacetate	$ROCOCH_2SO_3Na$
Acyl sarcosinate	$RCON(CH_3)CH_2COONa$
Acyl glutamate	$RCONHCH(COONa)CH_2CH_2COONa$
Alkyl ether sulfate	$RO(CH_2CH_2O)_{1-3}SO_3Na$
α-Olefin sulfonate	$RCH=CHCH_2SO_3Na, RCHOH(CH_2)_{2-3}SO_3Na$
Alkyl ether carboxylate	$RO(CH_2CH_2O)_3CH_2COONa$
Paraffin sulfonate	$RCH_2CH_2SO_3Na$
Acyl lactylate	$RCOO(CHCH_3COO)_{1-3}Na$

[a] $R=C_8-C_{22}$, preferably $C_{12}-C_{14}$ for lathering (mild) surfactant, and $C_{16}-C_{18}$ for less lathering, milder ones.
[b] Also appears as alkoxy hydroxy propane sulfonate.

Surfactant synthesis can be straightforward—getting a surfactant that is easily mixed into the formulation blend. But the more exciting and skillful cost-to-performance formulations are based on "in one pot" in situ manufacturing techniques, that by the nature of their by-products or reactant excess, will determine the final formulation. The most brilliant example is the "Dove" formulation that exploited the fatty acid excess, used to shift the reaction equilibrium of sodium cocoyl isethionate (SCI) synthesis, cosmetically presenting it as a moisturizing cream. The SCI synthesis is presented in Fig. 5.3.

The old fatty acid chloride route (Daimler & Platz, 1937; Bistline et al., 1971) has been made more economical with many different synthetic and catalytic routes covered by several patents; some are referenced (Anderson et al., 1960; Login et al., 1985; Urban et al., 1985; Gattir & Matthaei, 1975). Special care must be taken to inhibit transesterification between isethionate and stearic acid that are added later (Urban et al., 1985). Lower molecular weight fatty acid reactants volatilized during the reaction are continuously supplied to the reaction mass (Gattir & Matthaei, 1975). Recently a batch and continuous process was disclosed to prepare soap-acyl isethionate composition with ratios as high as 20:1; it requires a hot (180–200°F) caustic solution premix of sodium isethionate (Kutny et al., 1991). Catalytic processes introducing new metallic oxides and heavy metal salt catalysts that require temperatures lower than 200°C, have no yield or reaction rate reduction, improve energy usage, and improve odor and color have been claimed (Day et al., 1993). The synthesis of main sulfonates (LABS, α-sulfo fatty acid esters) and sulfates (alkyl sulfates) by SO_3 sulfonation are presented in Fig. 5.4. The preparation of alkyl and alkanolamide sulfosuccinate by sodium sulfite sulfonation is shown in Fig. 5.5.

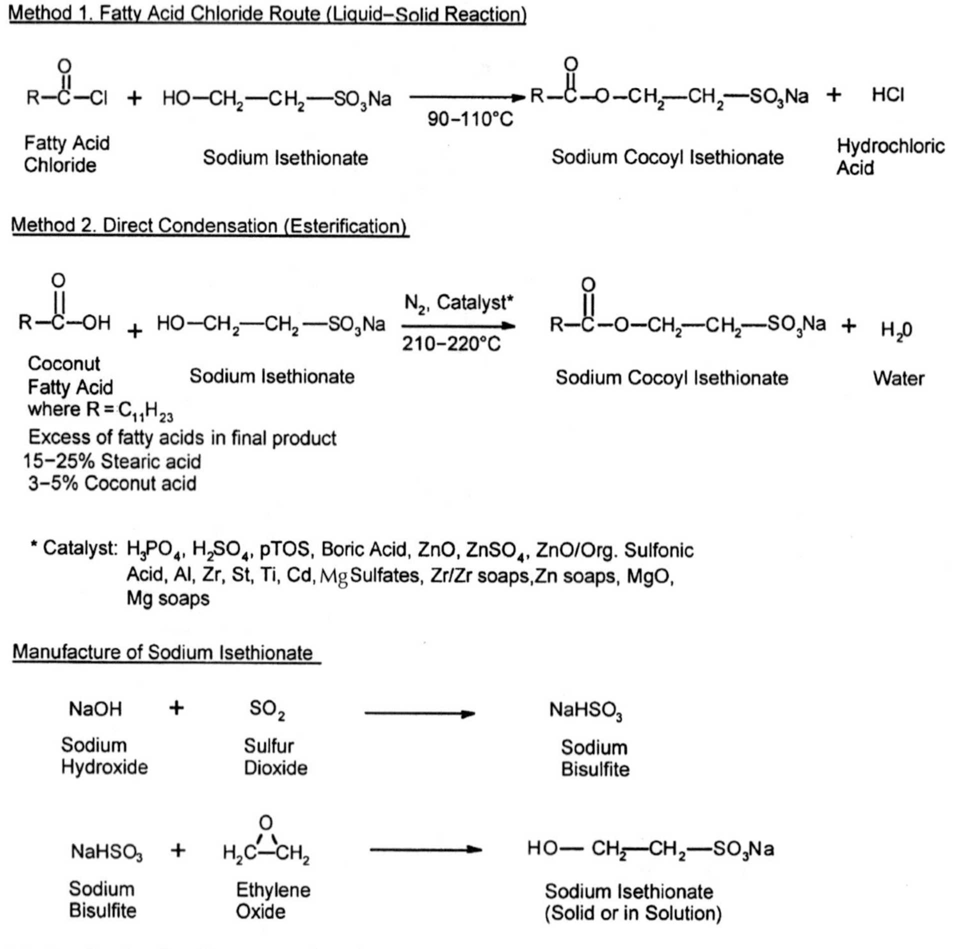

Fig. 5.3. Synthesis of sodium cocoyl isethionate.

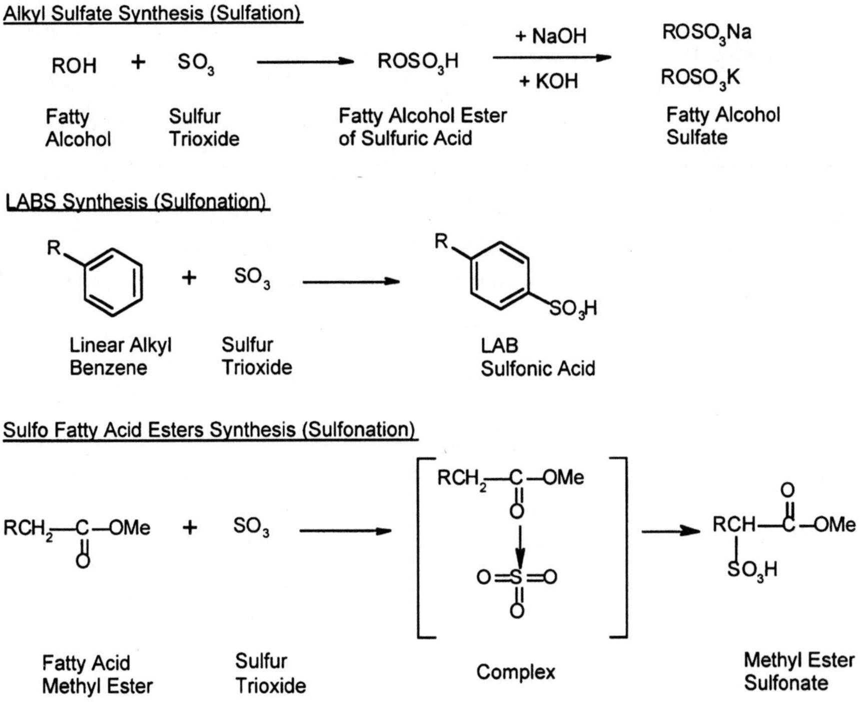

Chemistry, Formulation, and Performance of Syndet and Combo Bars • 157

Alkyl Sulfate Synthesis (Sulfation)

ROH + SO₃ ⟶ ROSO₃H —+NaOH→ ROSO₃Na
                                —+KOH→ ROSO₃K

Fatty Alcohol | Sulfur Trioxide | Fatty Alcohol Ester of Sulfuric Acid | Fatty Alcohol Sulfate

LABS Synthesis (Sulfonation)

R–C₆H₄ + SO₃ ⟶ R–C₆H₄–SO₃H

Linear Alkyl Benzene | Sulfur Trioxide | LAB Sulfonic Acid

Sulfo Fatty Acid Esters Synthesis (Sulfonation)

RCH₂–C(=O)–OMe + SO₃ ⟶ [complex] ⟶ RCH(SO₃H)–C(=O)–OMe

Fatty Acid Methyl Ester | Sulfur Trioxide | Complex | Methyl Ester Sulfonate

Fig. 5.4. Sulfonation/sulfation reactions by sulfur trioxide.

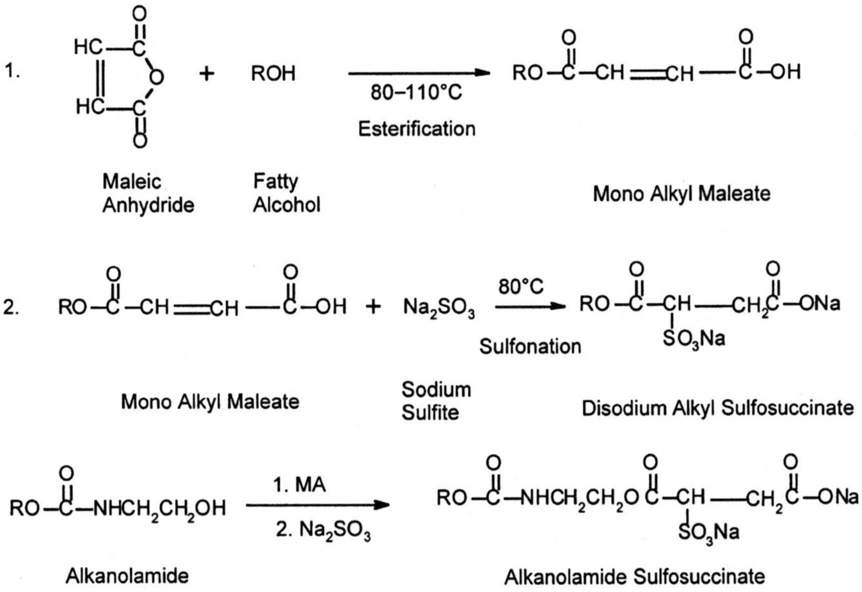

1. Maleic Anhydride + ROH —80–110°C, Esterification→ RO–C(=O)–CH=CH–C(=O)–OH

   Maleic Anhydride | Fatty Alcohol | Mono Alkyl Maleate

2. RO–C(=O)–CH=CH–C(=O)–OH + Na₂SO₃ —80°C, Sulfonation→ RO–C(=O)–CH(SO₃Na)–CH₂–C(=O)–ONa

   Mono Alkyl Maleate | Sodium Sulfite | Disodium Alkyl Sulfosuccinate

   RO–C(=O)–NHCH₂CH₂OH —1. MA, 2. Na₂SO₃→ RO–C(=O)–NHCH₂CH₂O–C(=O)–CH(SO₃Na)–CH₂–C(=O)–ONa

   Alkanolamide | Alkanolamide Sulfosuccinate

Fig. 5.5. Synthesis of alkyl and alkanolamide sulfosuccinate.

158 • M. Friedman

Sulfonation chemistry is well covered in the literature (Davidson & Milwidsky, 1987; Milwidsky, 1985). The hydrophobic chain length, neutralization media, and other parameters govern product suitability in solid-lathering formulations. For isethionates, the synthesis of nearly anhydrous sulfosuccinate in the presence of melted plasticizers and in situ preparation with anionic surfactants, plasticizers, and a small amount of water have been suggested as a straightforward method to manufacture a syndet base (Perla, 1975; Barker et al., 1983). The synthesis of some other surfactants such as sodium coco monoglyceride sulfate, that had been part of Colgate U.S. "Vel" brand (Weil & Stirton, 1976; Gray, 1959), acyl glutamate (extensively used in Japan), and acyl sarcosinate are illustrated in Fig. 5.6. Emphasis on chemical stability of the different hydrophylic moieties is important; it is necessary to compare carboxylate susceptibility to alkaline hydrolysis or sulfate hydrolysis under acidic conditions to the chemical stability of the sulfonates.

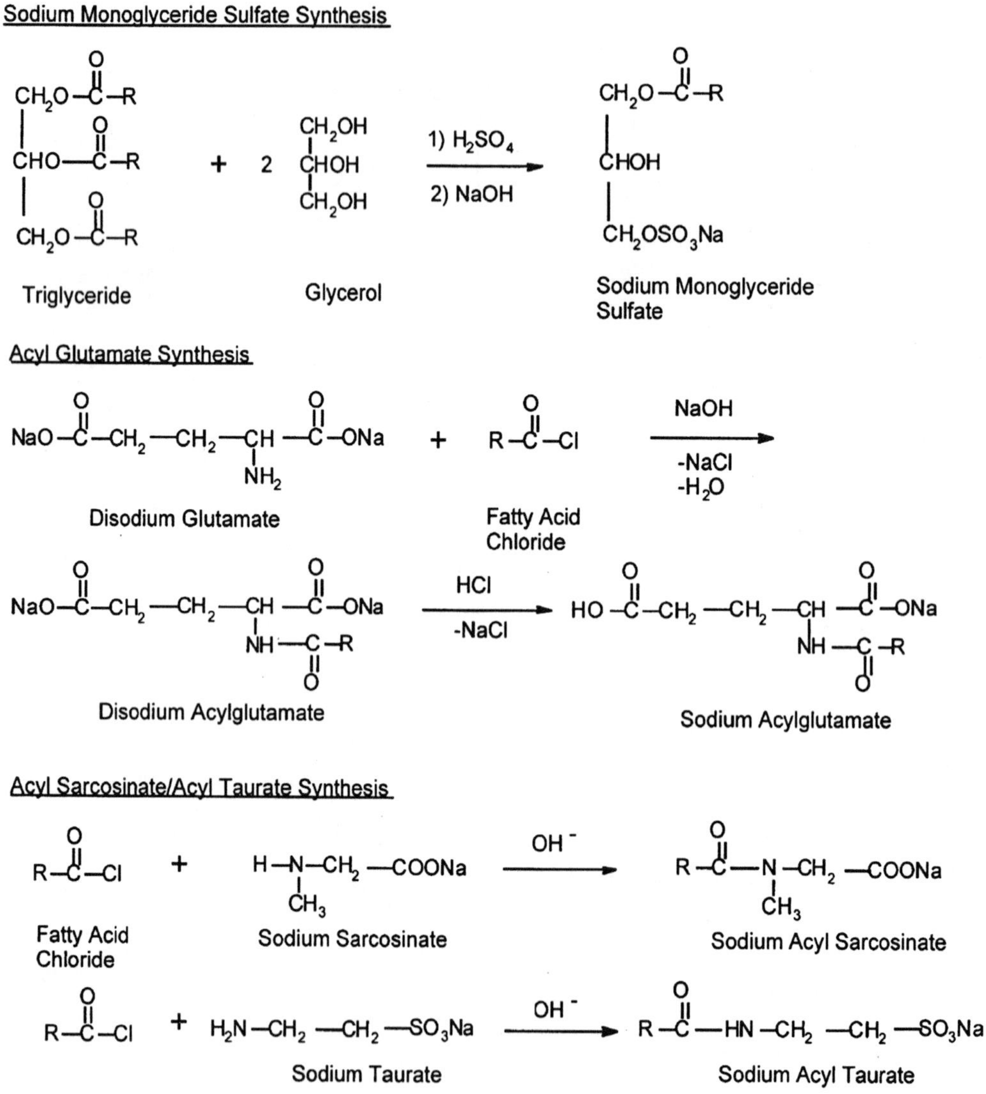

Fig. 5.6. Anionic surfactant synthesis.

Formulation of Cleansing Bars

The list of surfactants presented in Table 5.1 shows a few of the hundreds of synthetic detergents and innumerable plasticizers, binders, moisturizers, and fillers available to formulate syndets and combo bars. The formulation of cleansing bars always involves a skillful combination of scientific thought and almost artistic creativity when selecting appropriate ingredients. The challenge for a soap formulator goes far beyond the manufacture of an effective cleansing agent.

In order to fulfill consumer expectations, a broad range of qualities is demanded of a bar. Social, economic, and psychological demands have advanced along with cosmetic skin care and environmental awareness. This has been followed by substantial cosmetological, dermatological, and technological progress that allows the availability of the sophisticated multifunctional products that is found today. Following these trends, the formulation of syndet and combo bars has become a complex challenge, since final qualities and requirements, including their performance evaluation, are many and occasionally conflict (Table 5.2).

The different test conditions vary based on the test performed, but they have been standardized. Evaluation methods are elaborated according to the company's expertise or official methods and are always aiming for an objective measurement of realistic usage simulation (Wood, 1990; Piso & Winder, 1990). The state of the art in most of the recent patents is based on evaluation by human panels (Redd et al., 1992; Schwartz et al., 1992). These panels, ranging from expert evaluators in the laboratory through small- to large-scale consumer panels, are obvious for sensory evaluation; they are also used extensively for lathering, slipperiness, rinsing, or mush/smear appreciation (Gattir & Matthaei, 1975; Redd et al., 1992; Schwartz et al., 1992).

Table 5.2. Product Performance Evaluation

In addition to cleansing, color, and odor, which were always thought of as main attributes for consumer acceptance, the main bar performance evaluation parameters are:

1. Lathering (amount, stability, quality, density, speed of formation)
2. Mushiness/sloughing/smearing
3. Erosion/wear/use-up rate
4. Hardness
5. Rinsability (from skin and bathtub)
6. Physical stability (wet and dry cracking)
7. Chemical stability (odor, color, efflorescence)
8. Skin feel (during and after rinsing/drying)
9. Lather feel during washing
10. Mildness/dermatological compatibility
11. Processability/workability

Processing

Syndet bar processing is more difficult than combo bars and conventional soaps. The syndet products have a very narrow temperature range in which the plasticity permits proper refining and extrusion. Also due to the sticky and occasionally soft characteristic of the extruded product, stamping rates are adversely affected. There are several types of finishing lines, but for syndet products it is best to have a line that consists of a mixer, simplex refiner plodder, three-roll mill, and a duplex vacuum plodder. A detailed overview of bar soap finishing lines and equipment can be found in Baggini, Nizzero, and Spitz (Baggini et al., 1990). A main water chiller should supply cooling water to the line, and each machine should have its own independent cooling water temperature control unit. The soap press must have a low-temperature glycol/water die chiller and independent low-temperature controller unit for each stamping die group. A pilot-scale method to produce a syndet bar with long-chain alkyl sulfates for improved processability and bar characteristics is available (Schwartz et al., 1992).

The processing sequence comprises crutching, drying, amalgamating, milling (optionally), plodding, and stamping. A hygroscopicity test to evaluate processability is explained and examples are given. The higher the hygroscopicity, the stickier and more difficult to process a material tends to be (Schwartz et al., 1992).

Product Composition

To achieve all the requirements and properties, a suitable syndet/combar is essentially composed from a blend of ingredients as listed in Table 5.3. An ingredient can have multiple and interchangeable functions. So binders absorb some of the water, bind the various ingredients together, plasticize the mass, and act as an emollient. A good example of such functionality is high molecular weight polyethylene glycol that is also defined as a structurer (Fair & Farrell, 1995). Moisturizers and emollients are used synonymously.

Table 5.3. Ingredients of Syndet/Combo Bars

Ingredients	Range (%)
Surfactants	30–70
Plasticizers, binders, emollients	20–50
Lather enhancers	0–5
Fillers	5–30
Water	5–12
Fragrance	0–1.5
Mildness enhancers	0–5
Opacifying agents	0–0.3
Antibacterial, deodorant agents	0–2

Surfactants

Surfactants are primarily responsible for cleansing and lathering. Beside the anionic surfactants already reviewed in Table 5.1, nonionic and amphoteric surfactants used and seldom quoted in the patents are listed in Table 5.4 (Hollstein & Spitz, 1982; Orshitzer & Macander, 1977; Schwartz et al., 1992; Kacher et al., 1993; Kacher et al., 1993; Kacher et al, 1993; Wilson et al., 1993; Massaro et al., 1995; Medcalf et al., 1987; Pichardo & Kaleta, 1993; Hormes et al., 1995). The non-ionics also act as cleansers and plasticizers, but that depends on their HLB values.

A comprehensive list of surfactants, plasticizers, and binders was compiled by Hollstein and Spitz (Hollstein & Spitz, 1982). The state-of-the-art of patent literature, 15 years ago, showed that almost every existing surfactant is a potential active ingredient in syndets. In fact most of the patents cover the use of almost any surfactant, defining even the potential of using any "lathering mild synthetic surfactant that lathers at least about as well as the mild standard alkyl glyceryl ether sulfonate" (Redd et al., 1992).

Some compositions are given in Table 5.5, which partly illustrates the present state-of-the-art and has to be seen as complementary to previous lists (Hollstein & Spitz, 1982). During the last few years newly patented (Ospinal et al., 1999) sulfo methyl esters (SME) have been promoted commercially by Stepan Company (U.S.) for the formulation of combo bars (Ospinal, 1998).

Table 5.4. Nonionic and Amphoteric Surfactants

Surfactant Name	Reference
Alkyl polyglucosides	Hormes et al., 1995
Fatty alcohols	Hollstein & Spitz, 1982
Alkyl phosphate esters	Kacher et al. 1993b; Kacher et al., 1993c
Ethoxylated alkyl phosphate esters	Kacher et al. 1993b; Kacher et al., 1993c
Methyl glucose esters	Kacher et al., 1993a; Kacher et al. 1993b; Medcalf et al., 1987
Sucrose esters	Hollstein & Spitz, 1982
Imidazoline amphoterics	Hollstein & Spitz, 1982
Glyceryl monostearates	Hollstein & Spitz, 1982
Protein condensates	Kacher et al., 1993a; Kacher et al., 1993c
Alkyl phenol polyethylene glycol ether	Orshitzer & Macander, 1977
Polyethoxylated fatty alcohols	Schwartz et al., 1992; Wilson et al., 1993
Polyalkylene glycols (PEG)	Massaro et al., 1995
Alkyl betaines	Medcalf et al., 1987
Alkyl sultaines	Medcalf et al., 1987
Alkyl amine oxides	Medcalf et al., 1987
Polyhydroxy fatty acid amides	Pichardo & Kaleta, 1993

Table 5.5. Surfactant Composition in Combo/Syndets

Composition (%)	Reference
Magnesium soap (5–50%) + any surfactant (20–50%)	Redd et al., 1992
Soap (45–90%) + polyethoxylated fatty alcohol (1–8%)	Wilson et al., 1993
Alkyl glyceryl ether sulfonate (35–80%) + soap (1–25%)	Schwartz et al., 1992
Aldobionamide (20–75%) + polylalkylene glycol (15–65%)	Massaro et al., 1995
Soap (50–90%) + any surfactant (1–20%)	Medcalf et al., 1987
Soap (75–85%) + polyhydroxy fatty amide (1–10%)	Pichardo & Kaleta, 1993
Soap/fatty acid (10–50%) + any surfactant (20–65%)[a]	Kacher et al., 1993a; Kacher et al. 1993b [a]
Soap (70–85%) + alkyl ether sulphate (1–10%), alkyl polyglucosides (0.1–10%)	Hormes et al., 1995
Soap/fatty acid (5–50%) + any nonionic/anionic surfactant (15–65%)	Kacher et al., 1993c
Acyl isethionate (45–70%) + vegetable oil (0.5–2.5%)	Subramanyam, 1995; Subramanyam, 1994
PEG (70–80%) + lauryl sulphate/amphoteric (8–30%)[a]	Moran, et al., 1993 [a]
Sodium lauryl sulphate (70–90%)[a] + binder	Constantine, 1989 [b]
Long-chain alkyl sulfate (18–55%) + soap (1–20%)	Schwartz et al., 1992

[a] Refers to framed/molded bars.
[b] Presented as solid shampoo composition.

Some patents propose the use of new surfactant molecules. For instance, a nonionic glycolipid surfactant (more specifically an aldobionamide) at 20–75% in a composition structured by 15–65% polyalkylene gycol (preferably PEG of molecular weight 6000–10000) and containing no more than 15% fatty acids (Massaro et al., 1995). Unfortunately, the proposed preparation advantages (a renewable raw material source and a no-foam sacrificing structurer) are counterbalanced by the disadvantage of a costly synthesis of n-alkyl lactobionamides (obtained by reacting lactobiono-1.5-lactone with various alkylamines in anhydrous DMF, methanol, or net a lower yield). The choice of surfactants and their proportions not only determines the cleaning and lathering characteristics but also influences mushiness, plasticity, and skin compatibility.

Formulations

European and Israeli formulations are based mostly on a blend of alkyl sulfates and alkyl sulfosuccinates, reaching about 40–50% surfactant content. The pH of the bar is normally adjusted between 5.5 and 7.0. It has been found that the potassium salts give much better mushing and plasticity properties than the sodium salts, and there is an optimal K–Na ratio to minimize mushiness (Ramakers, 1992).

The U.S. mass-marketed brands are based on other surfactants. They also followed the soap–syndet concept already patented in the 1940s (soap + LABS + starch) (Hoyt, 1940), followed by nonsmearing bar formulations based on alkyl glyceryl ether sulfonate, sodium/magnesium soaps, and binder (Tokosh & Cahn, 1979). The leader, Dove, is based on fatty acid isethionates, specifically SCI. Since this is an expensive ingredient, the formulation includes about 25–27% fatty acid (specifically stearic acid and a minor proportion of coconut acid) and fatty acid soap, mostly neutralized to a final pH of 7.2–7.5 (Dederen, 1992). The fatty acid presence due to the reaction equilibrium shifting, produces an overall best-performing package, brightly presented as containing 25% moisturizing cream. The product contains some dodecyl benzene sulfonate (1–2%) as a lather enhancer. This surfactant blend reduces the final cost of the formulation but, due to the price of cocoyl isethionate, it is still relatively expensive. SCI bar formulations presented by Ho Tan Tai (2000) according to Unilever patents list a broad range of SCI content of 44–60% and a high electrolyte content of 5% sodium sulfate. The U.S. Dove formulations of the 1990s and the 2008 current formulations are presented in Table 5.6.

Table 5.6. Dove Bar Formulations (%)

Ingredient	1992 (USA)	Current
Sodium cocoyl isethionate (SCI)	47–49	48–50
FFA (stearic acid + coconut acid)	23–25	23–26
Soap	7–10	9–11
Sodium isethionate	4–6	4–6
Alkylbenzene sulfonate (LABS)	1–2	—
Cocamidopropyl betaine	—	2–3
Sodium chloride	0.5	0.5–1
Water	4–6	5–6

Since the mildness concept has been strongly advertised by Lever Brothers as an intrinsic property of isethionate bars, a serious change has occurred in bar formulation in Western Europe in the past few years. Products have been reformulated; more expensive isethionate bars, claiming even milder properties, have been introduced. This has changed the surfactant base from alkyl sulfate/disodium lauryl sulfosuccinate (DSLSS) to SCI/DSLSS base. A typical formulation of Sebamed (Germany) from the late 1990s contained about 25–30% SCI, 10–15% DSLSS, and a blend of fatty acids (C_{12}–C_{18})

and mineral oils of about 30–35%. Similar trends were followed by European syndet base suppliers (Ramakers, 1992, Dederen, 1992).

The drawback of the cocoyl isethionate bar (easily perceived by consumers) lies in its strong, characteristic odor. This odor recalls the coconut source of the fatty acid and needs a higher dosage of fragrance to be covered. It also creates a problem in the fragrance-free hypoallergenic formulations. An interesting contribution concerning this issue should be mentioned (Beerse, et al., 1995). The object of this contribution was to deliver a bar with reduced off-odor and a formulation that is mild to skin and easy to process. These objectives are achieved by using a cleansing bar comprising 10–70% of a sodium-distilled, topped cocoyl isethionate (STCI). The STCI of this invention contains little or no (0–4%) highly soluble acyl groups ($C_6 + C_9 + C_{10} + C_{18:1} + C_{18:2}$), about 45–65% C_{12}, and about 30–55% C_{14}, C_{16}, and C_{18}. The lack of the more volatile shorter chain odorous hydrophobics, as well as the unsaturated C_{18} chains are obviously the reason for reduced bar odors. It was also claimed that the STCI formulation allowed the use of higher levels of hygroscopic lather boosters, such as alkyl glyceryl ether sulfonate (AGS), without exhibiting processing drawbacks that would otherwise be experienced with regular SCI. STCI usage also allows the addition of larger amounts of water or liquid ingredients, such as glycerine and vegetable oils (Beerse et al., 1995).

However the significant relevance of these patents (Beerse, et al., 1995) to the subject lies in the inclusion of a broad range of functional ingredients that influence soap, combo, and syndet bar properties, such as off-odors, processability, stickiness, brittleness, mushiness, lather quality, and their combinations. The patents become an explicit formulation showcase, that is worth presenting in depth, that allows a close practical insight into the ingredient–performance relationship (Table 5.7). The STCI bars produced with the process are comprised of three key ingredients: STCI, a plasticizing agent, and a binder, in addition to a mild surfactant matrix (Table 5.7) (Beerse et al., 1995). Each will be discussed further.

Table 5.7. STCI Bar Composition (EP 0728186)

Components	Full range	Preferred	Preferred	Function
STCI	10–70	15–60	20–50	The key to the present invention. It is made from topped distilled coconut fatty acid.
Na-alkyl glyceryl ether sulphonate	0–50	5–30	10–20	Included as a lather-boosting synthetic surfactant. It is made from coconut fatty alcohols. Equivalent synthetic surfactants can be used.
Na-alkyl ether sulfate	0–10	1–8	2–6	A mild lather-boosting synthetic surfactant.
Na-cetearyl sulfate	0–40	4–30	8–20	A non-soil load filler and processing aid.
Na-soap	0–20	1–15	2–12	A lather booster and processing aid.
Mg-soap	0–50	4–30	8–20	A non-soil load filler and processing aid.
Fatty acid	0–35	3–25	5–20	A plasticizer.
Paraffin	0–30	3–25	5–20	A plasticizer.
NaCl	0–5	0.1–3	0.2–2	Provides bar firmness and improves bar smear.
Na_2SO_4	0–5	0.1–3	0.2–2	Provides bar firmness and improves bar smear.
Na-isethionate	0–15	1–10	2–8	Provides bar firmness and improves bar smear.
Water	3–20	4–15	5–10	A binder.
Fragrance	0–2	0.5–1.5	0.8–1.2	A binder and improves odor.

Abbreviation: Sodium topped cocoyl isethionate (STCI).

Other major surfactants used in U.S. formulations are sodium cocoglyceryl ether sulfonate contained in Procter & Gamble's Zest combo bar and in the Olay syndet bar. These last two products also contain SCI. In Japan, acyl glutamate is the basic surfactant in "Mignon," an expensive soapless cleansing bar. Classical formulations of high pH (Davidson & Milwidsky, 1987; Milwidsky, 1985) combo bars, like Lever 2000 and Zest, are described in Table 5.8.

Table 5.8. Combo Bar Formulations

Ingredient	Zest	Lever 2000 Current
Sodium glyceryl ether sulfonate	12–18	—
Sodium cocoyl isethionate (SCI)	—	—
Sodium methyl sulfo laurate	—	1–3
FFA	6–10	5–7
Soap (tallow/coconut)	50–60	65–75
Magnesium soap	5–10	—
Water	5–8	8–12

The formulations covered in the 2004 edition of this chapter (Friedman, 2004) have undergone various changes. In the United States, Zest (Aqua Pure) contained the classical Zest ingredient SCGES, but another Zest variant (Tropical Fresh) contains no surfactant at all. Zest in Mexico contains potassium lauryl sulfate as the surfactant of choice. During 2007, some of the Zest bars became regular soap bars without any surfactant. Table 5.8 illustrates the traditional Zest combo bar formula. The Lever 2000 combo bar replaced sodium cocoyl isethionate with sodium methyl 2-sulfolaurate.

A revised list of several suppliers that offer raw materials to formulate syndet and combo bars, as well as fully formulated ready-to-use soap-free or combo bases is presented in Table 5.9. The unpredictable fast changing soap/syndet/combo bar market during the last 3 years, makes the updating of such a list difficult and sometimes not reliable; even the company's names are difficult to follow. While the list is not current it has historical value.

For example, Procter & Gamble (P&G) promoted for a short time alkyl glyceryl sulfonate (AGS), and labeled it as sodium cocoglyceryl ether sulfonate (Procter & Gamble, 2000, 2001, 2003). AGS 1214 was presented as a surfactant–polymer mixture comprised of (on a dry basis) 72% of an AGS monomer (Table 5.1) and 13% of an AGS dimer, composed of 11% of disulfonate, 1% of chlorosulfonate, and 1% of hydroxysulfonate. The balance contained other molecules, such as 3% of unreacted alcohol, 6% of glyceryl ether (diol), 3% of sodium sulfate, and 3% of sodium chloride, which influence the overall properties of AGS (Procter & Gamble, 2000, 2001, 2003). The product was available as a paste (47% active) or in flakes (75% active). AGS was produced by reacting fatty alcohol with epichlorhydrin, followed by a reaction in a caustic solution to produce an intermediate; this was further reacted with a sulfite mixture, and finally oxidized and pH-trimmed to produce the AGS paste (Procter & Gamble, 2000, 2001, 2003).

A typical combo–bar formulation using AGS was provided by P&G. The formula included 52% of sodium soap, 17% of AGS, 14% of magnesium soap, 8% of water, 3% of salt, and 2% of superfatting agents. Ready-to-use Zest flakes of similar formulation were also offered (Procter & Gamble, 2000, 2001, 2003).

Table 5.9. Commercial Syndet Raw Materials and Ready-to-Use Bases

Company Name	Base Material Trade Name	Base Material Type
Ajinomoto Inc	Amisoft CS-11	Sodium cocoyl glutamate
	Amisoft LS-11	Sodium lauroyl glutamate
Akzo Nobel	Elfan AT 84	SCI[a]
Bradford	Jordapon SB II Syndet Base	SCI + SS
Innospec	Tauranol I-78-C	SCI
	Tauranol I-85-T	SCI
	Syndet Base 96-143-1	SCI
Hampshire Chemical Corp.	Hamposyl	Acyl sarcosinate
Clariant	Hostapon SCI 65	SCI
	Hostapon SCI 85	SCI
Galaxy	SN 8212	SCS + DSLSS
	SN 8501D	SCS + SPLS + DSLSS[b] + SCG
	SN 8102	SCS + SPLS + Soap
Rhodia	Geropon AS-200	SCI
	Geropon SDT	SCI + Soap
Tensa Chem	Tensianol N1LM/N1A	SPCS
	Tensianol 815/815C	SPLS + DSLSS
	Tensianol SF/ISL/STM/DEO	SPLS + SCI
	Tensianol 3456 B/3720/ 3713 /3714/400	SCI + DSLSS
Zschimmer and Schwarz	Zetesap 813A	DSLSS + SCS
	Zetesap 5165	DSLSS + SCI
	Zetesap 5213	DSLSS + SCA
	Zetesap ST 5251	DSLSS + SCS + SLES

[a]SCI raw materials contain free fatty acids (stearic and coconut) in proportions of 5–10% for 85% SCI and 20–30% for 65% SCI, and free sodium isethionate of 2–4%.
[b]SLS and DSLSS are currently supplied as commodities at 30–95%.
Abbreviations: Sodium/potassium lauryl sulfate (SPLS); sodium/potassium coco sulfate (SPCS); sodium coco sulfate (SCS); sodium cocoyl isethionate (SCI); disodium lauryl sulfosuccinate (DSLSS); sodium lauryl sulfoacetate (SCA); sodium lauroyl sarcosinate (SS); sodium cocoyl glycinate (SCG); alkyl glyceryl (ether) sulfonate (AGS); sodium lauryl ether sulfate (SLES), and sodium lauryl sulfate (SLS).

Recently, ready-to-use combo flakes were launched by Galaxy under the name "SN 8102". One can use this base as such or dry mixed, properly mixed with soap flakes in various proportions, from 80:20 to 20:80 (Galaxy Surfactants, 2008).

Valuable support is given by most raw-material suppliers in providing suggested formulations for optimal performance, and by the syndet- and combo- based noodles (pellets/flakes) to overcome processing problems. Tensa Chem offers probably the broadest range of syndet bases, defined as "fully formulated soap-free flakes for cleansing-bar manufacture" (Tensa Chem, 2006) (Table 5.9).

Clariant has done extensive development work to promote Hostapon SCI use in combo and syndet bars (Clariant, 2003). The "grittiness," a characteristic problem of SCI bars, was researched, and relevant parameters to be mastered were considered. This crystallization phenomenon, which sometimes occurs under storage, was influenced by the way the soap is made. Higher mixing temperature, longer and more intensive mixing, fluid (water, etc.) presence, and additional milling were recommended.

The crystal-size distribution of the SCI material is also an important factor mastered by the supplier. As recommended, one should make a grittiness assessment immediately and at 4 weeks and 3 months after production (Clariant, 2003). Due to the high mixing temperature (100–150°C) and the high moisture level (10–18%), SCI is susceptible to hydrolysis. The electrolyte level had some effect on limiting the amount of SCI lost through hydrolysis (Clariant, 2003; Unilever, 2000).

Recently, Zschimmer and Schwarz expanded its range of syndet compounds by launching Zetesap ST 5251, recommended for a variety of formulations. The solid emulsion system and the resulting liquid-crystal structures of the syndet base, combined with the advanced formulation, are considered to give the ideal base with a smooth touch and an exceptionally creamy foam (Zschimmer & Schwarz, 2007a,b)

Therefore, the consumer perception will evaluate dense creamy foam as superior to bubbly foam, ranking it as a more caring washing product.

Cleansing Efficiency

An interesting evaluation of the cleansing effect of various bars indicates that alkyl sulfate/sulfosuccinate blends have higher cleansing power, when compared to acyl glutamate and triethanolamine soaps, as presented in Figure 5.7 (Weber, 1987). The acyl isethionates were not tested for in Weber (Weber, 1987). The cleaning activity was determined with a skin-washing machine that cleans the artificially stained skin of the forearm under standard conditions (Molls & Schrader, 1984).

By another proposed evaluation method, a fat-based ointment that simulated dirt, was placed on the dorsum of the hand and washed in a rotating soap solution for 5 minutes. The cleansing efficiency of various soaps was assessed by comparing sebumeter reading before and after the washing (Wolf & Friedman, 1996). Better results, as shown in Figure 5.7, have been found for alkyl sulfate–based bar soap compared to leading isethionate brand, specifically 81% vs. 75% cleansing capacity; water gave about 30%. Deposition profiles checked in P&G patents and unpublished/commercial publications (New York Times, 2001; Unilever, 2002) also evaluate rinsing/cleansing efficiency.

AGS was found to rinse cleaner and deposit less compared to SCI, while sodium soaps were found to have a much poorer cleaning efficiency (Procter & Gamble, 2000, 2001, 2003). This data emphasizes again the high surfactant tolerance to hard water given by the higher solubility of Magnesium AGS and Calcium AGS compared to Sodium AGS; this results in AGS that rinses cleaner and deposits less. The rinsing efficiency and deposition profile of several surfactants is shown in Fig. 5.8 (Procter & Gamble, 2000, 2001, 2003).

Chemistry, Formulation, and Performance of Syndet and Combo Bars • 167

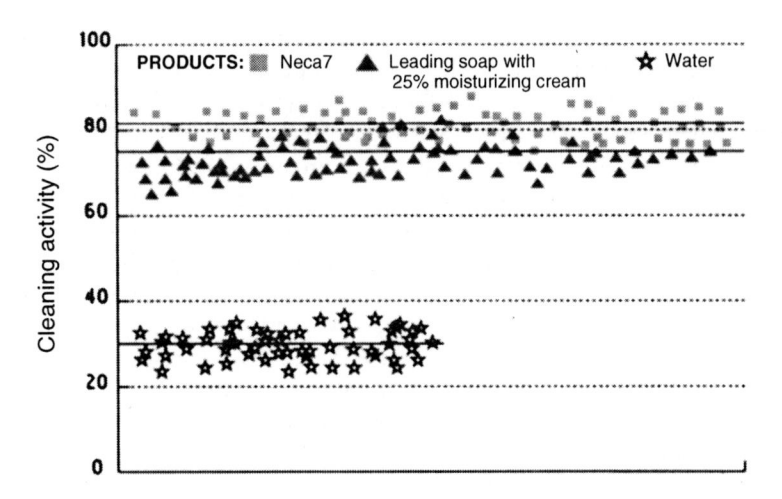

Fig. 5.7. Skin cleansing efficiency of two different cleansing products compared to water. (Wolf & Friedman, 1996)

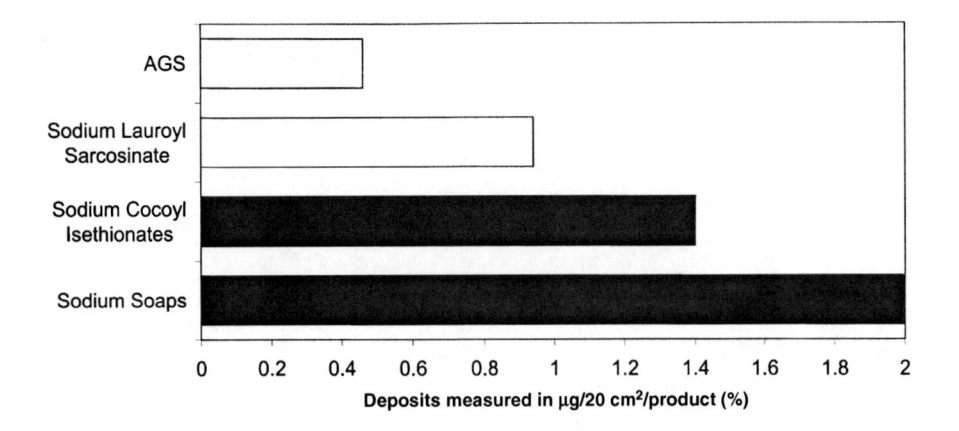

Fig. 5.8. Rinsing efficiency and depostion profile of different surfactants.

Foaming/Lathering

The foaming properties of different surfactants and finished syndet formulations can be checked by several methods whose description is beyond the scope of this chapter. A common and reliable test to check foam volume and height is the mechanical inversion test. When checked in hard water (320 ppm $CaCO_3$) a higher foam performance was found for alkyl sulfate and sulfosuccinate formulations compared to isethionate-based syndet/combo bars and natural soap (Friedman & Wolf, 1996) (Fig.5.9). However, this test doesn't give any information on the foam quality, as a more bubbly foam gives more volume than a dense creamy foam, the latter being better appreciated for moisturizing formulations. Therefore, the consumer perception will evaluate dense creamy foam as superior to bubbly oam , taking as a more caring washing product. Even the classic Ross-Miles method (Ross & Miles, 1953) is still being used in recent patents (Massaro et al., 1995); for instance, an initial foam height of 153 mm, collapsing to 5 mm after 10 min was found for C12 lactobionamide compared to an initial height of 145 mm that stays stable at 140 mm after 10 min for C14 lactobionamide. It is evident that the sensitivity of foam stability measurements depends on the hydrophobic chain.

Expert-rated panel-lathering tests are seldom preferred (Gattir & Matthaei, 1975), reporting especially on speed of formation, quality of foam, and lather feel during washing. Lather volume is also practically determined by accumulating the lather generated by rotating a bar ten times with a wetted glove under a stream of water at a given temperature (Massaro et al., 1995). Some results from the German DIN 53902 method are reported for ethoxylated isostearyl monoglyceride added to the syndet solution; they reduced foam from 500 mm to 150 mm (Hollstein & Spitz, 1982). Hydrophobic additives, selected for their emollient or plasticizing contribution, reduce foam considerably.

To enhance the lather performance of the primary surfactant in lather formation speed, stability, or cold water performance, some foam enhancers can be used. Dodecyl benzene sulfonate, sodium lauryl sulfate, alkanolamides, and even neutralized fatty acids perform this function well. Alkyl polyglucosides (APG) have been promoted recently by Cognis (Cognis, 2001) and Seppic (Seppic, 2001) as foam enhancers, especially in soap formulations containing 1.2–1.5% Plantaren/Plantacare (Cognis, 2001) or Oramix (Seppic, 2001). Both companies suggest that the same result was perceivable in syndet/combo formulations. Cognis reports on SCI:soap combo (ratios 50:8 to 70:8 on 100% active basis) with 8–9% APG showing improved processability and lathering as compared to the non-APG original base; 15% APG is also recommended as a co-surfactant in a syndet based on 40% DSLSS and 15% paraffin wax as a lipid-layer enhancer.

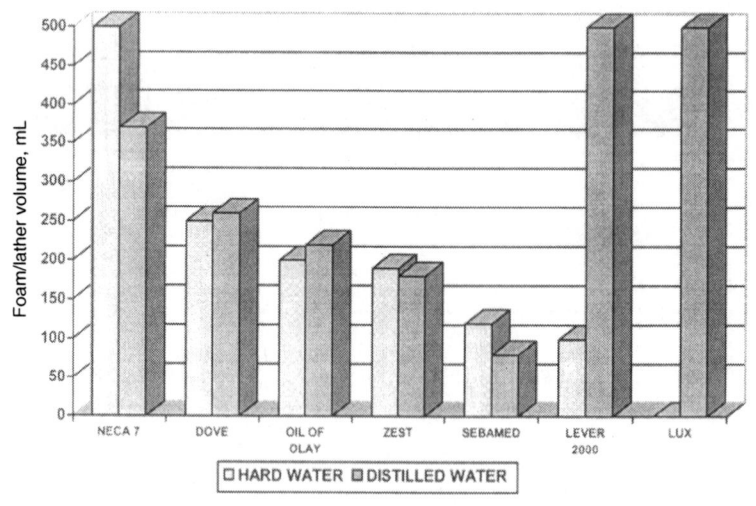

Fig. 5.9. Foam/lather performance. Water hardness, 320 ppm $CaCO_3$; Water temperature, 30°C; Product concentration, 0.16 wt %.

Plasticizers and Binders

To obtain good processability and usage properties, the formulation is stabilized with plasticizers and binders. Plasticizers are included to facilitate better extrusion and stamping of the syndet bar. The plasticizers act by lowering the viscosity of the material at the manufacturing temperature, providing flow under pressure.

Binders prevent separation of macroscopic aggregates, caused by local stress, and provide cohesion and anticracking behavior of the solid product. Natural soap does not have plasticity problems, as the soap itself is a classic plasticizer, beyond its cleansing surfactant role. Plasticizers and binders strongly influence lathering, mushing, and wear characteristics. Generally, the plasticizers and binders are used together, and one material can perform at least two roles. They are able to absorb some or all of the free water of the syndet, bind the various ingredients together, plasticize the mass, and act as emollient simultaneously. The plasticizers most commonly used are long-chain fatty alcohols (higher than C16), polyol esters (glycerol monostearate, sorbitan stearate, and glycerol mono- and distearate), polyethylene glycol, sodium stearate, stearic acid, fatty acid ethoxylate, hydrogenated castor oil, paraffin wax, fatty alkyl ketones, and a combination of hydrogenated triglycerides with fatty alcohols or acids (Hollstein & Spitz, 1982; Friedman & Wolf, 1996).

As pointed out in Table 5.7 (Beerse et al., 1995), at least 20% plasticizers are used; these are comprised of fatty acids and paraffins that are solid at room temperature but are malleable at bar-plodding process temperatures of about 35–45°C. The binder used in the same patent is any material that is by itself liquid, and is selected from water, liquid polyols, and even fragrance. Their levels in the bar are theoretically 3–20% water and 0–15% polyols (Beerse et al., 1995).

The plasticizers and binders have high melting points and high olecular weights. It seems that their binding activity is obtained when the melting point of the mass is simply raised. Ingredients, such as gums and gum resins, provide additional cohesion by acting as binders.

Some patents (Fair & Farrell, 1995) describe the binder as stucturer, specifically referring to polyalkylene glycol (25–60%). Recent patents (Kacher et al., 1993abc) define special binding as formation of a rigid crystalline phase skeleton structure, comprising an interlocking, open three-dimensional mesh of elongated crystals. The crystals in this case, essentially consist of 5–50% (Kacher et al., 1993ac) or 10–50% (25) fatty acids, of which 20–65% are neutralized, in addition to 15–65% (Kacher et al., 1993c) or 20–65% (Kacher et al., 1993ab) of anionic and/or nonionic bar firmness aid, including a relatively high water content of 15–55% (Kacher et al., 1993b) or 15–40% (Kacher et al., 1993ac). The surprising aspect of this invention is that bar hardness and low smear are obtained in the presence of soft hydrophobic materials (including petrolatum, paraffins, and most natural and synthetic waxes) nonionic solvents/co-solvents (propylene glycol, glycerine, etc.), and anionic surfactants that typically result in bar softening. This results in the interlocking crystalline skeleton network that contains substantial "void" areas being filled by soft and/or liquid aqueous phases; this drastically changes the colloidal structure, and consequently, the physical properties of a conventional bar. The only disadvantage of this almost ideal bar is that it is not extrudable under normal working conditions, and its preparation is made by pouring the molten mixture into a bar-shaped mold (Kacher et al., 1993ab) or by a freezer process in which the molten mixture is cooled to a semisolid in a scraped wall heat exchanger and then extruded as a soft plug, cooled and crystallized, fed on a moving belt (Kacher et al., 1993c). Worth mentioning are several patents (Moran, et al., 1993; Constantine, 1989) proposing solid shampoo formulations that are processed in a molded form, basically formulated with alkyl sulfates.

Cast (Poured) Transparent/Translucent Syndet/ComboCleansing Bars

The ultimate challenge of the syndet/combo market is to create an extrudable transparent/ translucent of a neutral pH. Technically this target has not yet been achieved but in the meantime the market segment of cast (also called poured, molded) transparent combars is steadily increasing. Beside the concept of purity and aesthetic appeal, this target provides a way to produce milder formulations with enhanced skin properties. With regular extrusion techniques, both soap and syndet bars have limitations to the amount of additives that can be used. These limitations can be overcome by using the cast–pouring techniques, while maintaining only a minimum degree of stickiness and mushiness. The special bar binding through formation of a crystalline skeleton network (Kacher et al., 1993abc) seems to give the required bar hardness and low smear for a pourable transparent combar containing SLS, SLES, SCI, and CAB (cocamidopropylbetaine) together with fatty acids, PEG, paraffin wax, and silicones (Whalley, 2000).

Several commercial combo bases are offered on the market by different suppliers, such as Stephenson, Galaxy, Tensachem, and Unichema. Crystal ST from Stephenson is a hot-pour base containing sodium stearate/stearic acid, sodium laurate/lauric acid, and SLS/SLES surfactant mixture besides a glycerine/propylene glycol/sorbitol clear matrix (Stephenson, 2003). The Galaxy SN-900 series used in transparent combo body wash bars contain 25–40% soap, 25–35% glycols, 10–15% anionics (SLES) and amphoterics (CAB), 10–15% sugars, and 10–15% moisture (Galaxy Surfactants, 2003). A recent patent (Jaworski & Park, 2002) claims a transparent bar using a synthetic detergent and a soap having an enhanced transparency, clarity, and mildness not achieved previously. The composition is composed of the following: 40–45% surfactant (of which 15–35% is an SLES type); 10–30% of a polyhydric alcohol (such as propylene glycol, glycerol, sorbitol, or PEG); 15–30% soap; and 5–20% alkanolamide as a foam stabilizer. A preferred pH of 9.6 and not below 8.0 (because of loss of clarity and hardness) is claimed. Surprisingly, the claimed product has mildness equal to Dove as checked by a modified soap chamber test even at this relatively high pH (Frosch & Kligman, 1979; Jaworski & Park, 2002).

Transparent and translucent combo bar formulations have been commercially promoted by adding nonoxynol-10-carboxylic acid and C12–15 alkyl benzoate to the soapsyndet base (Finetex & Novakovic, 1989). Commercial offerings of cleansing bars based on sarcosinates (Hampshire, 1994) or acyl glutamates (even of translucent appearance) (Ajinomoto, 1994) have been made during the last few years, in spite of the industrially inefficient hot-pour manufacturing method used. A beautiful, round-shaped acyl glutamate syndet has recently been launched by Ajinomoto in a Korean–Japanese venture that is being promoted abroad (Ajinomoto, 2001). This is a so-called gel-soap called Amino Crystal, emphasizing the main amino surfactant, TEA cocoyl glutamate; additional components are decyl glucoside, hydrolyzed collagen, dipotassium glycyrrhizinate, glycerine, urea, and isopropyl alcohol. Two factors are especially striking: the pH is about 6 (which is quite rare for a totally transparent syndet), and the price.

Performance- and Appearance-Improving Additives

One perceptible drawback of the syndets is their solubility and mushiness in water, which is known as bar smear or bar slough. This messy, unattractive and uneconomical property is due to the high solubility of some surfactants and inappropriate formulation. Under certain formulations, this paste-like mush is unable to return to a solid state by losing water from the mush layer. Using higher water levels in the formulation will result in the wanted manageable neat phase (also known as the G or lamellar phase) transforming into a viscous jelly-like middle phase. One major solution used for alkyl sulfate syndets is to use their potassium salts, which gives a low mushiness and an economical base.

Other low-solubility enhancers are inorganic salts, such as sodium sulfate and sodium chloride. A recently patented process produces a soap and combo bar with 0.5–10% auxiliary mild surfactants (Aronson et al., 2001); these bars have enhanced skin-care properties from the addition of protic acid salts (PAS) to soap formulations with fatty acid soaps, free fatty acids (FFA), and polyalkylene glycols. A protic acid is any acid that readily yields protons, for example, a Brønsted acid. The salts of such protic acids are selected from inorganic and organic acids, with preference being given to sodium chloride/sodium sulfate for the inorganic, and sodium lactate/sodium citrate for the organic. The required ratio FFA:PAS of 0.5:1 to 3:1 is highly emphasized (Aronson et al., 2001).

Aluminum triformate efficiently reduces water solubility in sulfosuccinate formulations (Hollstein & Spitz, 1982). Slipping has been obtained by adding zinc stearate (recently reappearing in the ingredient list of Extra Sensitive Dove brand, marketed in Europe) and ethoxylated sorbitan ester (Milwidsky, 1985; Hollstein & Spitz, 1982). Preferred slipping agents are high molecular weight ethylene oxide polymers (such as PEG 14M) (Aronson et al., 2001). Sodium isethionate is added in acyl isethionate formulations to reduce wear rate (Tokosh & Cahn, 1979).

Bar appearance (water retention and/or shrinkage prevention) aids are seldom selected from a long list of water-soluble organics, salts, clays, and suitable waxes (preferably paraffin) to impart skin mildness, plasticity, firmness, processability, glossy look, and smooth feel to the bar (Kacher et al., 1993abc). Antibacterial additives, such as trichlorocarbanilide and triclosan, are good examples of successful additives for deodorant and antibacterial bars, even if their use is currently being questioned.

Fillers

The fillers are, by definition, cheaper ingredients, used to reduce the bar cost. In the case of syndets, the fillers are not inert ingredients but participate in improving the internal structure and hardness of the finished product. Fillers can therefore be called additional binders. The best-known fillers are dextrin, starch, and modified starch (degraded, ethoxylated).

Talcum powder has also been used as a filler to aid against mushiness, while buffered borax is added to reduce specific gravity and lower wear. The drawbacks of the fillers are a rough surface texture, loss of slip, and loss of attractive overall appearance. For this reason, one should not exceed an optimal concentration of these substances. Recently, Luzenac (Luzenac, 2002) promoted new talc grades presented as innovative technology and applications for soaps and syndets (Luzenac, 2002; Arseguel, 2003).

The list of fillers quoted by Hollstein and Spitz (1982) also notes sodium sulfate, $CaHPO_4$, $MgHPO_4$, NaH_2PO_4, dextrin, and mannitol as being state-of-the art examples. A current state-of-the-art list will include calcium carbonate and talcum (1–40%), aluminosilicate clays and/or other clays (0.5–25%), and salt and salt hydrates (1–40%) of almost any existing cations and anions (Kacher et al., 1993abc). Sometimes wheat flour is preferred to cornstarch since it imparts a more acceptable wear rate to an acyl isethionate bar, which can contain between 25 to 65% filler (O'Roark, 1966).

Mildness Improvers/Skin Conditioners/Moisturizers

The formulation of a mild or even ultra mild soap–synbar has become a top focus in the field; a great deal of research and development was devoted to make soaps milder. Moisturizers, emollients, and superfatting additives were included to provide skin conditioning benefits and to improve mildness. Most of them are presented in Table 5.10 (Medcalf et al., 1987; Mausner, 1981). The "superfatting" term comes from fats and oils found in the soap boil in excess of stoichiometric needs. As in conventional soaps, the superfatting additives promote in the syndet/combo bars a denser, creamier lather, that leaves the skin smoother and softer. Classical superfatting agents that were used in alkali soaps, such as palm and olive oil (Palmolive), cocoa butter (Tone), and fatty acids (Coast/Shield), are also of practical use in syndet/combos (Piso & Winder, 1990). The addition of these additives has a plasticizing effect on the base, so that moisture content usually needs to be lowered and the hardening agent content (such as salt) has to be increased.

Table 5.10. Moisturizers/Emollients in Soap-Syndet (Combo) Bars

Moisturizer Type	Purpose	Type of Raw Material
Occlusive	Deposit on skin surface, reducing rate of evaporation	Long-chain fatty acids (25% moisture cream), cream), mineral oil, ethoxylated/propoxylated ethers of lanolin alcohol or methyl glucose, vegetable oils.
Non-occlusive	Hygroscopic substances to stratum corneum, retaining water, improving lubricity	Glycerine, propylene glycol, polyols, sorbitol, lactic acid, pyrolydone, carboxylic acid, urea, L-proline, guanidine, pyrolydone, hexadecyl, myristyl, isodecyl, isopropyl esters of adipic, lactic, oleic, stearic, isostearic, myristic or linoleic acids, isotearyl 2-lactylate, sodium capryl lactate, proteins, aloe vera gel, acetamide mea.

Patented skin conditioning/mildness aid agents seldom listed are hydrated cationic polymers selected from a broad range of cationic polysaccharides, cationic, co-polymers of saccharides, and synthetic cationic monomers of synthetic nitrogen polymers (Massaro et al., 1995; Pichardo & Kaleta, 1993). A preferred polymer is a hydrated cationic guar gum having a molecular weight range of 2,500–350,000. Incorporating 1% cationic guar gum (such as Jaguar C-14-S) to soap/syndet (about 67% soap) formulation is reported to reduce erythema/dryness score below that of a Dove type formulation, when checked by in vivo forearm test (Massaro et al., 1995). A new version of Dove, launched in 1995, unscented and recommended for sensitive skin, contains cocamidopropyl betaine that was seldom included in European formulations for its synergistic improvement of anionic surfactant mildness. Since then, betaine became the preferred foaming mild co-surfactant and replaced the harsher DDBS in all the new versions of Dove.

Some 2 in 1 soap and cream concepts have been solved not by soapless formulations but by including skin mildness and moisturizing additives with mild surfactants. Examples are FA 2 in 1 Soap and Cream with sodium myreth sulfate, decyl glucoside, sodium laureth sulfate, sodium lactate; Monsavon 2 in 1, Savon et Creme with 25% base hydratante, containing ceteareth 80, glycerine, guar hydroxypropyl trimonium chloride; and Palmolive 2 in 1 Wash & Creme with disodium lauryl sulfosuccinate, coconut acid, stearic acid, and lanolin.

A comprehensive ingredient list of the most well-known syndet and combo bars is given in Table 5.11. The ingredient lists for most formulations have been updated to the 2003 formulation list. Table 5.11 demonstrates the dynamic nature that formulas must possess to be able to cope with the market, consumer, environmental, and regulatory changes, while providing formulation improvements.

Striking differences are sometimes found for the same multinational brand manufactured in different parts of the world. For instance U.S. Zest is based on sodium glyceryl ether sulfonate (SGCS), while Zest Mexico contains potassium lauryl sulfate instead (Table 5.11).

Table 5.11a. Syndet/Combar Compositions (from Ingredient Labels)

VEL[a]	Dove Exfoliating[b]	Dove Nutrium[b]	Caress[b]	Dove[b]
SCI	SCI	SCI	SCI	SCI
Stearic acid	Stearic acid	Stearic acid	Stearic acid	Stearic acid
Sodium tallowate	Sodium tallowate	Sodium tallowate	Coconut acid	Sodium tallowate
Water	Coconut acid	Water	Sodium tallowate	Aqua (Water)
Coconut acid	Sodium isethionate	Sodium isethionate	Water	Sodium isethionate
Sodium isethionate	Water	Coconut acid	Sodium isethionate	Coconut acid
SC/SPK	Sodium stearate	Sodium stearate	Glycine soja oil	Sodium stearate
HCO	CAB	CAB	Sodium stearate	CAB
Fragrance	SC/SPK	Sunflower Seed oil	CAB	Parfum
Titanium dioxide	Fragrance	SC/SPK	Mica	SPK
Glycerine	Sunflower seed oil	Fragrance	SC/SPK	Sodium chloride
Sodium chloride	Tocopheryl acetate	Lanolin alcohol	Fragrance	Trisodium EDTA
PSP	Sodium chloride	Tocopheryl acetate	Sodium chloride	Zinc stearate
	Polyethylene	Glycerine	Tetrasodium EDTA	TSE
	Tetrasodium EDTA	Sodium chloride	TSE	
	TSE	Titanium dioxide	BHT	
	Titanium dioxide	Tetrasodium EDTA	Titanium dioxide	
		TSE		
		BHT		

Table 5.11b.

Cataphil[c]	Zest Mexico[d]	Zest[d]	Lever 2000[b]	Olay[d]
SCI	Sodium tallowate	Sodium tallowate	Sodium tallowate	SCI
Stearic acid	Magnesium tallowate	SCGS	SCI	Paraffin
Sodium tallowate	PLS	SC/SPK	Sodium cocoate/laurate	SCGS
Water	Sodium cocoate	Magnesium cocoate	Water	Glycerine
Sodium stearate	Water	Water	Sodium isethionate	Water
DDBS	Talc	Magnesium tallowate	Stearic acid	Magnesium silicate
Sodium cocoate	Sodium sulfate	Talc	Coconut/lauric acid	Magnesium stearate
PEG-20	SLES	Sodium chloride	Fragrance	Sodium isethionate
Sodium chloride	Magnesium cocoate	Triclocarban	Sunflower seed oil	Stearic acid
Fragrance	Fragrance	CA/PKA	Tocopheryl acetate	Magnesium cocoate
Sodium isethionate	Sodium chloride	Fragrance	WGAPD	Coconut acid
Petrolatum	Titanium dioxide	Titanium dioxide	Hydrolized wheat protein	Sodium stearate
SIL	Citric acid	Tallow acid	Titanium dioxide	Sodium cocoate
Sucrose laurate	Sodium citrate	PSP	Sodium chloride	Fragrance
Titanium dioxide	PSP	TSE	Disodium phosphate	Magnesium laurate
PSP			Tetrasodium EDTA	Titanium dioxide
TSE			TSE	Lauric acid
			BHT	Sodium laurate
				Tetrasodium EDTA
				TSE
				PEG-90M

Table 5.11c.

Eubose[e]	Satina[c]	Lanosan[f]	Nivea Milk Bar[g]	Sebamed[h]
DSLSS	SCI	Paraffin	SCI	DSLSS
Sodium coco-sulfate	Glyceryl distearate	SCI	Stearic acid	Triticum vulgare
Triticum vulgare	PEG-200	DSLSS	Sodium tallowate	Paraffin
Cetearyl Alcohol	Cornstarch	Glyceryl stearate	DSLSS	SCI
Paraffin	Paraffin	Sucrose	Sodium cocoate	Glyceryl Stearate
CAB	DSLSS	Water	Coconut acid	Stearic acid
PEG-CCG	Coceth 20	Lauryl polyglucose	Paraffin	Palmitic acid
Aqua	Petrolatum	Milk protein	PEG-150	Cetyl Palmitate
Glycerine	Panthenol	CAB	Parfum	Cetearyl Alcohol
Allantoin	HAP	Citric acid	Sodium chloride	Aqua
Aluminum formate	Bisabolol	Stearic acid	Tetrasodium EDTA	Lecithin
Titanium dioxide	Disodium EDTA	Aluminium triformate	Sodium etidronate	Sodium lactate
	Sodium hydroxide	PEG-GI	Glycerine	Tocopheryl acetate
	Fragrance	Betaine	Lanolin alcohol	Glycine
		PEG-45 M		Magnesium aspartate
		Cetearyl alcohol		Mixture (M)[i]
		Fragrance		Parfum

[a]Colgate-Palmolive; [b]Unilever; [c]Bayer; [d]Procter & Gamble; [e]Dr. Holbein; [f]Wella; [g]Beiersdorf; [h]Sebapharma; [i]Alanine pyroxidone HCl, lysine, leucine, urea phosphate, polyquarternium 22, disodium EDTA, PEG14 M.

Abreviations: Cocoamidopropyl betaine, CAB; Coconut/palm kernel fatty acid, CA/PKA; Disodium lauryl sulfosuccinate, DSLSS; Hydrolyzed animal protein, HAP; Hydrogenated castor oil, HCO; PEG-6 Caprylic capric glycerides, PEG-CCG; PEG 15 Glyceryl isostearate, PEG-GI; Pentasodium pentetate, PSP; Potassium lauryl sulfate, PLS; Sodium cocoate/sodium palm kernelate, SC/SPK; Sodium cocoyl isethionate, SCI; Sodium cocoglyceryl ether sulfonate, SCCS; Sodium dodecylbenzene sulfonate, DDBS; Sodium isostearoyl lactylate, SIL; Sodium laureth sulfate, SLES; Sodium palm kernelate, SPK; Trisodium etidronate, TSE; Wheat germ amidopropyl dimethylamine, WGAPD.

Design of the Dove Beauty Bar—A Development Showcase

An interesting paper on the experimental approach used in the development of the Dove Beauty Bar was recently presented by Hill and Post (2005). Defining the original product-design problem as the formulation of "a non-soap cleansing bar that did not leave a bathtub ring" the paper emphasizes, besides this primary attribute, several secondary properties essential for a successful product. These additional properties relate to consumer perception of firmness, rich and creamy lather, absence of grit, skin mildness, absence of unpleasant odors or colors, slippery-wet feel, low mush rate, and no cracking. Processing attributes are also included.

The logics of the product-design problem followed by the problem-solution strategy offer to the reader a prototype of good research and development practice. The presentation provides detailed data of the DEFI Process (used commercially by Lever Brothers Co. for Direct Esterification of Fatty Acid and Sodium Isethionate) also referred to as DEFI (Cahn et al., 1967) and Unilever screening tests for lather (Farrell & Nunn, 2005), odor stability, color, wet feel, mush (Van Gunst et al., 2001), firmness (Haass & Lamberti, 1968), and processibility (Post et al., 1997).

Product formulations affect the listed final-product properties:

- higher lather by pH adjustment and co-surfactant choice (DDBS preferred),

- higher stability of odor and color by appropriate choice of partially hydrogenated coconut fatty acid (to 3–5 IV) as preferred raw material for SCI,

- better binding and plasticity and lower cracking by using a combination of long-chain fatty acid (25–30 %) and sodium soap (5–10%),

- lower cracking, while assuring a lower water absorption by keeping the pH not appreciably above 7, and

- better extrusion and stamping by keeping water content at about 4–6 %.

All the above make the paper an explicit formulation showcase, revealing a deep understanding of the ingredients–performance relationship. In this respect, it is similar to previously presented patents (Beerse et al., 1995). The original solutions of the product-design problems, as proposed 30–40 years ago, stood the test of time of the most successful syndet bar on the market ever. Dove's composition remained largely unchanged to this day (Hill & Post, 2005).

Mildness Concept

Over the past 25 years many changes in the approach to soaps have occurred. The turning point was in 1979 when Frosch and Kligman described a new method to assess soap irritant properties (Frosch & Kligman, 1979). They demonstrated that the chief weakness of the existing tests was that under normal usage conditions the reactions were weak and did not discriminate between different soaps. They proposed the soap chamber test (SCT) to conduct tests on people known to have sensitive skin under extreme conditions, thus producing strong reactions that emphasize the differences between various soaps (Frosch & Kligman, 1979). After five weekday Duhring chamber exposures to 8% soap solutions (24 hour application in the first day, 6 hour applications in subsequent days), the skin reaction (erythema, scaling, and fissures) are read and rated, that characterize a soap as mild and non-irritant when the total score is less than 1 and harsh when it is close to 5, the maximum score.

Tests performed in Israel in 1992 confirmed Dove mildness, producing similar results for Neca 7 syndet with 25% moisturizing cream and Soft with 25% moisturizing cream (Fig. 5.10). The two local brands were based on SCI/DSLSS soapfree syndet blends. However, similar tests made in the U.K. during the winter of 1993 scored Lever 2000, considered to be mild, around 2.0, while Neca 7 and Soft scored 0.9 and 1.6, respectively (Internal Neca Publication, unpublished). The total overall mean score was primarily due to the erythema results, while the scaling and fissuring values were lower than 0.2 for all the products.

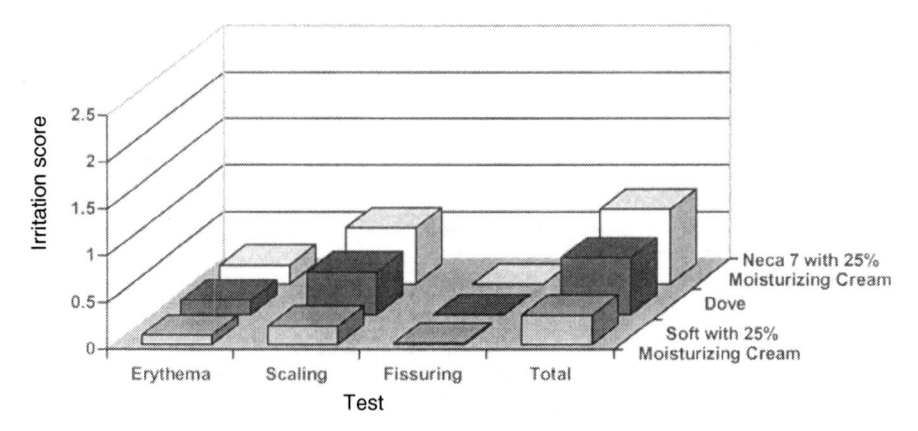

Fig. 5.10. Soap chamber test of 0.06 mL (8% w/v), overall mean score.

Climate conditions and skin panel sensitivity greatly influence the results, as already found and reported by Frosch and Kligman (Frosch & Kligman, 1979). Similar large variations in the SCT mildness scores for Dove over the course of a year have been reported, showing values from 1.05 to as high as 2.65 (Aronson et al., 2001). Thus, whenever certain bars were evaluated, the mildness of both Dove and Ivory (used as a control for a so-called "irritant" product) were also evaluated (Aronson et al., 2001).

The lack of correlation that sometimes exists between consumer tester reports and the proposed Frosch and Kligman "toxicity ladder" is intriguing. Consumer survey reports from 500 women who took part in usage tests showed that the best-liked soap was Neutrogena Normal to Dry, which ranked first in all categories, including luxurious skin feel, quick rinsing, super cream feeling, and produced extra skin softness (Boughton & Huges, 1981; Wolf, 1994). Dove ranked much lower. The results contradicted the soap chamber test in which Dove was the leader and Neutrogena ranked seventh (Frosch & Kligman, 1979). Similar discrepancies between facial wash and SCT results have been reported by Frosch (Frosch, 1982). "The question then arises: What test actually reflects soap's quality most accurately—the irritancy tests under extreme conditions, or a broad survey of consumers? A not less important question is: What is the most important property of a soap—mildness, or cleansing power and feeling of the skin after use?" (Wolf, 1994).

A comprehensive survey of the models for surfactant–skin interactions that detail the mildness from in vivo and in vitro evaluation methods was compiled by Paye (Paye, 1999) and is summarized in Table 5.12. As can be seen, the extreme usage conditions of SCT are obvious when compared to other in vivo tests that are normally used in state-of-theart patents and performed under realistic rinse-off usage conditions (application time not exceeding 90 seconds, compared to at least 6 hours in the SCT). For this reason the SCT method has been largely replaced in the industry by other in vivo tests, such as the forearm-controlled application techniques. This methodology was also selected for a recent Unilever patent (Aronson et al., 2001) that used controlled application wash tests, such as the Flex wash and Arm wash tests to quantify the relative potential to induce irritation, skin barrier damage, and dryness. These tests utilize a combination of subjective evaluations (visual skin condition assessment by expert graders) as well as objective instrumental biophysical measurements to quantify the induced changes to the skin barrier function and the skin's ability to retain moisture (Aronson et al., 2001).

Table 5.12. Mildness Evaluation Methods

Type of test	Method	Reference
in vivo	Flex wash test (3 washes daily, total 15 washes) (60 sec application)	Redd et al., 1992; Medcalf et al., 1987
	Arm wash test (4 washes daily, total 40 washes) (90 sec application) evaluate smoothness, erythema, and dryness	Medcalf et al., 1987
	Forearm wash test (4 washes daily, total 17 washes) (90 sec application) evaluate erythema, and dryness	Medcalf et al., 1987; Redd et al., 1992
	Soap chamber test (5 d, 24 h and 6 h × 4 applications) evaluate erythema, scaling, and fissures	Frosch et al., 1979
in vitro	Skin barrier destruction test Skin barrier destruction is measured by the relative amount of radio labeled water (^{3}H–H$_2$O) which passes from the test solution through a skin epidermis membrane into the physiological buffer contained in the diffusive chamber	Redd et al., 1992; Schwartz et al., 1991, 1992; Kacher et al., 1993ab; Wilson et al., 1993; Frantz, 1975; Small et al., 1987
	Zein test The Zein protein (similar to keratin) is solubilized in a surfactant solution. The Zein solubilized is determined by the released nitrogen content. The Zein number is measured as mg N/g surfactant. Red blood cell (RBC) test Hemoglobin denaturation or haemolysis test; measured by visual examination, transepidermal water loss (TEWL), or blood flow	

Mildness Evaluation Methods

Different protein denaturation in vitro tests have been developed to predict surfactant-induced eye or skin irritation (Paye, 1999). The Zein test is one of the widely used in vivo screening methods to evaluate local tolerance of surfactants (Table 5.12). Zein is a protein obtained from corn and resembles keratin. The Zein test developed by Gotte (Gotte, 1966) is based on the solubilization of the corn protein; the protein is normally insoluble in aqueous solutions unless denatured. It is incubated with the surfactant solution for 1 hour at constant temperature and under slight shaking. At the end of the incubation, the soluble fraction is separated from the insoluble one by centrifugation and filtration. As the surfactant-induced irritation increases, more Zein will be denatured and solubilized. Zein solubility in a surfactant solution is measured and given as mg nitrogen solubilized per gram of surfactant, known as the Zein Number (mg N/g). Values lower than 200 classify the product as mild and nonirritating. Zein values of different surfactants, as measured by Clariant (Clariant, 1999), are presented in Figure 5.12.

Combar mildness evaluation by the Zein test has been used in several patents, such as in a Unilever patent (Post et al., 1998). They found that a fatty acid soap containing 1–25% polyoxyethylene–polyoxypropylene surfactants (EO–PO polymer of ratio between 1.2:1 to 15:1 and a MW of 2000–25000) has enhanced mildness with no sacrifice in processability or lather. The claim was proved by showing that Zein values of combars containing about 34% soap, 14% SCI, 10% fatty acid, and 25% EO–PO polymers are lower by 40% when compared to non EO–PO bars (Post et al., 1998).

Chemistry, Formulation, and Performance of Syndet and Combo Bars ● 179

The red blood cell (RBC) test also investigates the protein-denaturing effect of surfactants by using red blood cells as a biological material substrate (Pape et al., 1987). The surfactant solution causes hemolysis of the blood cells and subsequently releases hemoglobin into solution and partially denatures it. After eliminating intact cells and cell debris by centrifugation, the amount of released hemoglobin and the proportion of denatured hemoglobin are assayed spectrophotometrically. The capacity of the product to induce cell hemolysis is currently used to predict the eye irritation potential of the material, while the skin irritation potential is predicted from the proportion of pigment which is denatured (Paye, 1999). RBC values of different surfactants, based on Clariant results (Clariant, 1999; Henning et al., 1999), are presented in Figure 5.12.

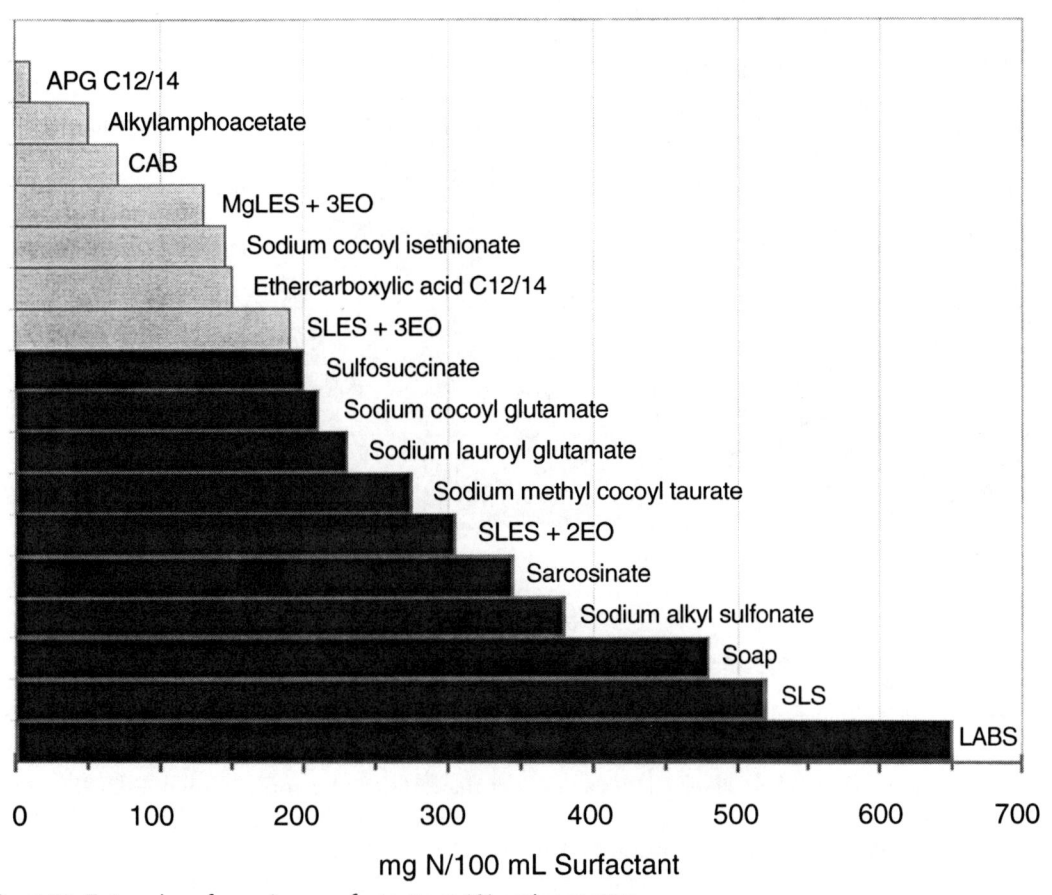

Fig. 5.11. Zein values for various surfactants at 1% active matter.

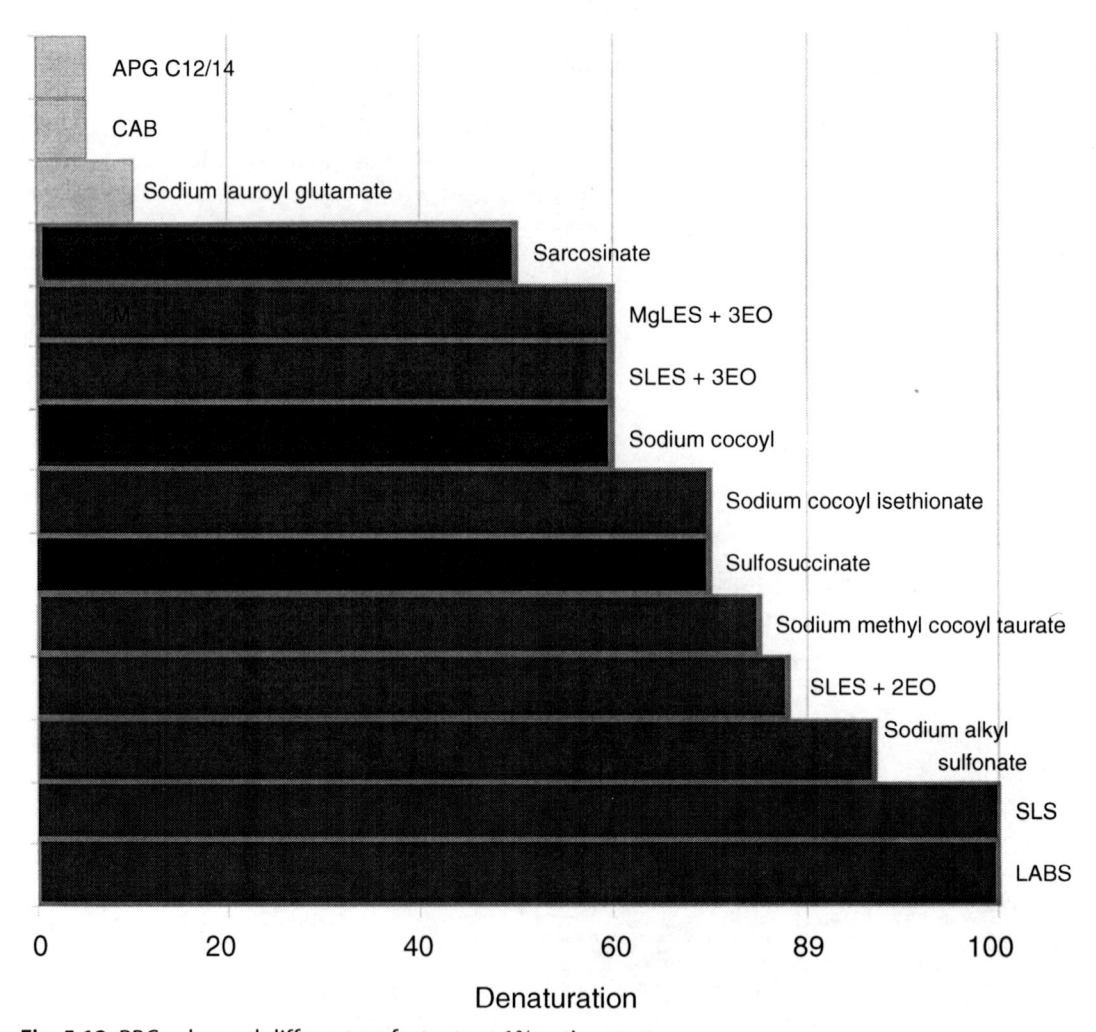

Fig. 5.12. RBC values od different surfactants at 1% active matter.

Most of the P&G patents (Beerse et al., 1995) measure the surfactant mildness by a skinbarrier destruction (SBD) test, developed by Franz (Frantz, 1975). One of the patent claims, based on SBD findings, is that the long-chain alkyl sulfates are milder than AGS and incorporation of 8–20% sodium cetearyl sulfate contributes to the required mildness of the specified product (Beerse et al., 1995). A surfactant mildness comparison, as measured by SBD test, has been also used by P&G to promote the newly commercialized AGS surfactant (Procter & Gamble, 2000, 2001, 2003). The results are presented in Figure 5.13. In Israel one particular neutral soapless soap (Neca) commands about 30% of the market. Although it has been on the market for about 40 years and has undergone many clinical tests throughout that period, it has never lost its place as the most widely sold soap in the country. Testing this soap with the new methods produced confusing results. In SCT, Dove was shown to be much better than the local soap. In the Zein test, which generally correlates well with the results of the chamber test (Wolf, 1994; Kaestner & Frosch, 1981), the two soaps were equal; both were only moderately irritating (Fig. 5.14). Another low pH soap, Softcare, has proven itself to be a very mild baby soap in clinical experiments in Israel. It was found to be much superior to Dove in the Zein test,

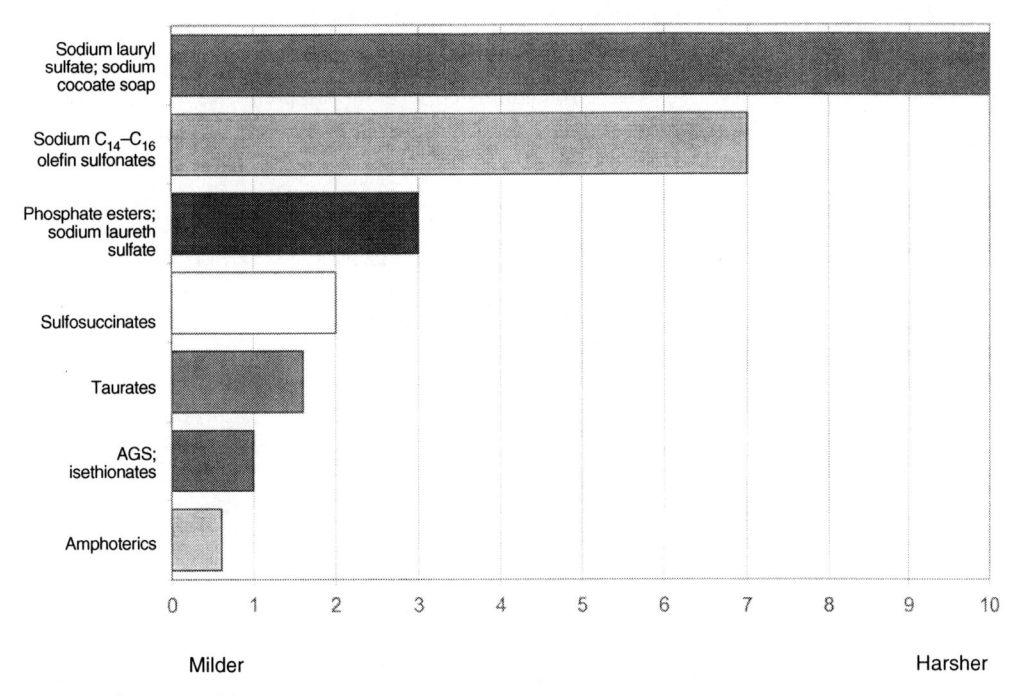

Fig. 5.13. Surfactant mildness comparison.

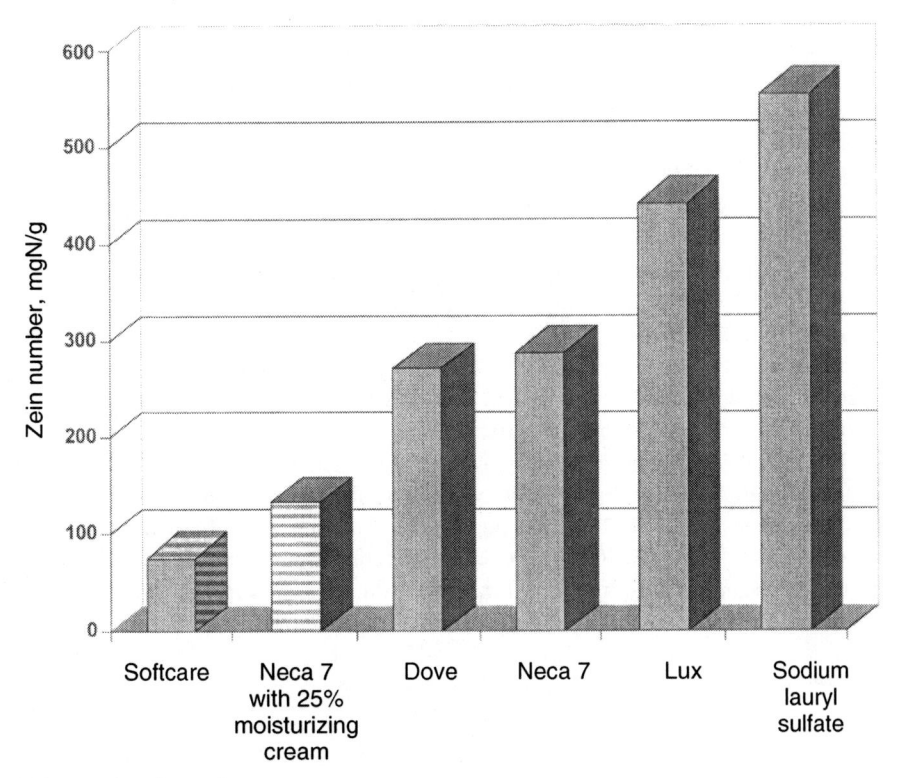

Fig. 5.14. Zein number for various soaps.

76 vs. 272, respectively, and in the SCT the opposite was the case (Fig. 5.14). The two local syndet brands are based on an alkyl sulfate/sulfosuccinate blends of different alkyl sulfate types and alkyl chain lengths as well as various AS/DSLSS ratios. The confusing results presented here raise the question of whether laboratory testing under extreme conditions is more reliable than clinical testing or consumer opinion that reflects decades of use. Should we rely on the sometimes confusing data that arise from very modern and sophisticated tests or on clinical tests and broad survey of consumers and usage tests (Wolf, 1994)? It seems that the answer can be found in the remark made by Frosch (Frosch, 1982) that "with regular use the soapless soaps . . .cause no adverse effects in people with normal skin . . . and are often beneficial to people of oily skins."

"The wide acceptability of the mentioned soap is consistent with this remark of Frosch, since 80% of the population have oily to normal skin and therefore, laboratory testing under extreme and unphysiological conditions is irrelevant for them" (Wolf, 1994). This statement confirms later objective self-criticism of Kligman on the patch test procedure (Kligman, 1996). The proposed approach is a syndet bar "for each skin type" recommending soaps for normal to oily, normal to dry, and for dry and sensitive skin. Different formulations and ingredients should suit different skin types. For instance, an alkylsulfate/sulfosuccinate base is suited to normal to oily skin, while an isethionate base is suited to dry and sensitive skin.

Market Development

Syndet bars have come a long way since their first commercial development about 50 years ago. It was then that mild cleansing bars were formulated by German dermatologists for patients who suffered from "soap eczema" due to sensitive skin. At that time, the bar was composed of surfactants similar to those popular and still widely used in shampoos. In fact, some soapless soap manufacturers defined their bars as "solid shampoo," promoting a successful marketing approach. However, it is interesting to note that even today's patents (Orshitzer & Macander, 1977; Moran, et al., 1993; Constantine, 1989) recall this concept, but they refer to moulded formulations. A similar approach was adopted by the Israeli market, the only one in which syndet cleansing bars cover over 50% of the total soap-bar market. This market share, favored by hard water considerations, is still greater than in any other market, although the current American and European markets have been growing tremendously.

At the beginning of the 1990s, Germany seemed to be the major European soapless soap cleansing bar market with 6% in soap volume but with 20% of the total value market, since syndet bars are much higher priced. The high price of the syndet bars was due not only to the more expensive formulation than the natural soaps, but was also due to the marketing positioning and strategy. Most of the successful brand names in Germany were positioned on a dermatological platform as premium-priced, upscale products; focusing on their skin benefits and selling primarily through pharmacies and drugstores resulted in a final consumer price of $2–3/100-gram bar. In France, where the soapless soaps are categorized as "pain dermatologique," leading brands account for about 2% of the total soap market and are priced similarly to their German counterparts.

Neca took an entirely different approach in Israel. Neca recognized the mass-market potential of a bar having the benefits of a synthetic soap but selling at a popular price. In 1964 the company introduced Neca 7, a soapless cleansing bar having a dermatological value similar to the European syndets but priced at $0.50–0.70/100-gram bar and marketed as an all-purpose cleansing bar.

Neca 7 quickly became the leading brand in Israel, and until now has commanded about 30% of the total soap market, surpassing the slightly lower market share of another local syndet bar, Hawaii, owned by Henkel. One can most likely consider Neca 7 and Hawai the world's best-selling syndets per capita in the highest syndet market-share country in the world.

The same marketing approach was successfully applied in the United States by Lever Brothers, which followed the same strategy of mass-market distribution, pricing, and promotion for its soapless and combination cleansing bars (especially Dove and, more recently, Lever 2000). In 1955, Dove was introduced as being different from other toilet bars because it "looks like a soap, it's used like a soap, but it is not a soap". Later, Lever began emphasizing that Dove "contains one-quarter moisturizing cream" and "won't dry your skin like soap," which remains the slogan of the present advertising campaigns. After a 1979 successful dermatological promotion supported by Kligman tests reporting that Dove soap was the best for the skin, Dove's market share grew. In 1982 Dove began successful marketing to doctors; it increased its market share constantly until it surpassed 9% by volume in 1990, when unscented Dove was introduced, while the last original Dove patent expired (*New York Times*, 2001). By 1993, Dove was the best seller, accounting for 16.4% of the U.S. bar-soap market. The success of soapless and combo bars in the United States was so dramatic and sustained that in 1991 Lever Brothers was able to oust P&G from the leadership position in the toilet-soap market. Two years later, in 1993, P&G responded in this "soap war," and regained the leadership with the success of its Whitewater Zest and, especially, Oil of Olay cleansing bath bars. The marketing strategy of the Oil of Olay bar, "a beauty bar at a competitive price," helped P&G regain the top position by 1993.

The key to the P&G success was primarily due to over a decade of research and development that enabled the creation of a "unique formula, which contains a specially engineered synthetic cleansing system." This again emphasized the importance of a properly chosen formulation for a targeted cost-effective product. The 1999 Dove Nutrium skin-nourishing bar was launched, visually emphasizing a dual formula (white–pink strips) that contained natural-skin needs. The white formula was advertised as a gentle cleanser, while the pink formula contained a nutrient-enriched lotion with vitamin E (*New York Times*, 2001). Dove advertising for 2002 emphasized "sensitive-skin innovations of unsurpassed mildness" comparing a relatively high erythema score (1.9) of a Cetaphil gentle-cleansing bar with a Dove Sensitive score that was only 1.4 (Unilever, 2002). The same advertising highlighted the proven compatibility for medicated applications containing Retin A and benzoyl peroxide (Unilever, 2002).

Dove Gentle Exfoliating Beauty bar was launched in 2003 and was formulated with small, soft, smooth blue beads that combined the brand's moisturizers with gentle exfoliation (Henson, 2003). During the last years, P&G was active too, focusing more on the emotional needs of consumers and launching a variety of new invigorating, fresh and wild scents for Zest (Henson, 2003).

The continuous development of the combars in the United States increased gradually their market share. In 2002 the combars reached a market share of 48.3% out of a total bar- soap sales of 1432 million USD. Five of the ten U.S. soap vendors of those days were combars (Dove, Lever 2000, Zest, Caress, Olay), with market shares of 23.7% (Dove) to 4.5% (Olay) (Euromonitor, 2002, 2003; IFF, 2002).

Nevertheless, the data continued to show in 2002 an annual decrease of 3.8% in total bar sales, continuing to surrender to liquids.

During the last years, bar-soap sales dropped further, and liquid-cleansing products increased significantly. The marketing titles of 2000, such as "liquids move up in the soap market" already emphasizing in 1999/2000 an annual increase of 7.7% in liquid-soap sales (Marchie, 2000), went on in 2002 when liquid soap rose 4.6% (Henson, 2002).

An overall view of the United States' and Western Europe's (Big 5) markets of solid bars and liquid-soap alternatives (hand, shower, and bath) was compiled for 2001 from *Euromonitor* data (2001; IFF, 2002).

The European segment of all the personal-wash-products market includes the body washes, mainly in the bath-care segment; this probably also contains some more sophisticated cleansing items. However, the data show that overall, the alternative- cleansing sources—such as liquid soaps, shower gels, and body washes—outpaced the bars significantly. The bar-soap market share dropped to about 37% in the United States, 25.6% in the United Kingdom, 22.4% in Italy, 19.5% in France, 12.2% in Germany, and 11.1% in Spain (*Euromonitor*, 2001; IFF, 2002). For comparison, the *Euromonitor* statistics of the global soap market showed a total sales of $8,975 million in 2001, of which Unilever, P&G, and Colgate-Palmolive had a market share of 33, 11, and 10%, respectively, while the key leading global brands were Lux (10%), Dove (9%), and Palmolive (4%). A slightly different estimate of the worldwide toilet-soap-bar market values the global retail value as U.S. $9,650 million, and the estimate is that about $1,400 million is claimed by syndet and combo bars (Hendrickx, 2002).

The 2007 figures follow the same trend globally, but differences are found by geographical areas. As many other things, the soap business is also a matter of geography. According to the *Euromonitor* statistics, the total retail- sales value of the global soap market summed up in 2007 to U.S. $10,002 million. The key leading brands were, as in 2001, Lux (12%) , Dove (11%), and Palmolive (4%). To this list, worthy of mention is Zest, the combo of our interest, with a global share of 2% (IFF, 2008a). The increase of the total global-soap-market sales, by 11.1% from U.S. $8,975 million in 2001 to $10,002 million in 2007, seems surprising for a shrinking soap-bar market.

However, if we consider the influence of the other geographical areas, the reason seems obvious. The Asia–Pacific area, that had in 2007 a retail sales value of U.S. $3,478 million, is characterized by a much higher consumption of bar soaps. For instance, the Indian bar soaps account for about 80% of the total body-wash sales, amounting to U.S. $1,369. The same figures of higher bar-soap consumption compared to total liquids and gels characterize, as well, the Eastern European, as compared to the Western European market. Updated available figures of solid and liquid soap consumption in Europe are presented in Table 5.13. All data compared *Euromonitor* figures between 2001 and 2007 (IFF, 2008a,b,c), except for Russia, where 2005 figures were available (Symrise, 2005).

The 2007 bar-soap consumption in Western Europe, compared to 2001, decreased 18% in France, 20% in Germany, 25% in Italy, and 38% in the United Kingdom against the use of liquid soaps and shower gels. In Russia and Turkey, bar-soap consumption has been steady for the time being. However, not like the classical bar-soap situation, the syndet/combo bar market in Western Europe reveals a more positive situation. This is due mainly to Dove's impressive market penetration and the use of specially formulated medicated-type syndet and combo bars.

According to *Datamonitor* data, during 2000–2001, Dove became the leader in France, Italy, and the United Kingdom with market shares of 16.7, 13.6, and 21.5%, respectively (Symrise, 2003). However, except for Sebamed, with a market share of 4.3% in Germany, no other syndet/combar was a leading brand (Symrise, 2003).

The 2006 market shares of the leading European bar soaps are presented in Table 5.14 (IFF, 2008a). Dove dominates in all the five listed countries. The French syndet-bar market is active, offering various formulations and functional variants; among them are: A-Derma (Laboratories Ducray), Klorane–Dermo (Laboratories Klorane), Avene (Laboratories Ducray), and Vendome (Laboratories Vendome).

Future Trends

Milder formulations with enhanced skin-care properties targeted to offer an aesthetic, consumer-appealing appearance seem to be a major trend in the future. The emphasis on environmental care and the use of natural renewable raw materials, made exclusively from vegetable-based fatty materials, will continue to drive development.

A transparent neutral syndet—perceived by customers as a pure product, free of any harmful ingredients—will also be targeted as a specialty bar. A transparent, extrudable neutral syndet, of reasonable cost–performance, should be a supreme challenging target.

During the last few years, bar soaps evolved into more complex products that must satisfy multiple functional needs (hygiene, antibacterial protection, skin care, mildness, and crossover) and emotional pleasures (well-being, attractive shapes, decorative look, and packaging) (Branna, 2001; Emsley, 2002; Marchie, 2001). The realization of these challenging targets, creating "new sensations in soap" and "wow" products, will drive new marketing strategies and continuous innovation, which nowadays are more important than ever for the future market of bar soaps.

Table 5.13. Solid and Liquid Soap Markets in Europe (2001–2007)

	Bar soaps		Liquid soaps		Shower gels	
	2001	2007	2001	2007	2001	2007
France	132	108	65	90	385	440
Germany	100	80	95	95	470	450
Italy	160	120	130	140	160	180
Spain	31	30	13	14	230	340
United Kingdom	210	130	95	145	330	450
Turkey	55	78				
Poland	61	59				
Russia	347	434				

(Sales Revenue in Million Euro) (IFF, 2008a, b; Post et al., 1997)

Table 5.14. Market Shares of European Bar Soaps (2006), %

	France	Germany	Italy	Spain	United Kingdom
Dove (Unilever)	16.4	11.8	23.0	16.6	27.6
Le Petit Marseillais (Lab Vendome)	11.7				
Monsavon (Sara Lee)	7.1				
Palmolive (Colgate)	5.1	5.6	12.7	7.4	4.1
Nivea (Beiersdorf)	4.0	10.5			
CD (Lornamead)		9.7			
Lux (Unilever)		7.9			
Sebamed (Sebapharma)		4.3			
FA (Henkel)		7.6			
Atkinson (P&G)			10.5		
Neutro Roberts (Manetti & Roberts)			8.8		
Fiori Roberts (Manetti & Roberts)			2.9		
Heno de Pravia (Perfumeria Gal)			2.4	21.7	
La Toja (Henkel)				10.5	
A-Derma (Pierre Fabre)	4.9			1.2	
Spuma di Sciampagna (Italsilva)				5.6	
Magno (Henkel)				2.7	
Imperial Leather (Cussons)					17.6
Simple (Accantia)					4.5
Pearl (Cussons)					3.5
Tesco (Tesco)					3.9
Neutromed (Henkel)					2.2
Johnson & Johnson pH 5.5 (J&J)					1.9

References

Ajinomoto, Japan. Internal publication, 2001.

Ajinomoto, Japan. Ibid. 1994.

Anderson, R.J. et al. U.S. Patent 2,923,724 (1960).

Aronson, M.P.; C.C. Nunn; S.G. Leopoldino; J.C. Chambers; C. Gorman; S. Azr Meehan. U.S. Patent 6,218,348 (2001).

Arseguel, D. The gentle touch. *Soap, Perfumery & Cosmetics* **2003**, *76*, 39–40.

Baggini, D.; F. Nizzero; L. Spitz. Drying and finishing, *Soap Technology for the 1990s*; L. Spitz, Ed.; American Oil Chemists' Society: Champaign, IL, 1990; pp. 154–208.

Barker, G.; L. Safrin; M.J. Barabash. U.S. Patent 4,335,025 (1982), Can. Patent 1,151,967 (1983).

Beerse, P.; J. Dunbar; E. Walker. WO 9513,357; European Patent 0728186(1995); SOFW 2003, 129, 68–72.

Bistline, R.G. et al. Surface Active Agents From Isopropenyl Esters. *J. Am. Oil Chem. Soc.* **1971**.

Boughton, P.; M.E. Huges. Soap cleansers, *The Buyer's Guide to Cosmetics;* Random House: New York, 1981; pp. 118–125.

Branna, T. New ideas are bubbling. *Happi* **2001**, *38*, 48–49.

Cahn, A.; H. Lemaire; R. Haass. Preparation of Sulfonated Surface Active Agents, Fatty Acid Ester. U.S. Patent 3,320,292 (1967).

Clariant, Hostapon Sci: A Versatile Mild Surfactant, Internal publication, 2003.

Clariant. *In vitro* and *In vivo* Tests, Internal publication, 1999.

Cognis. APG in Bar Soaps, Internal publication, 2001.

Constantine, M.S. European Patent 0330435 B1 (1989).

Daimler, K.; K. Platz. U.S. Patent 1,881,172 (1937).

Davidson, A.S., and B. Milwidsky, *Synthetic Detergents*, John Wiley & Sons, New York, 1987, pp. 132–178.

Day, J.F.; W. Mueller; R.H. Muth. European Patent 0585071A1 (1993).

Dederen, J.C. Skin Cleansing and Mildness: A Comparison, ICI Surfactants internal publication RP71/92E, Everberg: Belgium, 1992.

ECM Marketing Report, January 1997.

Emsley, J. A fresh start. *Soap, Perfumery & Cosmetics* **2002**, *75*, 24–27. *Euromonitor,* 2001.

Ibid. 2002.

Ibid. 2003.

Fair, M.J.; T. Farrell. U.S. Patent 5,540,854 (1995).

Farrell, T.J.; C.C. Nunn. Fatty Acid Soap/Fatty Acid Bars Which Process and Have Good Lather. U.S. Patent 6,846,787 B1 (2005).

Finetex; Esposito, M.A.; M. Novakovic. U.S. Patent 4,851,147 (1989).

Frantz, T.J. *J. Invest. Dermatol.* **1975**, *64*, 190.

Friedman, M. *Chemistry, Formulation and Performance of Syndet and Combo Bar, SODEOPEC.* L. Spitz, Ed.; AOCS Press: Champaign, Illinois, 2004: pp. 147–188.

Friedman, M.; R. Wolf. Chemistry of soaps and detergents: various types of commercial products and their ingredients. *Clin. Dermatol.* **1996**, *14*, 7.

Frosch, P.J. Irritancy of soaps and detergent bars, *Principles of Cosmetics for the Dermatologist;* P.J. Frosch, S.N. Horwitz, Eds.; Mosby: St. Louis, 1982; pp. 5–12.

Frosch, P.J.; A.M. Kligman. The soap chamber test: a new method for assessing the irritancy of soaps. *J. Am. Acad. Dermatol.* **1979**, *1*, 35–41.

Galaxy Surfactants. (www.galaxysurfactants.com), Internal publication, Mumbai, India, 2003.

Galaxy Surfactants. Internal publication, Mumbai, India, 2008.

Gattir, L.; R.J. Matthaei. U.S. Patent 3,879,309 (1975).

Gotte, E. *Aesth. Med.* **1966,** *10,* 313.

Gray, F.W. U.S. Patent 2,868,812 (1959).

Haass, R.; V. Lamberti. Synthetic Detergent Bar. U.S. Patent 3,376,229 (1968).

Hampshire, England, Internal publication, 1994.

Helliwell, J.F. Int. Patent WO 9,403,151 (1994).

Hendrickx, J. The beauty of multifunctionality. *Soap, Perfumery & Cosmetics* **2002,** *75,* 33–34.

Henning, T. et al. Milde Tenside, 46, Sepawa Congress, Bad Durkheim, Germany, 1999.

Henson, M. Making a splash in personal cleansers. *Happi* **2003,** *40,* 80–86.

Henson, M. It's a wash: Cleansers 2002. Ibid. **2002,** *39,* 84–92.

Hill, M.I.; A.J. Post. *Design of the Dove Beauty Bar.* A.B. Editor et al., Eds.; Elsevier B.V./Ltd; 2005.

Ho Tan Tai, L. *Formulating Detergents and Personal Care Products;* AOCS Press: Champaign, IL, 2000; pp. 237–239.

Hollstein, M.; L. Spitz. Manufacture and properties of synthetic toilet soaps. *J.Am. Oil Chem. Soc.* **1982,** *59,* 442.

Hormes, M.; W. Schneider; W. Scholz; U. Hennen. WO 95/11959 (1995).

Hoyt, U.S. Patent 2,432,169 (1940).

IFF. A Walk Through the World of Bar Soaps, Internal publication, November 2002.

IFF Bar Soap Review Europe 2007, Internal publication (based on *Euromonitor*), 2008a.

IFF Liquid Soap Review Europe 2007, Internal publication (based on *Euromonitor*), 2008b.

IFF Shower Gel Review Europe 2007, Internal publication (based on *Euromonitor*), 2008c.

Jaworski, R.J.; D.A. Park. U.S. Patent 6,395,692 (2002).

Jungerman, E. The formulation and properties of syndet bars. *Cosmet. Toilet.* **1982,** *97,* 77–80.

Kacher, M.K.; J.E. Taneri; D.G. Schmidt; M.W. Evans; C.S. Koczwara; S.K. Hedges; T.F. Leslie. Int. Patent WO 93/19158 (1993a).

Kacher, M.K.; J.E. Taneri; D.G. Schmidt; D.J. Quiram; M.W. Evans. Int. Patent WO 93/19154 (1993), U.S. Patent 5,262,079 (1993b).

Kacher, M.K.; J.E. Taneri; D.G. Schmidt; T.K. Wong. Int. Patent WO 93/19159, WO 93/19157 (1993c).

Kaestner, W.; P.J. Frosch. Hautirritateionen verschiedener Anionen-Aktiver Tenside in Duhring-Kammer-Test des menschen in Vergleich zu *In-vitro*-und die experimentellen Methoden. *Fette Seifen Anstrichm.* **1981,** *81,* 433 (1981).

Kligman, A.M. A personal critique of diagnostic patch testing. *Clin. Dermatol.* **1996,** *14,* 35.

Kutny, Osmer et al. U.S. Patent 5,041,233 (1991).

Login, et al. U.S. Patent 4,515,721 (1985).

Luzenac. (www.luzenac.com), Internal publication, Paris, France, 2002.

Marchie, M.K. Liquids move up in the soap market. *Happi* **2000,** *37,* 72–84.

Marchie, M.K. A new attitude: Soaps get serious. Ibid. **2001,** *38,* 67–78.

Massaro, M.; G. Gruden; G. Rattinger. Int. Patent WO 95/12382 (1995).

Mausner, J. *Cosmet. Toilet.* **1981,** *106,* 5–30.

Medcalf, R.J.; M.O. Visscher; J.R. Knochel; R.M. Dahlgren. European Patent 0227321 (1987).

Milwidsky, B., Syndet Bars, *Happi 31:* 58–70 (1985).

Molls, W.; K. Schrader. Lecture at Tegewa Congress, Proceedings of CESIO World Surfactant Congress, Munchen, Vol. 4 (1984), pp. 202–207.

Moran, T.F.; B. O'Brian; D. Molan. Int. Patent WO 93/07245 (1993).

New York Times, July 22, 2001.

Nielsen data. Israel, 2003.

O'Roark, J.R. U.S. Patent 3,248,333 (1966).

Orshitzer, P.; A. Macander. U.S. Patent 4,012,341 (1977).

Ospinal, C.E. Sulfomethyl Ester-Based Combo Bars, Stepan Internal Publication, 1998.

Ospinal, C.E.; J.S. Nelson; J.R. Sporer; M.J. Nepras. U.S. Patent 5,965,508 (1999).

Pape, W.J.W.; U. Pfannenbecker; U. Hoppe. *Molec. Toxicol.* **1987,** *1,* 525.

Paye, M. Models for studying surfactant interactions with the skin. *Handbook of Detergents, Part A: Properties;* G. Broze, Ed.; Marcel Dekker Inc.: New York, 1999; pp. 469–509.

Perla, G.; G. Mattielo. U.S. Patent 3,926,863 (1975).

Pichardo, F.A.; J.A. Kaleta. U.S. Patent 5,254,281 (1993).

Piso, Z.; C.A. Winder. Soap, Syndet, and Soap/Syndet Bar Formulations, *Soap Technology for the 1990s;* L. Spitz, Ed.; American Oil Chemists' Society: Champaign, IL, 1990; pp. 209–229.

Post, A.J.; F.S. Osmer; M.F. Petko. Mild Bar Compositions Comprising Blends of Higher Melting Point Polyalkylene Glycols and Lower Melting Point Polyalkylene Glycols as Processing Aids. U.S. Patent 5,683,973 (1997)

Post, A.J.; E. van Gunst; M.H. Wayne; M. Fair; M. Massaro. U.S. Patent 5,786,312 (1998).

Procter & Gamble. AGS, Technical data sheet, Cincinnati, Ohio, 2000.

Procter & Gamble. Zest flakes, Ibid. 2001.

Procter & Gamble. Ibid. 2003.

Ramakers, H.P.E. Syndet Bars: Conception and Characteristics, Surfactants (internal Publication), Everberg: Belgium, RP73/92E (1992).

Redd, B.H.; E.C. Walker; R.E. Hare; D.A. Niederbaumer; J.A. Dunbar; T.A. Bakken. Int. Patent WO 92/16609 (1992).

Ross, J.; G.D. Miles. Foaming properties of surface active materials (Method D), *Annual Book of ASTM Standard;* American Society for Testing and Materials: Philadelphia, Pennsylvania, 1953; Vol. 15.04, pp. 1153–1173.

Schwartz, J.R.; W.R. Cassidy; T.A. Gehrig; E.J. Miller. Int. Patent WO 91/13137 (1991), U.S. Patent 5,108,640 (1992).

Schwartz, J.R.; W.E. Eccard; T.A. Bakken; L.A. Gilbert. Int. Patent WO 92/07931 (1992).

Seppic, Oramix, More Than Just Foaming Agent, Internal publication, Paris, France, 2001.

Small, et al. U.S. Patent 4,673,525 (1987).

Stephenson (www.stephensongroupuk.com), Internal publication, Bradford, England, 2003.

Subramanyam, R. Int. Patent WO 95/16022 (1995), U.S. Patent 5,284,598 (1994).

Symrise. European Soap Market, Internal publication, May 2003.

Symrise. New Bar Soap 2005, Ibid. (based on *Euromonitor*), 2005.

Symrise. 2008 Trends for Soap, Ibid., 2008.

TensaChem. Product Range, Internal publication, 2006.

Tokosh, R.; A. Cahn. U.S. Patent 4,180,470 (1979).

Unilever advertising. Pharmagraphics, 2002.

Unilever. European Patent 0508006 (2000).

Unilever. U.S. Patent 2,970,116.

Urban, et al. U.S. Patent 4,536,338 (1985).

Van Gunst, E. et al. Low Synthetic Soap Bars Comprising Organic Salts and Polyethylene Glycol. U.S. Patent 6,255,265 B1 (2001).

Weber, G. *Acta Derm.* **1987,** *134(Suppl.),* 33–34.

Weil J.K.; A.J. Stirton. *Anionic Surfactants;* W.M. Linfield, Ed.; Marcel Dekker Inc.: New York, 1976; Chapter 6.

Whalley, G.R. Better formulations for today's bar soaps. *Happi* **2000,** *39,* 86–88.

Wilson, D.B.; C.D. Tereck; P.A. Niederbaumer; R.G. Bartolo; F.A. Pichardo; T.J. Welch. Int. Patent WO 93/17088 (1993).

Wolf, R. Has mildness replaced godliness next to cleanliness? *Dermatology* **1994,** p. 217.

Wolf, M.; M. Friedman. Measurement of the skin cleaning effect of soaps. *Int. J. Dermatol.* **1996,** *35,* 598.

Wood, T.E. Analytical methods, evaluation techniques, and regulatory requirements, *Soap Technology for the 1990s;* L. Spitz, Ed.; American Oil Chemists' Society: Champaign, IL, 1990; pp. 260–291.

Zschimmer; Schwarz. Beyond the Surface, Internal publication, 2007a.

Zschimmer; Schwarz. Zetesap–Syndets for the 21st Century, Ibid. 2007b.

Transparent and Translucent Soaps

Teanoosh Moaddel[1] and Michael I. Hill[2]

[1]Unilever HPC, Trumbull, Connecticut, USA; [2]Columbia University, New York, New York, USA

Introduction

Transparent and translucent soaps have a surprisingly long history. It was discovered at least as far back as 1789 that a transparent soap bar could be made by dissolving soap in hot ethyl alcohol, pouring the mixture into open molds to solidify, and then allowing the soap to age (Cristiani, 1881). Well over two centuries later, transparent and translucent soaps make up a good proportion of soap bars sold in the marketplace. They are sold in a variety of shapes and degrees of optical clarity, and some are hand-crafted by artisans, whereas others are mass-produced by the large soap manufacturers. However they are made, these soap bars have much consumer appeal, as clarity in a soap bar connotes health, purity, mildness, and freshness.

Although transparent soaps have far greater optical clarity than translucent soaps, they do share some commonalities. All compositions contain mostly tallow and coconut soap, and they can also include varying levels of soaps made from castor oil, safflower oil, or synthetic detergents. In addition, unlike ordinary soap bars, which are generally exclusively sodium soap, transparent and translucent soap bars both typically contain mixtures of sodium, potassium, and/or triethanolamine soaps. Finally, both transparent and translucent soap bars generally contain polyols, such as glycerine, propylene glycol, sorbitol, or sucrose.

The manufacturing process for transparent soaps differs from that for translucent soaps. Translucent soaps are manufactured by energetic working, which easily lends itself to mass production on a standard soap manufacturing line that has been modified to allow for this. In contrast, transparent soaps are manufactured by preparing and casting a melt, followed by cooling and solidification and sometimes by additional aging, a process that cannot be used in a standard soap finishing line.

In this chapter we provide a coherent understanding of the relationships and interdependencies between optical clarity, bar composition, processing route, and processing parameters. First we build on basic concepts of soap phases and structure to identify sources of light scattering, and we show how molecular interactions amongst the components of a soap bar are critical to the development of optical clarity. Then we discuss the importance of the process path and how it affects the efficiency of molecular interactions.

Soap Structure and Transparency Development

To understand why soap bars may be more or less transparent, we need to consider how the various features of a soap bar can impact the transmission of light. For example, surface imperfections will cause surface light scattering, decreasing the intensity of transmitted light. Internal objects of size greater than 200 nm will also scatter light and thereby reduce the transmitted light intensity. In addition, the presence of any objects that can absorb the incident light beam will also reduce the transmitted light intensity. These possible sources of interaction of soap with light are depicted in Fig. 6.1.

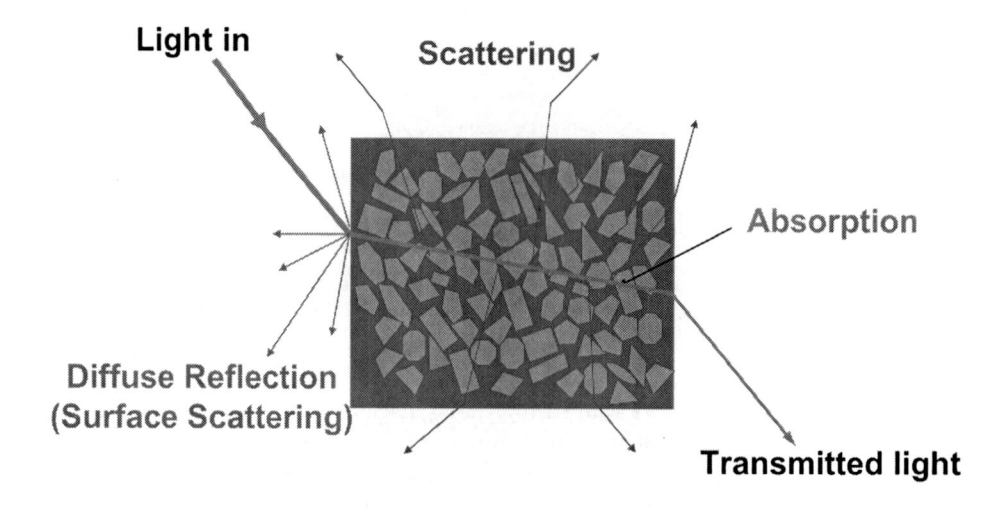

Fig. 6.1. Possible sources of interaction of a soap bar with light.

Applying these general principles of light transmission through soap requires an understanding of the various phases present in a soap bar. As indicated in Chapter 3 of this book by Hill and Moaddel (Hill & Moaddel, 2004), commercial soap bars generally contain multiple components divided among multiple phases, and the particulars of the distribution of components across the phases depends on the oil blend and the process route that are used to get to the final structure. In conventional extruded toilet soap, a mixture of two separate crystal types forms at thermodynamic equilibrium. One crystal type, referred to as *delta phase*, is composed of the less soluble saturated long-chain soaps (e.g., C_{16} and C_{18} soaps) and is dispersed in a continuum of another crystal type composed of the more soluble saturated short-chain soaps and unsaturated soaps (e.g., C_{12} and $C_{18:1}$ soaps), referred to as *eta phase*. The configuration of less soluble soaps dispersed in a continuum of more soluble soaps can be compared to "bricks and mortar," as depicted in Fig. 6.2.

The continuous phase (the "mortar"), which is composed of the more soluble soaps, will also contain more water than the dispersed phase (the "bricks"), which is composed of the less soluble soaps. Further, because solid soap and water have different refractive indices ($n \approx 1.5$ for solid soap, $n = 1.0$ for water), these two phases will have different refractive indices. Thus, incident light can be scattered as it passes through the different phases in the soap bar. Large dispersed soap crystals, entrapped air, and surface roughness will also scatter light, and dark objects present in the soap bar will absorb light.

This suggests that to maximize light transmission, the soap formulator should endeavour to raise the refractive index of the continuous phase, reduce the size of solid soap crystals, minimize entrapped air, reduce surface roughness, and keep the color light. The latter three can be regarded as trivial, and we will not focus on them here. The former two, however, relate to soap solubilization and soap crystal size. Since a soap bar can consist of multiple components distributed amongst multiple phases, controlling these two parameters is nontrivial.

Typical Soap Phases, Their Properties, and Methods for Characterization

The more commonly encountered soap phase structures in a commercial soap bar can be illustrated in the generic binary phase diagram of soap and water, depicted in Fig. 6.3. The various concentrated soap phases can be classified into two main groups: solid crystals and liquid crystals.

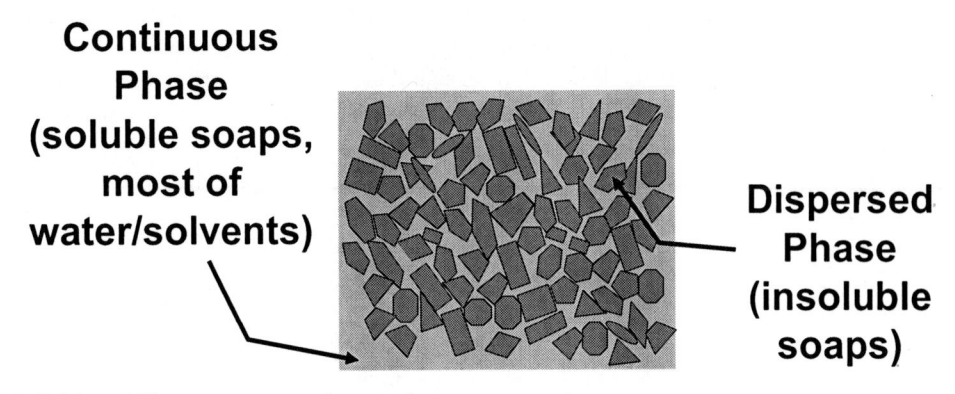

Fig. 6.2. Brick and Mortar structure of a soap bar.

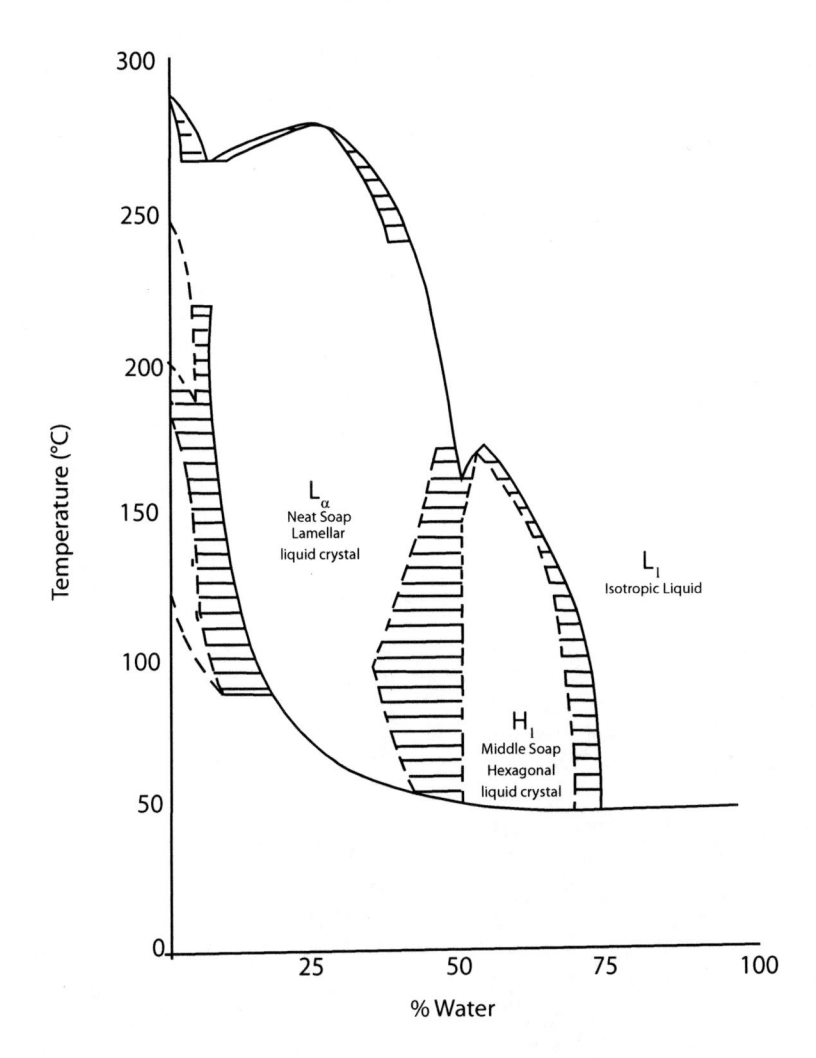

Fig. 6.3. Generic binary phase diagram of soap and water. Reprinted from Hill and Moaddel, 2004.

The structure of solid soap crystals is depicted in Fig. 6.4., and corresponds to packed bilayers of soap molecules. As Chapter 3 describes in greater detail, the more common soap phases have been described using two sets of nomenclature; see Table 6.1. Typical liquid crystal phases encountered in commercial soaps include the lamellar liquid crystal phase and the hexagonal liquid crystal phase, depicted in Fig. 6.5a and 6.5b, respectively.

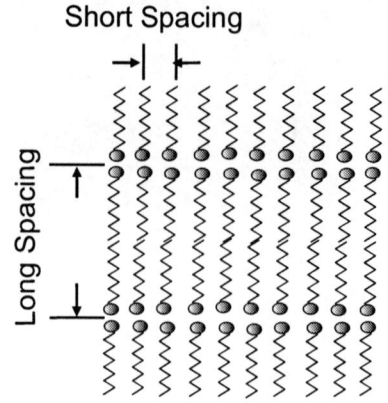

Fig. 6.4. Structure of solid soap crystals.

Table 6.1.

Ferguson	Buerger	
Omega	Kappa	Disordered mixed crystal with all chain lengths, sat'd
	Eta	Mixed crystal rich in unsat'd and short chain sat'd soap
Beta	Zeta	Small mixed crystals rich in long chain sat'd soap
Delta	Delta	Large mixed crystals rich in long chain sat'd soap

Sources: Buerger et al., 1945, Ferguson et al., 1942; Ferguson, 1944.

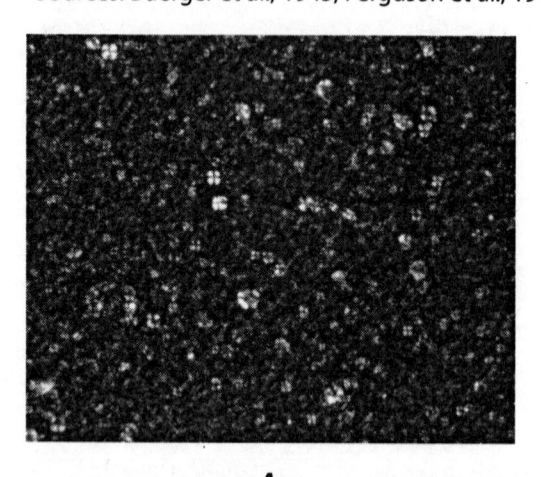

Fig. 6.5. Photomicrograph of lamellar liquid crystal (A) and hexagonal liquid crystal (B)

The lamellar liquid crystal phase, L_α, is similar to that of solid soap crystal, as both have soap molecules arranged in planar sheets. However, the hydrocarbon tails in the liquid crystal are in a "fluid" state, whereas the hydrocarbon tails in the solid crystal are in a "rigid" state. On the other hand, the more viscous hexagonal liquid crystal phase, H_1, consists of close-packed long cylindrical micelles with the soap molecules aligned so that the hydrophilic heads are on the cylinder surface and the hydrophobic tails point toward the center. We refer the reader to Chapter 3 for more details. Optical microscopy with cross-polarizers, differential scanning calorimetry, small angle X-ray diffraction and rheology are typical characterization tools used to investigate the structural nature and flow behavior of these different phases (Nemeth et al., 1998; Funari et al., 1992; Ahir et al., 2002; Rosevear, 1968; Borne et al., 2000; Laughlin et al., 1994).

Translucency Development during Neat Soap Drying and Finishing
Vacuum Spray Drying

Vacuum spray drying of neat soap rapidly converts the lamellar liquid crystalline phase to solid soap crystals. The thermodynamically preferred distribution of the various soaps has all the insoluble (long-chain saturated) soap in the bricks and all the soluble (short-chain and long-chain unsaturated) soap in the mortar at equilibrium. However, the viscosity of the neat soap imposes a mass transfer limitation onto the individual soap molecules during the drying process, preventing their migration to their preferred equilibrium phases. Thus, soluble and insoluble soaps are often trapped together as one solid soap crystal, a metastable situation. This is depicted in Fig. 6.6. The extent of this entrapment depends critically on the final water level of the spray-dried pellet, and for optimum translucency development using the minimal amount of energetic working, the final moisture in the soap noodles should typically be no less than 16%. Above 16% moisture a phase transformation takes place that renders a microstructure more amenable to translucency development.

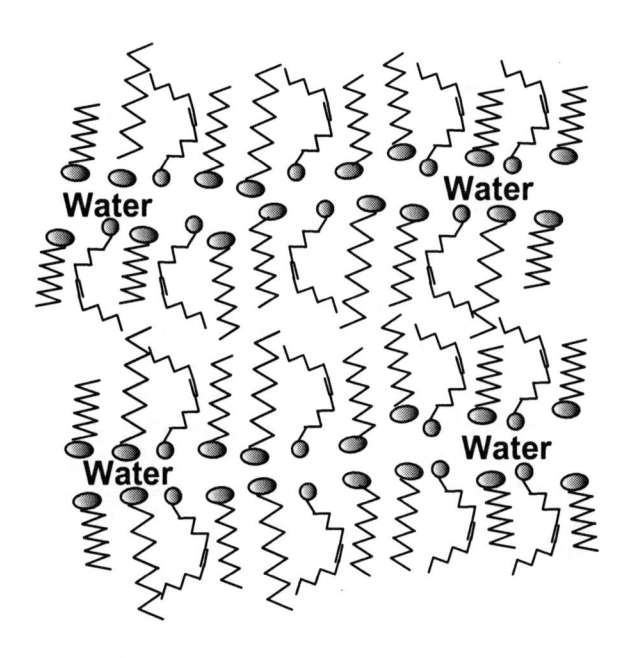

Fig. 6.6. Metastable solid soap crystals.

Intensive mechanical mixing of the spray-dried soap pellets can create sufficient surface renewal to overcome the mass transfer limitation, allowing the soluble soaps to redissolve or melt while leaving behind the insoluble soaps to recrystallize as zeta (beta) phase soap crystals. This is desirable in the manufacture of translucent soaps for two reasons. First, increased levels of soap dissolved in the continuous phase will raise its refractive index to more closely match that of the dispersed phase, and second, zeta (beta) crystals are smaller than other solid soap crystal forms. Both effects increase the transmission of light through the bar.

However, this recrystallization will occur only if the zeta (beta) phase is favoured at the temperature and water levels experienced during mixing. For example, under conditions of low moisture and temperature (below 13% and below 40°C), short-chain and unsaturated-chain soaps will not redissolve, and this recrystallization will not occur. At the other extreme, if the moisture and temperature are too high (above 18% and 50°C, respectively), as can be encountered during downstream soap-making operations (i.e., roll mills and plodders), the short-chain and unsaturated-chain soaps will not crystallize until the system cools sufficiently, and the long-chain soaps will crystallize as either kappa (omega) or delta phase soap rather than zeta (beta) phase. Thus, the *operating window* for translucency development spans conditions that are hot and wet enough for the unsaturated and short-chain saturated soaps to melt or dissolve into the liquid crystal, but cool and dry enough for the long-chain saturated soaps to remain insoluble. This is depicted in Fig. 6.7.

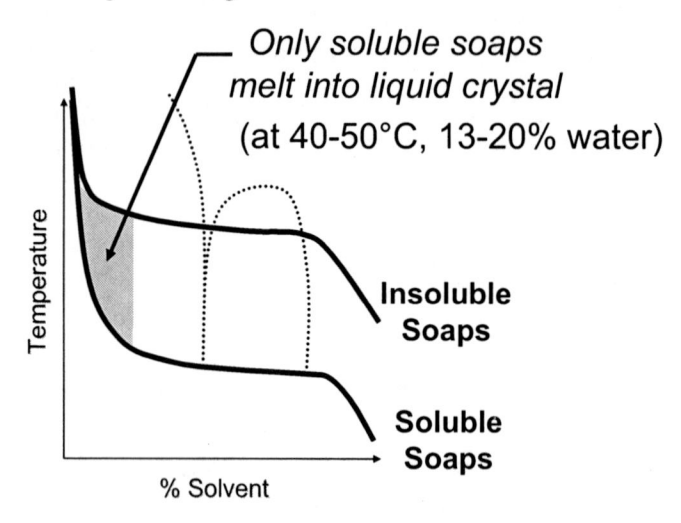

Fig. 6.7. Operating window for translucency development.

The development of translucency largely depends on how well formulation ingredients and processing equipment facilitate the recrystallization of the kappa (omega) phase into the zeta (beta) phase. One approach to achieving this recrystallization involves adding solubilizers. Solubilizers raise the solubility of the short-chain and unsaturated-chain soaps, there by facilitating recrystallization into zeta (beta) phase soap. Since solubilizers also have a high refractive index, their presence raises the refractive index of the continuous phase, which is desirable for translucency. Effective solubilizers include glycerine, propylene glycol, sucrose, sorbitol, and triethanolamine.

The choice of soap counter-ion and tail group can also affect the development of translucency by promoting disruption of crystal packing, leading to smaller and fewer solid crystals. For example, going from sodium to potassium to triethanolamine, the counter-ion becomes larger and crystal packing is more easily disruptable, due to the increasing area occupied by the hydrophilic head group, as shown in Fig. 6.8. Similarly, if the fatty tails of the soap molecules are unsaturated or branched, this also disrupts crystal packing by increasing the volume of the hydrophobic portion, as shown in Fig. 6.9.

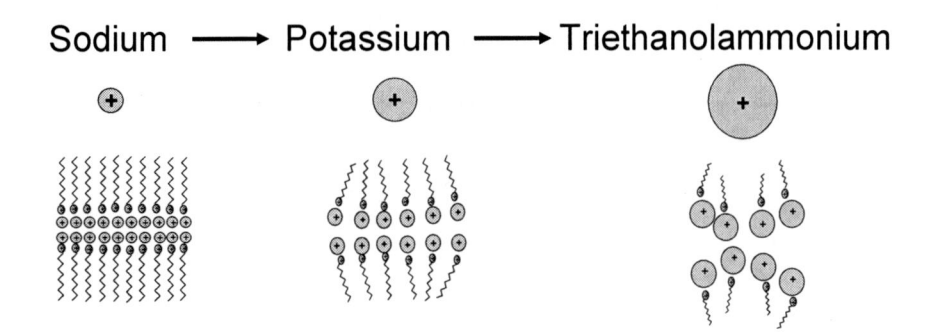

Fig. 6.8. Effect of soap counter-ion on crystal packing.

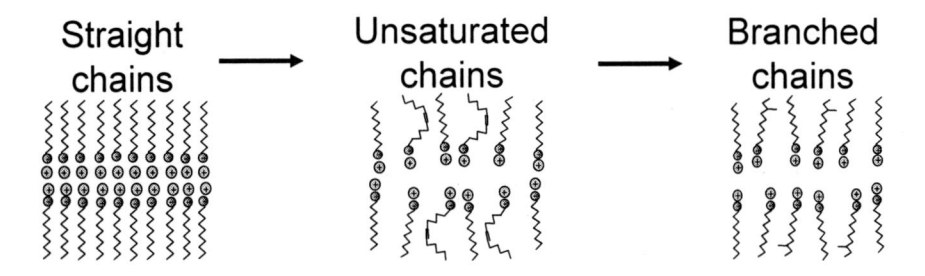

Fig. 6.9. Effect of soap hydrophobic tail group on crystal packing.

One final and sometimes overlooked factor necessary for translucency development is a proper ratio of short-chain to unsaturated-chain soaps. The solubility of short-chain and unsaturated soaps depends greatly on this ratio. Figure 6.10 depicts the phase behavior of a mixture of sodium laurate and sodium oleate as a function of the ratio of oleate to laurate. The water solubility of the mixture is greatest at a molar ratio of 1:1. Since translucency requires recrystallization into zeta (beta) phase, and this is facilitated by increasing the solubility of the continuous phase, translucency should be best at an oleate:laurate molar ratio of 1:1. This corresponds to a tallow-coconut ratio of between 90–10 and 65–35, respectively, or an IV of between 30 and 50.

Thus, production of translucent soap with excellent optical clarity requires 70–80% soap with an IV between 30 and 50; 5–12% of a combination of polyols, which can include select mixtures of glycerine, propylene glycol, and sugars; and 13–20% water, combined with sufficient intensive mixing within the right window of temperature and water levels (typically 40–50°C and 13–20% water).

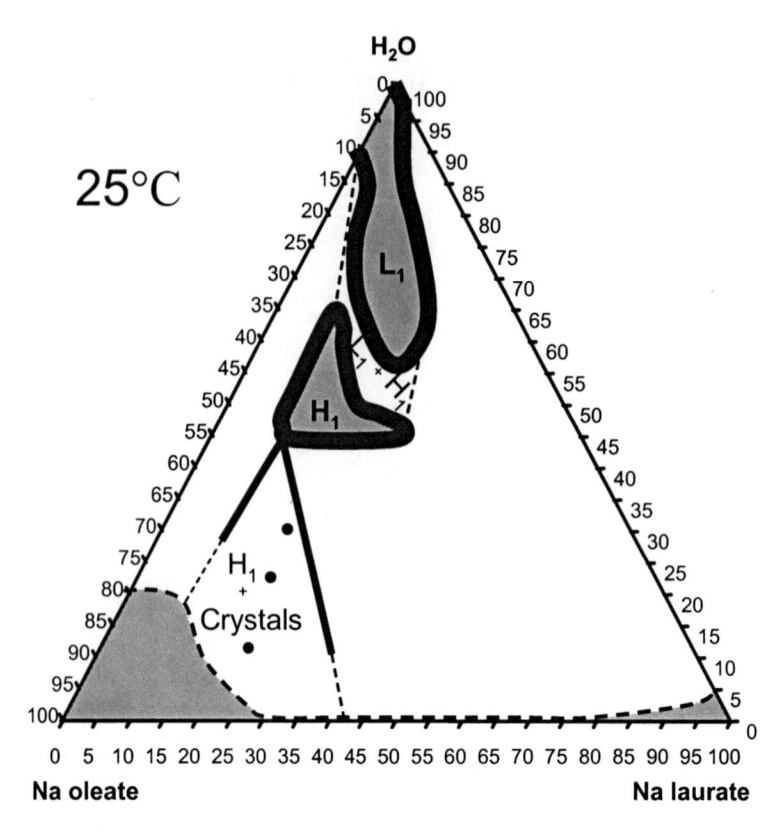

Fig. 6.10. Phase diagram of Sodium Oleate, Sodium Laurate, and water. Reprinted from (Mongondry et al., 2006). L$_1$ – Isotropic micellar phase; H$_1$ – Hexagonal liquid crystalline phase.

Intensive Mixing and Refining

The choice of intensive mixing equipment is also critical for achieving translucency. Double-Arm Mixers are the most widely used intensive mixing equipment. A very interesting unit that was used in the past, but has not gained wide acceptance, is the Cavity Transfer Mixer (Fig. 6.11).[1] Each has its advantages and disadvantages.

In all cases, all nonsoap ingredients can be added to standard opaque soap pellets, provided adequate distributive and intensive mixing is achieved. The Double-Arm Mixers and special plodders (see Chapter 12) can meet both these needs and using them is logistically the simplest approach. Adding the specific polyol blend to the neat soap prior to spray-drying is a more efficient means to incorporate polyols into soap. By this route translucency development is greatly enhanced as a result of better molecular solubilization of polyols and soap. The degree of translucency of the spray-dried pellets (noodles) produced depends on the number of plodders of the vacuum spray drying system.

---

1 In the Cavity Transfer Mixer the intensive work occurs between rotor and stator, it provides for a continuous operation, and is enclosed and jacketed which provides good temperature and moisture control.

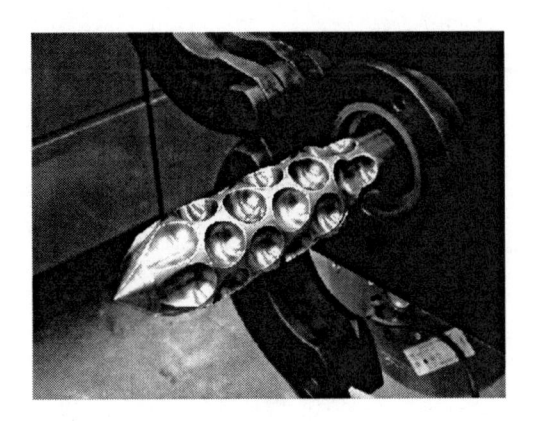

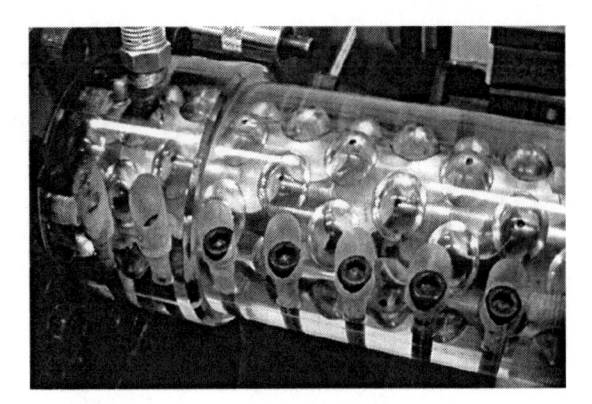

Fig. 6.11. Cavity Transfer Mixer.

The combination of added polyols and the mechanical work imposed on the system generates the right phase sequence and colloidal structure for effective translucency development for extruded bars. By this route no further intensive mixing or work is required for translucency development as spray dried pellets are already translucent. This route, although providing better translucency development, is logistically more complex. And so, it is the combination of added polyols with intensive mixing which dissolves the soluble soap fraction of the metastable solid soap crystal that is generated at the spray-drying stage, leaving the insoluble soaps to recrystallize as zeta (beta) phase soap and resulting in a colloidal structure amenable to translucency development. The more efficiently this separation is accomplished, the better the translucency development. These results reflect the strong interplay between the various polyols and the soap that drives the development of translucency (Wang et al., 2004). Figure 6.12 is a phase diagram of the soluble soap fraction of a typical soap bar with water and glycerine at 45°C,[2] and illustrates this association between soap and polyol.

Clearly evident is the formation of association structures between glycerine, water, and soap and the extension of the isotropic micellar phase, L_1, from the water corner of the phase diagram to the glycerine corner of the phase diagram. This specific interaction of glycerine with soap is fundamental to understanding of why and how soaps can be made translucent or transparent.

It follows from the previous discussion that the fat source directly affects translucency development. In particular, changing the fat source from animal fat to vegetable fat changes the inherent fat composition. This incurs the following major changes:

- A decrease in the level of C_{18} saturated soap at the expense of increasing C_{16} saturated soap
- Increased levels of $C_{18:2}$ (linoleic acid)

---

[2] The L_1 phase has been determined to be within 5%. The exact phase boundaries were not accurately determined for the more viscous phases and are only estimates.

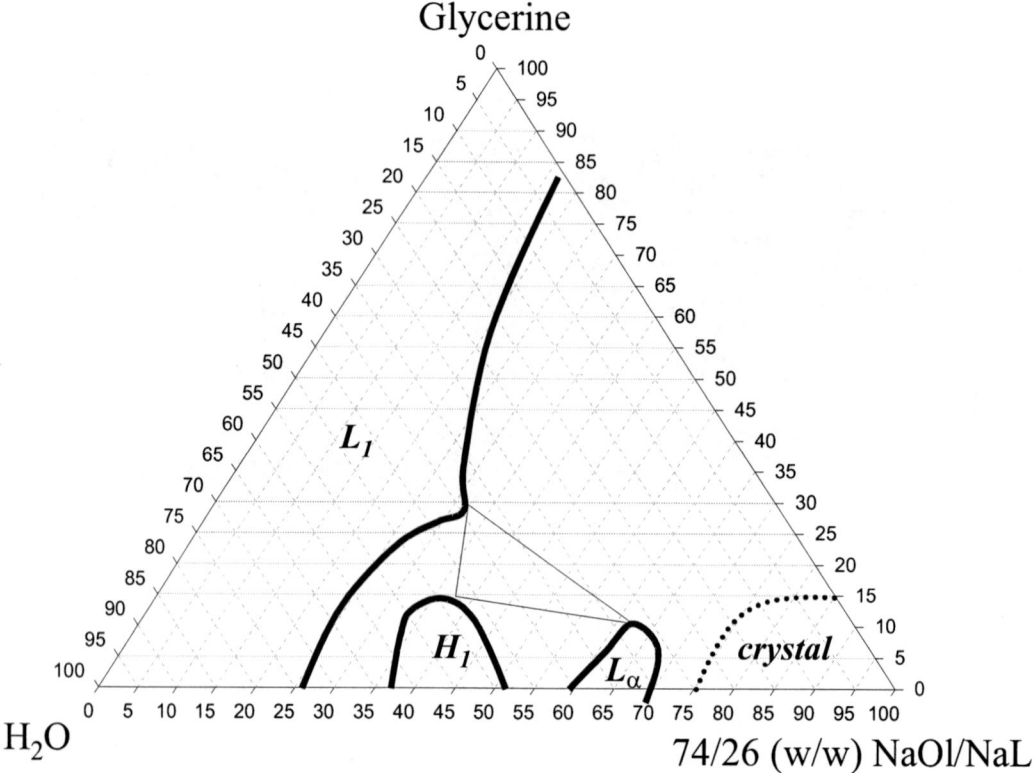

Fig. 6.12. Phase diagram of glycerine, water, soluble soap fraction of soap. L₁ – Isotropic micellar phase; H₁ – Hexagonal liquid crystalline phase; L$_\alpha$ - Lamellar liquid crystalline phase; NaOl - Sodium oleate; NaL- Sodium laurate.

Cast-melt Process and Transparency Development

Pears, the World's First Transparent Soap, was introduced in 1789. It contained glycerine, natural oils, rosemary, cedar, and thyme. The process by which these bars were made involved pouring saponified oils and additives into molds. These were then cut and allowed to mature for a period during which the alcohol evaporated and was recovered for reuse. The final bars where then stamped. The key ingredient helping to render these bars transparent was ethanol, which acts as a hydrotrope. That is, the addition of ethanol increases the solubility of the soap and prevents the liquid crystal phase from forming when the soap solution is cooled. In this case, when the soap solution cools it forms a gel structure rather than the usual bricks-and-mortar structure. It is still debated whether the structure of these bars is a gel consisting of a rigid network of small solid crystals with isotropic solution in the interstices, or instead consists of a supercooled solid (a "glassy state") with dispersed solid crystals (Wang et al., 2004). Other approaches to making transparent melt-cast soap bars involve the use of high levels of polyols, such as glycerine, propylene glycol, various sugars, and triethanolamine, to increase soap solubility and prevent liquid crystal formation on cooling. In contrast to the methods in which ethanol is used, there is no need for maturation. The patent literature also shows that highly soluble syndets, such as sodium lauryl ether sulfate, sorbitan oleate, nonyl phenol ethoxylates, and sodium cocoyl isethionate, are often used to make transparent soaps. In these cases, however, soap remains the primary structurant. Transparent soap bars of excellent optical clarity can be made by combining 35–50% soap (with minimal unsaturation), 30–40% specific blends of polyols, and 15–25% water. The soap used in these compositions must have a minimum level of unsaturation so as not to promote formation of the liquid crystalline phase on cooling.

References

Ahir, S.V.; P.G. Petrov; E.M. Terentjev. Rheology at the Phase Transition Boundary: 2. Hexagonal Phase of Triton X100 Surfactant Solution. *Langmuir* **2002**, *18*, 9140–9148.

Borne, J.; T. Nylander; A. Khan. Microscopy, SAXD, and NMR Studies of Phase Behaviour of the Monoolein-Diolein-Water System. *Langmuir* **2000**, *16*, 10044–10054.

Buerger, M.J.; L.B. Smith; F.V. Ryer; J.E. Spike, Jr. The Crystalline Phases of Soap. *Proc. Natl. Acad. Sci. U.S.A* **1945**, *31*, 226–233.

Cristiani, R.S. *A Technical Treatise on Soap and Candles*; Henry Carey Baird and Co.: Philadelphia, 1881; p. 423.

Ferguson, R.H.; F.B. Rosevear; R.C. Stillman. Solid Soap Phases. *Indust. Eng. Chem.* **1942**, *35*, 1005–1012.

Ferguson, R.H. The Four Known Crystalline Forms of Soap. *Oil Soap* **1944**, *21*, 6–9.

Funari, Sergio S.; Michael C. Holmes; G. J. T. Tiddy. Microscopy, X-ray Diffraction, and NMR Studies of Lyotropic Liquid Crystal Phases in the $C_{22}EO_6$/water System: A New Intermediate Phase. *J. Phys. Chem.* **1992**, *96*, 11029–11038.

Hill, M.; T. Moaddel. In *Soap Structure and Phase Behaviour*, Luis Spitz, Ed.; SODEOPEC 2004, p. 73–95, AOCS Press.

Laughlin, Robert G.; R.H. Ottewill; R.L. Rowell, Eds. *The Aqueous Phase Behaviour of Surfactants*; Academic Press: London, 1994.

Mongondry, P.; C. W. Macosko; T. Moaddel. *Rheol. Acta* **2006**, *45*, 891–898.

Nemeth, Z.; L. Halasz; J. Palinkas; A. Bota; T. Horanyi. Rheological behaviour of a lamellar liquid crystalline surfactant water system. *Colloids Surf., A* **1998**, *145*, 107–119.

Rosevear, F.B. Liquid Crystals: The Mesomorphic Phases of Surfactant Compositions. *J. Soc. Cosm. Chemists* **1968**, *19*, 581–594.

Wang; Joshi; Kumar; Yarovoy; Moaddel. Lyotropic Mesophases Formed by Solutions of Sodium Stearate in Glycerol and Water, Amer. Phys. Soc. March 2004.

Kettle Saponification: Computer Modeling, Latest Trends, and Innovations

Joseph A. Serdakowski
AutoSoft Systems, 2 Round Hill Court, East Greenwich, Rhode Island, 02818, USA

Introduction

This review demonstrates the value of utilizing a computer method to achieve a high degree of accuracy in process control for a full boiled kettle soap making process.

Definitions, Terminology, and Illustrations

The definitions, terminology, and some illustrations are derived from the original chapter on this subject published in 1994 (Spitz, 1996) in a book which is out of print.

The symbols in the curly brackets { } represent the shorthand notation used in the algebra.

Processing Steps

The processing steps are represented by sequential numbers spanning 0 to $k + 1$, with 0 being the loading and k being the number of washes. The processing step is represented as subscripts when applicable.

Ingredients

The ingredients are the materials which are either added to or removed from the kettle. The ingredients are represented as subscripts when applicable. They include:

- $\{f_1, f_2, ... f_i\}$ = Fats and oils (total number = i)
- $\{a_{i+1}, a_{i+2} ... a_{i+j}\}$ = Fatty acids (total number = j)
- $\{c\}$ Caustic = 50% solution of NaOH and H_2O
- $\{b\}$ Brine = saturated solution of NaCl in H_2O
- $\{l_0, l_1, l_2, ... l_k\}$ Lyes generated by process steps (k = # of washes, $k = 0$ is the spent lye for glycerine recovery)—solutions of glycerine, NaCl, NaOH, and H_2O

  Spent lye = a by-product of the kettle process which is high (>15%) in glycerine, and low (<0.5%) in NaOH.

  Wash lye = a lye which is generated and consumed by the kettle process

- $\{y_0, y_1, y_2, ... y_{k-1}\}$ = Lyes added to process steps
- $\{u_0, u_1, u_2, ... u_k\}$ = Curd (k = # of washes, $k = 0$ is the curd resulting from loading) = an intermediate remaining after lye removal
- $\{n\}$ Neat = the finished product of the kettle soap process

- {r} Seat [or Nigre] = remains in the kettle after neat soap removal.
- {w} Water = the liquid phase of H_2O
- {t} Steam = the vapor phase of H_2O

Components

Components are the chemical compounds present in the ingredients. The components are represented as superscripts when applicable. They include:

- {s} Soap
- {ω} H_2O
- {g} Glycerine
- {d} Sodium chloride [NaCl]
- {h} Sodium hydroxide [NaOH]

Physical Properties

Physical properties are quantitative characteristics of the components and/or ingredients. They include:

- {M} Mass (lbs)
- {X^χ} Mass fraction of component χ (lbs/lbs)
- {W} Molecular weight (lbs/lb-moles)
- {T} Temperature (°F)
- {ρ} Density (g/cc)
- {P} Heat capacity (BTU/lb °F)
- {Γ} Heat of reaction (BTU/lb)

Miscellaneous Parameters

- {$\overline{T}$} Reaction temperature of kettle (220°F)
- {D} Day of the year
- {$E_0, E_1, E_2, ... E_{k+1}$} Electrolyte settling ratio (where $k + 1$ is the finish step)— the ratio of the different electrolytes as they settle through different phases, specifically, F[NaCl][NaOH]
- {δ} Separation efficiency—the fraction of the available lye which separates from the curd phase.
- {G} Glycerine concentration factor—this is a measure of glycerine's preference to concentrate in the lye phase during phase separation.

Cooling constants determine how fast the kettle cools. These values are site specific, and are a function of the kettle geometry, insulation, and environment. They include:

- {T_∞} Equilibrium temperature (°F)
- {D_T} Half-life (Days)

Evaporation constants determine how fast the kettle loses water due to evaporation. These values are site specific, and are a function of the kettle geometry, insulation, and environment. They include:

- $\{\omega_\infty\}$ H$_2$O loss at infinite time (°F)
- $\{D_\omega\}$ Half-life (Days)

Conservation Equations

Since neither no one nor anything can create or destroy matter and energy, we can use that principle in our analysis of the kettle process. We apply this in three distinct ways. They include:

- Conservation of mass
- Conservation of mass of each component
- Conservation of energy

Kinetics

The saponification reaction is not spontaneous. As described in Woollatt (1985, p. 154), "... the reaction with neutral fats ... does not start readily. It is autocatalytic, that is[,] catalyzed by the product of the reaction, soap. Hence, the reaction rate accelerates greatly until most of the fat is reacted, when it slows down again." The secret to successful computer simulation is to keep things as simple as possible, but not too simple. The reaction time is much less than the batch time. One simplifying assumption we can make is that everything happens instantly.

Phase Diagram Theory

The kettle soap process has five components, and strictly speaking, a five component phase diagram is required to represent it. Since this is too complicated, the diagram of a simplified three component system is presented.

The components are soap, total electrolyte, which is a linear combination of the sodium chloride and the sodium hydroxide present, and solvent, which is a linear combination of glycerine and water. See Fig. 7.1.

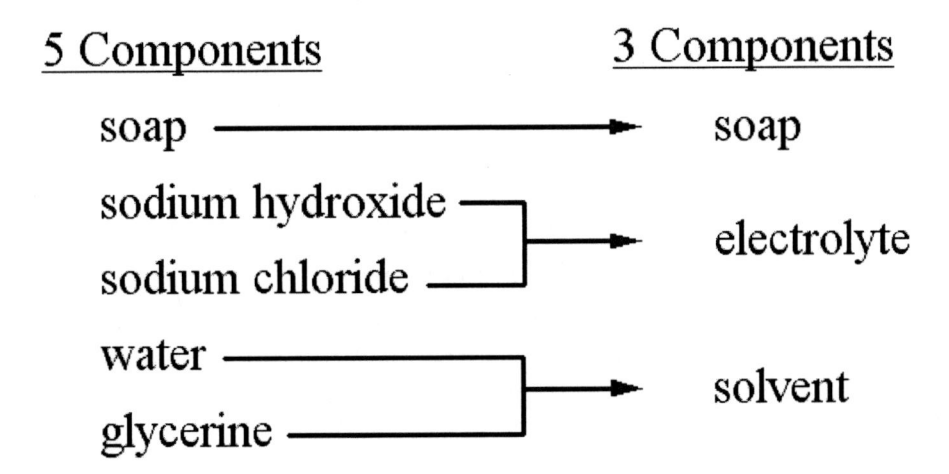

Fig. 7.1. This diagram illustrates the mapping from 5 components to 3 components.

The component list is then simplified to include:

- {s} Soap
- {v} Solvent
- {e} Electrolyte

with the linear combinations defined as (Spitz, 1996):

$$M^e = z^d \times M^d + z^b \times M^b$$

$$M^v = M^\omega + M^g - (1 - z^d) \times M^d - (1 - z^b) \times M^b$$

Here, {z}, defined as the "graining efficiency," a traditional soap making term, is a measure of how much of the particular electrolyte must be added to move the resultant mixture a certain distance in the *x* direction on the phase diagram. One can also use other electrolytes as described in Spitz (1996), p. 119. Typically, the "z" factors are normalized such that $z^d = 1$. The first equation above determines the total amount of electrolytes present. The second equation above determines the total amount of solvent present. The final two terms in the second equation above are necessary to assure that the conservation of mass components is maintained.

The phase diagram of a typical 80:20 tallow/coco soap as illustrated in Woollatt (1985) on p. 153 is illustrated in Fig. 7.2.

Also note the inclusion of several X axes. The values for the X (electrolyte) axis depend on the chain-length distribution of the soap. The graining index data presented in Spitz (1991) on p. 118 allows us to determine the phase diagrams for a number of different soaps. The relative graining indices are aligned, and the electrolyte is scaled in proportion to yield the phase diagrams for all listed soaps. As a first order approximation, the X axis is scaled in proportion to the graining index. For example, the coordinates of the point of intersection of the D and Q regions occur at 6.3% of electrolyte for the 80% of tallow/10% of coco soap as illustrated in Fig. 7.2. This soap has a graining index of 13. A pure coconut oil soap with a graining index of 22.5 will have the point of intersection of the D and Q regions at:

$$\frac{22.5}{13} \times 6.3\% = 10.9\%$$

In this fashion, one can determine phase diagrams for soaps of all chain length distributions.

The phase diagram is further approximated for computerization. Only the two phase regions M and N are required for modeling. Both regions are approximated by straight edged quadrilaterals (i.e., linear approximations), which have proven to be sufficient. Higher order approximations (quadratic) were tested. The higher order approximations complicate the mathematics, but do not provide any improvement to the model. A specific linearized phase diagram is discussed below.

Kettle Soap Boiling—General Discussion

Since the publication of the prior work, the rising cost of energy and raw materials and the plunging value of glycerine resulted in a paradigm shift in the types of kettle processes employed.

We consider the following processes:

Countercurrent or full boil: The traditional way of processing a kettle of soap. In this process, one generates a low glycerine (<3%) neat soap, a seat, and spent lye with 15% or more of glycerine. The steps involved with generating lyes are relatively forgiving, settle quickly, and are easy to manage. The finishing step requires a "fitting" of the kettle. The fitting brings the kettle to a state where the neat soap

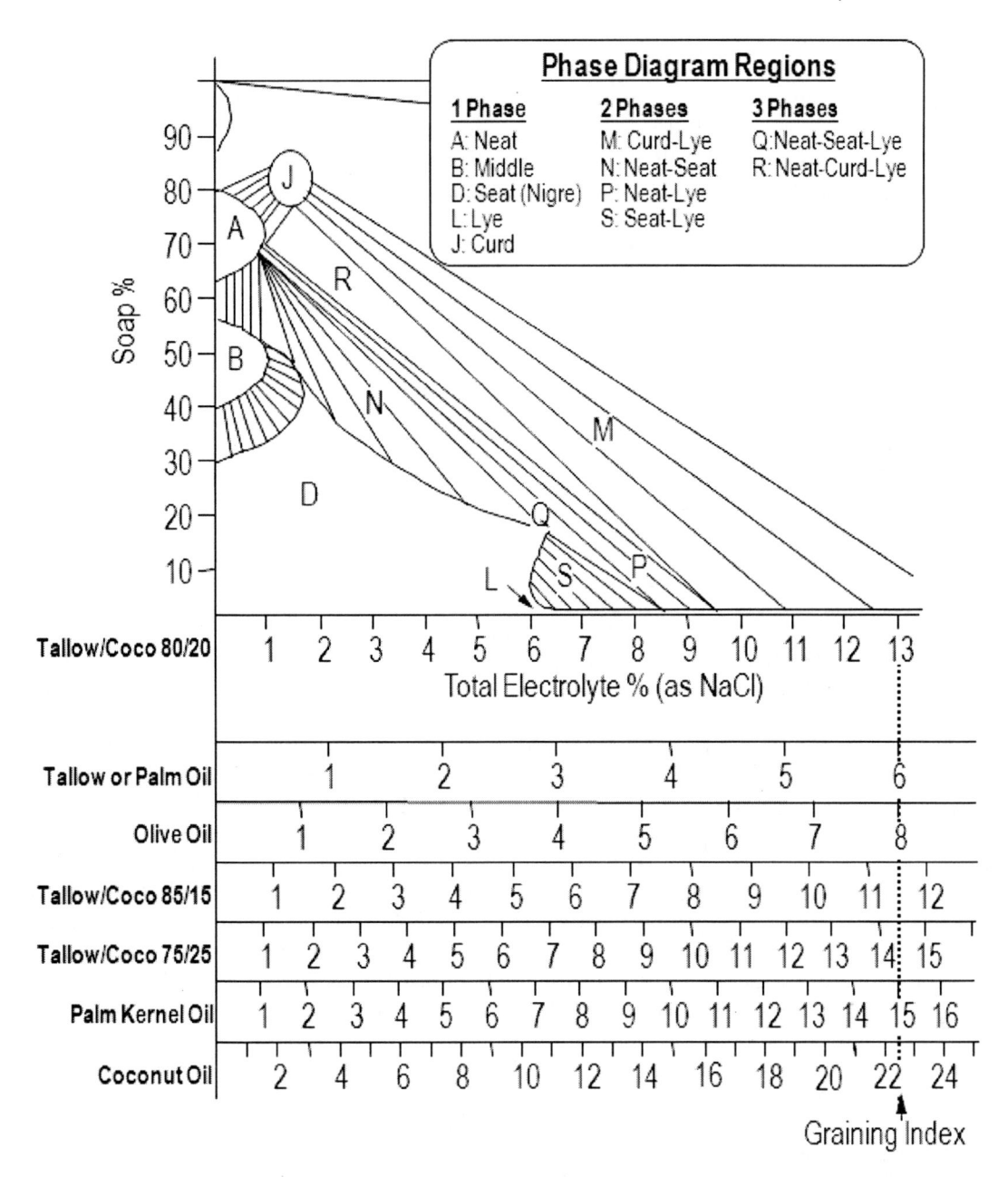

Fig. 7.2. Soap Phase Diagram with Multiple X-Axes. This allows the same image to represent many soap blends

separates from the seat over a 24 to 96-hour period. This fitting is difficult to achieve, and leaves open the possibility that an acceptable neat soap will not be available after the prescribed settling period, resulting in process interruption and considerable rework. Even if neat soap is successfully produced, the quantity of neat soap may vary significantly from batch to batch because of the difficulty in reaching the best fit. This process is fully outlined in many sources cited in the bibliography.

Semi boiled: This chapter refers to this process as "seatless". In the seatless process, an empty kettle is loaded, saponified, and finished in one step, leaving all of the glycerine (≈8%), color, and odor in the neat soap. Neither seat nor lye is generated, and the soap is ready for drying in as little as 4 hours after the loading commences. This high level of glycerine provides considerable "Nomar" qualities, but sometimes results in a base that will "sweat" and stink under high humidity conditions, and will display more cracking than a full boiled kettle.

Lyeless: This process loads an empty kettle to allow the kettle to be finished directly. The resulting neat soap is available in 24 hours at around 5% of glycerine content, leaving behind a seat containing many of the color and odor bodies. This process is advantageous if one has an outlet for the seat in a lower grade base. The physical properties of this base will be intermediate to the bases outlined in 1. and 2. above.

Oil Finish: This process loads either an empty kettle or a seat, and generates one or more lyes in the same fashion as the full boil process. Unlike the full boil process which has the finicky and time-consuming fitting, the resulting curd of the Oil Finish Process has a very low sodium chloride content, allowing for the addition of a high quality fat (e.g., edible tallow); oil (e.g., edible coconut oil); fatty acid (e.g., coconut fatty acid); and/or citric acid to consume the excess free alkalinity, and results in a kettle containing only a low glycerine (<3%) neat soap.

Steps to Kettle Soap Boiling

Step 1—Loading

Typically, a kettle with a capacity of 20,000 to 200,000 pounds is used. The seat often remains in the kettle from the prior batch. The seat is brought to a boil by the introduction of live steam into the bottom of the kettle, through a specialized nozzle called a rosebud (because of its appearance), and through a series of open steam coils.

Precise amounts of fats, oils, and/or fatty acids are combined with caustic, brine, and water. In the case of the countercurrent process, recycled lyes from the first wash of a prior kettle are also added. The materials are added such that the rate of saponification is maximized.

Since spent lye is the desired output of this kettle, the electrolyte or X axis of the phase diagram should be composed of only NaCl, with only enough NaOH added to the kettle to saponify the fats, oils, and fatty acids. This poses a problem for the soap maker, since high excess levels of NaOH drive the saponification reaction to completion; however, no excess should exist, and perhaps even a slight deficit of NaOH should occur at the conclusion of the loading process to assure the formation of a spent lye.

The Seatless Process magnifies this problem. Since in this process the kettle is loaded and directly pumped, the soap maker must assure that all of the fat and oil was completely saponified, and the free NaOH must be very low. Attempts at loading a seatless kettle without sophisticated mass flow meters to precisely measure the fat, oil, and caustic additions were not successful. However, with precise calculations and measurements, an experienced and motivated soap maker can be very successful in loading and finishing a seatless kettle in a timely fashion. By using only whole fats and oils in the Seatless Process, one can simplify the problem of achieving a fully saponified kettle with a low free NaOH (<0.05%) if one has some fatty acid or citric acid available to neutralize the last bit of free NaOH after all of the fats and oils are saponified.

We now turn our attention to identifying the region of maximal saponification on the phase diagram. It is slightly lower in electrolyte than Region M (the two phase curd–lye region). The exact region in which this area is located is subject to some debate. Most published phase diagrams illustrate three distinct regions, those being M, P, and R; however, our experience is that for all practical matters those regions are indistinguishable during the production process. That being the case, the point of maximal saponification will occur in Region Q or perhaps even Region N.

The actual "location" of this point of maximal saponification regarding Regions R, P, Q, N, etcetera is inconsequential when one's principal priority is optimizing production. An experienced soap maker inherently knows this region by the appearance of the kettle contents. To identify this crucial point in the soap making process, simply sample the kettle at the point when the experienced soap maker "knows" the kettle "looks" best. A maximum of 10 to 12 kettles should be more than enough data to define this point for the fat and oil blend being used. Once the first fat and oil blend is identified, one can extrapolate other similar blends by using the relative graining index method outlined above.

The strategy outlined in the preceding paragraph deserves additional attention. Traditionally, two separate and distinct approaches to optimizing the kettle process (as well as all processes) existed. Since the dawn of time, manufacturers have relied on trial and error to optimize any process (observation). More recently, the application of the laws of physics and technology was applied to fully understand and optimize the process. Both approaches can be time consuming. I have always advocated and implemented a hybrid approach, breaking the large problem down to a series of smaller ones, and deciding step-by-step if the answer can be more quickly ascertained through observation or application.

The percentage of soap, or Y axis, has a limited working range, since levels in excess of 55% of soap result in a mixture which is too viscous to permit good agitation by using only live steam, and levels below 40% result in excessive amounts of spent lye, reduced kettle capacity, and low glycerine concentrations in the spent lye.

Remember that the loading starts not with an empty kettle but with a seat, which should have a composition on the border between Region D (the one phase seat region) and Region N (the two phase neat–seat region). The loading should proceed to bring the partial contents of the kettle to the saponification point as soon as possible, and then keep the kettle composition at the saponification point for the remainder of the loading process.

Step 2—Graining

After all of the fats, oils, and/or fatty acids are saponified, the kettle needs to be positioned on the phase diagram at a point which results in an unstable mixture of curd and lye. This area is in Region M (the two phase curd–lye region). Only a limited area in Region M will effect good separation of the lye from the curd, this area being just over the border from Region R. Complicated interactive forces at the molecular level exceed gravitational forces; thus, the lye and curd do not completely separate. The percentage of total separation is the "separation efficiency," in which 83% seems to be a realistic maximum for industrial kettle soap processes. Movement away from this border results in an "over-graining" condition where even though the lye and the curd are two distinct phases, quite visible to the naked eye, they do not separate. In these cases, separation efficiencies can drop below 50%, yielding a process which cannot be economically viable since the resulting curd will not be high enough in soap percentage to allow for effective fitting.

Movement from Region R to Region M is done by the addition of brine. In theory, this could also be done by the addition of rock salt if the amount of generated lye needs to be minimized, or by the addition of NaOH if the presence of excess NaOH in the (now not) spent lye is acceptable. This process is called "graining" the kettle since the kettle's appearance changes from being smooth to being very grainy. Numerous traditional soap maker checks are made to assure that the proper grain is achieved. These tests are discussed in the traditional references outline by Tom Woods (Spitz, 1996).

Note: Again, the exact point of "best" settling is known by the experienced soap maker. Sampling a small number of kettles defines this point, and allows the computer to bring the soap maker to this point on a routine basis.

The efficiency of kettle agitation can be enhanced by the installation of a recirculation pipe as depicted in Fig. 7.3.

Kettle Recirculation Pipe
-improves mixing during loading and washing
-allows soapmaker to sample lye

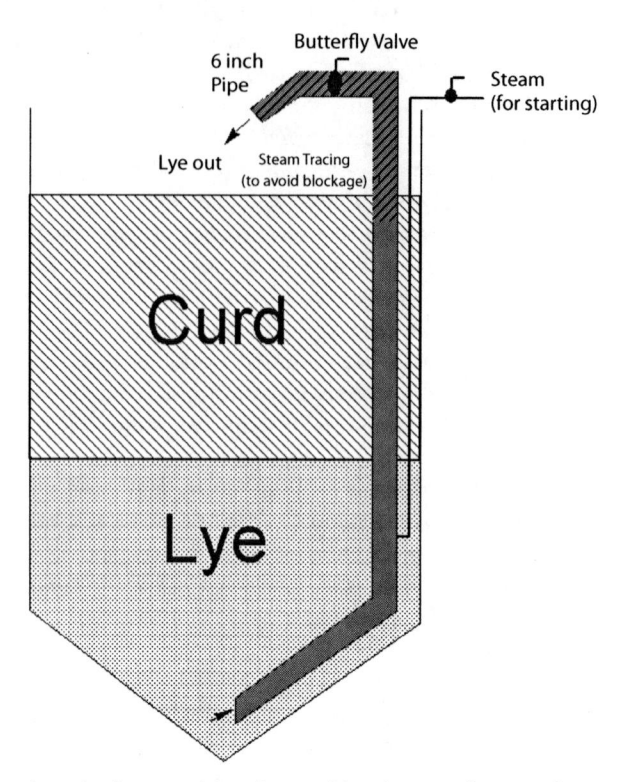

Fig. 7.3. The efficiency of kettle agitation can be enhanced by the installation of a recirculation pipe as depicted in above.

This recirculation pipe allows lye that accumulates on the bottom of the kettle to flow up the pipe and be disbursed on the top of the kettle. This process allows for more rapid saponification and a full consumption of the NaOH. One can sample and test the recirculated lye for both free alkali and salt levels. Once the desired levels are achieved, the graining process is considered complete. The desired levels are determined from the phase diagram by constructing a "tie line" which passes through the graining point. The intersection of this tie line with Region L (the one phase lye region) determines the electrolyte concentration. The absence of free alkali in the recirculated lye sample indicates that the saponification reaction is complete.

Step 3—Settling and Spent Lye Removal
The kettle is allowed to settle, which results in an accumulation of lye at the bottom of the kettle. The composition of this lye is predicted by the use of the tie line as described above. The total quantity of lye is determined by the ratio calculation standard to all phase diagrams. The available lye is determined by multiplying the available lye by the separation efficiency, remembering that 10% or more of the available lye cannot be removed without the aid of increased gravitational forces (e.g., a centrifuge).

A properly grained kettle can have lye removal occur almost immediately. This immediate removal of lye does not come without a price, however. The solubility of soap in the lye is a partial function of the lye temperature. Lye removed immediately upon graining has temperatures in excess of 220°F, and carries with it more than 1% of soap. Upon storage and subsequent cooling, this soap precipitates out of the lye and floats to the surface, eventually creating a solid mass inside the lye storage tanks. One can add this soap back to subsequent kettles, but this requires management to assure that the lye storage capacity is not clogged with precipitated soap. Kettles allowed to settle for longer time periods yield cooler lyes and less precipitated soap problems.

Countercurrent processing nets glycerine concentration in the lye in excess of 15%. Concurrent processing (i.e., lack of countercurrent processing) yields spent lyes with less than 12% of glycerine.

One can calculate the exact concentration of glycerine in the lye once one considers the mechanisms at work. The solvent in the simplified phase diagram consists of glycerine and water. Upon completion of the graining process, a definite ratio of glycerine to water is noted in the solvent. One of three mechanisms can occur: the ratio of glycerine to water can increase in the solvent rich or lye phase relative to the entire mass of the kettle, the ratio can stay the same, or the ratio can decrease. The first mathematical models of the system assumed that the ratio was constant throughout. Comparing actual results to the model netted slightly higher glycerine concentrations in the lye than predicted. A glycerine concentration (or "fudge") factor was defined. A value of 1.1 matched the model to the actual results, meaning that glycerine had a slightly higher tendency to migrate into the solvent rich or lye phase in preference to the water.

Step 4—Kettle Washing

Washing is the process of adding additional amounts of caustic, brine, and water to a settled curd. Remember that the settle curd is located in Region M, close to Region J (the single phase curd region). Washing moves the kettle composition down the tie line toward Region L. Again the same constraints apply regarding overgraining the kettle.

The washing is performed for several reasons. First, the color and odor of the soap are improved. Second, the concentration of glycerine in the soap is reduced. Third, the free alkali to salt ratio is controlled. The loading and graining steps require the generation of a spent lye, in which the electrolyte is composed purely of NaCl. Since the spent lye is removed from the kettle process and used as the feedstock for a glycerine evaporator, one must minimize the free alkali content to reduce the treatment costs associated with the glycerine recovery. Thus, the free alkali to salt ratio is effectively zero. Such high levels of salt, if carried through to the neat soap, will produce a rise so high in salt that subsequent processing and bar pressing will be severely compromised, if not impossible. By washing the kettle with precise amounts of caustic and brine, one can shift the free-alkali-to-salt ratio to provide a soap base with superior handling characteristics.

This is carried to the limit with the Oil Finish Process. Here, the NaCl level of the settled curd has to match the finished neat soap specification. Applying caustic washing, without any additional brine added during the washing steps, rapidly lowers the NaCl levels to allow for oil finishing. Please refer to the next section for more on this topic.

The washing is performed in such a way as to bring the kettle to a point of instability (as described in the graining step above). Again, the recirculation pipe is utilized to effect better mixing and to allow sampling of the lye. Once the recirculated wash lye achieves the desired free alkali and salt concentration, the washing is complete.

Step 4A—Washing an Oil Finish (OF) Kettle

The principle for washing an OF kettle is identical to that described above. However, the goal of the OF wash is to leave 0.5% of NaCl in the curd. Since the total electrolyte level is dictated by the physical chemistry of the phase diagram, one has to grain the kettle out with just caustic instead of the traditional brine and caustic mix. This "caustic wash" has another unique property: the resultant unsettled curd has a very unique appearance. The unsettled curd has a very small grain that looks like wet sand. If properly balanced, the lye drops out very quickly, allowing one to proceed directly to the

OF step and the completion of the kettle. The challenge here is to effect a complete separation, so that the settled curd is low enough in moisture (<32%), and when the kettle is finished, the soap is indeed all neat soap and not a mixture of neat and middle soap. One can somewhat mitigate this problem with the use of citric acid during the finishing.

Step 5—Settling and Wash Lye Removal

The kettle is allowed to settle, which results in an accumulation of lye at the bottom of the kettle. Again, lye removal can proceed almost immediately if the kettle is properly grained. This lye is stored and is used during the loading of subsequent kettles. A properly designed kettle process yields an amount of lye from a wash to match the amount of lye to be recycled back into the prior step of the subsequent kettle. The computer model can greatly simplify this problem as is demonstrated later.

At the conclusion of lye removal, the kettle is in Region M, with the best location being as close to Region J as possible, meaning most of the lye is removed.

One can repeat Steps 4 and 5 as many times as necessary to achieve the proper color, odor, glycerine concentration, and free alkali to salt ratio. Diminishing marginal returns occur after two well defined washes.

Step 6—Finishing or Fitting the Kettle

This step is the step which traditional soap makers appear to hold as most mysterious and skillful. However, a properly designed kettle soap process results in very consistent finishes. At the start of this step, the kettle was drained of all available wash lye, and the desired free alkali to salt ratio was achieved. The kettle is in Region M, close to Region J. Water is used to finish the kettle. The addition of water to the kettle moves the kettle's composition directly toward the origin on the phase diagram. The kettle passes through Region R (the three phase curd–seat–lye region) and into Region P (the two phase neat–lye region). If effective settling were possible in Region P, only neat soap and lye; however, this is not observed, most probably due to the insufficient gravitational forces generated on Earth (one has to wonder if future generations of soap makers will ply their trade on Jupiter to take advantage of increased gravity there). Further addition of water will move the kettle's composition into Region Q (the three phase neat–seat–lye region). Good separation is found in this region; however, the addition of the seat-lye phase increases the variability of the process and complicates processing. Best fitting occurs just over the border into Region N (the two phase neat–seat region). Here, terrene gravity can just overcome the molecular level forces and permit the neat soap and seat to separate. The addition of excessive water results in relatively large amounts of seat and subsequent smaller kettle yields. A minimum of 8 hours is required before one can remove the neat soap from the kettle, and longer times, if available, provide a more consistent product.

Step 6A—Finishing an Oil Finish (OF) Kettle

At the point where an oil finish (OF) kettle is to be finished, it needs to have a chemical composition of no more than 0.7% of NaCl and 32% of water, the upper bound in neat soap for these two components. Of course, the NaOH content is much higher than the <0.1% levels required for neat soap. Typically, the settled OF curd has a NaOH content in the 0.75–1% range. This excess NaOH content is removed by the addition of one or more of the following: a fatty acid, a fat, and/or citric acid. If a combination is desired, then add the fat first because it is the most difficult to react and requires an excess of NaOH to saponify in a timely fashion. Great danger is possible in adding too much fat if one is not patient enough to allow the saponification to be complete. Additionally, the fat added at this stage retains all of its glycerine, color, and odor in the kettle, which could be a problem. A fatty acid reacts quickly; however, most fatty acids have their own odor and color issues depending up storage and handling; however, no glycerine is added at this step. Using citric acid to consume some or all of the excess NaOH is a recent development with surprising results. Although citric acid and NaOH produce sodium citrate, another electrolyte, the graining power of sodium citrate is quite weak, and the kettle remains smooth. Additionally, a dramatic reduction in the viscosity of the finished neat soap occurs. This is of critical importance in the frequent occurrence of an incomplete settling during the final wash step. If indeed the settled curd retains some lye (which is often the case), the moisture of the settled curd remains around 34%; thus, when the fat or fatty acid is added, the finished soap has considerable middle soap content, making the soap unpumpable. Sodium citrate levels in excess of 0.25% dramatically reduce the viscosity of the neat–middle soap mixture well below even the most fluid, properly composed neat soaps. The danger lies in sodium citrate levels approaching 1%, which results in a dried soap which is difficult to press into a bar, being too crumbly. One needs to target sodium citrate levels at the 0.25% level to achieve pumpable viscosities in high moisture soaps without pressing problems in the finished rice.

Secondary H_2O Considerations

To accurately calculate the process defined above, a high level of accuracy and precision is required because the critical areas of maximal settling are relatively small. Traditional methods may overlook the following contributions to kettle soap H_2O.

Steam used for agitation and heating condenses into the kettle mass. One must calculate this amount by using the temperatures, heat capacities, and heat of reactions of the various ingredients. Evaporation occurs during the settling process, and must be considered. Finally, the kettle cools during settling and requires condensed steam to reheat.

Countercurrent Illustration

The countercurrent nature of this process is illustrated with a four kettle system by in Fig. 7.4. The processing steps are listed across the top: load, first wash, second wash, third wash, and finish. The loading of every kettle generates a neutralized or spent lye which is then sent to glycerine recovery. The wash lye removed from the first wash in kettle 1 goes into the loading of kettle 2. The wash lye removed from the second wash of kettle 1 goes into the first wash of kettle 2. The wash lye removed from the third wash in kettle 1 goes into the second wash of kettle 2. Kettles 2, 3, and 4 follow the same pattern. The seat generated during the finishing and the fitting of the first kettle is used for the loading of the second kettle. The other seats are handled in the identical fashion. The countercurrent flow becomes evident. The lines representing the production of soap go from left to right, and the lines representing the flow of glycerine go from right to left.

Variations on a Theme

Many possible variations are available to the process outlined above. The process defined by Thomas Wood in Appendix A (Spitz, 1996) of my prior writings adds the coconut oil during the first wash. Other options include graining the seat "off-line" and re-introducing this concentrated and washed seat into a washing step, thus loading on an empty kettle. One can also hold back a relatively large amount of NaOH during the loading step, thus saponifying only a fraction (say 85%) of the fats during the first step of the process. These variations and others all have their features and benefits. However, all deal with the same phase diagram and the same concepts of graining and settling; thus, one can calculate all of them in the same fashion.

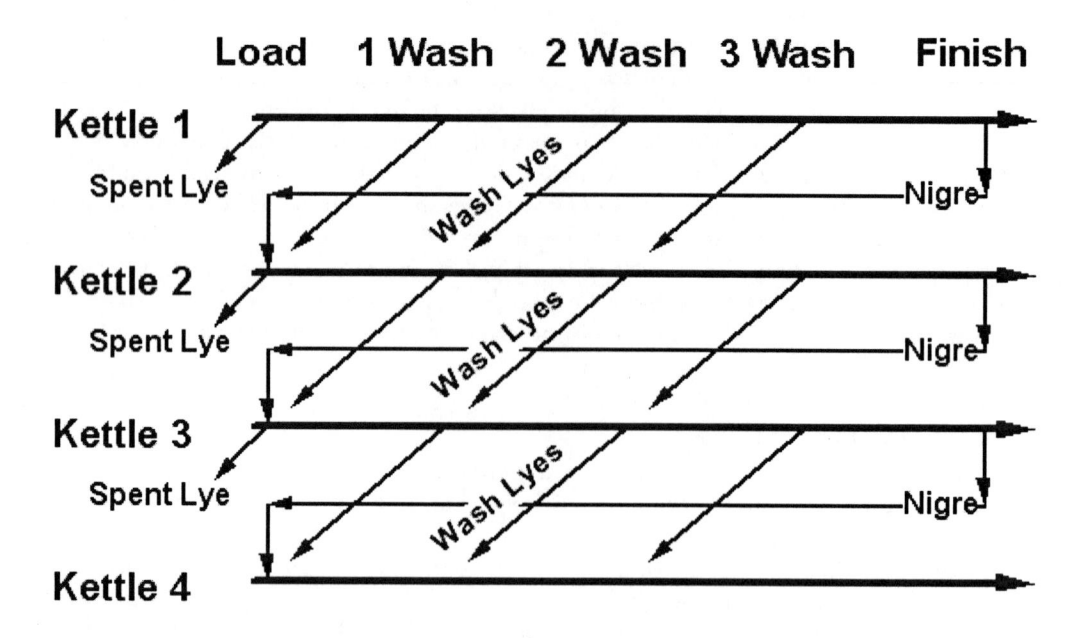

Fig. 7.4. Material Flow in the Counter-Current Process

The Mathematics of a Kettle of Soap

As in my prior work, I used Microsoft Excel for the construction of my model. From 1982 to 2005, I wrote and modified the Kettle Soap Process Simulator (KSPS) using Excel and Excel's built-in macro programming language. In 2005 I decided that a total rewrite of the program was warranted. The KSPS was on Version 19, and had so many patches and modifications that it was very difficult to follow. Excel Version 11(2003) no longer fully supported the Excel Macro language. Excel's macro language migrated to Visual Basic for Applications (VBA), and offered much more power and flexibility. I also had developed a new strategy for loading and fitting a kettle which the KSPS could not support.

I started with a blank spreadsheet in the newest version of Excel, and created the Kettle Soap Process Controller (KSPC), which incorporates all of the newest technologies. Interested parties are welcome to request a copy of the KSPC.

Step 1—Loading

In this section, my current approach to the kettle loading calculations is explained. The same exact calculation scheme is valid for washing a kettle. Finishing a kettle is a different problem, and is discussed separately.

A kettle is loaded with the ingredients as outlined above. The total mass of the kettle is the sum of the ingredients:

$$M = M_f + M_c + M_b + M_y + M_r + M_w + M_t$$

We need to determine how much of each ingredient is required.

The kettle mass is also equal to the sum of its components:

$$M = M^s + M^\omega + M^g + M^d + M^h$$

Kettle Saponification • 215

The mass fraction of all of the components must equal 1:

$$1 = X^s + X^\omega + X^g + X^d + X^b$$

We also know that the mass of the kettle multiplied by the mass fraction of a component equals the mass of the component:

$$M^s = X^s M$$
$$M^\omega = X^\omega M$$
$$M^g = X^g M$$
$$M^d = X^d M$$
$$M^b = X^b M$$

In a similar fashion, each mass component has its own equation:

$$M^s = M^s_f + M^s_c + M^s_b + M^s_y + M^s_r + M^s_w + M^s_t$$
$$M^\omega = M^\omega_f + M^\omega_c + M^\omega_b + M^\omega_y + M^\omega_r + M^\omega_w + M^\omega_t$$
$$M^g = M^g_f + M^g_c + M^g_b + M^g_y + M^g_r + M^g_w + M^g_t$$
$$M^d = M^d_f + M^d_c + M^d_b + M^d_y + M^d_r + M^d_w + M^d_t$$
$$M^b = M^b_f + M^b_c + M^b_b + M^b_y + M^b_r + M^b_w + M^b_t$$

Combining equations gives us:

$$X^s M = M^s_f + M^s_c + M^s_b + M^s_y + M^s_r + M^s_w + M^s_t$$
$$X^\omega M = M^\omega_f + M^\omega_c + M^\omega_b + M^\omega_y + M^\omega_r + M^\omega_w + M^\omega_t$$
$$X^g M = M^g_f + M^g_c + M^g_b + M^g_y + M^g_r + M^g_w + M^g_t$$
$$X^d M = M^d_f + M^d_c + M^d_b + M^d_y + M^d_r + M^d_w + M^d_t$$
$$X^b M = M^b_f + M^b_c + M^b_b + M^b_y + M^b_r + M^b_w + M^b_t$$

We also know that the mass of a component in an ingredient is equal to the mass fraction of the component in that ingredient multiplied by the mass of that ingredient. Our five mass component equations then become:

$$X^s M = X^s_f M_f + X^s_c M_c + X^s_b M_b + X^s_y M_y + X^s_r M_r + X^s_w M_w + X^s_t M_t$$
$$X^\omega M = X^\omega_f M_f + X^\omega_c M_c + X^\omega_b M_b + X^\omega_y M_y + X^\omega_r M_r + X^\omega_w M_w + X^\omega_t M_t$$
$$X^g M = X^g_f M_f + X^g_c M_c + X^g_b M_b + X^g_y M_y + X^g_r M_r + X^g_w M_w + X^g_t M_t$$
$$X^d M = X^d_f M_f + X^d_c M_c + X^d_b M_b + X^d_y M_y + X^d_r M_r + X^d_w M_w + X^d_t M_t$$
$$X^b M = X^b_f M_f + X^b_c M_c + X^b_b M_b + X^b_y M_y + X^b_r M_r + X^b_w M_w + X^b_t M_t$$

This is getting a bit messy. However, some hope does reside here. Many of these terms are zero. For example, no soap exists in brine. Many others are known. For example, the NaOH content of caustic is typically 49.6%. Also, one typically knows the amount of seat that is available for the kettle load. The mass fraction of H_2O in water and steam is 100%, so $X^\omega_w = X^\omega_t = 1$. The zero terms are eliminated, and the unknown values are bolded in our next set of equations:

$$X^s M = X^s_f \mathbf{M_f} + X^s_y \mathbf{M_y} + X^s_r M_r$$
$$X^\omega M = X^\omega_f \mathbf{M_f} + X^\omega_c \mathbf{M_c} + X^\omega_b \mathbf{M_b} + X^\omega_y \mathbf{M_y} + X^\omega_r M_r + \mathbf{M_w} + \mathbf{M_t}$$
$$X^g M = X^g_f \mathbf{M_f} + X^g_b \mathbf{M_b} + X^g_y \mathbf{M_y} + X^g_r M_r$$
$$X^d M = X^d_c \mathbf{M_c} + X^d_b \mathbf{M_b} + X^d_y \mathbf{M_y} + X^d_r M_r$$
$$X^b M = X^b_f \mathbf{M_f} + X^b_c \mathbf{M_c} + X^b_y \mathbf{M_y} + X^b_r M_r$$

We still have a ways to go, because we have eight unknowns but only five equations. We have to get this to a system where the number of equations equals the number of unknowns. This series of equations should be kept "linear" so that one can apply matrix inversion techniques to achieve an exact solution.

We can perform an energy balance around the kettle to capture the mass of the steam that condenses as a function of the other ingredients:

$$0 = P_f(\overline{T} - T_f) M_f + P_c (\overline{T} - T_c) M_c + P_b (\overline{T} - T_b) M_b$$
$$+ P_y (\overline{T} - T_y) M_y + P_r (\overline{T} - T_r) M_r$$
$$+ P_w (\overline{T} - T_w) M_w + P_t (\overline{T} - T_t) M_t + \Gamma M_f$$

Depending on the climate, the temperatures may fluctuate with the season. I have measured temperature fluctuations as great as 20 °F in the northeastern United States, and I routinely adjust for it.

Here, the last term is the energy released during the saponification reaction. Now six equations exist.

We revisit our mass reaction component summation equation. for our seventh and final equation.

$$1 = X^s + X^\omega + X^g + X^d + X^b$$

We now have only seven equations and eight unknowns! We cannot solve this problem.

We actually can by solving four separate problems. Recall that we are adding recycled lye to this kettle. We first solve the problem of loading the kettle with the constraint of no lye added, or $M_y = 0$. This then gives us a linear system of six equations and six unknowns— something which we can solve exactly with only one solution. After the solution is achieved, we have to confirm that indeed all of the ingredients are positive. You could imagine a situation where a very large and wet seat is used, resulting in a negative water addition. In this case, one has to reduce the amount of the seat until all ingredients are non-negative. This solution must have all non-negative ingredients to proceed.

So we now have a solution in hand for a kettle loading with no added lye. That is solution number 1.

Solution number 2 sets the caustic addition to zero, or $M_c = 0$. We again solve the problem. This solution could very well have a very large amount of lye added to the kettle, thus forcing a negative water addition to achieve the correct loading target. This is of no consequence.

Solution number 3 sets the brine addition to zero, or $M_b = 0$. We again solve the problem. Again, this solution could very well have a very large amount of lye added to the kettle, forcing a negative water addition to achieve the correct loading target. This too is of no consequence.

Solution number 4 sets the water addition to zero, or $M_w = 0$. We again solve the problem. Again, this solution could very well have a very large amount of lye added to the kettle, forcing a negative caustic and/or brine addition to achieve the correct loading target. This too is of no consequence.

We now have four mathematically valid solutions to loading this kettle, although three of them may be physically impossible because of negative ingredient additions. Since we have valid solutions, any linear combination of these four solutions will also be a valid solution. In many cases we seek to consume as much lye as possible during the loading stage. Of the three solutions that consume lye, one or more solutions may be available where all ingredients are non-negative. In this unlikely occurrence, simply choose the solution with the largest lye consumption, and your job is over. Most probably, all the lye containing solutions will have one or more negative ingredients. Pick the one with the least amount of lye, and perform a linear combination with the no-lye solution to find a solution that maximizes the lye addition with all other ingredients being non-negative.

If one desires to consume a fixed amount of lye which is less than the maximum calculated amount, then perform a linear combination of the two solutions weighted to achieve the desired addition of lye.

Fortunately, the power of Microsoft Excel and its associated Visual Basis for Applications permits the above series of calculations to occur in fractions of a second. This is the "heart" of the KSPC, and is available to those who request it.

Rates of Addition

As discussed earlier, the success of the kettle is a strong function of maintaining the proper point on the phase diagram to assure maximal saponification. The loading target as defined is this maximal point of saponification. However, at the start of the loading process, the kettle's composition is identically the seat's composition. In the simplified case of pre-blended fats, the first stage of loading is to move the kettle to the point of maximal saponification.

Fluctuating Fat Ratios

Often, a pre-blend tank is not available, forcing the soap maker to add the fats either serially or sequentially. In the serial case, the tallow/coco ratio varies as the kettle's ingredients are charged. To maximize saponification in this case, one has to "hit a moving target" since the maximal saponification point is moving. By using the various X axes in the phase diagram of Fig. 7.2, one can use a computer to predict this optimal point as a function of the tallow/coco ratio; however, that calculation is quite involved.

I solved this problem by providing myself with a graphical representation of the progression of the loading process, as illustrated in Fig. 7.5. I built my loading program to identify up to fifteen different intermediate targets. I know the starting point of the load (the seat composition), and I know the end point of the load (the final loading target). The best way to determine the intermediate loading targets is by trial and error, with a rendering of the path through the phase diagram imaged in an Excel chart. Fig. 7.5 shows a kettle with a smooth loading path from seat to curd.

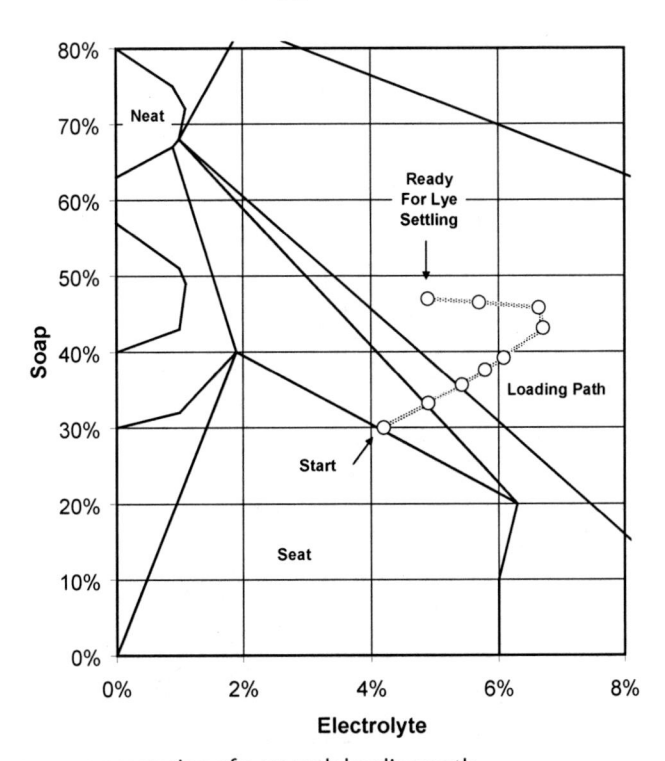

Fig. 7.5. Phase diagram representation of a smooth loading path.

Kettle Settling

The kettle is loaded, and the lye is dropping out. How much lye drops out, and what is its composition? Here is how we figure this out. We approximate the lye–curd region as a quadrilateral.

Fig. 7.6 shows a screen shot from the KSPC. This is a mathematical representation of the curd–lye region of the phase diagram. The X axis is the percentage of total electrolyte; the Y axis is the percentage of soap. The points 1, 2, 3, and 4 define this two phase region. L is the point where the kettle is loaded. Now here is where the math begins. First we define Point I, which is the intercept of the two lines defined by line segments 1–2 and 3–4. We then draw a line through Points I and L. This represents the phase diagram tie-line on which the loading point L resides. This tie line intercepts the line segment 2–3 at point U, and intercepts the line segment 1–4 at point Y. Point U defines the composition of pure curd that evolves when the kettle fully settles. Point Y defines the composition of pure lye that evolves when the kettle fully settles. The relative mass of the curd and lye equals the relative length of the line segments defined by L–U and L–Y. These calculations follow the rules of the phase diagram theory as outlined in any book on the topic. All of the math that defines this is simple algebra.

Our work is not yet done. The curd and lye do not fully separate. I define a "lye-drop factor," which is a value from 0 to 1 to define what fraction of the lye actually drops out of a kettle. This lye-drop factor varies with the loading target, and is typically between 0.78 and 0.9, meaning 78 to 90% of the total available lye is removed from the kettle.

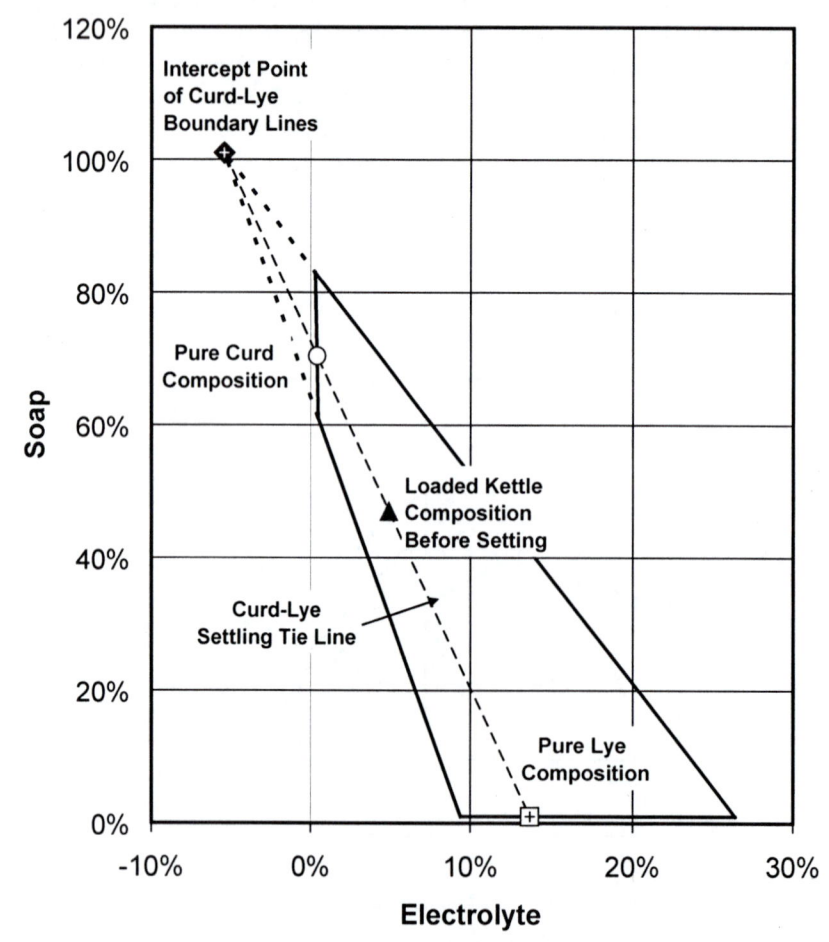

Fig. 7.6. Computerized phase diagram of lye-curd region displaying the "Tie" line which determines the curd and lye compositions.

We also have to determine the amount of NaOH and NaCl in the lye and curd phases. I found that the ratio of NaOH to NaCl stays constant during the settling. This is also true for the water to glycerine ratio.

Of course, the kettle has cooled during this process. Steam must be used to reheat the kettle for the next processing step. This steam condenses and increases the water content of the kettle. One must determine this to maintain an accurate record of the kettle contents.

This summarizes all of the aspects of loading and drawing lye from a kettle. Obviously, someone attempting to implement this technology has a lot of work ahead of him or her. Hopefully, this effort provides some useful guidance. As mentioned earlier, washing a kettle follows the exact same scheme. Fitting a kettle is a different problem.

Kettle Fitting

Unlike loading or washing a kettle, fitting a kettle involves the addition of only water to move the kettle from the curd–lye region on the phase diagram to the neat–seat region. Consider the phase diagram. The addition of water to a kettle is represented on the phase diagram as moving toward the origin. Recall that the origin is 100% of solvent, 0% of soap, and 0% of electrolyte. Also recall that the key to successfully settling a kettle between two phases (curd–lye or neat–seat) is strategic placement on the phase diagram. During the loading and washing steps, a specific point on the phase diagram is specified and achievable because two or three components (soap, solvent, electrolyte) are being added. During the fitting step, only one component is being added (water), so in most cases to achieve an exact point on the phase diagram is impossible. One can, however, identify the desired line segment that represents the ideal tie line to achieve optimal neat–seat settling. This is exactly what is done.

Fig. 7.7 is from the KSPC and illustrates how this is achieved. The composition of a curd is identified on this graph as well as a quadrilateral representation of the neat–seat region. Infinite dilution of the kettle with water is represented by a line segment drawn from the curd to the origin. The target line segment is illustrated as well. Simple algebra calculates the interception of the line to the origin and the ideal tie line. This interception point is the point where the kettle is to be fitted.

Once this fitting point is identified, the software then determines the amount of water to be added to the kettle to achieve that point. Since this is a very exact measurement, one has to make steam adjustments. Two steam adjustments are considered: first the amount of steam required to bring the added water to a boil and second a substantial amount of steam that condenses while heating the kettle. Once the kettle is boiling, the additional steam that is injected for continued agitation simply passes through the kettle. A small amount of steam continues to condense while the kettle is being mixed. This is due to the fact that the kettle is not perfectly insulated, and some heat is lost through the kettle sides and bottom.

Oil Finish

This new and novel approach requires careful manipulation of the kettle. Again, the kettle is loaded and washed in the traditional fashion; however, the wash targets are skewed to include a significantly greater amount of free NaOH and much less NaCl. After the final lye is removed from the final wash, the total NaCl content of the kettle equals the desired NaCl content of the neat soap. Therefore, one must neutralize an extremely high NaOH content.

A well behaved OF kettle has between 0.75 and 1.0% of NaOH content. As mentioned earlier, one must avoid a high solvent content in the curd or else middle soap exists and makes the resultant neat–middle mixture too viscous to pump. For a 85:15 palm/coco soap, I found a wash target to be 49.9% of soap, 2.27% of NaOH, and 1.53% of NaCl. This generates a lye composed of 7.23% of NaOH and 4.85% of NaCl. Removing 80% of the lye (remember, we are on Earth, not Jupiter) has a resulting curd of 64.7% of soap, 0.76% of NaOH, and 0.51% of NaCl, leaving the total solvent level at

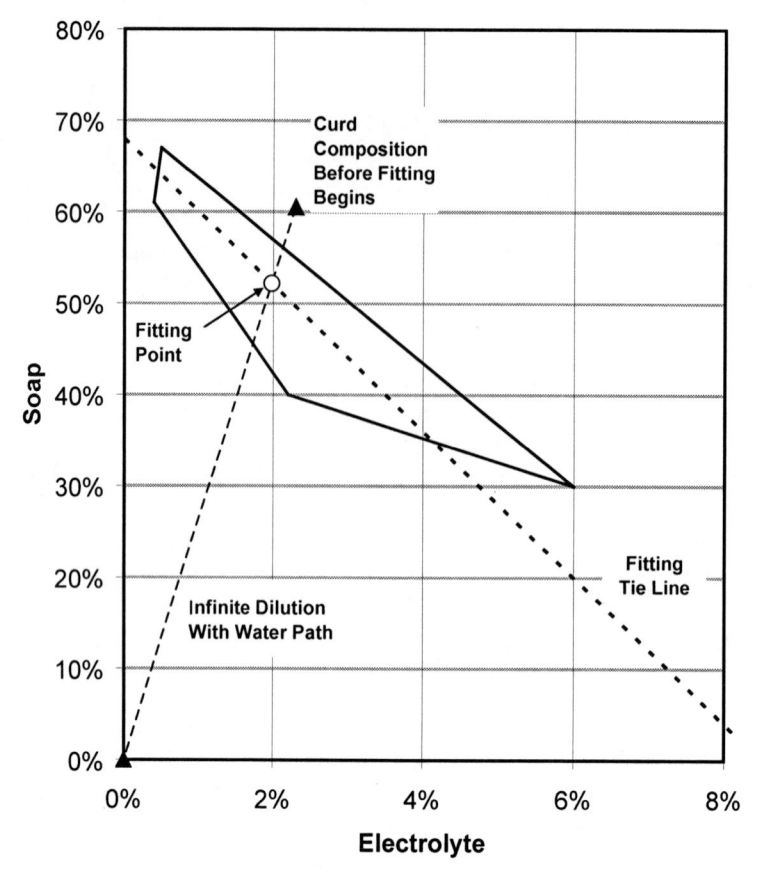

Fig. 7.7. Computerized phase diagram of neat-seat region displaying how the fitting point is determined.

33.9%. Remember, the total solvent level is the sum of the water and glycerine. The glycerine content is a partial function of the amount of recycled materials added in prior steps and can "float," so the total solvent level should be the focus of attention.

Again, I must caution that attempting to saponify fats or oils to consume this excess NaOH is difficult and time consuming. However, one could reduce or add some fat and/or oil to the kettle to consume a fraction of the excess NaOH. I found that one should maintain a minimal excess of 0.5% of NaOH to assure that any fat or oil added at this step is completely saponified.

Once the 0.5% of the NaOH level is achieved, we are left with two options to neutralize the balance. Just fatty acids sometimes results in stiff soap which is difficult if not impossible to pump. However, neutralizing one half of the remaining NaOH with citric acid actually substantially lowers the viscosity of the neat soap and is highly recommended. One has to do some trials to determine the maximal amount of citric acid that a particular formula can tolerate and still maintain the proper physical properties of the finished bar.

Levels as low as 0.1% of sodium citrate have a tremendous benefit to the neat soap viscosity without impacting the final bar.

Future Plans

The KSPC has reached a high level of technical development. It was successfully used to achieve high-quality soap of many different compositions. All interested parties are encouraged to contact me for a copy. Hopefully others can expand upon my work, and a new wave of kettle soap development will spread across the world.

Acknowledgments

I would like to thank the AOCS for the opportunity to present my approach to kettle soap making, and Luis Spitz for his support and assistance.

References and Bibliography

Davidsohn, J.; E.J. Better; A. Davidsohn. *Soap Manufacture;* New York: Interscience, 1953; Vol. 1.

George, E.; J. Serdakowski. Computer modeling in the full boil soap making process. *HAPPI* **1985.**

George, E.; J. Serdakowski. Correlation of fat and oil quality with soap base color. *Cosmetics and Toiletries* **1993,** *108.*

Spitz, L. (Ed.) *Soaps and Detergents: A Theoretical and Practical Review;* AOCS Press: Champaign, IL, 1996.

Spitz, L., (Ed.) *Soap Technology for the 1990s;* AOCS Press: Champaign, IL, 1991.

Woollatt, E. *The Manufacture of Soap, Other Detergents and Glycerine;* John Wiley & Sons: New York.

Continuous Saponification and Neutralization Systems

Timothy Kelly
VVF North America, Kansas City, Kansas, USA

The objective of this chapter is to provide an updated overview of continuous saponification and neutralization systems. Included are fundamental principles of operation, an overview of equipment and systems technology, and a review of commercially available systems.

Introduction

The commercial soap making systems available today are the result of a long evolution of technology advances. According to ancient Roman legend, soap got its name from Mount Sapo, where it was first made by accident as the result of the tallow from animal sacrifices reacting with the ashes from the burned wood to produce a mixture capable of cleaning clothes better. As the centuries progressed, soap making evolved into a batch process where animal fats and vegetable oils were boiled with ashes, soda ash, or caustic soda in small kettles to produce crude soap. To satisfy increasing demand, larger batches were produced by what today is commonly called the kettle soap making process.

Continuous soap making began in the mid-1940s with the development of processes for converting fats and oils into fatty acids via a hydrolysis reaction carried out under high pressure and temperature. The fatty acids, with the glycerine from the hydrolysis reaction removed, were reacted with a caustic soda, water, and salt blend to produce soap in a continuous fashion. This method of manufacture is referred to as continuous neutralization. To save energy and to improve yields, technology further advanced with the development of continuous processes for reacting neutral fats and oils directly with caustic soda to produce soap and glycerine. An important part of these processes was the technology employed to remove the glycerine from the soap. This method of manufacture is referred to as continuous saponification.

Whether soap is produced via fatty acid neutralization systems or continuous saponification systems, the major soap making equipment manufacturers have continued to improve these technologies. In recent years, commercially available systems were modified to reduce energy consumption, to reduce changeover time and losses, to reduce plant maintenance, to increase flexibility, and to improve plant layouts. The information that follows provides an overview of the principles behind these technologies and the manufacturing systems available in the market today.

Continuous Saponification

Overview

The continuous saponification method for making soap is also called neutral fat/oil saponification or full boiled saponification. This is globally the most widely used process for soap making. In this process, a blend of fat and oil is reacted with caustic soda (NaOH) to produce soap and glycerine. The glycerine and other water soluble impurities are removed from the soap. Once the soap is adjusted for final alkalinity, it is sent to storage to await further processing. A simplified block diagram of the process is shown in Fig. 8.1. The reader is referred to the Appendix for a summary of the definitions and terminology used in this diagram and throughout the remainder of this chapter.

The simplified block diagram shows the major unit operations required in the continuous saponification process. These operations and their purposes are:

- Metering/Dosing—This operation provides for an accurate addition of all raw materials into the system.

- Saponification—This operation provides for a complete reaction of the fats/oils with caustic to produce soap and glycerine.

- Cooling and Spent Lye Separation, Washing and Half Spent Lye Separation, and Lye and Neat Soap Separation—These three operations combine to remove glycerine and other water soluble impurities from the soap and to establish the NaCl and NaOH content in the soap to neutralization.

- Neutralization—This system reduces excess NaOH in the finished soap, and also provides for the addition of minor ingredients such as antioxidants into the soap prior to storage.

A simplified flowsheet for a typical plant is shown in Fig. 8.2. In this process, raw materials are metered into the reactor, the washing column, and the neutralizer. The reacted soap leaving the reactor is cooled with recycled spent lye from the static separator. Upon entering the static separator, soap and spent lye separate, with the soap exiting the top of the static separator and the spent lye exiting the bottom. The soap exiting the static separator has the glycerine removed by using wash lye in a countercurrent extraction column. Due to density differences, wash lye leaves the bottom of the wash column as half spent lye, and is recycled to the reactor. Soap exits the top of the wash column, and is then centrifuged to remove entrained wash lye. The centrifuge lye is recycled back to the wash column. The centrifuged soap flows into a neutralizer where the alkalinity is adjusted by using a neutralizing agent, and any required minor ingredients such as an antioxidant are added.

Metering/Dosing

The accurate addition of raw materials into the saponification system is critical to soap composition control and problem free system operation. The purpose of the metering system is to continuously dose the proper amount of each raw material into the saponification, washing, and neutralization systems.

Raw Materials

- Fats—The most common fat used globally is tallow. This tallow is typically either edible tallow or bleached and filtered lower grade tallow. Other suitable fats include palm oil or a combination of palm oil and palm oil stearin. The fat produces a longer chain length (16 to 18 carbons) soap used to provide good soap structure and slow but stable lather.

- Oils—Globally, coconut oil and palm kernel oil are the most commonly used oils. To assure good soap quality, these oils are most often of a refined, bleached, and deodorized grade. The oil produces shorter chain length (12 to 14 carbons) soap used to provide a fast, creamy lather.

- Caustic Soda—A caustic soda solution with a strength of 50% of NaOH is utilized. Most commonly the caustic is a low salt (<1%), low iron version. Lower iron content helps to protect the color stability of the finished soap. The caustic soda provides the NaOH required to saponify the fats/oils into soap and glycerine. The NaOH, along with the NaCl from the brine, provides the electrolyte required to achieve the neat soap/lye two phase separation critical to glycerine removal in the washing system. Soap with a slight excess of caustic improves the stability of the soap in storage tanks.

Continuous Saponification and Neutralization Systems ● 225

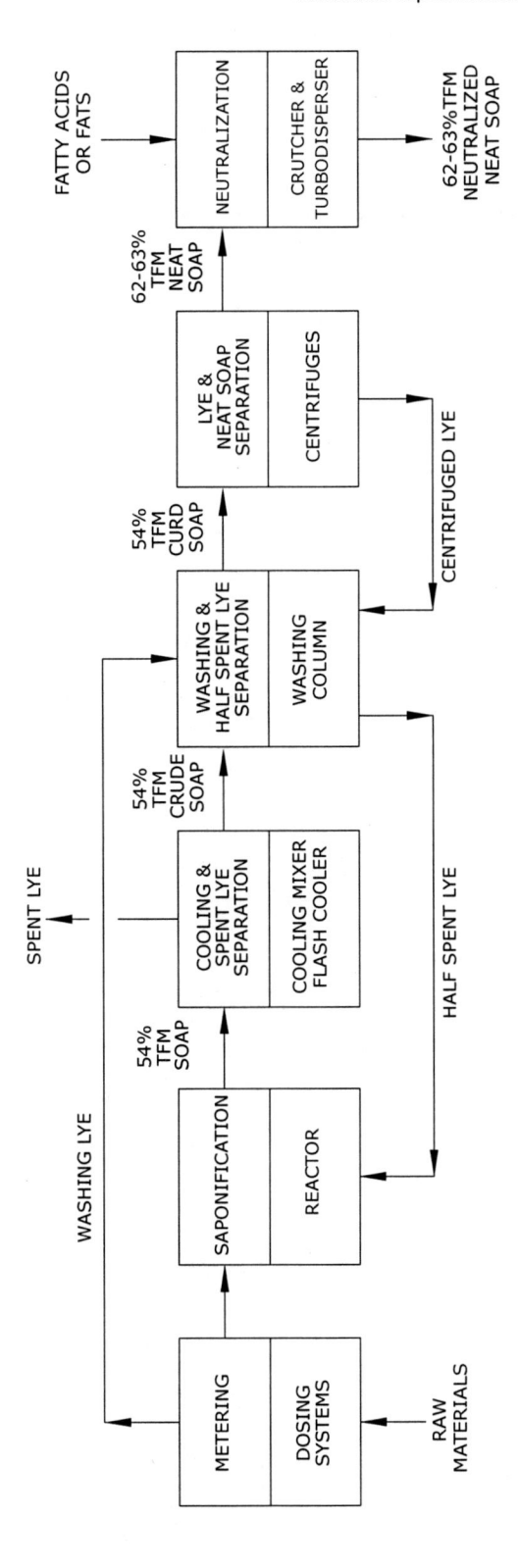

Fig. 8.1. Continuous Saponification System Diagram.

226 ● T. Kelly

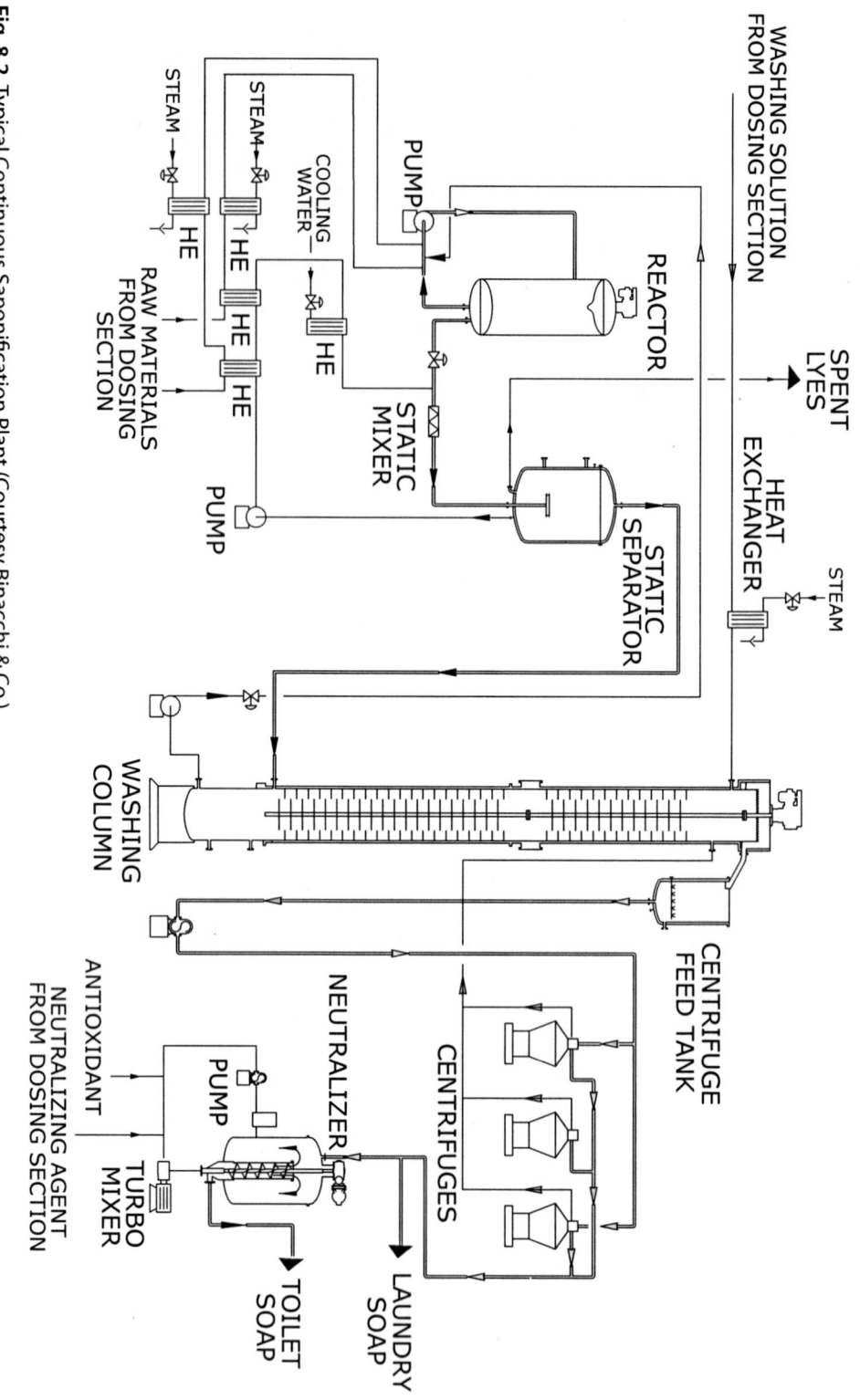

Fig. 8.2. Typical Continuous Saponification Plant (Courtesy Binacchi & Co.).

- Water—Water is used, in combination with caustic soda and brine, to provide the washing lye necessary to achieve the neat soap/lye two phase separation critical to glycerine removal in the washing system. The water is usually softened to remove excess water hardness. Water hardness components can cause the scaling of heat exchanger surfaces in the wash lye pre-heater.

- Brine—Usually 20% of NaCl in a water/brine solution is utilized. However, some operations utilize a saturated brine solution (approximately 26% of NaCl). With either brine solution, the softening of the water used to make the brine helps to prevent heat exchanger scaling due to water hardness. The NaCl in the brine provides the electrolyte required to achieve the neat soap/lye two phase separation critical to glycerine removal in the washing system. It also provides the finished neat soap with NaCl level important for optimal soap viscosity and flowability.

- Fatty Acid—Fatty acid can be utilized for neutralizing excess NaOH in the soap neutralization system. Many plants will react the excess NaOH with the same fat or oil used in the saponification system.

Critical Metered Streams
The raw materials are combined to form several important flow streams into the operation.

- Fat/oil—The fat and oil are combined together into a single stream added into the saponification system. The ratio of fat to oil is important to producing the proper quality soap. Formulations of fat/oil vary from 50:50 to 85:15 tallow/coconut oil.

- Reaction Lye—Caustic soda, brine, and water are combined to produce the reaction lye utilized in the saponification system. In some systems, the half spent lye from the bottom of the washing column is also combined in this stream. Avoid mixing the brine directly with the caustic soda to prevent the precipitation of salt from the mixture.

- Washing Lye—Caustic soda, brine, and water are combined to produce the washing lye utilized in the washing column. Again, avoid mixing the brine directly with the caustic soda to prevent the precipitation of salt from the mixture.

Metering Equipment
Several types of equipment are available for metering systems. Many older system designs used a multi-headed piston style metering pump with a common motor and drive shaft. Raw materials were supplied to the pump heads from individual level tanks. The primary advantage of this system was the ability to change all flow rates with one pump speed change. Individual stream flows were adjusted by changing the stroke length of the piston. With improvements in flow measurement capability, commercial equipment vendors developed improved metering performance by utilizing flow meters and either centrifugal or positive displacement pumps:

- Centrifugal Pump Metering System—Figure 8.3 shows a typical metering system utilizing a centrifugal pump. Raw materials enter the pump after being strained through duplex strainers. A mass flow meter and a pressure control valve are used to accurately control the flow from the pump. The primary advantage of this system is the ability to meter multiple streams with one centrifugal pump. This minimizes capital required for the plant.

- Positive Displacement Metering System—Figure 8.4 shows a typical metering system utilizing a positive displacement pump. Again, raw materials enter the pump after being strained through duplex strainers. A mass flow meter and a variable frequency drive are used to control the pump speed to accurately control flow from the pump. The primary advantage of this system is it will allow a larger minimum and maximum flow rate range than the centrifugal pump metering system. The disadvantage of this system is the higher capital cost and larger equipment layout required.

228 ● T. Kelly

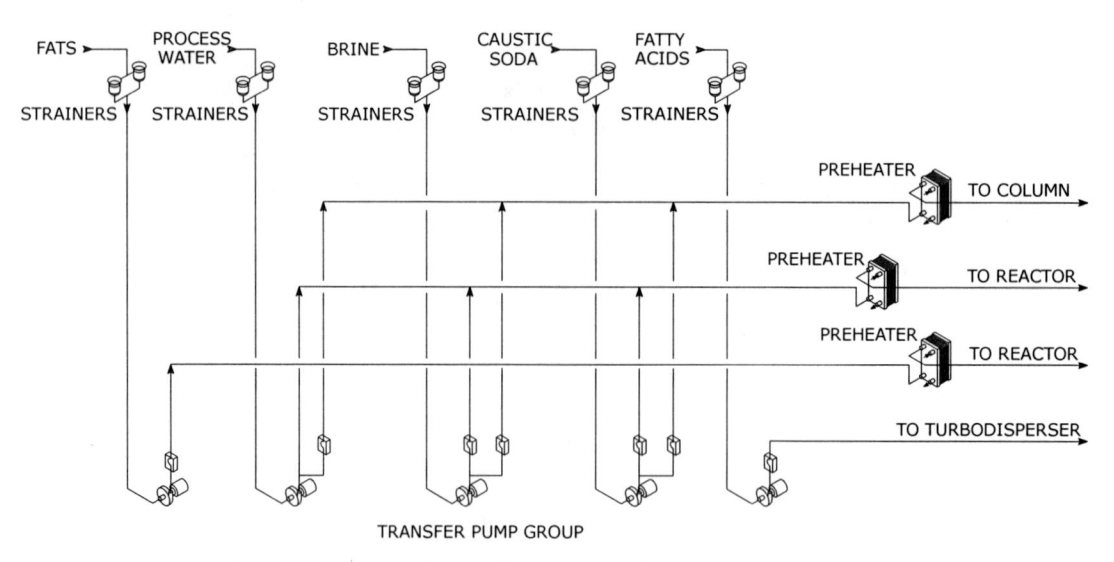

Fig. 8.3. Mazzoni LB "SCNT" Continuous Saponification Centrifugal Pump Metering System.

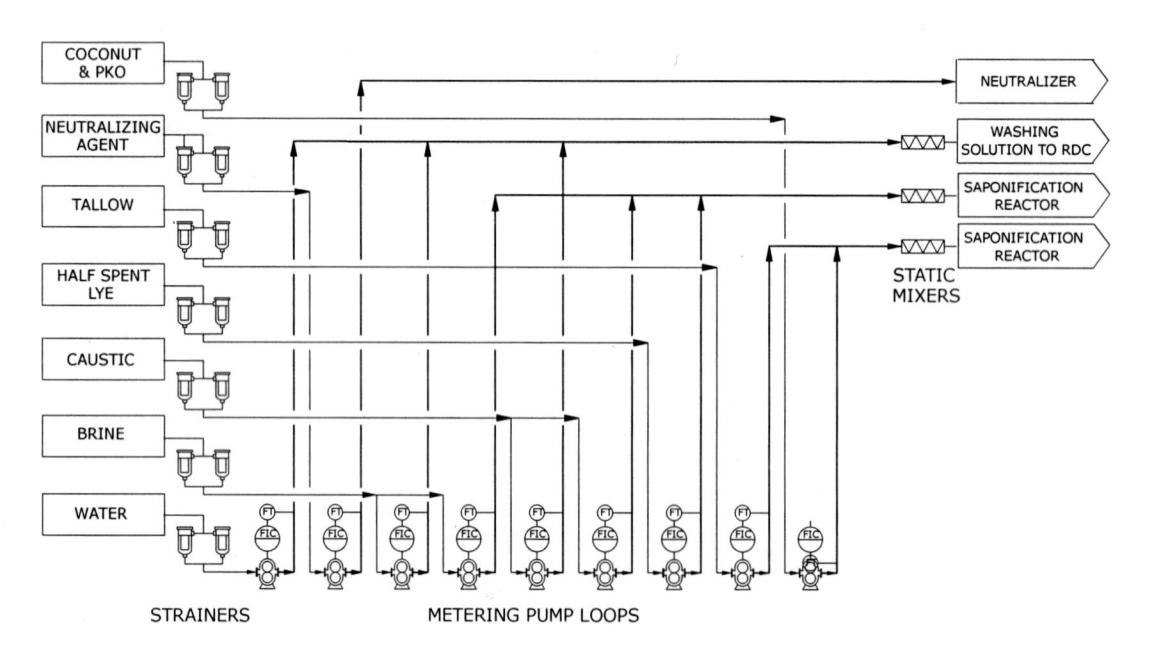

Fig. 8.4. Binacchi CSWE-3 Continuous Saponification Positive Displacement Pump Metering System.

Saponification

Saponification is the reaction of fats and oils with caustic soda to produce soap and glycerine. Completely converting all of the fats and oils into soap and glycerine is critical to the subsequent steps in the washing process. An incomplete reaction will not allow the separation of soap and lye in the washing system.

Saponification Reaction

The saponification reaction is a chemical reaction between the fats/oils and the NaOH present in the caustic soda. The equation for the reaction is:

$$\text{Fat/oil} + 3\text{ NaOH} \rightarrow 3\text{ Soap} + \text{Glycerine}$$

From the reaction equation 1 mole of fat/oil reacts with 3 moles of NaOH to produce 3 moles of sodium soap and 1 mole of glycerine. While the reaction may seem simple, it is actually a step wise reaction with 1 mole of NaOH reacting with the fat/oil triglyceride to form a diglyceride, liberating 1 mole of sodium soap. The diglyceride then reacts with another mole of NaOH to form the monoglyceride, releasing another mole of sodium soap. The final step is the monoglyceride reacting with the third mole of NaOH to form the glycerine and 1 final mole of sodium soap. The saponification reaction is exothermic, releasing 60 cal/mole of fat/oil reacted. The heat liberated from this reaction is used in the process to maintain the temperature in the saponification reactor.

Rate of Reaction

The saponification reaction is a heterogeneous reaction since the initial reactants are not soluble in each other. Due to this heterogeneous nature, the reaction proceeds in three stages: a slow initial stage, a fast autocatalytic stage, and a slow final stage. Figure 8.5 shows the degree of saponification as a function of time.

- Initial Stage—During the initial stage, the NaOH containing water and oil is slowly forming an emulsion. As the emulsion forms, it promotes better contact between the oil and water phases, increasing the rate of reaction. The initial stage of the reaction is slow because the relatively high electrolyte content of the water portion of the reaction mixture forces the soap into an insoluble form which is not able to emulsify the oil. During this part of the reaction, the reaction rate is effectively limited by the agitation present in the reaction vessel.

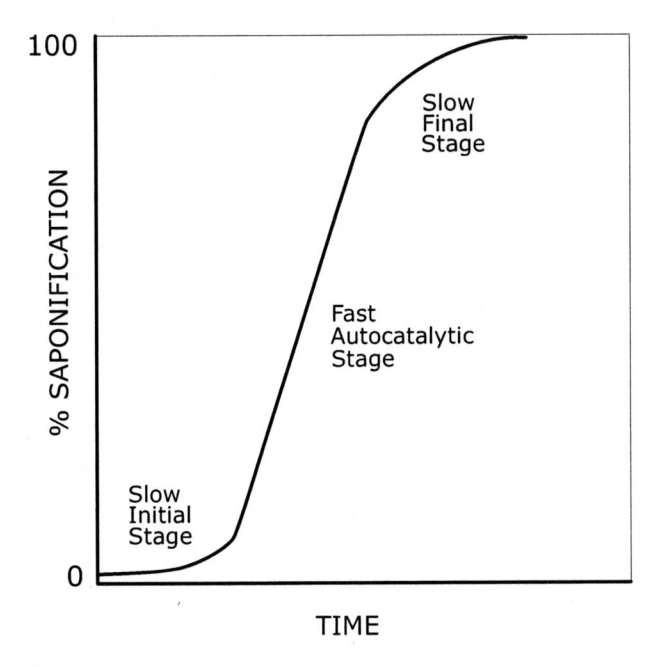

Fig. 8.5. Rate of Saponification.

- Autocatalytic Stage—In this stage, rapid reaction is taking place because the soap concentration has reached the point where soap micelles are beginning to form. The micelles solubilize unsaponified fat, and thus promote contact between the unreacted fat and the NaOH.
- Final Stage—In this stage, the reaction slows down as the concentration of reactants decreases. During this part of the reaction, the reaction rate is effectively limited by the agitation present in the reaction vessel.

Critical Reaction Factors
To achieve complete saponification, reaction system designs must take into account several critical factors.

- Mixing/Shearing—Intimate contact between the fat/oil and the NaOH containing water phase is critical to forming a fine emulsion to minimize the slow initial stage of the reaction. In addition, good mixing also promotes reaction during the slow final stage.
- Temperature—As a rule of thumb, the reaction rate of saponification will double for every 10°C increase in temperature. Most saponification systems operate at 120 to 140°C to achieve the fastest reaction times.
- Composition—A slight excess of caustic is required to complete the reaction. In addition, the correct electrolyte concentration is needed to achieve the proper soap phase. Too high an electrolyte content will slow the reaction rate due to a hard soap grain.
- Residence Time—Sufficient residence time must be provided to complete the reaction. With all of the other factors, noted above, controlled properly, a residence time of 10 to 15 minutes is sufficient.

Saponification Equipment
Several different commercial equipment systems are available for the saponification process. Each of these systems attempts to utilize the factors critical to good saponification, to achieve complete saponification in the shortest period of time required. Two types of reaction systems are common. One design pre-shears reactants in a high shear device, followed by reaction completion in a subsequent vessel. The other design is based on a stirred tank reactor principle where reactants are introduced into a recycled saponified soap stream to pre-solubilize the fats/oil in the soap phase. The fat/soap blend returns to a vessel where a continual stream of reacted soap is removed.

Figure 8.6 is a simplified flowsheet of a design where the fats/oils and the reaction lye mixture are introduced together into a high shear mixer to promote the formation of a fine emulsion critical to minimizing the slow stage of the reaction. The high shear mixer is followed by an agitated reaction vessel where the saponification reaction is completed. Details of the internals of the reaction vessel are shown in Fig. 8.7. This reaction system is capable of producing fully saponified soap in a short period of time.

Figure 8.8 is a simplified flowsheet of a design where the fats/oils and reaction lye are introduced into a recycled saponified soap stream. The blend then proceeds through a non-agitated tubular reaction vessel where saponified soap exits continuously. This vessel is operated at high temperature (130°C) and 3 bar pressure to prevent soap flashing, until the reacted soap mass is flashed in the separate flash cooler.

Continuous Saponification and Neutralization Systems ● 231

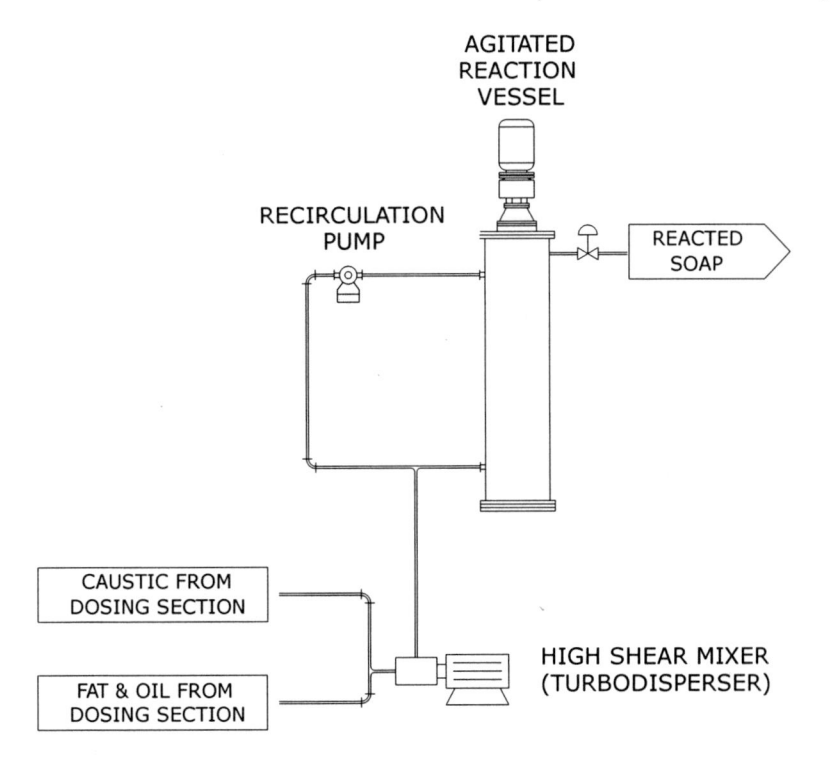

Fig. 8.6. Typical High Shear Reaction System.

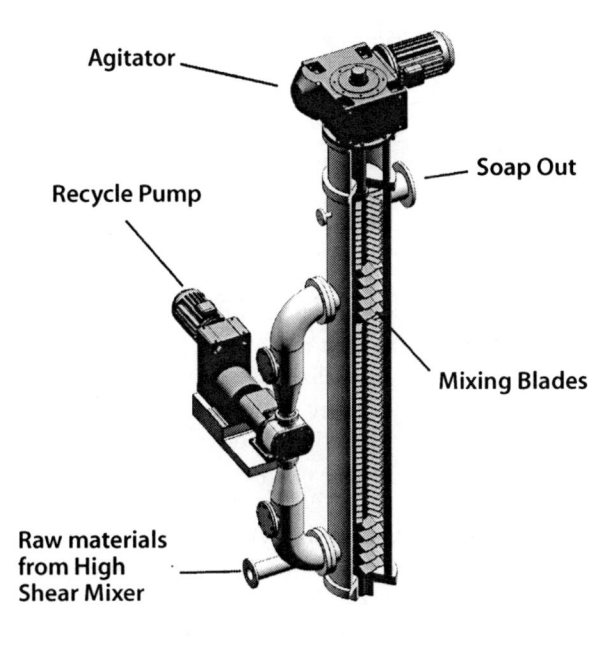

Fig. 8.7. Reaction Vessel Details (Courtesy Soaptec srl).

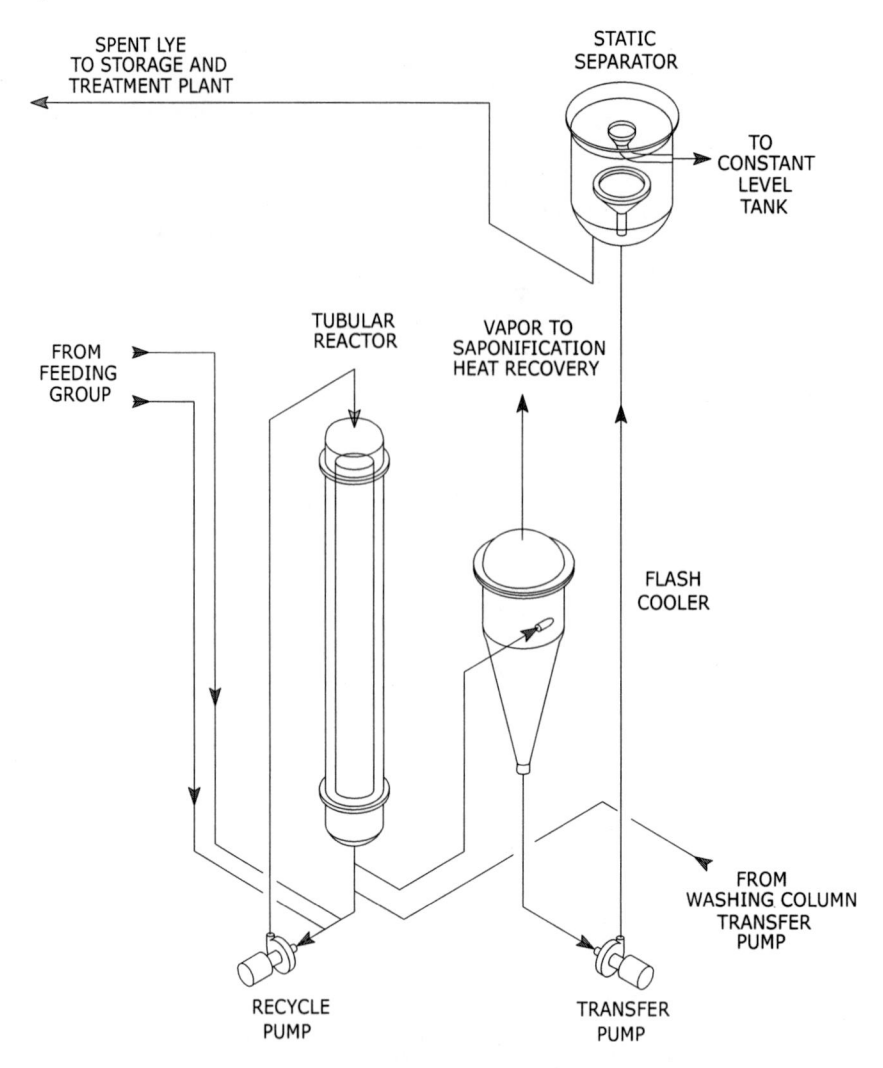

Fig. 8.8. Mazzoni LB "SCNT" Continuous Saponification Plant Reaction/Cooling System.

Soap Washing/Extraction

The removal of glycerine and water soluble impurities takes place in the soap washing/extraction system. In addition to the removal of glycerine, this system is also responsible for establishing the NaCl and NaOH content of the soap exiting to the neutralization system. As depicted in the simplified block diagram of the process shown in Fig. 8.1, the key unit operations in this system are cooling and spent lye separation, washing and half spent lye separation, and lye and neat soap separation. The washing system begins with the cooling of the reaction soap mixture by a cooling mixer, flash cooler, or spent lye recycle stream, followed by spent lye separation. The key part of the system is a rotating disc contactor (RDC) liquid/liquid extraction column, where wash lye is used to extract the glycerine from the liquid soap. The final step is the centrifuging of entrained wash lye carried over with the soap from the top of the wash column.

Two key principles are critical to understanding the operation of the washing system. The first is the concept of soap phase chemistry. The second is Wigner's Model for the distribution of NaCl and NaOH between the soap and lye phases. Each of these is discussed briefly.

Soap Phase Chemistry
From a chemistry point of view, once the saponification of fats and oils is completed, the continuous saponification process is a five component system consisting of soap, NaCl, NaOH, water, and glycerine. A five component phase diagram is required to represent this system. However, this is a complicated system so it is simplified to a three-component system consisting of soap, electrolyte (NaCl and NaOH), and solvent (water and glycerine).

Soap Phase Diagram
Figure 8.9 is a phase diagram for 75:25 tallow/CNO soap. The phase diagram depicts phases present at specific system compositions. The vertical axis represents the percentage of soap composition in the mixture. The horizontal axis represents the percentage of electrolyte in the mixture. The electrolyte shown on the phase diagram is reported as the percentage of NaCl. Since electrolytes have different graining efficiencies, the equation for calculating electrolyte value for NaCl and NaOH is:

$$\%electrolyte\ (as\ \%NaCl) = \%NaCl + (1.2) \times (\%NaOH)$$

The third component is water. The bottom left-hand part of the phase diagram represents 100% water.

Of particular interest for the continuous saponification process is the neat soap single phase area and the neat soap–lye two phase area. In the neat soap–lye two phase region, the majority of the continuous saponification process operates. The removal of glycerine from the reacted soap depends on the washing system composition being controlled in this two phase region. Once soap exiting the centrifuges has the final alkalinity adjusted in the neutralization system, the neat soap single phase is achieved. The approximate operating points of the continuous saponification system are shown on the phase diagram.

Limit Lye Concentration
The left hand side of the neat soap–lye two phase region is known as the limit lye concentration. At the limit lye concentration, if electrolyte decreases, soap begins to dissolve in the lye, and the three phase neat soap–nigre–lye region is present. Further electrolyte decreases will enter into the two phase neat–nigre region. The location of the limit lye concentration on a phase diagram is a function of the fat/oil blend composition. Fats and oils have a unique graining index which determines the limit lye concentration. Table 8.1 shows the graining index for typical fats and oils and their blends. The limit lye concentration is also a function of the system temperature. As temperature is increased, the limit lye concentration increases, meaning an increase in the temperature of the system will result in a loss of neat soap–lye separation. Another way to think of this is that the neat soap is more soluble in the lye at higher temperatures.

Soap Grain
Within the neat soap–lye two phase region, the grain of the soap varies with electrolyte content. Figure 8.10 shows three types of soap grain. A normal grained soap would be found near the limit lye concentration line. At lower electrolyte levels, the grain softens until no separation is apparent. This is shown in the soft grained photo. As electrolyte is increased to the right hand side of the neat soap–lye region, the soap grain becomes harder. The harder grain of soap has whiter peaks and is more bunched. This hard grain of soap entraps more lye in the soap grains. Glycerine removal efficiency is best when normal grained soap is present.

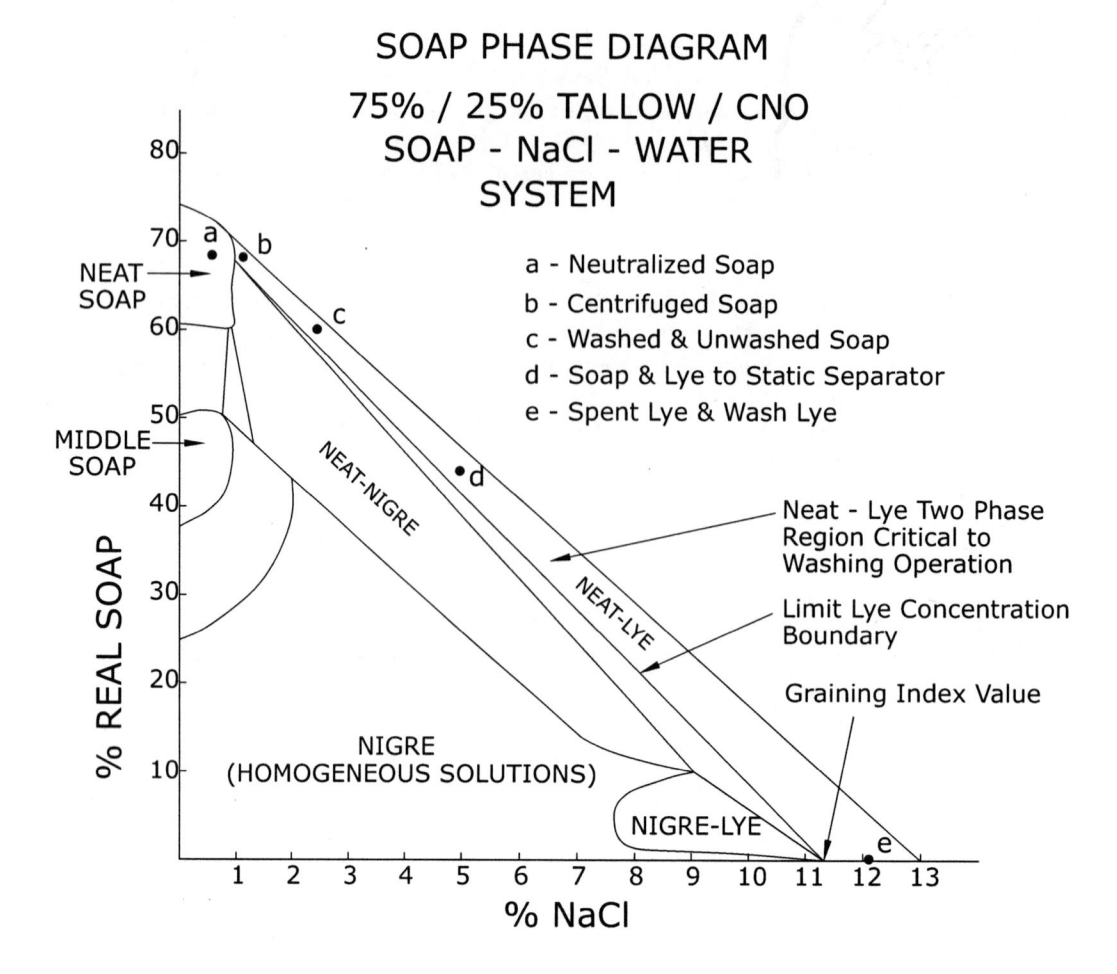

SOAP PHASE DIAGRAM
75% / 25% TALLOW / CNO
SOAP - NaCl - WATER SYSTEM

a - Neutralized Soap
b - Centrifuged Soap
c - Washed & Unwashed Soap
d - Soap & Lye to Static Separator
e - Spent Lye & Wash Lye

Neat - Lye Two Phase Region Critical to Washing Operation
Limit Lye Concentration Boundary
Graining Index Value

NEAT SOAP
MIDDLE SOAP
NEAT-NIGRE
NEAT-LYE
NIGRE (HOMOGENEOUS SOLUTIONS)
NIGRE-LYE

% REAL SOAP vs % NaCl

Fig. 8.9. Illustrative Soap Phase Diagram.

Table 8.1. Graining Index

Material	Graining Index (%NaCl)
Tallow	5 - 7
Palm Oil	5 - 7
Palm Kernel Oil (PKO)	13 - 17
Coconut Oil (CNO)	20 - 25
80/20 T/PKO	10 (use weighted average of high end value plus a 10% safety factor)

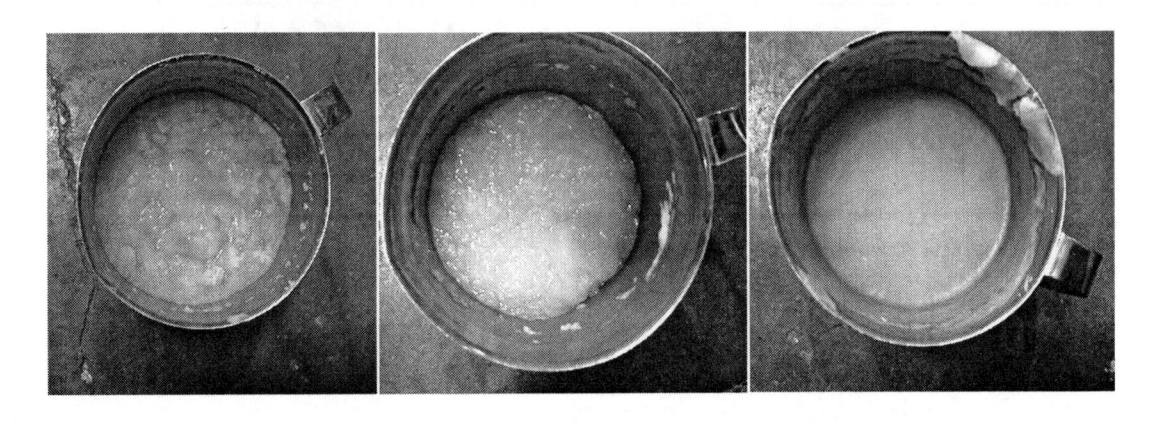

Hard   **Normal**   **Soft**

Fig. 8.10. Soap Grains.

Wigner's Model
The distribution of NaCl and NaOH in soap separated from a lye phase is critical to the proper operation of the continuous saponification system. The composition of the wash lye determines the NaCl and NaOH content of the centrifuged soap. In 1940, J.H. Wigner proposed a model to describe the distribution of NaCl, NaOH, and glycerine between the separated soap and lye phases. An excellent discussion of Wigner's model was done by Villela and Suranyi (Spitz, 1996). A graphical representation of the model is shown in Fig. 8.11. The model states:

- Soap curds consist of a soap hydrate containing 66% of total fatty matter (TFM) mixed with enmeshed lye. The 66% of TFM soap hydrate can never be separated as a discrete phase, and the soap curds will always contain some enmeshed lye.

- Electrolytes are present in the enmeshed lye and in the separated lye, but not in the hydration water of the soap hydrate. Both the electrolyte content of the enmeshed lye and the separated lye are identical.

- The sum of the hydration water plus enmeshed lye is the lye-in-soap. Glycerine is present in all of the aqueous phases at the same concentration, including the water of hydration.

To illustrate the power of this model, a simple calculation example is presented.

	Grained Soap		
Soap Curd			Separate Lye
66% TFM Hydrate		Enmeshed Lye	
Anhydrous Soap	Hydration Water		
	0% NaCl	X% NaCl	X% NaCl
	0% NaOH	Y% NaOH	Y% NaOH
	Z% Glycerine	Z% Glycerine	Z% Glycerine

Fig. 8.11. Graphical Representation of Wigner's Model.

Example 1

Assume 1000 kg of 62% of TFM neat soap with 0.5% of NaCl was centrifuge separated from a lye solution. Calculate the NaCl content of the separated lye.

Solution

First the amount of 66% of hydrated soap must be calculated. This is done as follows:

Hydrated soap = (1000 kg) × (0.62 TFM)/(0.66 TFM in hydrated soap)
= 939.39 kg soap hydrate

If 939.39 kg of the neat soap is hydrated soap, the remainder is enmeshed lye. The amount of enmeshed lye is:

Enmeshed lye = 1000 kg – 939.39 kg
= 60.61 kg enmeshed lye

All of the NaCl in the neat soap is contained in the enmeshed lye. The concentration of the NaCl in the enmeshed lye is:

%NaCl in enmeshed lye = (1000 kg) × (0.5% NaCl)/(60.61 kg enmeshed lye)
= 8.25% NaCl in the enmeshed lye

Since the composition of the enmeshed lye is the same as the separated lye, the NaCl content of this lye is 8.25%.

The above example shows the detailed calculations for NaCl. Similar calculations can be made for NaOH. In general, divide the total amount of NaCl in the neat soap by 0.06 to obtain the maximum amount of NaCl in the washing lye. From a practical experience standpoint, the following values apply to the relationship of NaCl and NaOH in the soap and separated lye:

- For every 1% of NaCl in the washing lye, 0.066% of free NaCl exists in the neat soap.
- For every 1% of NaOH in the washing lye, 0.07% of free NaOH exists in the neat soap.

Glycerine Washing Equipment

The washing system begins with the cooling of the reaction soap mixture followed by spent lye separation. The key part of the washing system is an RDC liquid/liquid extraction column, where wash lye is used to extract the glycerine from the liquid soap. The final step in the system is the centrifuging of entrained wash lye carried over with the soap from the top of the wash column. Each of these systems plays a critical role in removing the glycerine from the saponified soap.

Cooling and Spent Lye Removal

Since most saponification systems operate at temperatures between 120 and 140°C, reducing the saponified soap temperature is necessary to achieve the neat soap and lye separation required for the removal of glycerine. Traditionally, this temperature reduction was accomplished in a vessel called a cooling mixer. Saponified soap entering the cooling mixer is cooled by water flowing through tubes inside the agitated vessel. The 85°C temperature of soap is controlled by regulating the flow of cooling water through the tubes. The cooled soap/lye stream then flows into a static separator where residence time is provided to allow the soap to separate from the spent lye. From the static separator, spent lye leaves the washing system for further glycerine recovery processing. The soap proceeds on to the washing column.

While the cooling mixer system was a standard offering in the industry for many years, recent technology advances have replaced it. Two types of systems are currently produced by equipment manufacturers.

- Flash Cooler—Mazzoni LB offers a flash cooler system, shown previously in Fig. 8.8. This system uses the temperature available in the soap from the saponification system to flash cool the soap/lye mixture prior to the static separator. The water that flashes off as steam is available to preheat raw materials. The flashed soap is then pumped to a static separator for the removal of the spent lye. The advantages of this system over the cooling mixer are higher glycerine in the spent lye (40%), increased energy efficiency, and reduced mechanical reliability issues.

- Cooled Spent Lye Recycle—Binacchi and others offer a cooled spent lye recycle system, shown in Fig. 8.12. In this system, a recycled stream of spent lye from a static separator is pumped through a plate and frame heat exchanger where it is cooled. The cooled spent lye is injected into the saponified soap leaving the reactor. The spent lye and soap mixture is mixed in a static mixer, and then sent to the static separator for the removal of the spent lye. The advantages of this system over the cooling mixer are reduced mechanical reliability issues and reduced plant layout.

Washing and Half Spent Lye Removal

The key part of the glycerine removal system is the wash column shown in Fig. 8.13. The wash column is an RDC countercurrent liquid/liquid extraction column. Unwashed soap and entrained lye from the static separator enter the bottom of the wash column. Wash lye enters near the top of the column. Since the soap is less dense than the wash lye, soap rises to the top of the column and overflows as washed soap to the centrifuges. Wash lye proceeds to the bottom of the column, where it exits as half spent lye. Lye from the centrifuges is recycled to the top portion of the column.

Figure 8.14 shows the internal details of the top and bottom of the wash column. The column consists of a series of mixing and settling zones known as extraction stages. Each stage contains a stator ring of metal attached to the inside diameter of the vessel and a flat rotating disc attached to an agitator shaft which runs most of the length of the column. Most wash column designs have 40 of these extraction stages. As unwashed soap moves up the column, it is subjected to sequential mixing and separating from the rotating discs and stator rings, respectively. This series of mixing and separating stages provides for the removal of glycerine from the soap.

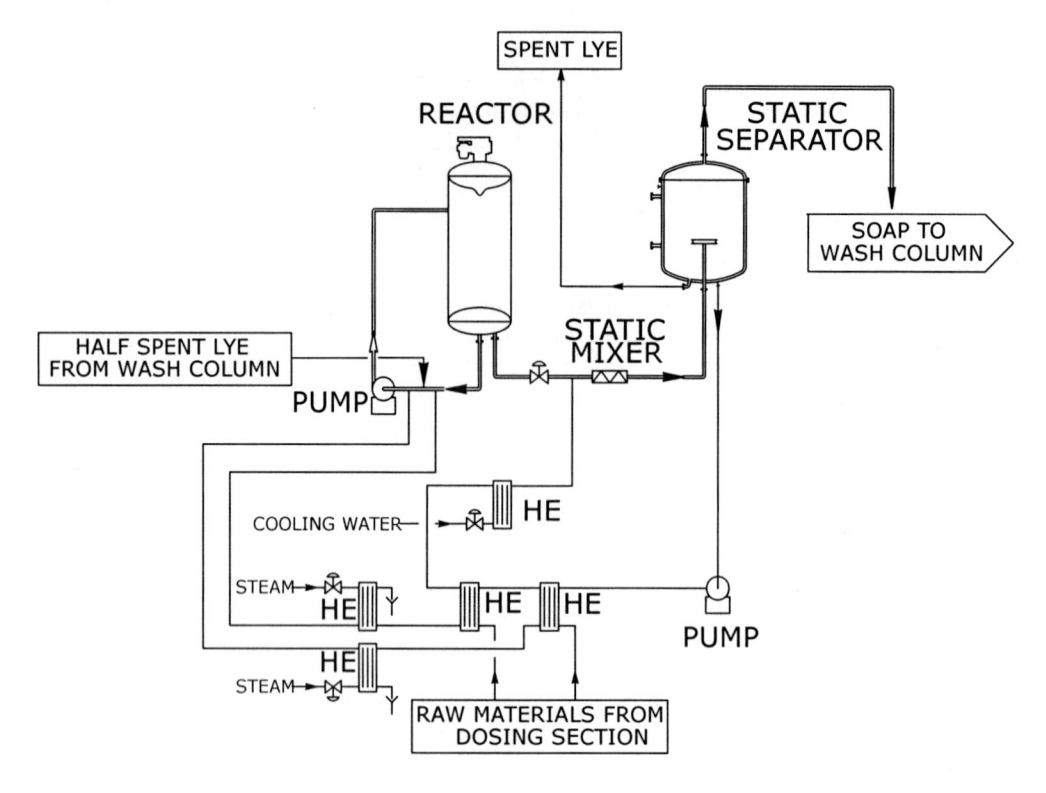

Fig. 8.12. Binacchi "CSWE-3" Continuous Saponification Cooled Spent Lye Recycle System.

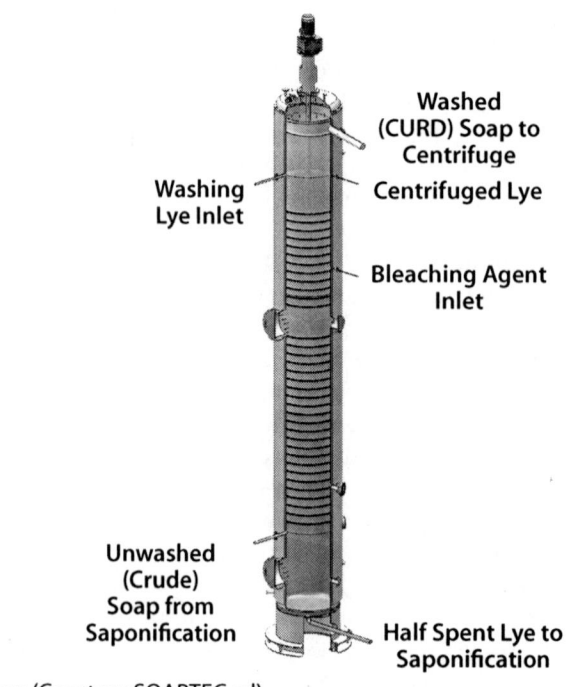

Fig. 8.13. RDC Column (Courtesy SOAPTEC srl).

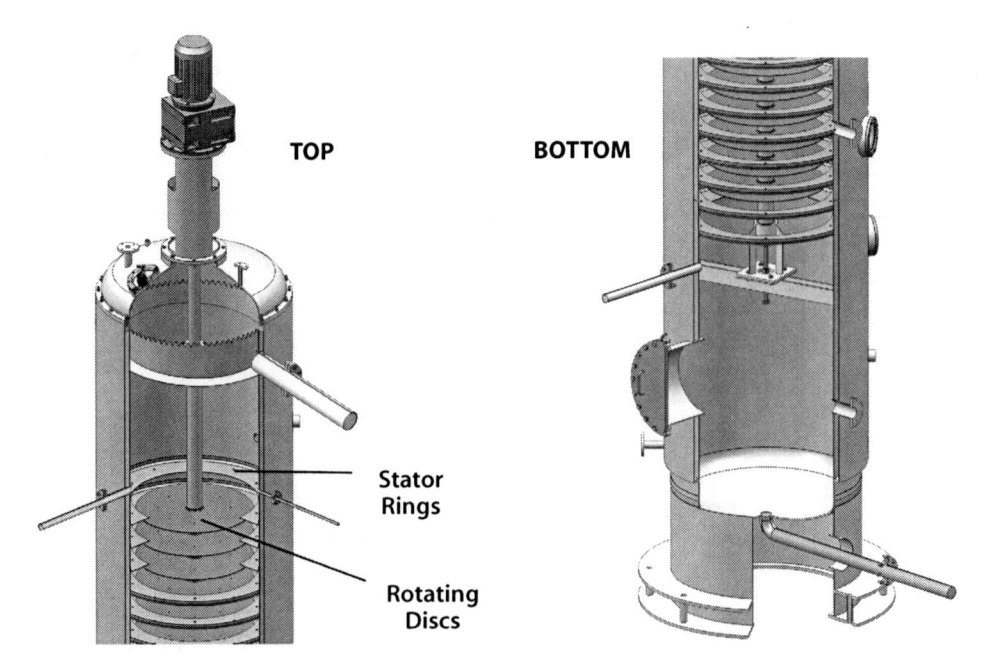

Fig. 8.14. RDC Details (Courtesy SOAPTEC srl).

Lye and Neat Soap Separation
The soap exiting the top of the wash column entrains with it approximately 15–20% of wash lye. The final step in the soap washing system is centrifuging the entrained wash lye and recycling it back to the top of the wash column. The neat soap, with the lye removed, proceeds to the neutralization system. The centrifuge is specifically designed for soap and lye service. A typical centrifuge arrangement is shown in Fig. 8.15. The soap and entrained wash lye enter the centrifuge through a feed pipe which directs the flow to the bottom of the bowl. The bowl, spinning at approximately 5000 rev/minute, separates the soap and lye. Lye travels the outside part of the bowl to an impeller paring disc where it exits and returns to the top of the wash column. Neat soap travels to the center of the bowl and proceeds to another impeller paring disc where it exits to the neutralization system. Back pressure valves on the soap and lye outlets are used to maintain the proper separation of lye from the soap. Centrifuge designs were improved to increase mechanical reliability and to enable capacities of 8000 kg/hour of neat soap at 62% of TFM. Most continuous saponification systems are equipped with an extra centrifuge to allow for cleaning and repair.

Glycerine Washing Efficiency
A properly controlled washing system will leave approximately 0.1–0.2% of glycerine in the neat soap. Several key factors are critical to good glycerine removal:

- The electrolyte in the washing lye should be close to the limit lye concentration. The electrolyte must be adjusted for the fat/oil blend being utilized. With this normal grain soap (versus hard grain), maximum glycerine removal is achieved.

- Increasing the number of extraction stages in the wash column increases glycerine removal. Most wash columns in systems today contain 40 stages of extraction.

- The wash column agitator speed should be properly set via experimentation. Slow agitator speed provides inadequate mixing. High agitator speed causes back mixing between stages and poor separation.

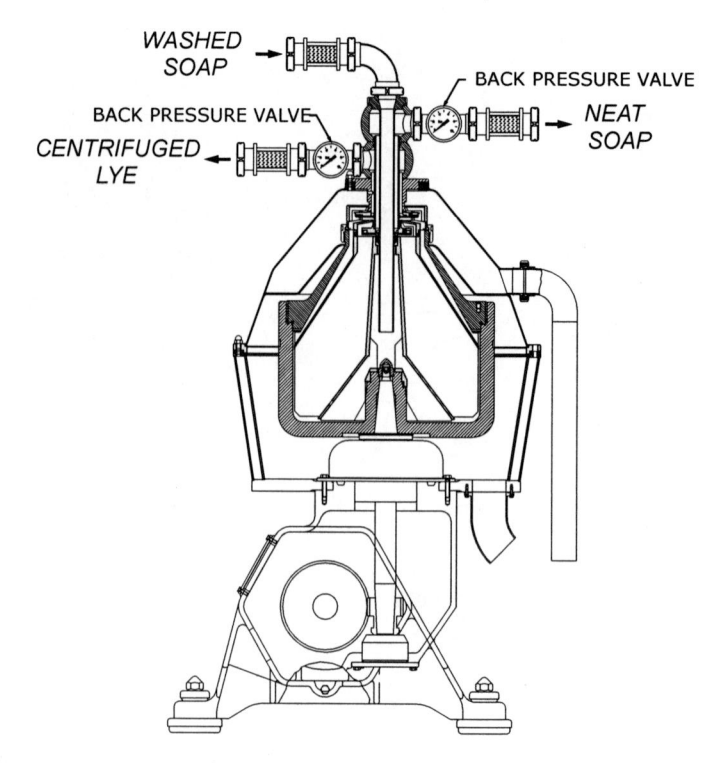

Fig. 8.15. Soap/Lye Centrifuge Arrangement.

- Incoming washing lye glycerine content must be below 0.3%. Salt is recovered in the glycerine recovery process and is recycled back to the continuous saponification system. One must control glycerine content in the salt to avoid excess glycerine in the wash lye.
- The washing ratio/bulk of lye can be varied to change glycerine removal. Both of these measures are indications of the amount of wash lye used per unit of soap produced.

Neutralization

The final step in the continuous saponification process is the adjustment of the alkalinity of the neat soap leaving the centrifuges. Typically, the centrifuged neat soap has a caustic level of 0.2–0.3% of NaOH, with the exact amount being determined by the wash lye NaOH concentration. The NaOH is reduced to a final level set by the finished soap customer specifications, usually 0.01–0.08%. The reduction of the excess NaOH is accomplished by adding either a fatty acid or the same lauric oil used in the saponification process.

Neutralization Equipment

The equipment systems used in the final neutralization step employ designs similar to the saponification systems. Figure 8.16 is a drawing of a typical neutralization system. Soap from the centrifuges enters the neutralization vessel. A recycled soap stream from the neutralizer is pumped through a high shear mixing device known as a turbodisperser. Neutralizing agent is added into the high shear mixer. The purpose of the high shear mixer is to provide intimate mixing between the NaOH in the soap and the neutralizing agent. The intimate mixing speeds the reaction rate. Product leaving the high shear mixer is returned to the neutralizer vessel. A continuous stream of finished specification soap is removed from the soap neutralizer vessel. Control of the finished soap NaOH level is achieved by means of a pH control system. The pH control system varies the amount of fatty acid metered into the system to maintain a constant NaOH level in the finished soap. Most pH probes are of a self-cleaning design with provisions to remove the probes for maintenance/calibration while the plant is still operating.

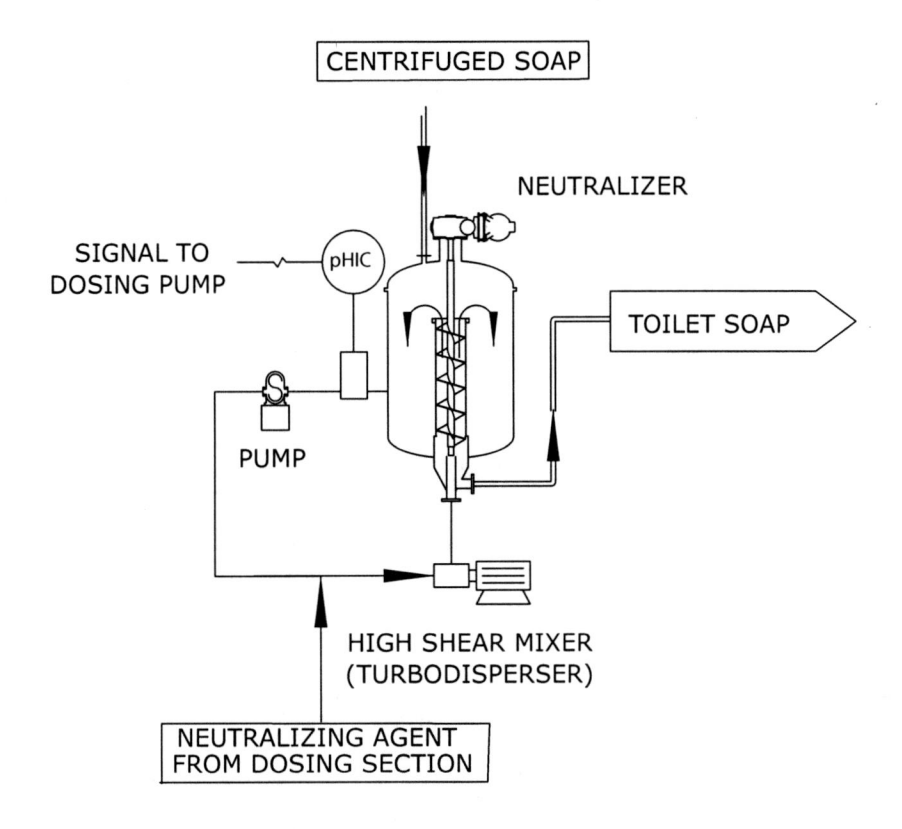

Fig. 8.16. Binacchi "CSWE-3" Continuous Saponification Neutralization System.

Commercially Available Systems

Throughout the previous discussion concerning the fundamental principles of operation and the overview of equipment and systems technology employed in continuous saponification processes, many examples shared come from commercially available systems. Today's available equipment has incorporated design improvements aimed at increasing energy efficiency, improving flexibility by reducing changeover times and product holdup times, reducing plant maintenance by simplifying equipment designs, reducing plant layout area requirements, and providing flexibility to operate systems without glycerine removal to allow for operation as a semiboiled soap process. Commercially available systems today are described briefly below.

- *Mazzoni LB S.p.A.*—SCNT-N Plant
  This system, shown in Fig. 8.17, was introduced to the market in 2001. The key design features include centrifugal metering pumps with mass flow meters, a tubular reactor system capable of reducing saponification holding time to 8–10 minutes, a flash cooler system to allow up to 40% of glycerine in the spent lye and reduced energy consumption (100 kg of 3 bar steam per 1000 kg of neat soap), and a tubular neutralizer design which can be used as a stand alone fatty acid neutralization system. Plant capacities are available from 3 to 20 tons/hour.

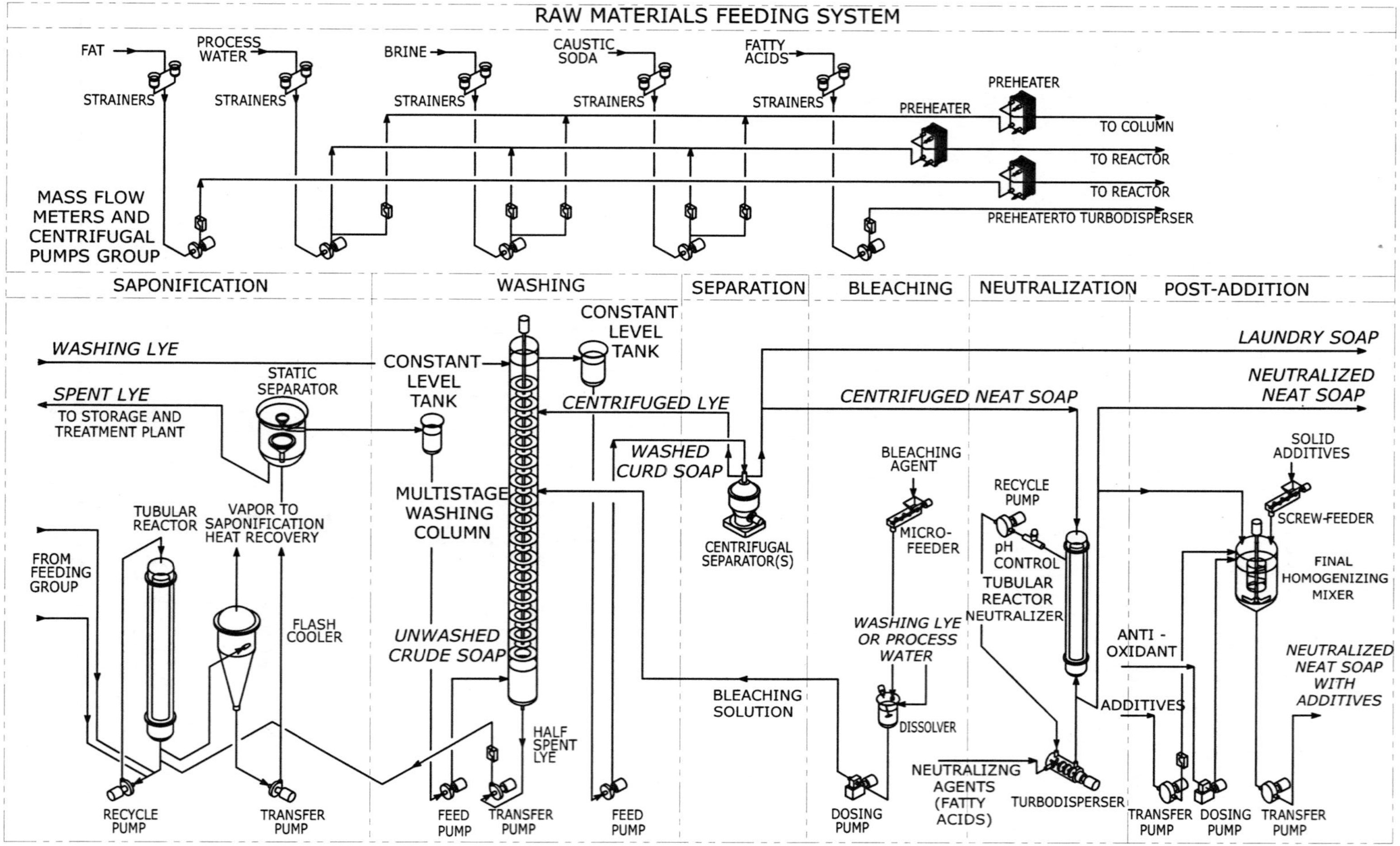

242 • T. Kelly

Fig. 8.17. Mazzoni LB "SCNT-N" Continuous Saponification Plant.

- *Binacchi & Co.*—CSWE 3 Plant
  This plant was depicted previously in Fig. 8.2. The key design features include positive displacement metering pumps with mass flow meters for improved metering accuracy, reducing plant layout size by the utilization of a vertical reactor system, and the recycling of cooled spent lye to reduce the saponification mixture temperature, allowing for a reduced plant layout and a reduced product hold-up. Energy consumption is estimated at 120 kg of 3-bar steam per 1000 kg of neat soap. Plant capacities are available from 2 to 16 tons/hour.
- *Sela GmbH*—KVN Plant
  Similar in design to Binacchi's CSWE 3 system, plant capacities are available from 0.3 to 13 tons/hour.

Continuous Fatty Acid Neutralization

Overview

As was mentioned at the beginning of this chapter, continuous soap making processes began in the mid-1940s with the development of processes for converting fats and oils into fatty acids. The fatty acids, with the glycerine from the hydrolysis reaction removed, are reacted with a caustic soda, water, and salt blend to produce soap in a continuous fashion. This method of manufacture is referred to as continuous fatty acid neutralization.

Neutralization Reaction

The neutralization reaction is a chemical reaction between the fatty acids and the NaOH present in the caustic soda. The equation for the reaction is:

$$\text{Fatty Acid} + \text{NaOH} \rightarrow \text{Soap} + \text{Water}$$

From the reaction equation 1 mole of fatty acid reacts with 1 mole of NaOH to produce 1 mole of sodium soap and 1 mole of water. This reaction proceeds at a very fast rate compared to the saponification reaction. The important factors for a good neutralization reaction are accurate raw material metering, high shear mixing of fatty acid and NaOH to promote good interfacial area, operation with a slight excess of NaOH, and good temperature control. The neutralization reaction is exothermic, releasing 14 cal/mole of reacted fatty acid. The heat liberated from this reaction is used in the process to maintain the temperature in the neutralization reactor.

Neutralization Equipment

Compared to the continuous saponification method for making soap, this process is very simple because no glycerine removal equipment is necessary. As with continuous saponification technology, fatty acid neutralization technology has continued to evolve. Improvements in metering accuracy, reactor designs, and process controls have continued to reduce utility requirements, improve flexibility, reduce maintenance, and reduce plant layout space requirements.

Mazzoni LB SCT-SSCT Process

Shown in Fig. 8.18 is the Mazzoni SCT-SSCT process for continuous soap making. When used as the SCT process, this system provides for fatty acid neutralization. When used as the SSCT process, this same system operates as a neutral oil saponification system without glycerine removal, commonly referred to as a semiboiled soap process. The semiboiled soap process is discussed in greater detail in a subsequent chapter.

In a fatty acid neutralization process, raw materials are metered into the neutralization reaction loop by a series of centrifugal pumps. Raw materials enter the pump after being strained through duplex strainers. A mass flow meter and a pressure control valve are used to accurately control the flow from the pump. The proper amount of brine, water, caustic soda, and fatty acid blend are injected into a high shear mixer known as a turbodisperser. A recycled soap stream from the tubular reactor is also injected into the turbodisperser. Both the temperature of the caustic soda/water and of the fatty acid are controlled via plate and frame heat exchangers, with operating set points established to maintain the desired neutralizer temperature. The turbodisperser provides the intimate mixing of the reactants to facilitate a complete and uniform reaction. The soap leaving the turbodisperser flows into the tubular reactor where a continuous stream of soap exits. A pH meter is located in the recycled soap loop. This pH meter controls a trim pump which allows for fine pH adjustment of the finished soap by using fatty acid.

The tubular reactor is designed to operate under high pressure and temperature. One advantage of this process is the ability to close-couple the SCT neutralizer system with a vacuum dryer known as the SCT-C system. This close-coupled process is discussed in greater detail in Chapter 9.

Other Commercial Systems

Binacchi also offers fatty acid neutralization systems. The CSFA plant, shown in Fig. 8.19, is a fatty acid neutralization system with positive displacement metering pumps and a series of two agitated reaction vessels. The CHBS system, shown in Fig. 8.20, is a flexible dual reactor system capable of producing both fatty acid neutralized soap and semi-boiled soap.

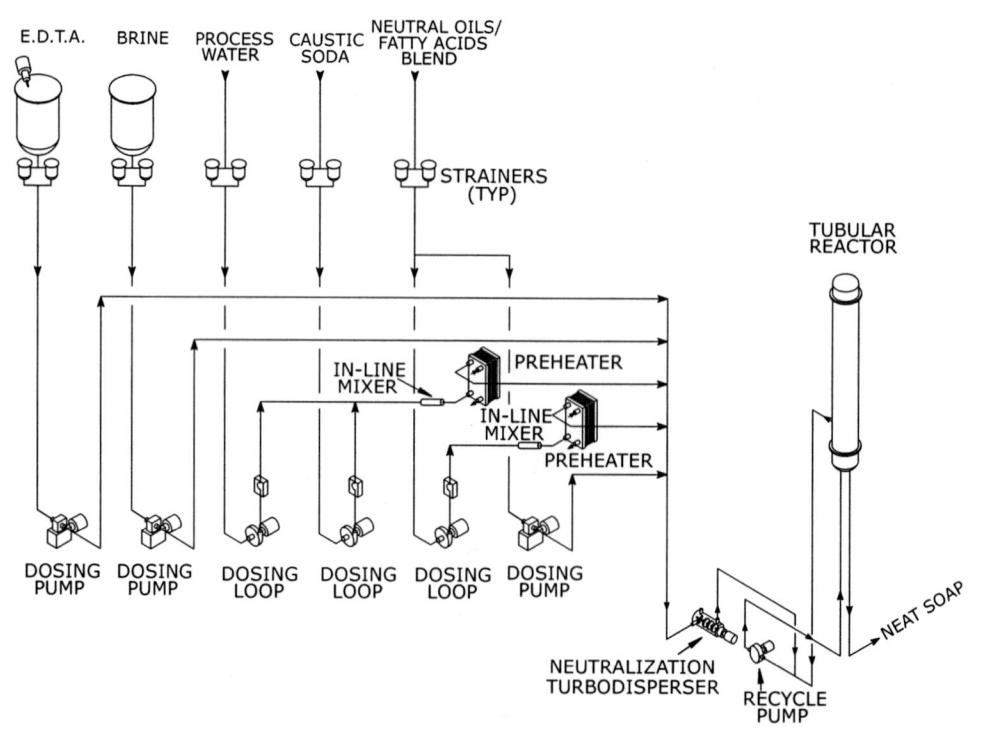

Fig. 8.18. Mazzoni "SCT" Continuous Fatty Acid Neutralization or "SSCT" Neutral Fats/Oils Saponification without Glycerin Extraction.

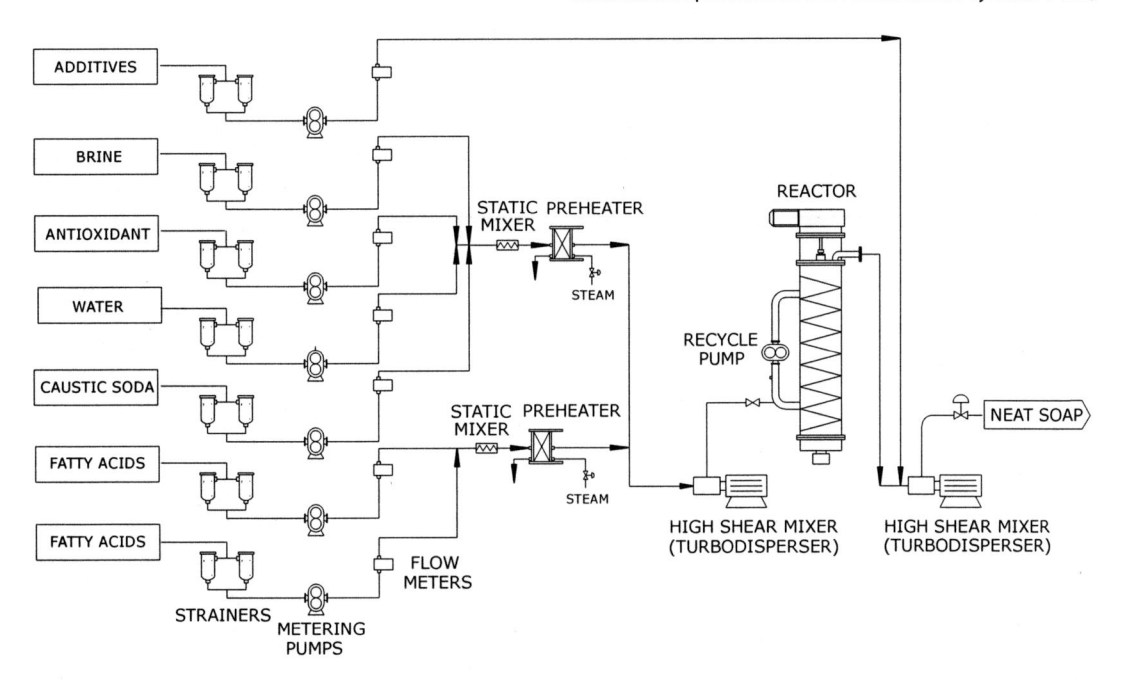

Fig. 8.19. Binacchi "CSFA" Continuous Saponification for Fatty Acids.

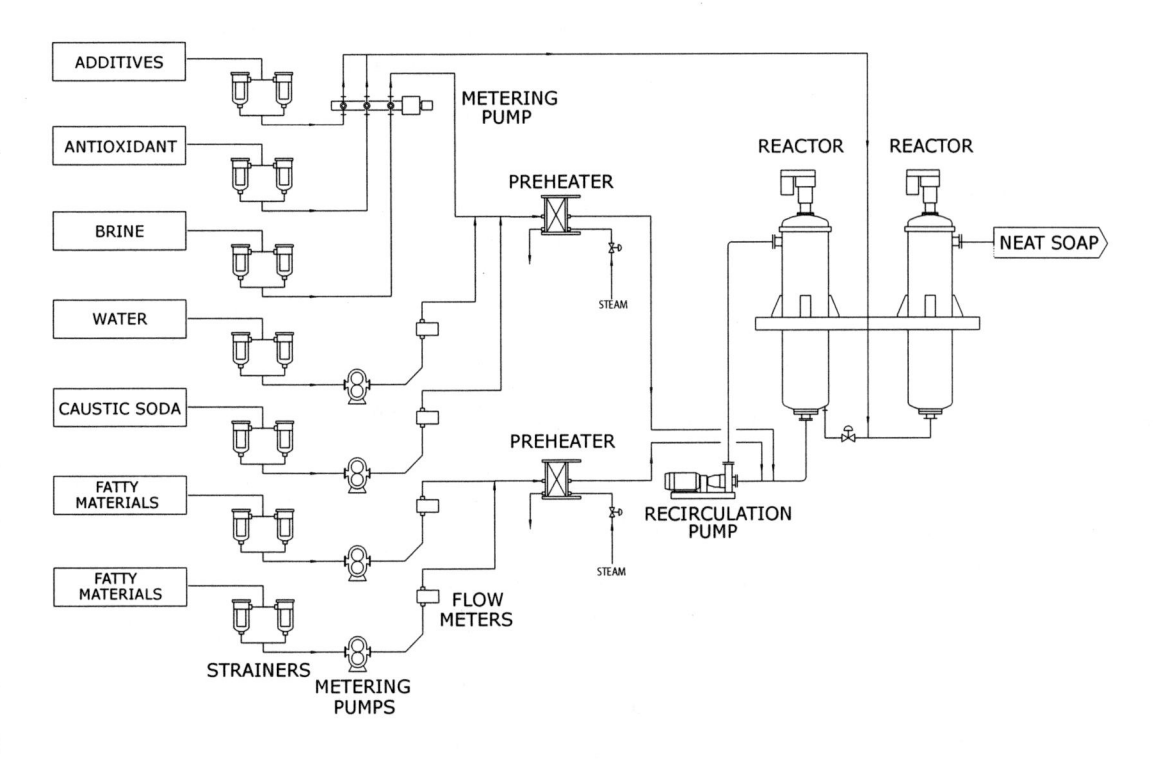

Fig. 8.20. Binacchi "CHBS" Continuous Saponification for Fatty Acids – Neutral Fats.

Summary

Whether the choice of soap making system is a saponification process, fatty acid neutralization process, or a semi boiled process, continuous methods of production have proven advantages over older batch processes. Continuous systems improve yields, reduce energy consumption, require less in-process inventory, reduce production cycle times, improve finished product quality/consistency, increase flexibility, and require smaller plant layouts. Today's commercially available systems have continued to improve these advantages. A summary of recent advances includes:

- Improved metering system designs which use the latest flow measurement technologies to increase metering accuracy and to improve finished product quality.

- Improved reactor designs which enable complete reaction with shorter residence time. Reaction times in some systems were reduced to 8–10 minutes. The reduced hold up times allow for simpler changeovers with less rework material.

- Increased flexibility of system designs. Many systems are now designed to be capable of producing both fatty acid neutralization soap and neutral oil saponification soap on the same system. This is critical as both raw material and finished by-product glycerine prices continue to fluctuate. These systems can also be directly coupled to vacuum drying systems to provide further flexibility and energy savings.

- Increased energy efficiency through the use of flash coolers and heat interchanges of raw material and in-process streams. Steam consumption has improved from 150 to 100 kg per 1000 kg of neat soap produced on some system designs. The flash cooler system produces spent lye glycerine content up to 40%, reducing energy requirements in the glycerine recovery process.

- Reduced plant maintenance through cooling mixer elimination, use of standard centrifugal and positive displacement pumps, and improved centrifuge mechanical designs.

As raw material and energy prices increase, the design of soap making systems will continue to improve to respond to changing market needs.

Acknowledgments

I wish to thank Luis Spitz for his help and support in the development of the material presented. In addition, I wish to thank each of the companies who supplied the information on their continuous soap making equipment. Their names are acknowledged in the illustrations they provided.

References

Spitz, L. (ed.) *Soaps and Detergents: A Theoretical and Practical Review;* AOCS Press: Champaign, IL, 1996; pp. 141–147.

Wigner, J.H. Soap manufacture, *The Chemical Processes;* Chemical Publishing Co.: New York, 1940.

Further Reading

Palmquist, F.T.E. *Soap Technology-Basic and Physical Chemistry,* Zander & Ingerstrom AB: Stockholm, Sweden, 1983.

Woollatt, E. *The Manufacture of Soaps, Other Detergents and Glycerine;* Ellis Horwood Limited: London, England, 1985.

Appendix

Definitions and Terminology

Saponification
Saponification is the chemical reaction of fats and oils with caustic soda and brine to form soap and glycerine.

Washing
Washing is the removal of glycerine from soap by extraction with a brine, caustic soda, and water mixture in an extraction or washing column. The brine, caustic soda, water mixture is typically referred to as a lye.

Neutralization
Neutralization is the addition of fatty acid or fats to neat soap to reduce alkalinity.

Total Fatty Acid (TFA)/Total Fatty Matter (TFM)
Total fatty acid (TFA) is a measure of the fatty acid content present in a soap mixture. TFA does not indicate that free fatty acid is present, but is the amount of fatty acid present as soap molecules. TFA is also called TFM. The relationship between soap content and TFM content is: %TFM = % of sodium soap x 0.93.

Grain
When a sufficient amount of electrolyte (caustic soda and salt) is added to a soap-and-water mixture, the soap is converted to an insoluble form. Soap in this form is referred to as grainy neat soap, and has a grainy, lumpy texture similar to oatmeal. Sometimes graining is referred to as salting-out or splitting-out the soap. At high electrolyte content, soap is "grained out" from the lye with little soap dissolved in the lye. As electrolyte content decreases, the grain becomes softer and soap in the lye increases. At low electrolyte levels, no separation of soap from the lye occurs.

Electrolyte
"Electrolyte" is the term used to refer to solutions of NaCl and NaOH. In soap making, electrolyte is generally reported in terms of percentage of NaCl. Since different electrolytes affect soap graining differently, the following equation is used to relate them: %electrolyte (as NaCl) = %NaCl + 1.2 × %NaOH.

Limit Lye Concentration (LLC)
LLC is the total electrolyte concentration at which the "grained out" soap begins to dissolve in the lye. It is also the minimal concentration of total electrolyte in which soap cannot dissolve any further in the lye. LLC is also called limit lye, limit lye solubility, and is most commonly referred to as graining index (GI).

Washing Lye
Washing lye is a mixture of water, 50% of caustic soda, and 20% of brine which enters the top section of the countercurrent washing column.

Half Spent Lye
Half spent lye is lye that leaves the bottom of the washing column and is recycled to the saponification reactor.

Spent Lye
Spent lye is glycerine enriched lye which leaves the system after the saponification reactor or washing system.

Centrifuged Lye
Centrifuged lye is lye separated by the centrifuge, and recycled back to the upper section of the washing column.

Crude Soap (Unwashed Soap)
Crude soap is the saponified soap mixture from the reactor entering the bottom of the washing column.

Curd Soap (Washed Soap)
Curd soap is the 52–54% of TFM soap leaving the top of the washing column and sent to the centrifuges.

Neat Soap
Neat soap is the 62–63% of TFM soap leaving the centrifuges.

Neutralized Neat Soap
Neutralized neat soap is neat soap neutralized with fatty acids or fats in the neutralizer to final NaOH content.

Washing Ratio
Washing ratio is the quantity of spent lye removed per quantity of neat soap produced.

Bulk of Lye
Bulk of lye (BOL) is the quantity of spent lye removed per quantity of fats and oils saponified. BOL is roughly 1.5 times the washing ratio.

Semi-Boiled Soap Production Systems

Boris Radic[1] and Luis Spitz[2]
[1]Soaptec srl, Italy; [2]L. Spitz, Inc., Highland Park, Illinois, USA

Introduction

Semi-boiled saponification is a glycerine removal free process which produces "neat soap" with all the glycerine that is contained in the starting fats and oils.

Semi-boiled saponification does not require the washing and fitting steps of the traditional full-boiled kettle-soap process or the washing and neutralization of the continuous saponification systems which produce neat soap with only a small quantity of glycerine.

The traditional market for Semi-boiled soaps was mainly for different laundry soaps.

Currently, Semi-boiled soap manufacturing is also gaining importance for toilet soaps due to:

- the fluctuating cost of refined glycerine
- the increasing cost of many raw materials, especially those for spent-lye treatment
- the availability of complete glycerine processing plants

Semi-boiled soap making is a simple, single-step operation which offers these advantages:

- savings in energy consumption
- reduced capital investment compared to other systems
- lower environmental impact (no by-products)
- requirement of less-skilled personnel due to process simplicity
- less and simpler quality-control requirements.

The high-glycerine content of Semi-boiled soap alters the soap's appearance, and results in a harder finished product at given moisture content due to the increased viscosity of the liquid phase of the soap.

To account for this, Semi-boiled soaps are usually produced with 58–60% of Total Fatty Matter (TFM) content, in contrast to the full-boiled or continuously made neat soap with 62–63% of TFM content.

For comparison between the two types of soaps, the term "virtual TFM" is introduced to predict the behavior of the Semi-boiled soap in full-boiled terms (more details in the Definitions and Calculations section of this chapter).

In the full-boiled process, some or most of the glycerine and impurities are removed from the neat soap by washing and/or fitting. These procedures include the addition of water and electrolytes, some kind of mixing to assure contact between newly added material and preformed soap, and then settling to separate the neat soap from the "waste" (in the case of washing, spent lye; and in the case of fitting, seat soap and lye). For more details, refer to Chapters 7 and 8 in this book.

The purpose of washing is twofold: (i) to recover the glycerine liberated by the saponification reaction and (ii) to remove the lye-soluble impurities. Washing in the kettle process is time-consuming and less efficient than in the continuous systems with the commonly used countercurrent, high-efficiency, multistage washing columns [rotating disc contactor (RDC)]. The phase diagram shows that washing is conducted inside the neat soap–lye region.

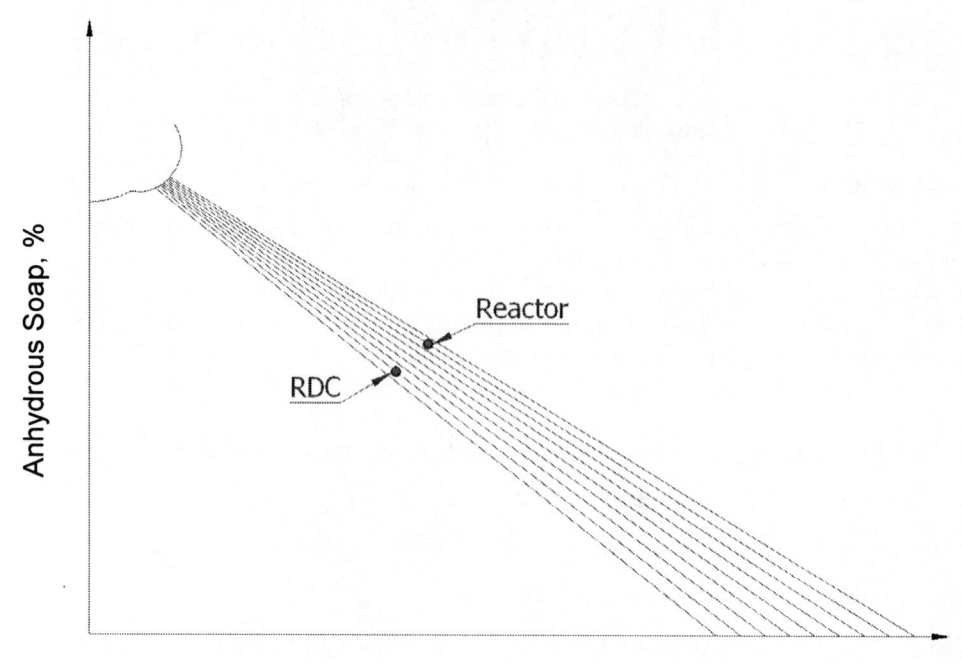

Fig. 9.1. Washing.

The purposes for "fitting" the last processing step in the kettle process are to:

- remove soap-soluble color bodies
- remove the minor impurities
- lower the electrolyte content
- raise the fatty-acid content.

When producing semiboiled soap, the process cannot be manipulated by physical means, only by chemical means; therefore, important points to bear in mind are:

- Semi-boiled soap is produced directly inside the neat soap region: no separation or washing of any kind.
- All of the ingredients present at the beginning of the process remain inside the finished product.
- The quality of the finished soap directly depends on the quality of the raw materials used.

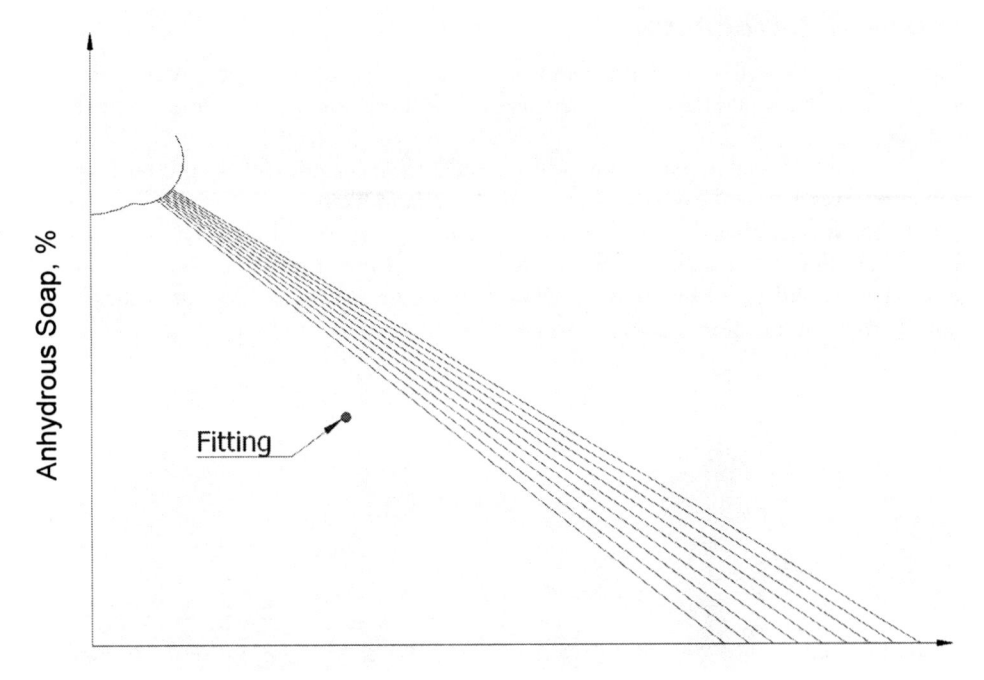

Fig. 9.2. Fitting.

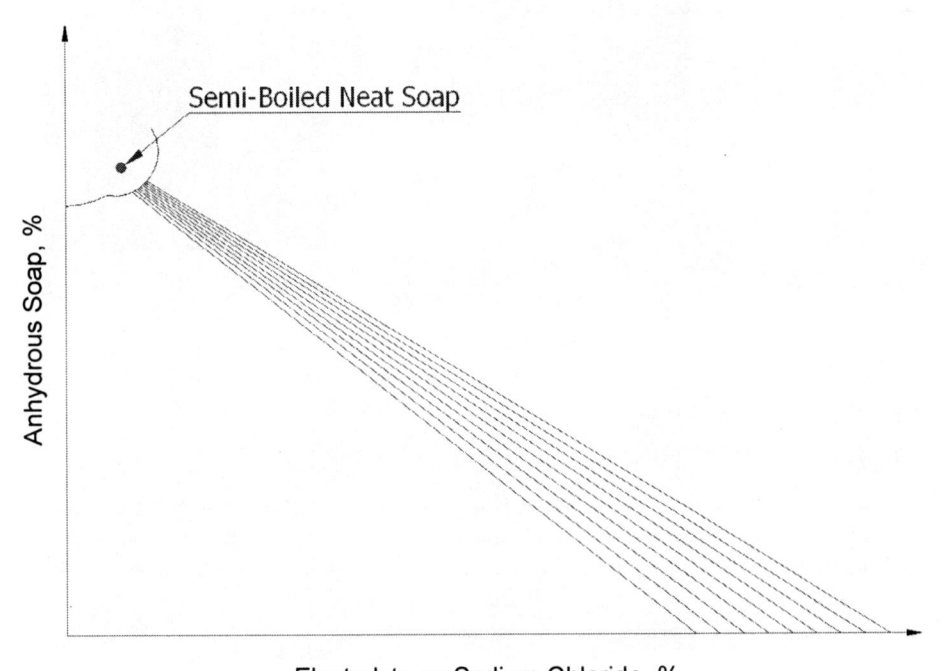

Fig. 9.3. Semi-boiled Neat Soap Phase Region

Crutcher—Mixer/Reactor

Semi-boiled soap production requires a fair amount of mechanical mixing. Open-steam mixing, typical in kettle saponification, would lower the percentage of TFM and introduce the risk of the formation of middle or seat soap.

A crutcher, a very efficient mixer/reactor, was and still is the first choice for Semi-boiled saponification systems. It also serves very well for mixing minor additives and fillers into all types of soap.

The crutcher is a jacketed vessel with a screw-type agitator in a draft tube. The rotating screw induces a vigorous flow of the materials across the vessel, resulting in very good mixing.

The crutcher's versatility is limited by its inability to handle high viscosities: for this reason it is used mostly for atmospheric-pressure applications.

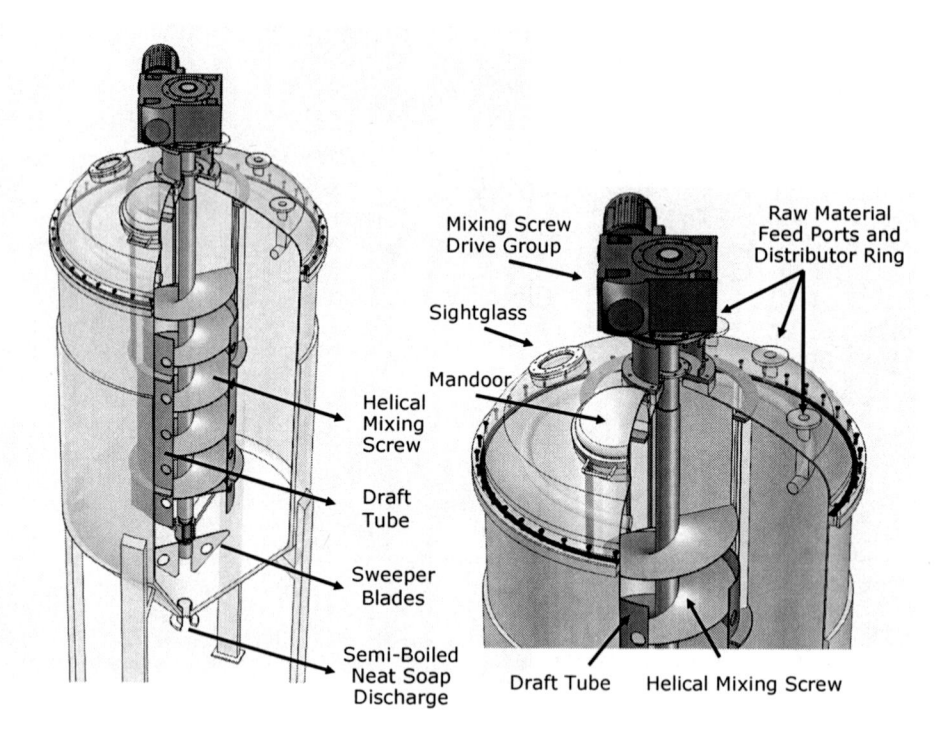

Fig. 9.4. Crutcher.

Atmospheric Systems

Batch Process

All the fats and oils are fed into the crutcher (Fig. 9.4.). Then caustic soda is fed gradually through a distributor ring (on top of the fats and oils) under constant agitation. A rule of thumb is that about three hours are needed for the preparation of a batch of Semi-boiled soap. This time includes the filling of all materials, the reaction time, and the discharge of the product.

A detailed sample for the preparation of one batch follows:

- Feed the fats and oils into the crutcher, and heat to 70°C by heating the water in the crutcher's jacket to 85°C: 30–45 minutes. When this temperature is reached, two-thirds of the caustic soda is added slowly at room temperature and at a predetermined strength.

- Keep the batch under constant agitation because the mixture will begin to thicken, and then add the rest of the caustic soda until the saponification reaction is completed in 90 to 120 minutes.
- Analyze for free alkalinity to determine if more caustic soda or fats/oils are needed. Adjust if required: 10 to 20 minutes.
- Discharge the finished Semi-boiled neat soap to storage: 20 to 30 minutes.

Total time is from 150–215 minutes.

Semi-continuous Process

The most commonly used Semi-boiled soap-production system consists of two crutchers. This plant consists of two batch systems (two crutchers) working in tandem to allow for the continuous operation of a downstream drying system. While in one crutcher a batch is saponified, the other crutcher is discharging an already finished batch of Semi-boiled soap.

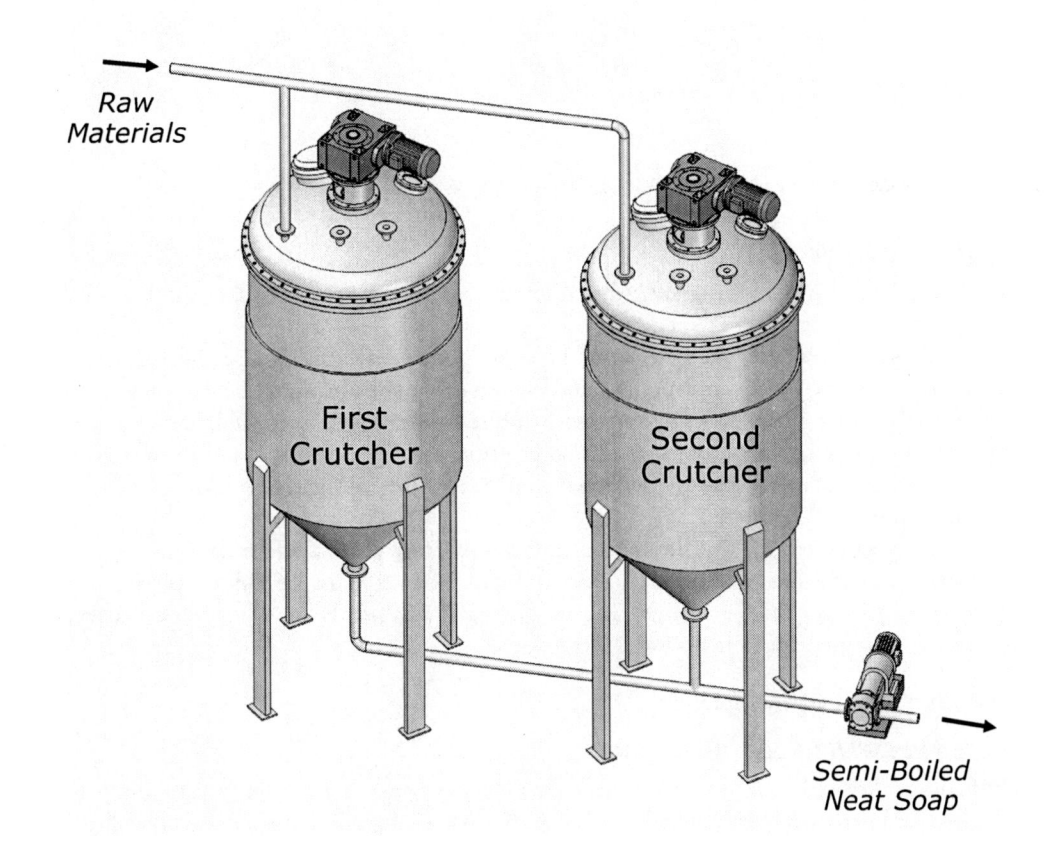

Fig. 9.5. Atmospheric Semi-continuous System.

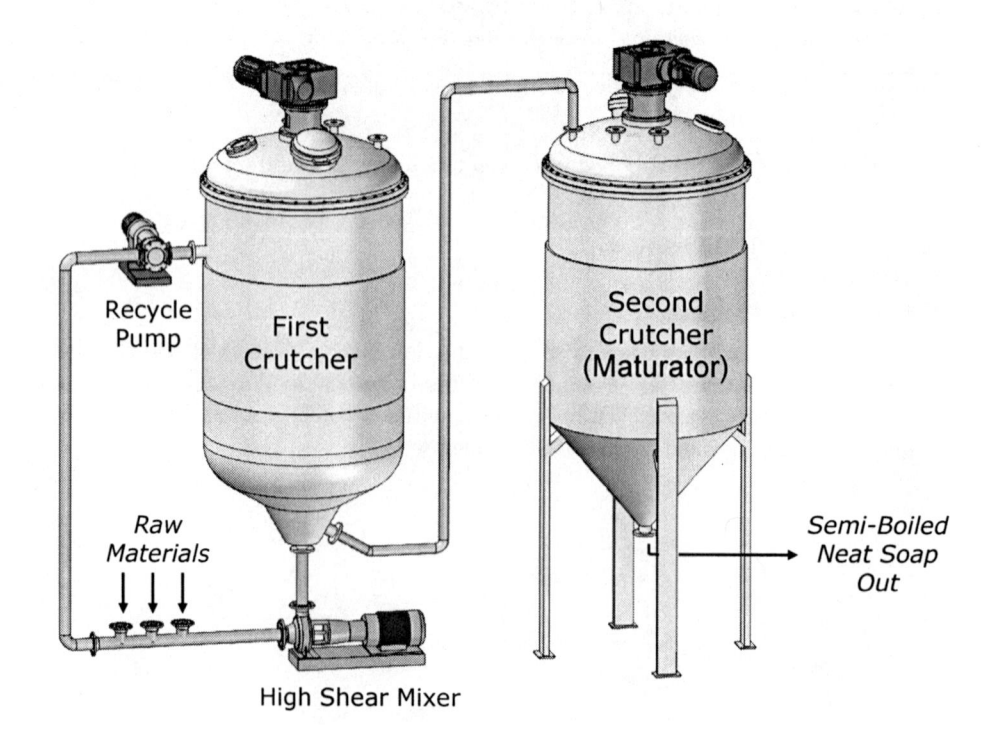

Fig. 9.6. Atmospheric Continuous System.

Continuous Process

The addition of a high-shear mixer (turbodisperser) and a recycle pump to the crutcher converts a batch system into a fully continuous process.

The high-shear mixer assures very good contact between all the reagents. Further vigorous mixing by the crutcher (combined with the enhanced action of the recycle pump) speeds up and completes the reaction. Also, since a reasonably small quantity of fresh reagent is injected in a stream of already formed soap, no separation occurs so that fats/oils and caustic remain in contact for the time necessary.

A nominal production rate is discharged continuously by the incoming fresh raw materials through a "siphon" pipe.

In many cases, especially when no intermediate storage occurs before drying, the first crutcher is followed by another one, called the "maturator," which serves to assure the completion of the reaction by allowing a longer residence time. The maturator is also handy for the addition of various minor ingredients that are used in many soap formulations.

Pressurized Systems

Stirred Vessel Reactor (Autoclave)

Stirred-vessel reactors, also called autoclaves, were developed for continuous saponification systems of full-boiled neat soaps, and only recently have been used for the Semi-boiled saponification of toilet soaps.

Figure 9.7 shows a typical stirred-vessel reactor inside a continuous saponification plant (full-boiled type).

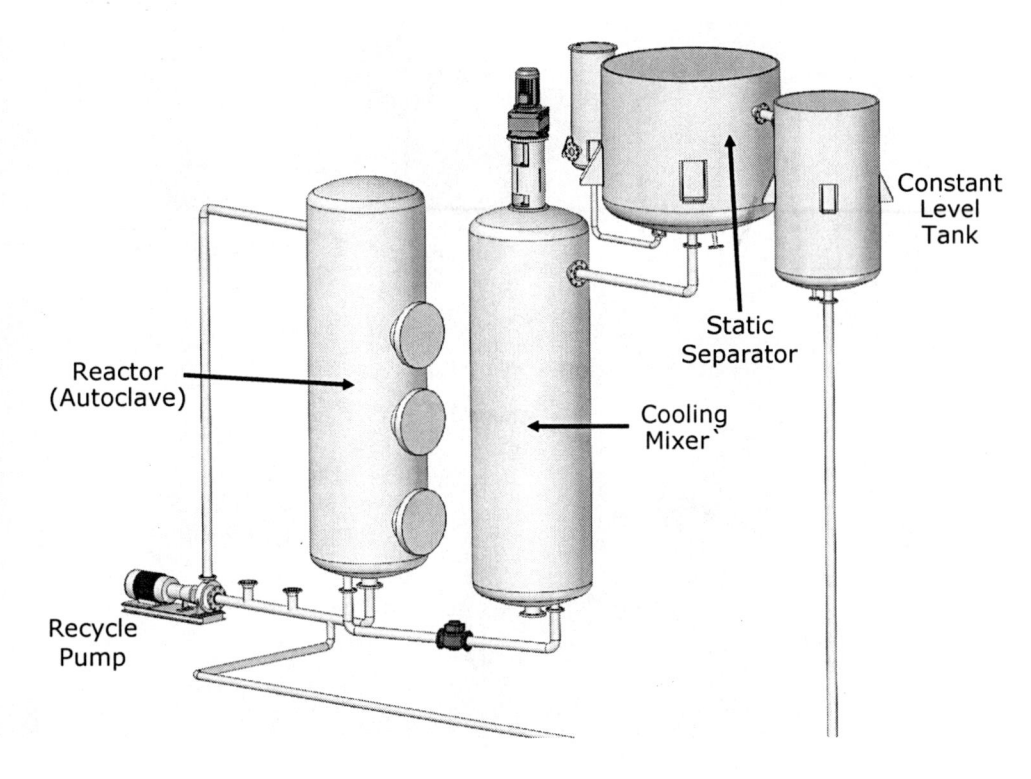

Fig. 9.7. Stirred Vessel Reactor (Autoclave) Pressurized System.

Vertical- or horizontal-type stirred-vessel reactors come with internal baffles, tubes, or mixing blades. These reactors are used in combination with a centrifugal recycle pump.

All previously described effects take place here as well. The good mixing action of the recycle pump and the preformed soap acting as an emulsifier will, given enough residence time, result in a complete reaction between the reagents. Important to notice is that the newly formed soap is not simply displaced by incoming raw materials, but has to go through a pressure-control valve. This allows a degree of control over pressure and temperature inside the reactor which, in turn, allows one to control the speed of the reaction and the viscosity of the finished neat soap.

These reactors are used for the production of neat soaps of regular viscosity: their design and the use of a centrifugal pump for recirculation are not well-suited for handling concentrated materials.

High Shear Mixer/Reactor

The high-shear mixer/reactor has an internal multiblade mixer shaft with large contact surfaces which assure quick and complete saponification/neutralization (Fig. 9.8).

It can be offered with or without a recycle loop. The system with a recycle loop is more efficient, more flexible, and can process materials of higher viscosity.

These units can handle both regular and semiconcentrated neat soaps.

This type of reactor utilizes the thixotropic characteristic of soap to its advantage by a constant, intense mixing action, allowing one to obtain high TFM content neat soap that would otherwise be too viscous to produce.

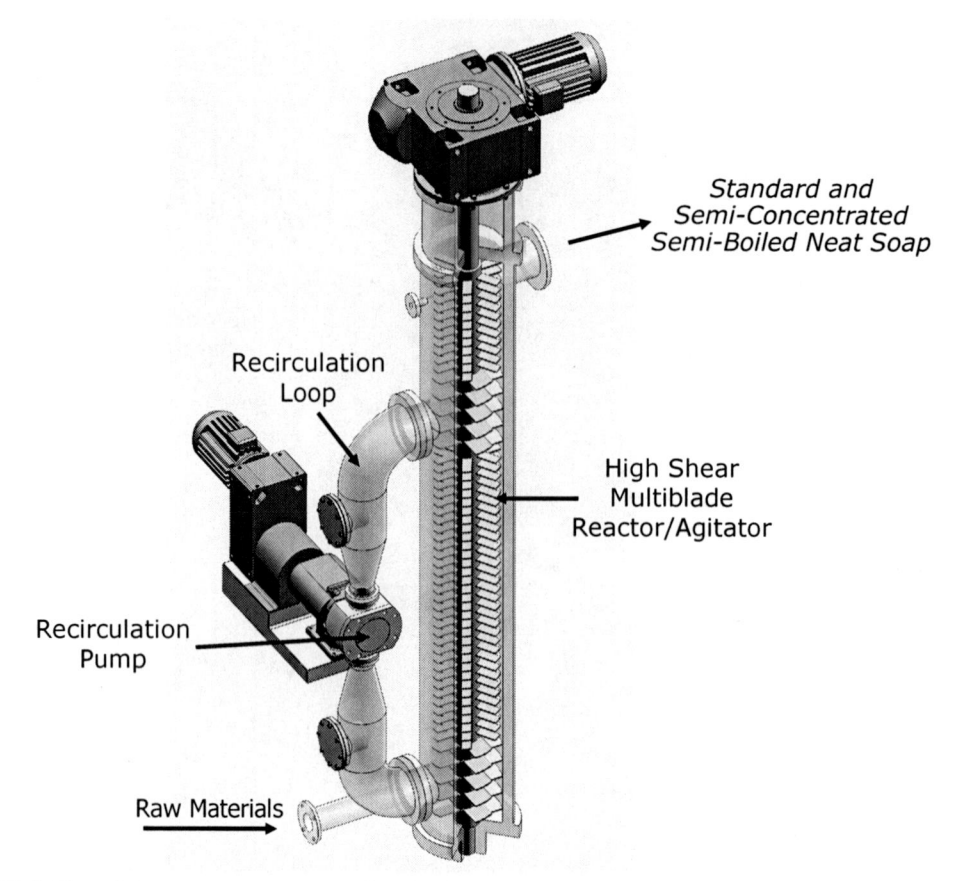

Fig. 9.8. High Shear Mixer-reactor Pressurized System.

Semiconcentrated Saponification Systems

Various pressurized systems exist that are designed to produce semiconcentrated soaps up to 70–72% of TFM content by using concentrated caustic soda.

The efficiency of the reaction is greatly enhanced by a higher concentration of reagents, allowing for shorter residence times and/or resulting in very complete saponification. Furthermore, this extra efficiency of reaction, when combined with longer residence times, allows for the successful processing of a wide variety of raw materials; hence, increased flexibility of operation results.

This implies that commercially available plants are a trade-off between the compactness of the installation (easier cleaning, smaller investment, smaller footprint but a more narrow range of processible raw materials) and residence time (a wider range of processible raw materials but more equipment or a bigger installation, a bigger investment).

Figures 9.9 and 9.10 show two typical commercially offered plants: the compact Mazzoni LB "SSCT" plant with a tubular reactor (Fig. 9.9) and Binacchi's "CHBS" plant (Fig. 9.10) utilizing two high-capacity (high residence time) reactors capable of processing a very wide variety of raw materials.

Combination (Integrated) Saponification/Drying Plants

The most interesting development of the semiconcentrated system is a combination saponification/drying plant where the saponification plant is directly coupled to the spray chamber of the vacuum spray-dyer atomizer without the use of an intermediate storage or feed tank and a heat exchanger (Fig. 9.11).

Semi-Boiled Soap Production Systems • 257

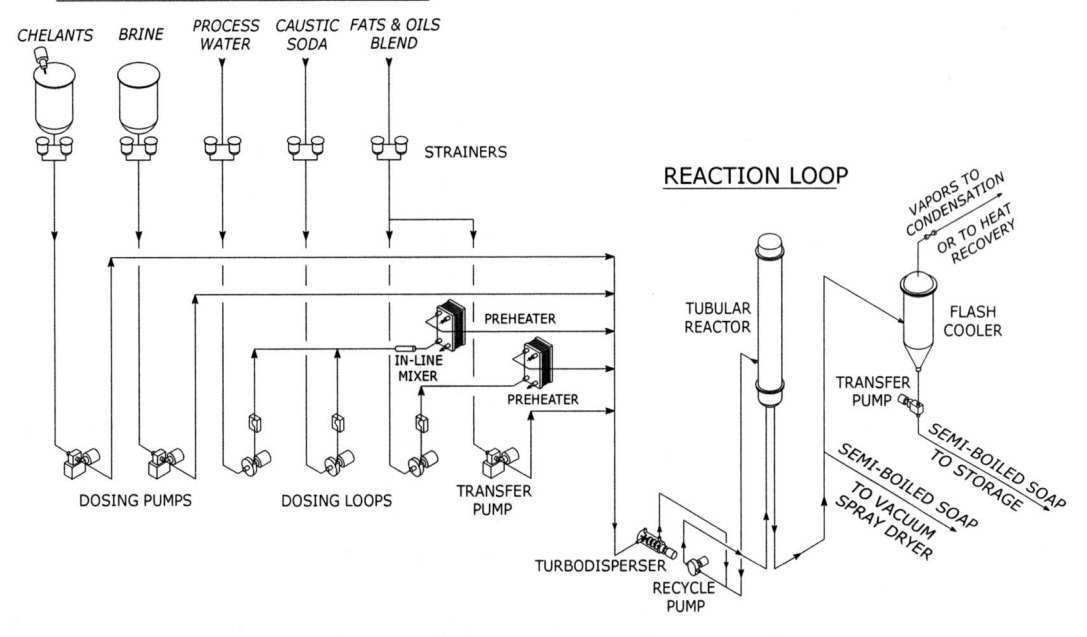

Fig. 9.9. Mazzoni LB "SSCT" Semi-boiled Soap Production Plant. Courtesy of Mazzoni LB, S.p.A.

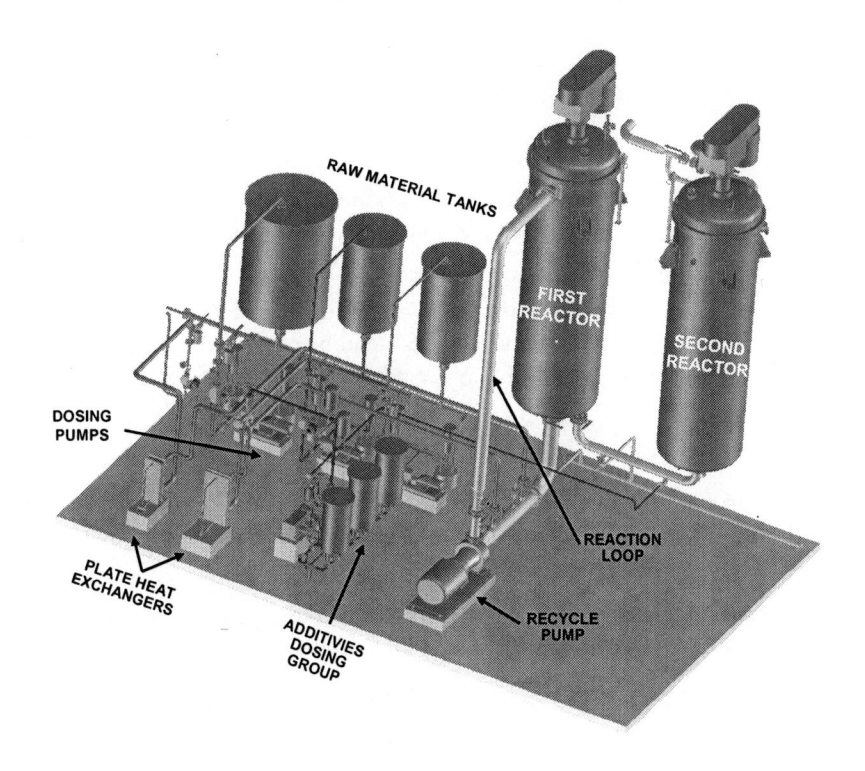

Fig. 9.10. Binacchi "CHBS" Semi-boiled Soap Production Plant. Courtesy of Binacchi & Co.

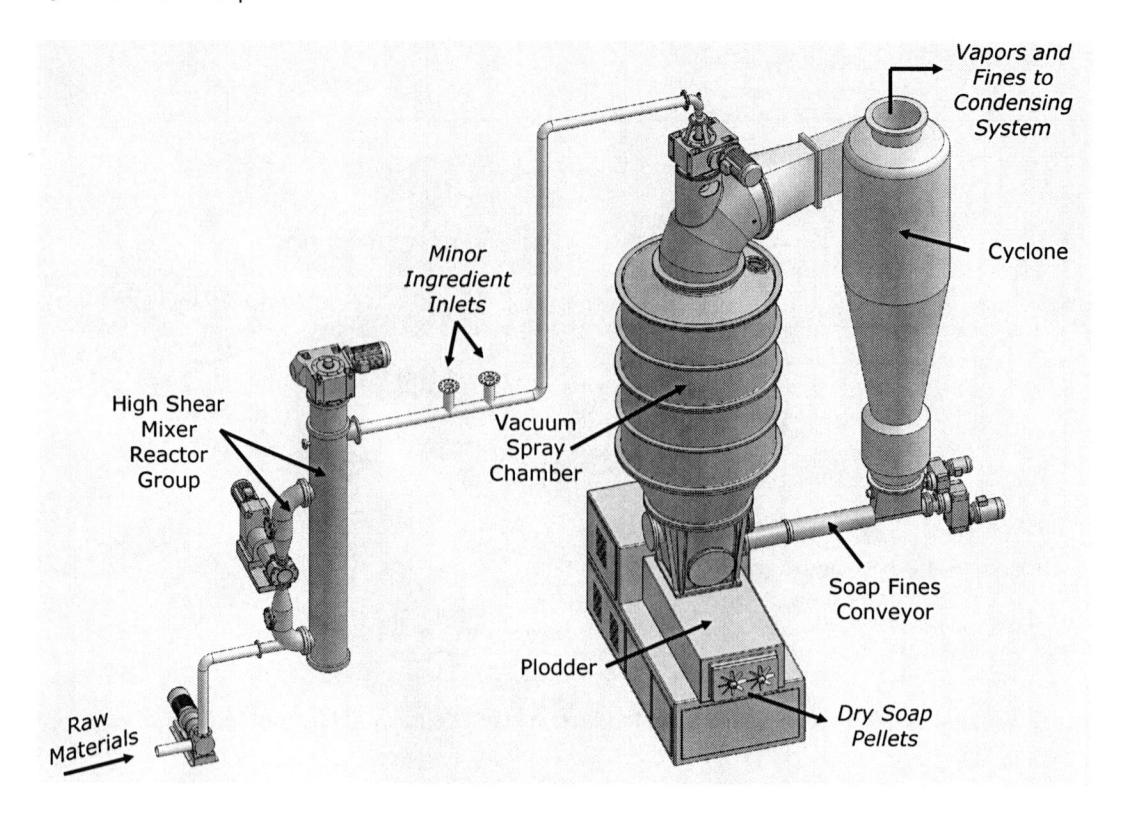

Fig. 9.11. Combination Semi-concentrated Saponification and Drying Plant.

Apart from obvious investment savings due to the reduced number of equipment pieces involved (at least a couple of pumps, a filter, a service tank, and a heat exchanger), this kind of setup delivers a series of further advantages:

- Exothermic nature of reaction is used to heat the soap without the need for external energy sources. The hot neat soap produced is sprayed directly into the vacuum-spray chamber of the dryer where the exact amount of water is flashed off to obtain the desired final moisture content of the dried soap pellets.

- Steam-saving of up to 60% with respect to the traditional plants.

- Reduced fines formation due to the reduced quantity of moisture to be removed (single liquid phase flow through the system). This permits one to utilize only one soap-fines recovery cyclone for medium-size plants.

- No intermediate storage tank and no heat exchanger eliminate the exposure of the soap to air and local overheating, resulting in improved quality of the final product.

Figures 9.12 and 9.13 show two typical commercially available combination (integrated) plants.

Semi-Boiled Soap Production Systems • 259

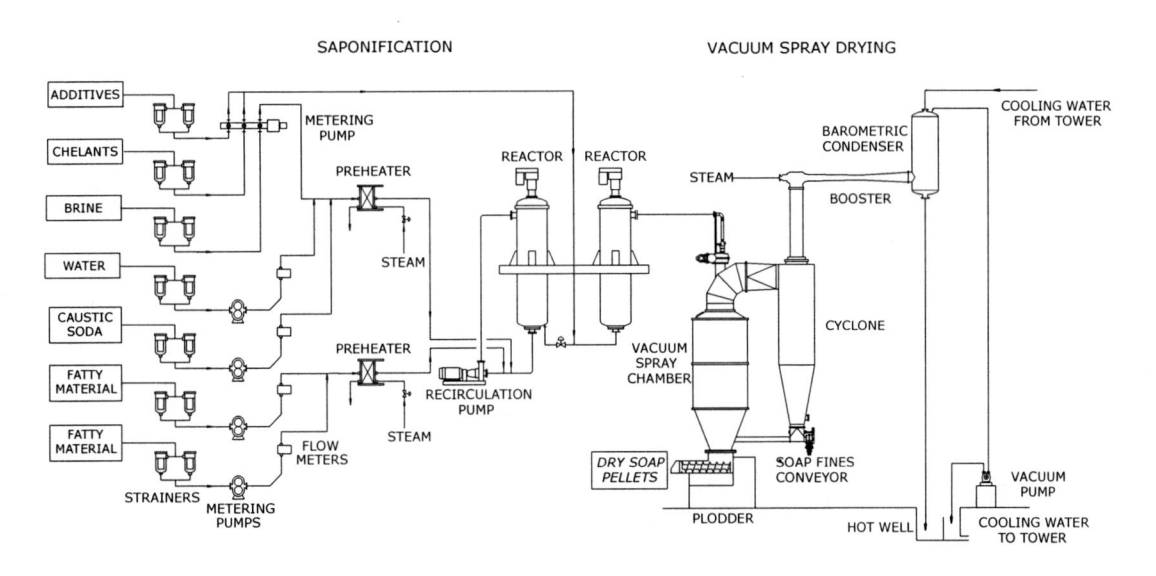

Fig. 9.12. Binacchi "CHBS-VSD" Combination Semi-concentrated Saponification and Drying Plant. Courtesy of Binacchi & Co.

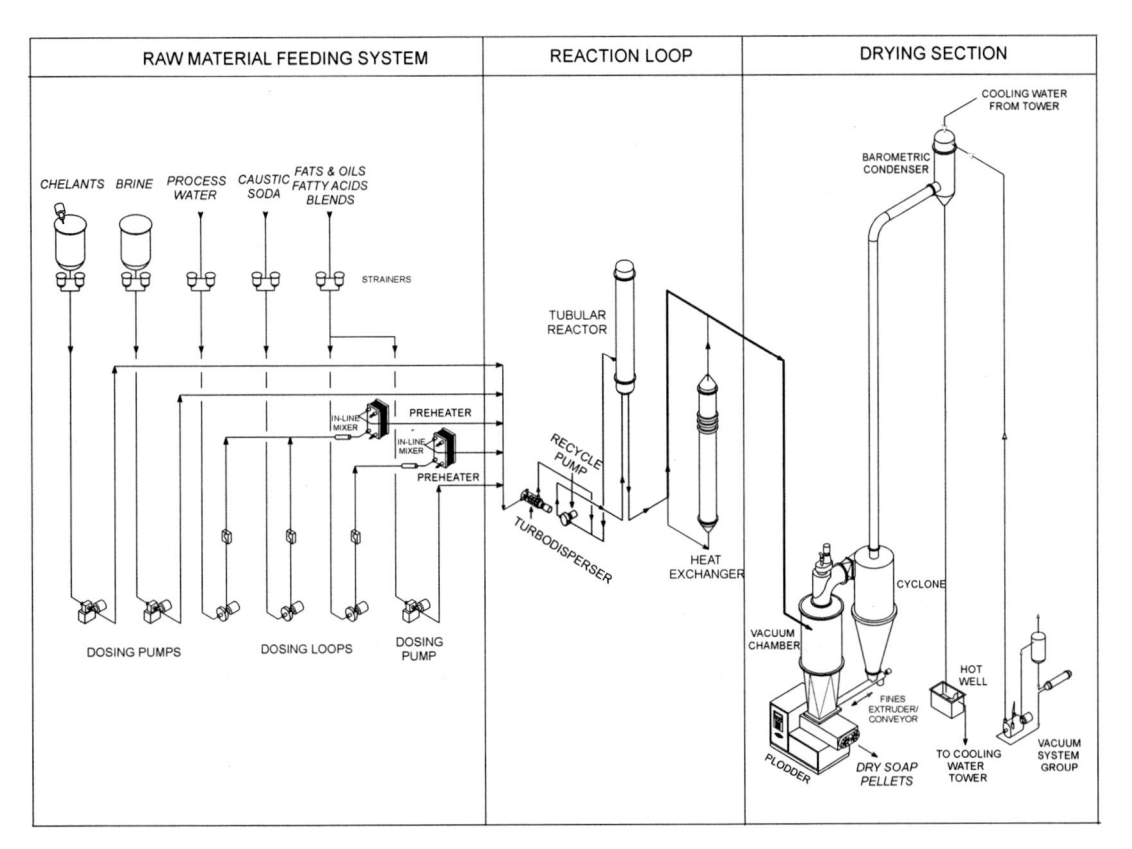

Fig. 9.13. Mazzoni LB "SSCT-C" Semi-concentrated Combination Saponification and Drying Plant. Courtesy of Mazzoni LB, SpA.

An Atmospheric- and a Pressurized-system Comparison

Atmospheric Batch Systems

These were the first systems, and are still widely used due to their operational simplicity. One does not need to bother with pressure control.

Mixing is continued until the reaction is completed. The soap is sampled periodically by the laboratory, and minor adjustments are made based on the analysis and the desired final-product specifications.

Atmospheric Continuous Systems

The atmospheric continuous systems are more complex, mostly due to the importance of avoiding incorporation of air during mixing. A vent valve on the crutcher eliminates the air during the filling stage and the produced soap leaves the crutcher through the "siphon" pipe, ensuring that the vessel is always full.

Keeping the vessel full at all times requires good temperature control of the raw materials. The newly formed soap inside the crutcher must not boil; its temperature must be kept under 100°C.

The residence time given to the soap is fixed in the function of a production rate. If a problem occurs with the efficiency of saponification and the temperature cannot be increased, the residence time can, but this will result in a decreased production rate.

Important to note is that atmospheric systems cannot produce semiconcentrated soap.

Pressurized Systems

Pressurized systems allow the use of increased pressure and temperature, resulting in a faster and a more complete reaction. Increasing temperature and pressure can also fix viscosity problems.

Pressurized systems are more complex and are normally more expensive.

The use of pressurized systems—the new combination (integrated) of saponification and drying plants for semiconcentrated soap production— represents the latest and most exciting advance in soap-processing technology in a long time.

From the above descriptions, the main characteristics of the atmospheric and pressurized systems are summarized as:

Atmospheric Systems

- Low investment cost (batch version)
- Ease of use and maintenance
- Best suited for batch-type production.

Pressurized Systems

- Require tighter control
- Very flexible and efficient
- Best suited for continuous production.

Continuous Saponification Plant Use for Semi-boiled Soap Production

The main components of the widely used, continuous, neutral fats-and-oils saponification process for the production of 62–63% of TFM neat soap are shown in Fig. 9.14. For more details on the continuous saponification plants, please refer to Chapter 8.

Semi-Boiled Soap Production Systems ● 261

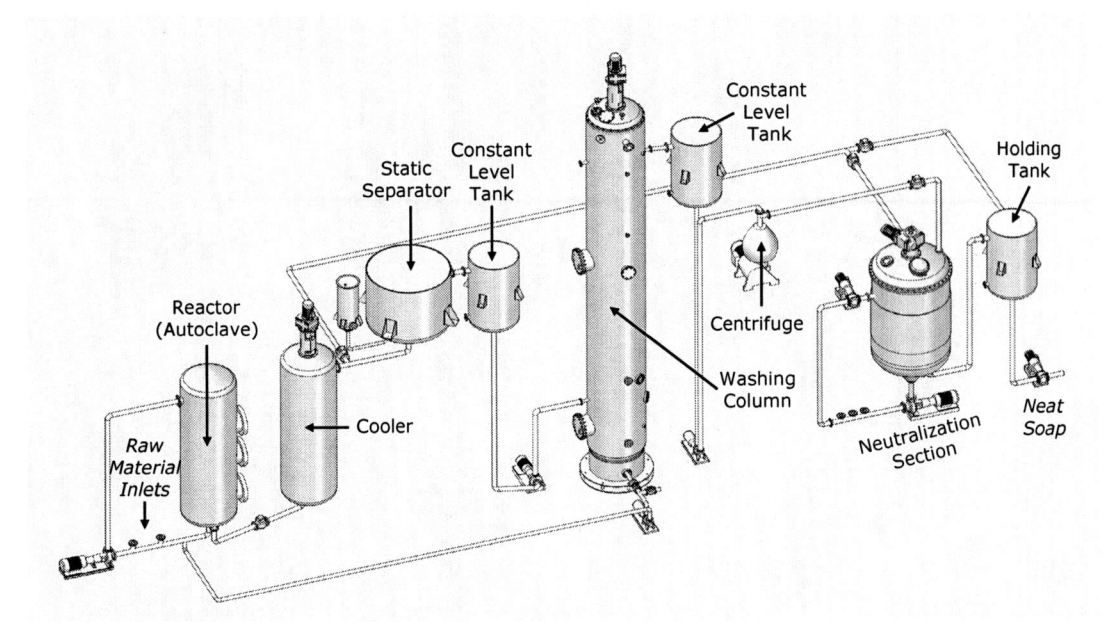

Fig. 9.14. Continuous Neutral Fats and Oils Saponification Plant.

One can utilize a continuous saponification plant for Semi-boiled soap production by using either the reactor route (pressurized system) or the neutralizer route (atmospheric system).

Reactor Route—Pressurized System

A pressurized system requires the use of the reactor. One can apply any type of horizontal, vertical, stirred, or not stirred reactor that performs well for a full-boiled soap operation for a Semi-boiled soap production.

A bypass must be installed from the cooler to the holding tank from where the finished Semi-boiled soap is transferred by using the existing transfer pump and piping.

Normally, continuous, neutral fats and oils saponification works with a mass balance that avoids pumping the water into the reactor. Therefore, the usual dosing set-up does not have a "water to reactor" pump, or if it does, it is only a small "emergency pump." For Semi-boiled soap production, water is required. A half-spent lye pump has the capacity required, and is therefore used for dosing the water into the reactor.

Alternatively, Semi-boiled neat soap produced in the reactor and passed through the cooler is fed into the neutralizer. This operation allows for a longer residence time.

Also, already installed pipes, valves, transfer pumps, etcetera of the plant can perform the eventual addition of minors.

Neutralizer Route—Atmospheric System

In the atmospheric pressure-type operation, the neutralizer's group components are the key units.

Raw materials are pumped into the recycle loop of the neutralizer, and are reacted there until the completion of saponification.

Two main issues reside with this solution: (i) one needs to repipe the dosing section and bring all the raw materials to the neutralizer (water by using Half Spent Lye (HSL) pump included) and (ii) one will suffer some loss of the production due to the inferior efficiency of the neutralizer (when compared to the reactor).

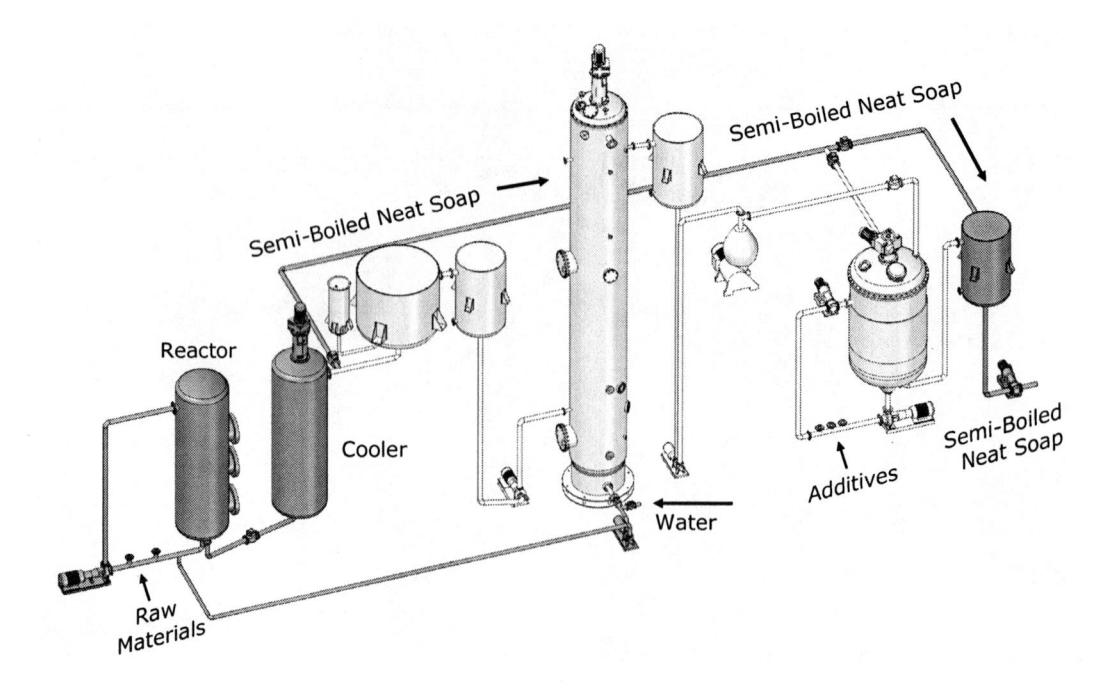

Fig. 9.15. Continuous Saponification Plant Utilization for Semi-boiled Soap Production Using the Reactor, Cooler, and Additional Water.

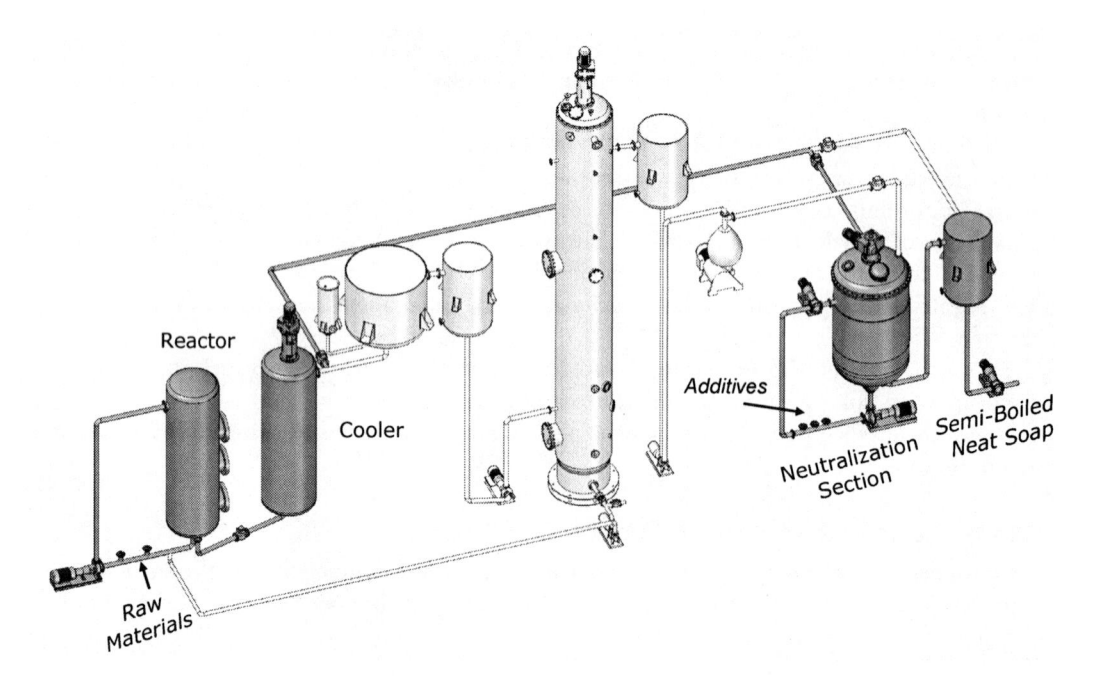

Fig. 9.16. Continuous Saponification Plant Utilization for Semi-boiled Soap Production Using the Reactor, Cooler, and the Neutralization Section.

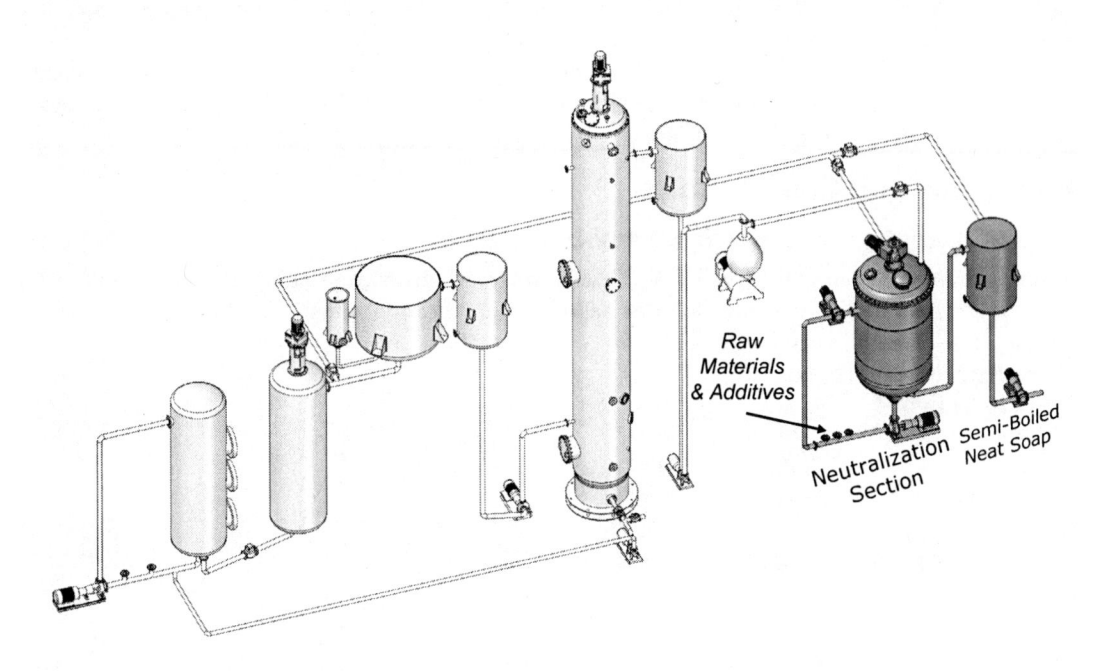

Fig. 9.17. Continuous Saponification Plant for Semi-boiled Soap Production via the Neutralizing Section.

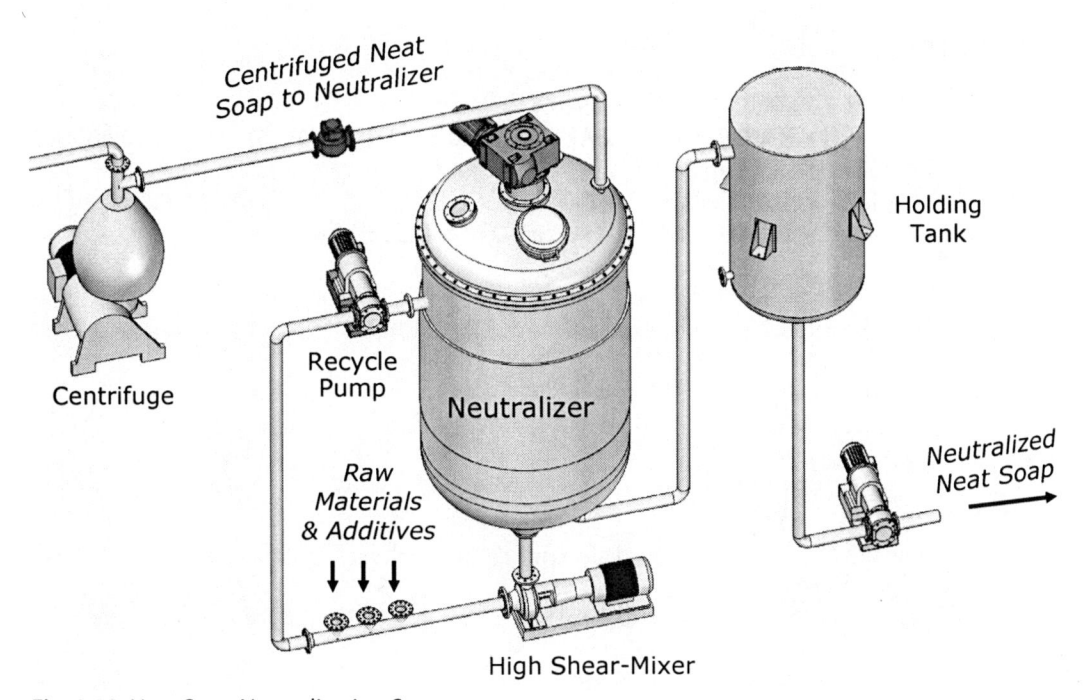

Fig. 9.18. Neat Soap Neutralization System.

Other than that, the system is already piped properly, and no bypasses are to be added. The inlet that receives the neat soap from the centrifuges should be blanketed off. The neutralization section's details are shown in Fig. 9.18.

The comparison with the atmospheric continuous system (Fig. 9.6) shows the similarity between these two systems—crutcher with a recycle loop, recycle pump, high-shear mixer, siphon pipe, and a second vessel to receive the finished product.

Definitions and Calculations

Sample Calculation for a Batch of Semi-boiled Soap

The production of 10 tons of 60% of TFM, 80:20 tallow/coco-blend neat soap with a residual of 0.05% of free NaOH, and of 0.5% of NaCl follows: Tallow—saponification value (SV): 200
Coconut oil— Saponification Value (SV): 250
48% of NaOH (caustic soda)
20% of NaCl (brine)

Desired:	Value	ID Letter
Quantity (kg)	10,000	A
TFM (%)	60	B
Residual NaOH (%)	0.05	C
NaCl (%)	0.5	D
At our disposition:		
Oil blend 80:20 SV	210	E
NaOH% in caustic	48	F
NaCl% in brine	20	G
TFM% in oils	94.63	H
Glycerine% in oils	11	I

Oils required = A × B/H = 10,000 × 60/94.6 = 6340.5 kg (= J)
Caustic soda required = [(J × E × 0.0713) + C × A]/F = [(6340.5 × 210 × 0.0713) + 0.05 × 10,000]/48
= [(94936.3) + 500]/48 = 1988.25 kg (= M)
Brine required = A × D/G = 10,000 × 0.5/20 = 250 kg (= K)
Water required = A-J-M-K = 10,000 - 6340.5 -1988.25 -250 = 1421.25 kg (= L)

The final formula is:
6340.5 kg of oil blend + 1988.25 kg of caustic soda + 250 kg of brine + 1421.25 kg of water.

The soap produced will contain:
Anhydrous soap.................. 65.22%
Glycerine 7.28%
Water 26.95%
Virtual TFM 63.64%
(The real TFM value is 60%, but the increased glycerine content will make this soap behave as if it had 63.64% of TFM, thus "Virtual TFM.")
Residual NaOH and NaCl should be as per-desired specs, that is, 0.05% and 0.5%, respectively.

All the calculations must be controlled via analysis.
When comparing the analysis with calculated results, one can make the necessary corrections.

%TFM	
Real	Virtual
56	59.40
57	60.46
58	61.52
59	62.58
60	63.64
61	64.70
62	65.76
63	66.82

Valid for 80:20 tallow/coconut mixture with 200 SV for tallow and 250 SV for coconut.

Glycerine Content Finder Diagram

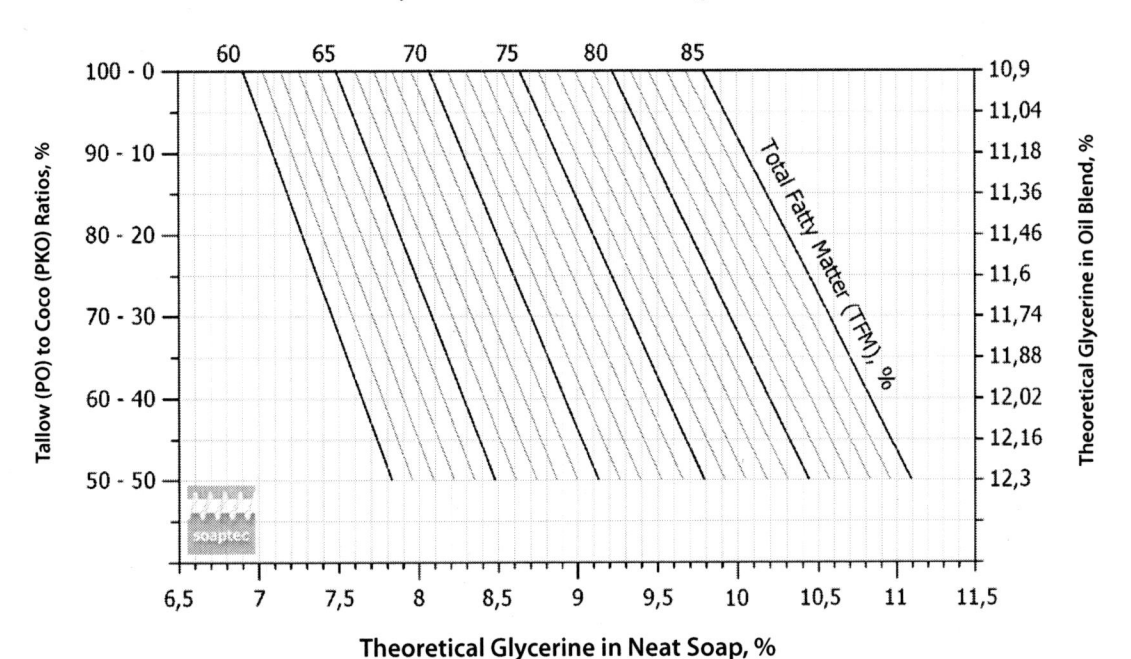

Note: Calculated for Tallow (PO) with SV of 250 and Coco (PKO) with SC of 250

Fig. 9.19. Glycerine Content Finder Diagram.

Soap Drying Systems

Luis Spitz[1] and Roberto Ferrari[2]

[1]L. Spitz, Inc., Highland Park, Illinois, USA; [2]Mazzoni LB. S.p.A.,Busto Arsizio, Italy

Soap Drying Systems

Drying is a very important processing step in soap manufacturing. The drying system used affects the physical properties of the dried soap (details are given in Chapter 3). Optimum final moisture content is a critical variable for obtaining the best finished bar performance and finishing line productivity. Liquid neat soap with water content ranging from 28–34% can be dried to several different grades in terms of moisture content (MC):

- Toilet soap pellets (noodles): normally 12–15% MC
- Special soap pellets (translucent or multipurpose): 18–22% MC
- Laundry soap pellets/bars: 22–30%MC

Vacuum Spray Drying

Vacuum spray drying, patented by Giuseppe Mazzoni, was introduced for industrial applications following World War II. The use of vacuum as a medium to simultaneously dry and cool the liquid neat soap is responsible for the beginning of the modernization of soap manufacturing.

Plants using this system were designed to convert liquid neat soap into dry toilet soap pellets and could also be equipped to produce continuously extruded laundry soap bars. The elimination of the laborious and costly laundry soap framing process used until the advent of power detergents was a revolutionary step in soap manufacturing.

Vacuum spray drying is still is the most widely used process for toilet and laundry soap applications. It is a very flexible and simple-to-operate system for the production of various types of soaps at different moisture levels.

Other Drying Systems

Hot-air Cabinet Dryers

Before the invention of vacuum spray drying, hot-air cabinet dryers were used. Neat soap was fed onto a steam-heated chill roll and then onto a steel belt of a long and large cabinet in which hot air completed the drying of the soap in flake form.

Hot-air Spray Towers

In a few factories, hot-air spraying towers similar to detergent-production plants are used for soap drying, but application of this type of system is limited by high energy requirements, large space requirements, and, last but not least, the high temperature and the physical characteristics of the dry soap obtained this way.

Expansion Dryers

After the Mazzoni vacuum spray drying process, only a few soap drying systems were introduced. The Tubular Drying System appeared in 1955 (Mazzoni, 1960), and the Parkson Atmospheric Dryer with a Plate Evaporator and a Chill Roll in 1966 (Palmason, 1963). Alfa-Laval also offered a similar system. Only a few of these dryers were installed.

In 1960, Miag developed, in collaboration with the 4711 Soap factory in Cologne, Germany, a "Double Expansion Drying" system and patented it in 1964 (Miag GmbH, 1964).

The tubular, plate, and double expansion dryers did not gain acceptance.

Currently, Sela GmbH offers two systems similar to the no longer offered Miag plants, a double-stage atmospheric system for drying toilet soap pellets (Fig. 10.1) and a single-stage system for drying and extruding high-moisture laundry soaps (Fig. 10.2).

Chill Rolls and Flakers

The combination of a chill roll and a flaker is currently regaining interest for synthetic and combo soap base production. This operation, which is not strictly a drying operation, but rather a simple solidification of a molten mass, is beyond the scope of this chapter.

Toilet and Laundry Soap Vacuum Spray Dryers

During recent years, most of the effort to optimize vacuum spray drying plants has focused on reducing energy requirements, improving pollution control, and the design of large industrial units capable of processing soap pellets or laundry bars at up to 10–12 tons/h.

An interesting new development is the so-called "integrated" plant. This consists of a continuous saponification/neutralization unit with direct processing of the neat soap into the dryer without any storage tanks in between. This design is shown in Fig. 10.8.

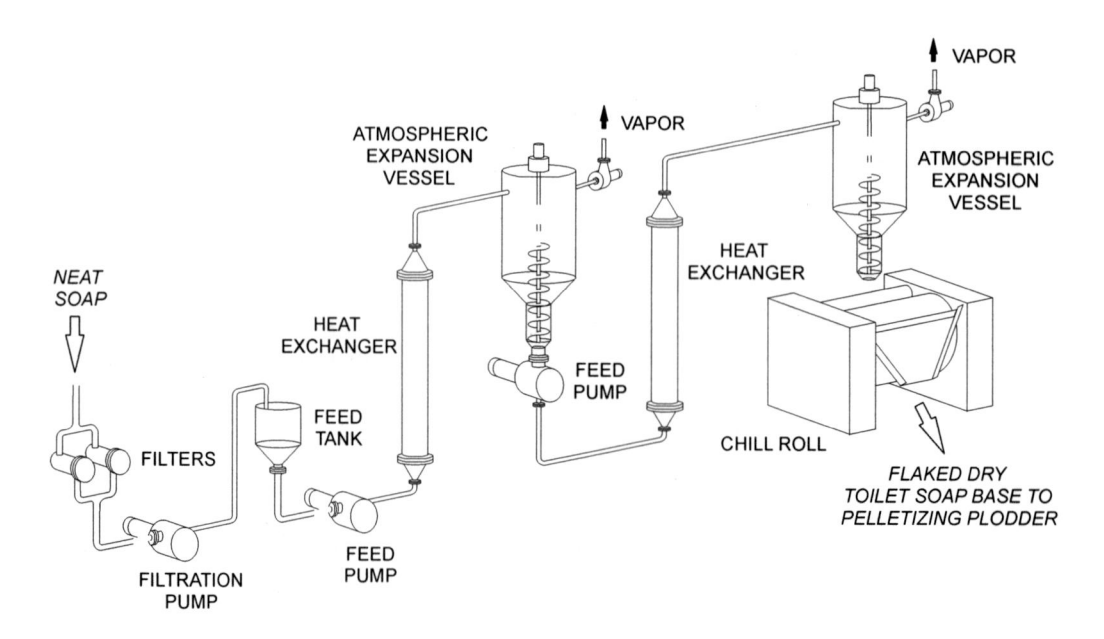

Fig. 10.1. SELA "TAAF" Atmospheric Double-stage Toilet Soap Drying Plant.

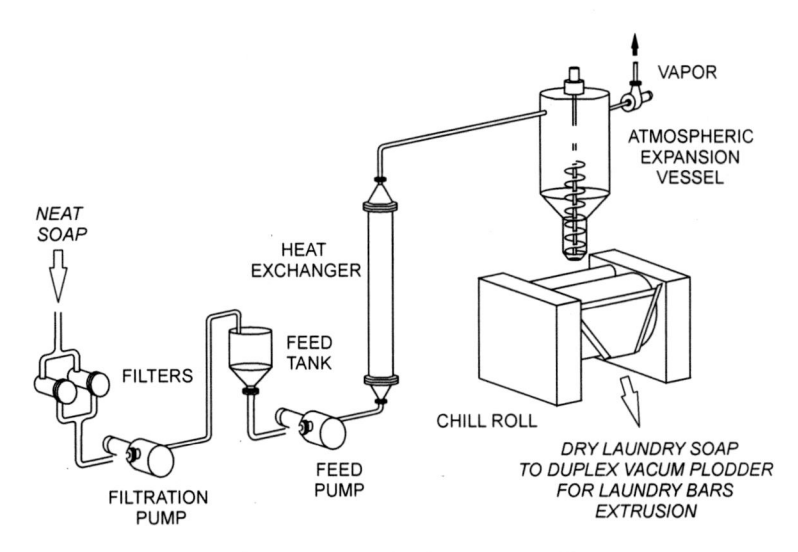

Fig. 10.2. SELA "TAH" Atmospheric Single-stage Laundry Soap Drying Plant.

Toilet Soap Dryers

The following table compares the processing steps for a traditional toilet soap dryer and the new, updated systems in current use.

Operating Step	Traditional	Current
1	Neat soap feeding	Neat soap feeding
2	Preheating and pre-evaporation	Preheating and pre-evaporation
3	–	Vapor separation
4	Vacuum drying	Vacuum drying

For Step 3 a vapor–liquid separator or vapor eliminator (hereinafter called "VLS") is used. Step 4 can be done in a conventional vacuum chamber with a rotating nozzle and scrapers, or in the new Mazzoni LB vacuum chamber with a stationary inlet nozzle.

The VLS concept was first patented in 1984 by SELA GmbH, and the first system was installed in 1987. The application of the VLS system with new dryers and easy retrofitting into existing plants has gained acceptance during the last decade. The main advantages of using a VLS system are as follows:

- Increased production capacity
- Utilities-saving and energy recovery options
- Reduced quantity of soap fines
- Improved vacuum level

The key processing variables that affect and control vacuum spray drying of liquid toilet soaps and laundry soaps are illustrated in detail in Fig. 10.3 for 78–80% TFM toilet soap drying. The variables shown are steam pressure, soap pressure and temperature, spray nozzle pressure, operating vacuum, and dry soap temperature leaving the system. The heating and evaporating phase distribution is also shown.

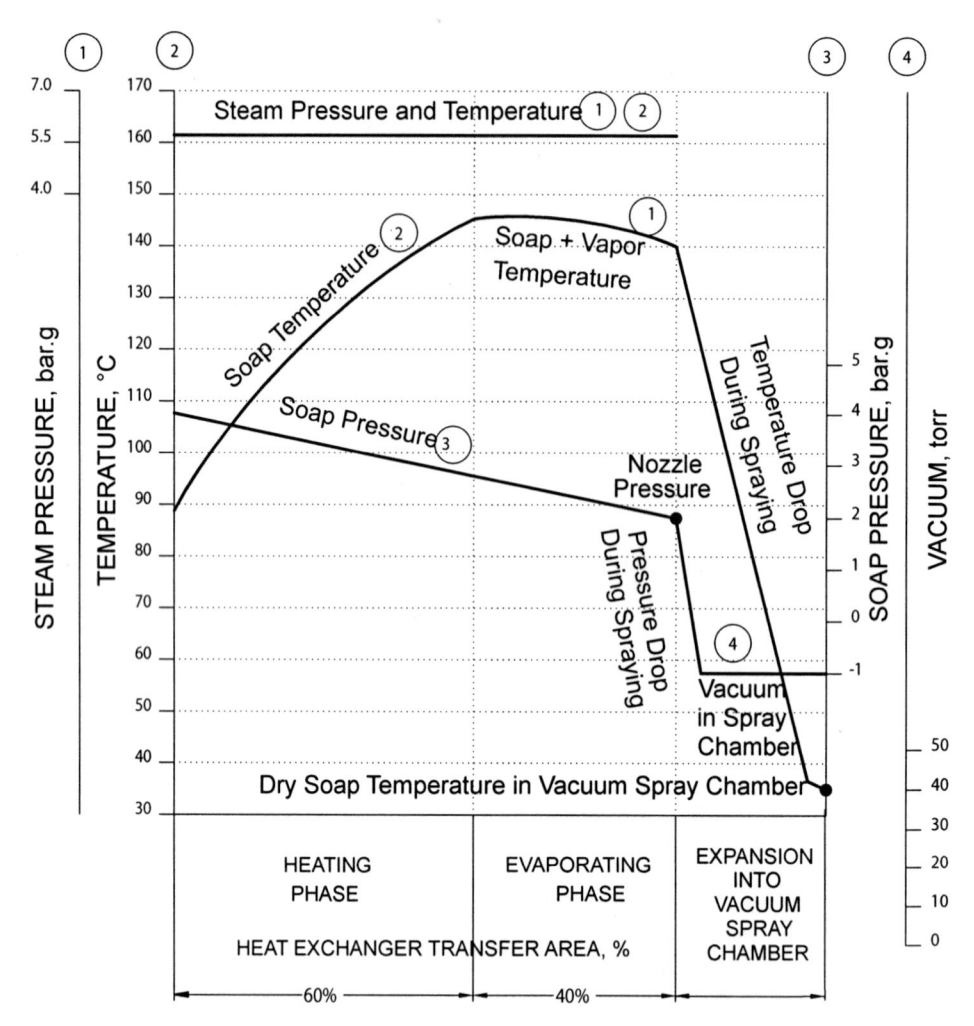

Fig. 10.3. Drying system for 78–80% RFM toilet soaps.

The function of toilet soap vacuum spray dryers is to convert liquid neat soap into dry soap pellets (noodles) by removing moisture.

To evaporate the necessary amount of water to reach the desired final moisture, the neat soap/water mixture has to be heated to its boiling point. Vacuum is used as a medium to obtain soap drying and cooling simultaneously.

Standard, superfatted, and translucent bases with different moisture levels can be produced with ease by simply changing operating conditions.

The preheated soap and vapor mixture coming from the heat exchanger is normally first sent to a VLS, which consists of a static vessel where pre-evaporated water vapor is separated by gravity from hot liquid soap under controlled temperature and pressure (Fig. 10.4).

The normal operating pressure for the VLS is 1.4–1.7 bar which is also the optimal back pressure fro the heat exchanger operation and the vaper-liquid separation. In some VLS designs a screw type agitator is used to facilitate pushing the liquid neat soap into the pump.

Soap Drying Systems • 271

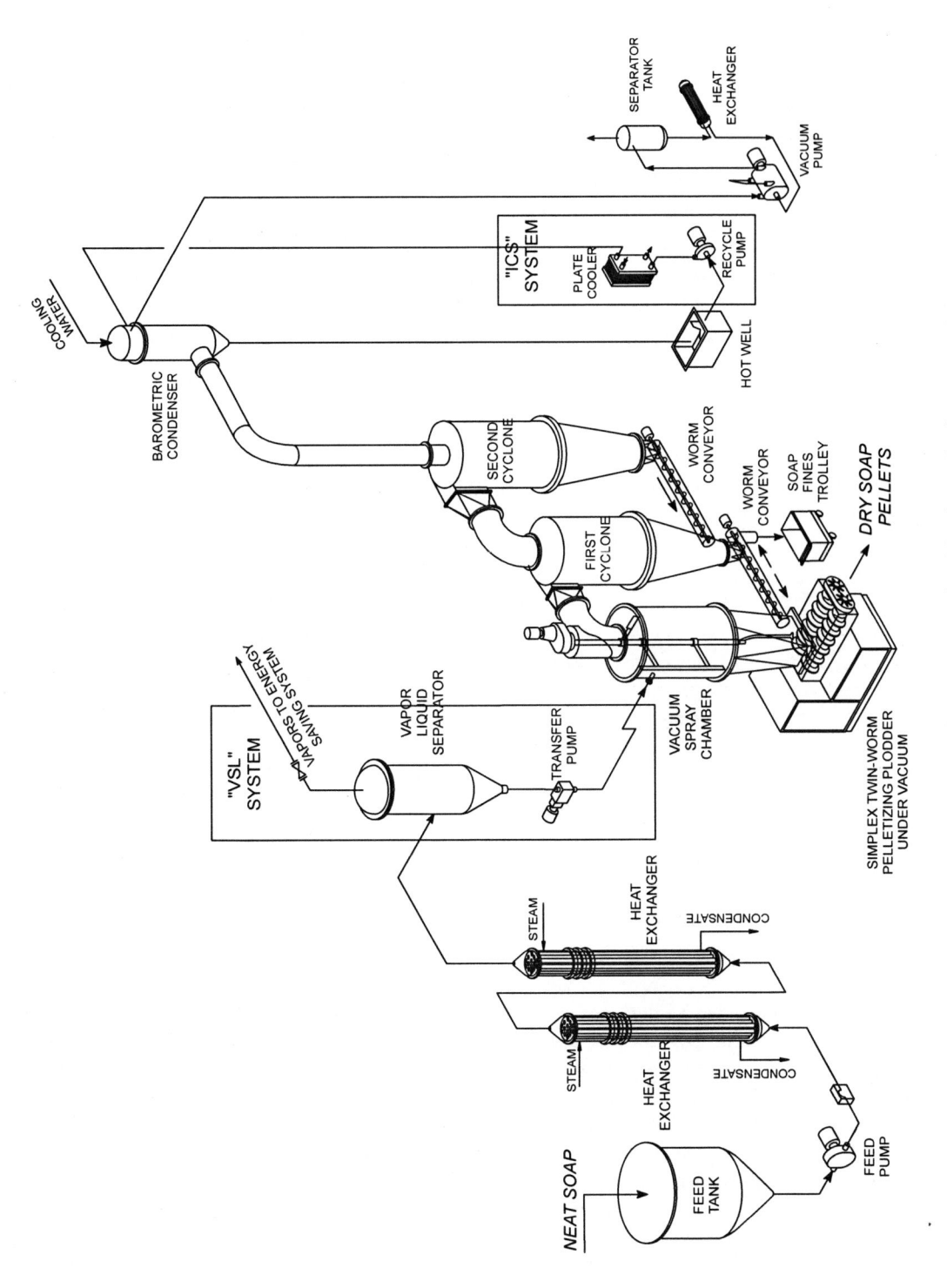

Fig. 10.4. Toilet Soap Vacuum Spray Dryer with "VLS" Vapor Liquid Separator and "ICS" Indirect Cooling Water System.

The vapors from the VLS can be used for various energy conservation purposes, as discussed separately.

The soap from the bottom of the VLS is pumped under level or flow control to a vacuum chamber. There are two types:

- Rotating spray nozzle (conventional design): This is equipped with rotating scrapers and a spray nozzle assembly.

- Stationary inlet nozzle ("no-spray"): This new Mazzoni LB vacuum chamber is equipped with a tangential stationary soap inlet nozzle, rotating scraper, and anti-entrainment internal baffle.

As the vapors are flashed off because of the vacuum, the soap dries to its final moisture. The rotating scrapers remove the dried and cooled soap, in flake form, from the chamber walls. The unevenly shaped dry soap flakes fall onto plodders to be pelletized.

During spraying, a certain amount of soap fines are formed and are carried off with the high-velocity vapors into cyclone-type fines recovery systems before being condensed in a condenser. A vacuum pump, steam ejector, or combination of the two maintains the vacuum and removes the noncondensables.

Laundry Soap Dryers

Fewer innovations have been applied to laundry soap dryer designs compared to toilet soap dryers. The VLS system cannot be employed for laundry soaps.

The processing steps for laundry soap dryers are as follows:

Operating Step	Pure Laundry Soap	Filled Laundry Soap with Solid Fillers	Filled Laundry Soap with Liquid Fillers
1	Crutching (solids mixing and/or other additives)		
2	Neat soap feeding		
3	Preheating only for higher than 68% TFM soaps	–	Preheating only for higher than 68% TFM soaps
4	–	Liquid filler injection (if applicable)	Liquid filler injection
5	Spraying into vacuum chamber		
6	Extruding finished laundry bars or pelletized laundry soap base		

Perfume and color are added by dosing systems in the plodders, where the soap is already compact, in order to minimize perfume losses in the vacuum.

The principles of vacuum spray drying of the various types of laundry soap products into a dry base in bar form at various moisture levels are shown in Fig. 10.5 for 68–70% TFM laundry soaps and Fig. 10.6 for lower than 68% TFM laundry soaps.

The figures show how the vacuum spray drying process is affected and controlled by key variables, such as steam and soap pressure and temperature, spray nozzle pressure, operating vacuum, dry soap temperature leaving the system, and the heating and evaporating phase distribution.

The function of laundry soap vacuum spray dryers is to convert liquid neat soap into dry soap bars with various additives and fillers by removing moisture. Vacuum is used to obtain soap drying and cooling simultaneously.

Pure unfilled, filled, and other type of bars of different moisture levels can be produced with ease by simply changing operating conditions.

Preheating is required only to produce soap bars with higher than 68% TFM content, starting from a regular 62% TFM neat soap. Simple flashing-off under vacuum drying without any preheating allows a final product with 67–68% TFM content to be reached. In any case the heat exchangers are used only to increase the soap temperature without pre-evaporating water; this makes the application of the VLS useless for laundry soap dryers.

If preheating is needed for liquid fillers, an inline mixer must be used after the heat exchanger. The preheated soap is directly sent to the vacuum chamber, which can be of the same type used for toilet soaps.

As the vapors are flashed off because of the vacuum, the soap dries to its final moisture. The rotating scrapers remove the dried and cooled soap, in flake form, from the chamber walls. The unevenly shaped dry soap flakes fall onto a duplex plodder to be pelletized in the first stage plodder and then extruded in the form of a continuous slug (billet) from the second (last) stage plodder. The extruded slug can be directly engraved, cut, and packed, or directly sent to a laundry soap finishing line that has a soap press. Laundry soap pellets can also be produced for sale.

During spraying, a small amount of soap fines are formed, which are carried off with the high-velocity vapors into cyclone-type fines recovery systems before being condensed in a condenser. A vacuum pump, steam ejector, or combination of the two maintains the vacuum and removes the noncondensables.

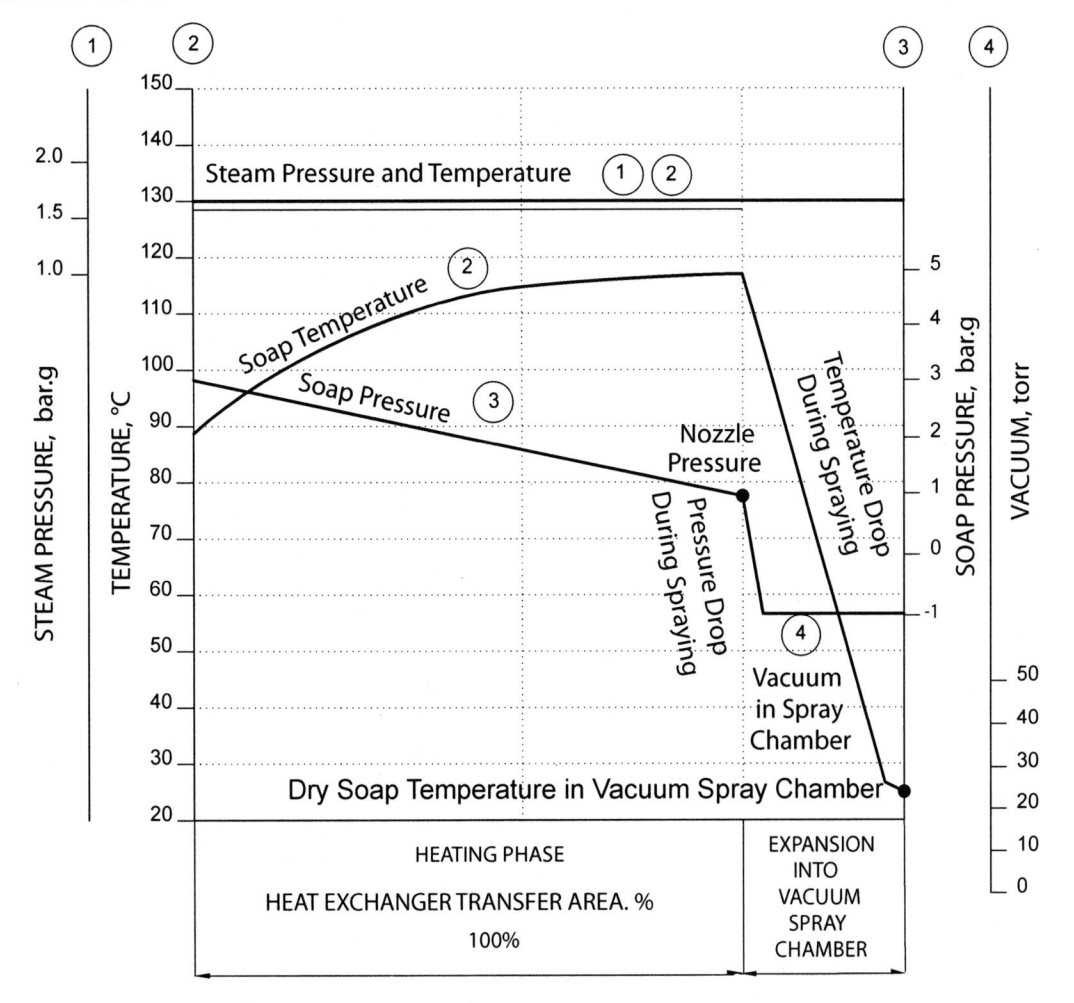

Fig. 10.5. Drying System for 68–70% TFM Laundry Soaps.

DRYING SYSTEM FOR LESS THAN 68% TFM LAUNDRY SOAPS

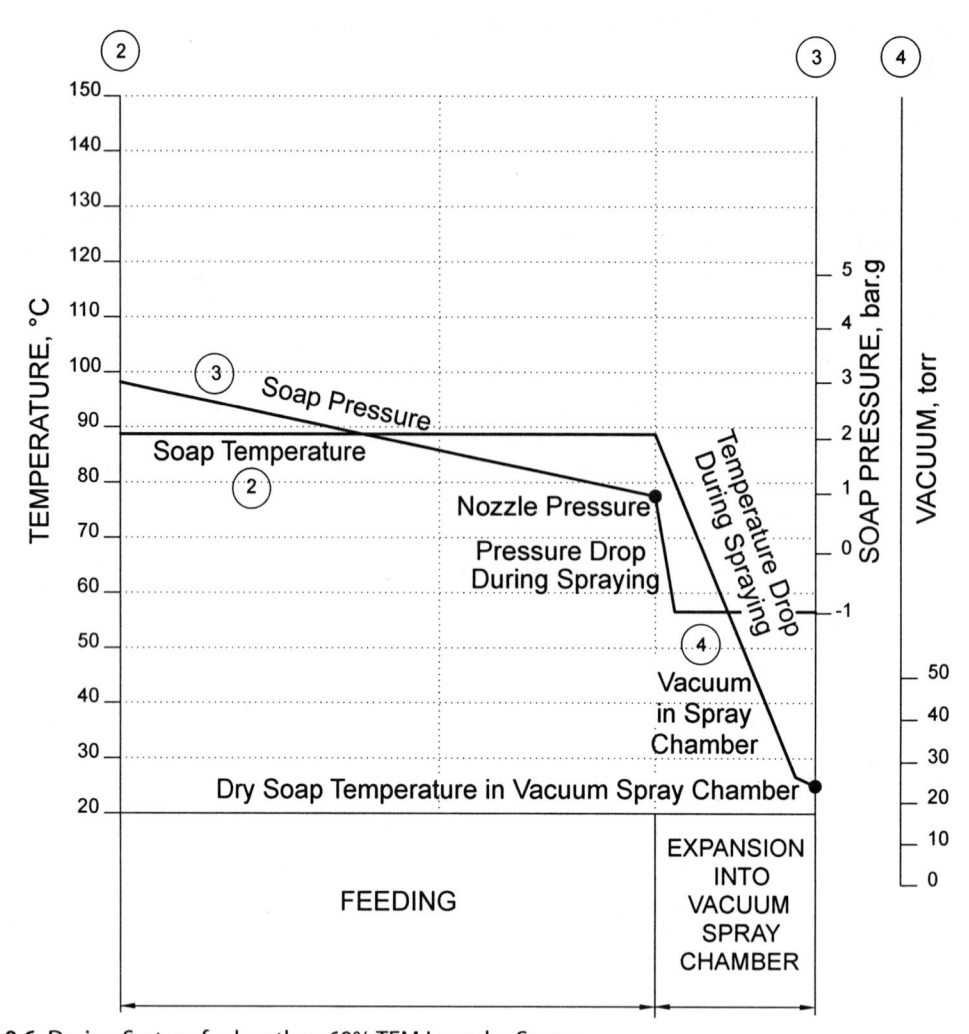

Fig. 10.6. Drying System for less than 68% TFM Laundry Soaps.

Dryer Operating Data for Toilet Soap Pellets and Laundry Soap Bars

The operating parameters for the vacuum spray-dried production of the most typical dried toilet soap bases in pellet form and laundry soaps in bar form are summarized in Table 10.1. A suggested plodder selection guide is also listed.

Dryer Equipment and Components
Filtration Pumps and Filters

Continuous saponification plants produce clean neat soap, whereas soap made via a kettle, or a semiboiled process in a crutcher, can contain impurities. Filters protect the feed pump, the heat exchanger, and the spray nozzles from damage due to foreign bodies.

Hollow disc type filtration pumps can handle impurities and still provide sufficient pressures of 3–4 bar to send soap to the dryer feed tank.

Soap Drying Systems • 275

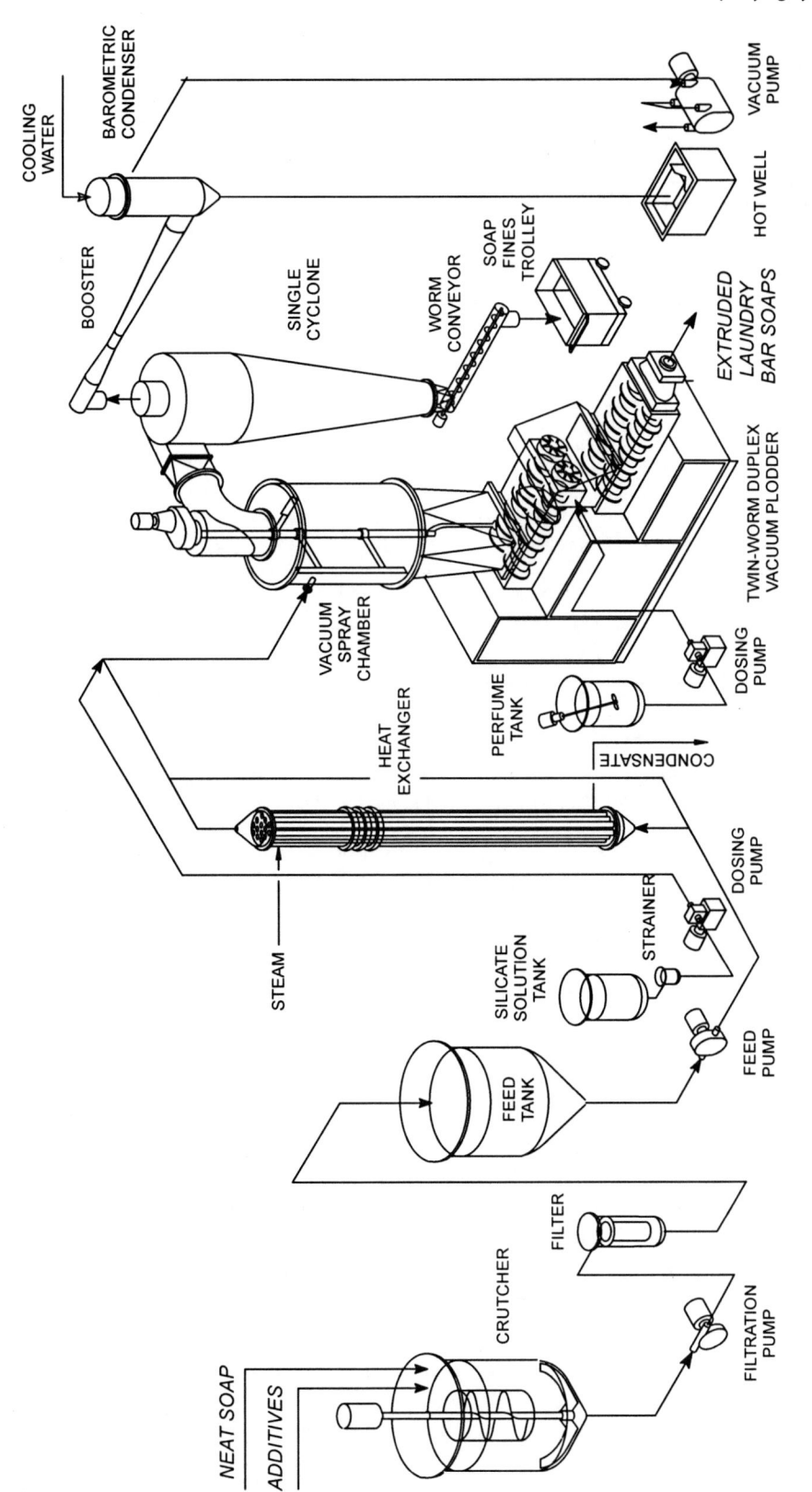

Fig. 10.7. Laundry Soap Vacuum Spray Dryer.

Table 10.1. Typical Spray Dried Soap Products and Operating Data

Product Types ▶	Toilet Soap Base Pellets (Noodles) Standard	Superfatted	Translucent Soap Pellets / Bars Translucent	Laundry Soap Bars Opaque Pure		Opaque Filled	
Product Characteristics ▼							
Total Fatty Matter content, % TFM	Normal range 78 – 80 Wide range 78 – 84	80 – 84	68 – 72	68 – 72	62 – 63	55	45 – 50
Moisture content, % H2O	Normal range 13 – 15 Wide range 14 – 8	12 – 10	18 – 22	25 – 21	32 – 31	32 – 28	32 – 28
Operating Conditions ▼							
Heat exchanger steam pressure, bar.g	5 – 6		0.5 – 1.5	0 – 2	Not Used		
Soap temperature after heat exchanger, °C	130 – 140		105 – 115	110 – 115	85 – 90	85 – 90	85 – 90
Pressure in Vapor Liquid Separator, bar.g	1.3 – 1.7		Not used	Not used			
Absolute pressure in vacuum chamber, torr	30 – 50		20 – 30	20 – 25	15 – 20	10 – 15	10 – 15
Plodder Selection Guide	One or Two Single-Worm Plodder or One or Two Twin-Worm Plodder(s)		Two Single or Twin-Worm Plodders with a Vacuum Chamber at the Final Extrusion Stage	Two Single or Twin-worm plodders with a Vacuum Chamber at the Final Extrusion Stage		Two or Three Twin-Worm Plodders with a Vacuum Chamber at the Final Extrusion Stage	

A complete filtration section with pump, dual filters, and an automatic control system is illustrated in Fig. 10.8. Basket filters with drilled steel-plate filtering equipment, with or without screens, are used in horizontal or vertical position.

Typical filtering elements are as follows:

- For toilet soap: drilled plate with 2 mm diameter holes, covered with 30–50 wire mesh screen
- For laundry soap: drilled plates with 0.5–0.8 mm holes
- A fast changeover filter is installed for continuous operations. Filter clogging is detected by a differential pressure transmitter, and the signal is used to stop the filtration pump or to switch over to the clean second filter. The use of bag filters is recommended when polyethylene contamination from fats packaging occurs.

Feed Pumps

The most widely used soap feed pumps are shown in Fig.10.9. External and internal gear pumps provide excellent positive flow with nonpulsating discharge. Internal gear pumps offer the advantage of having only a single shaft and one mechanical seal or stuffing box. External gear pumps can be made with spur, helical, or fishbone gears.

Soap Drying Systems • 277

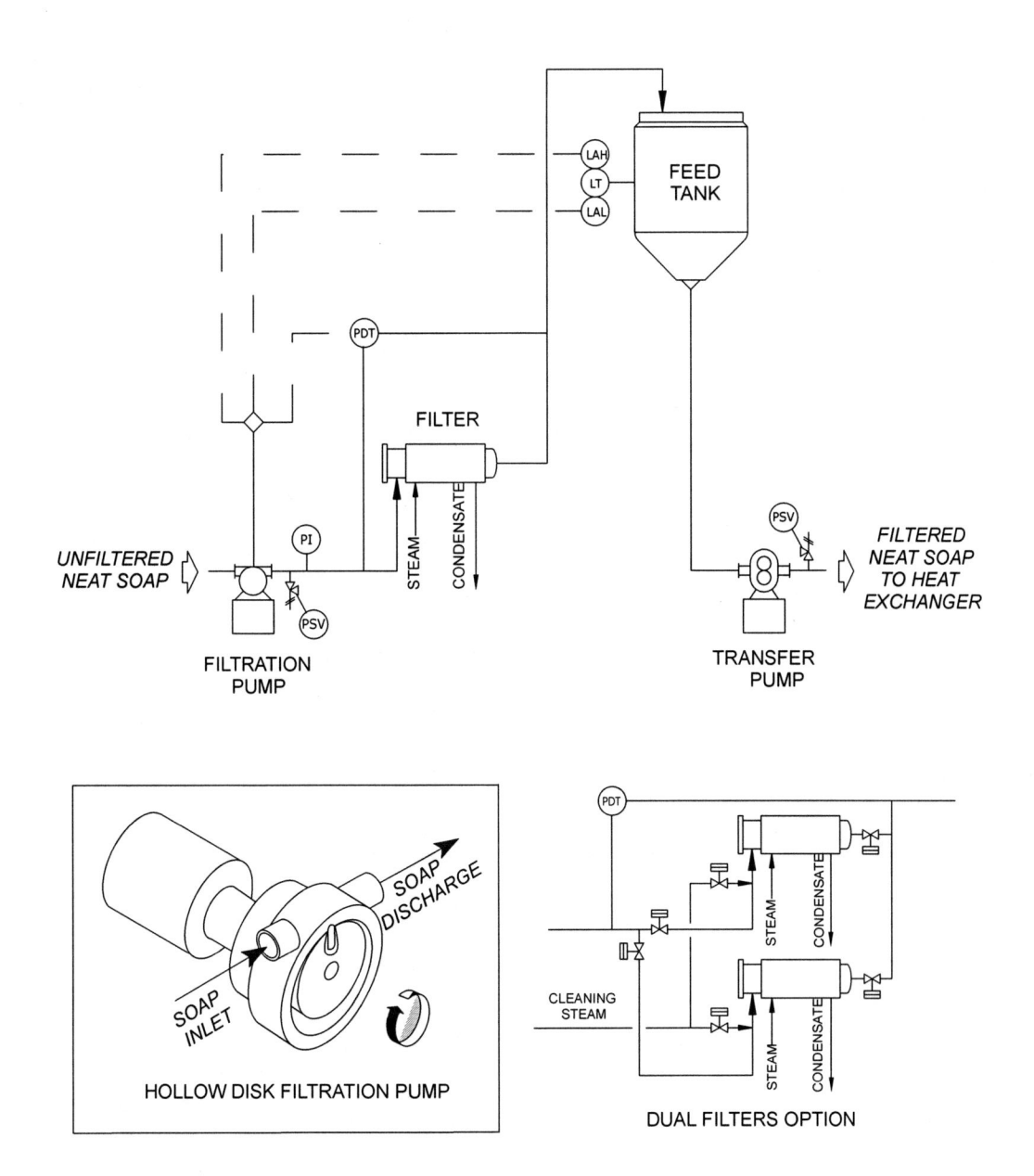

Fig. 10.8. Filtering Section: Filtration Pump and Filters Automatic Control.

Lobe pumps, today the most widely used, are also excellent for pumping shear-sensitive fluids such as neat soap. They are available in bi-lobe or tri-lobe design. A particular feature of lobe pumps is that the rotors and seals can be accessed easily by removing the font cover, without disconnecting the process line. It is suggested to limit the rotating velocity to 200 rpm in order to select a pump with sufficiently large inlet pipe sizes to allow easy pump feeding.

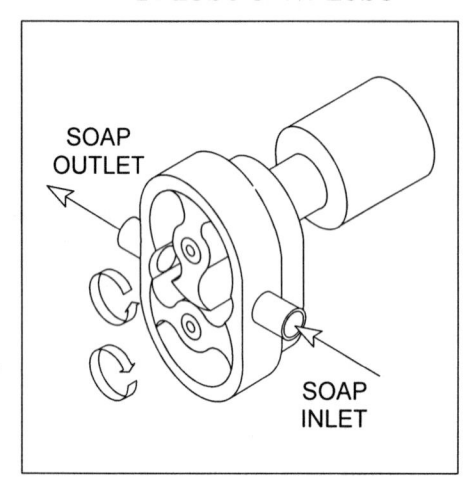

Mono screw pumps also provide excellent flow and pressure stability, but their application is limited due to wearing of synthetic material stators with consequently high maintenance costs. To extend the life of these pumps they are selected with very low rotation speed, such as 50 rpm maximum. They can be a good choice when abrasive components are added to neat soap. Whereas a lobe pump would suffer due to lobe abrasion, for the mono pump, the stator is the sacrificial element that preserves the stainless steel casing and worm.

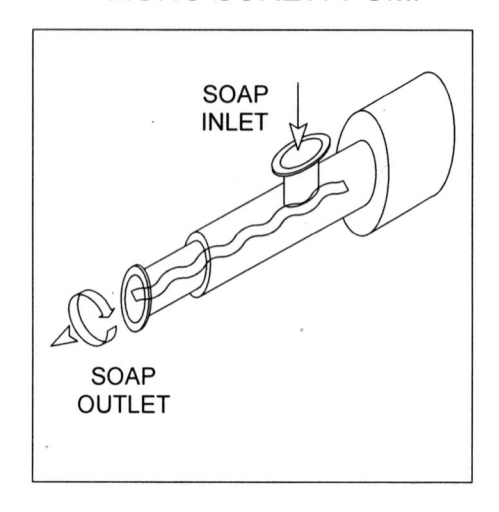

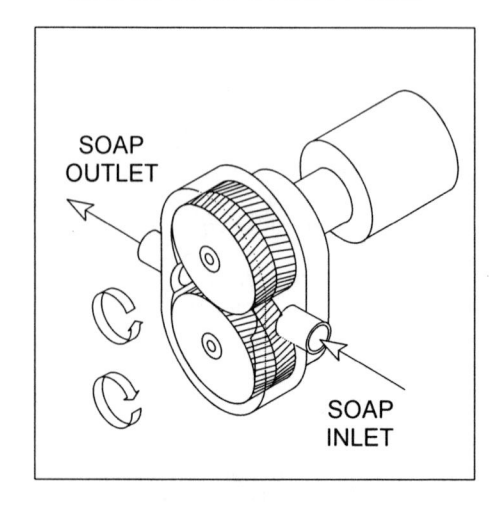

Fig. 10.9. Feed Pumps.

Heat Exchangers

Shell-and-tube Heat Exchangers

Single-pass shell-and-tube heat exchangers for soap applications are simple in design. A number of straight tubes are sealed between two perforated tube sheets. The tube plates and cones are made of 304, 316, or 316L stainless steel. To avoid stress corrosion cracking problems with soap formulations containing higher than 0.6–0.7% salt content in the neat soap, high-nickel-content materials, such as Alloy 825 (Unified Numbering System UNS-N08825), are recommended. Expansion joints should be used to prevent mechanical stresses.

The optimum tube sizes used are 10 mm ID/12 mm ID or 12 mm ID/14 mm ID.

Shell-and-tube exchangers have the lowest capital cost per square meter of heat transfer area and can be installed in various configurations:

- A single vertically placed heat exchanger for any plant capacity
- Two equally sized heat exchangers in series

The current practice is to design heat exchangers with a tube length of 6 meters, so that for a given transfer area, fewer tubes are used. This permits lower cost in terms of heat exchanger/transfer area, because there is no need to cut commercial pipes better inlet soap velocity, resulting in better heat transfer and reduced possibility of blocking.

Typical heat transfer surface areas for tube-and-shell heat exchangers used for vacuum spray dryers are, for toilet soap, 20–30 m^2/ton of dry soap, and, for laundry soap, 10–15 m^2/ton of dry soap. The actual selected area depends on the initial and final soap moistures of the processed soaps.

Plate-and-frame Heat Exchangers

Plate-and-frame heat exchangers (PAF) consist of a number of thin, corrugated metal plates and gaskets clamped together and fitted with a frame. The turbulence of the flowing material induced by the surface design of the plates results in a 3–5 times higher heat transfer coefficient U (see the appendix to this chapter).

Previously the use of PAF exchangers for soap drying was limited due to the gasket materials and their replacement costs. Today the use of ethylene propylene diene monomer (EPDM) gaskets designed for 200°C maximum temperature and 25 bar pressure has eliminated the problems encountered with less resistant gaskets.

Drying of high moisture content, heat sensitive, soap/synthetic (combo) products requires high pre-evaporation rates in the atmospheric drying stage. For this application, PAFs provide homogeneous two-phase flow with a higher steam rate.

PAF heat exchangers cost more per square meter of heat transfer area than shell-and-tube units, but their high heat transfer rate, compact size, and use for special applications can make them cost-effective.

The higher maintenance cost of PAF exchangers compared to shell-and-tube exchangers has limited application of PAFs to the special cases just described.

Vapor–Liquid Separators

After the installation in 1987 of the SELA vapor-liquid separating/extracting system (VLS), Mazzoni LB, Binacchi & Co. and Soaptec srl developed different designs to be used with vacuum spray dryers. Like all novel systems, it took years for the industry to accept the VLS, not only with new dryers but as a retrofit option. The VLS is positioned between the heat exchanger and the vacuum spray chamber, and disengages the vapors formed in the heat exchanger before the soap-and-vapor mixture reaches the vacuum spray chamber.

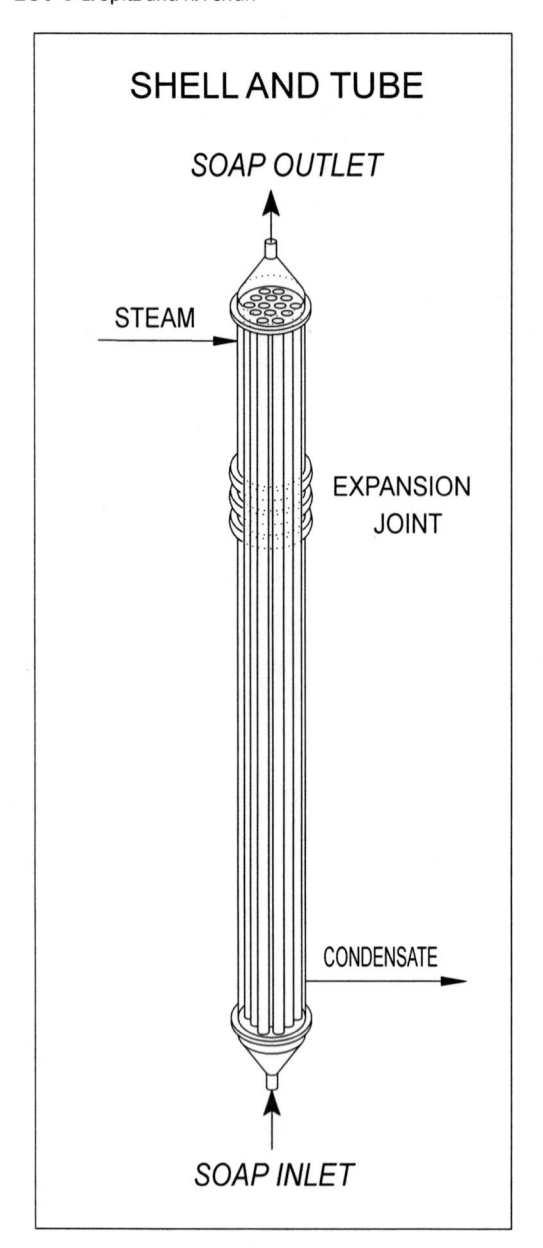

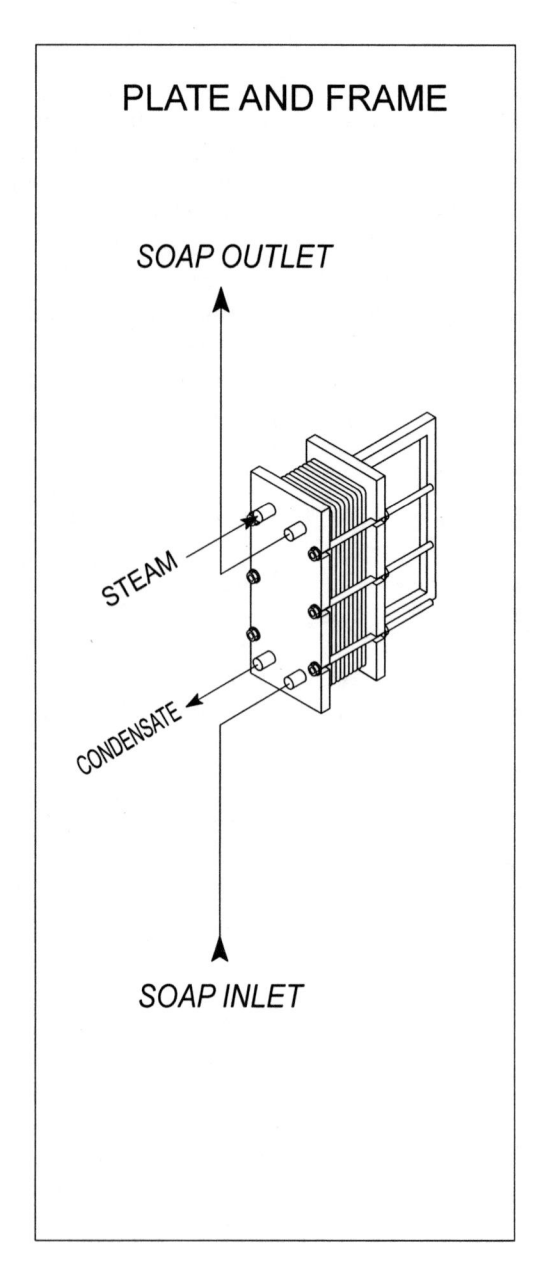

Fig. 10.10. Heat Exchangers.

Mazzoni LB VLS Design
The Mazzoni VLS unit consists of a vertical static vessel operating under a pressure of 1–2 bar, where the vapors that were formed in the heat exchanger separate from the liquid hot soap due to gravity.

The separation is governed by the pressure-controlled outlet of vapors from the top part and the level-controlled extraction of soap from the bottom. The pressure is maintained by a control-modulating valve on the vapor line, whereas the soap level is controlled by varying the speed of the pump under the VLS. This pump, which feeds the soap to the drying chamber, is a lobe or gear pump, preferably vertically mounted so that soap can directly fall into the pump without need for pipes and elbows. Lobes pumps are should preferably have an enlarged top rectangular port for easier soap entrance into the pump.

Soap Drying Systems ● 281

The vapors generated from VLS systems, in spite of their low pressure, can be reused for heat recovery purposes, depending on the plant configuration and the types of soaps produced.

1. In this application, the vapors from the VLS can directly drive a properly configured steam booster (Fig. 10.11), thanks to the favorable mass rate between vapors from the VLS and balance vapors from the vacuum chamber. It is possible to achieve a consistent 10–15 torr vacuum improvement compared to the same plant working without the booster, without consumption of fresh makeup steam. Fresh steam is used normally only for startup and/or transitory conditions.

2. The liquid neat soap can be preheated up to 110–120°C in a dedicated heat exchanger placed before the heat exchanger of the vacuum dryer. Since not all the vapors from the VLS are used, the remaining can be used to heat up the makeup water for the boiler, the raw materials, and the soap storage tanks.

3. It is possible to use the VLS vapors to preheat the raw materials used in a continuous saponification plant, if one is installed nearby. Refer to Alternate B shown in Fig. 10.11.

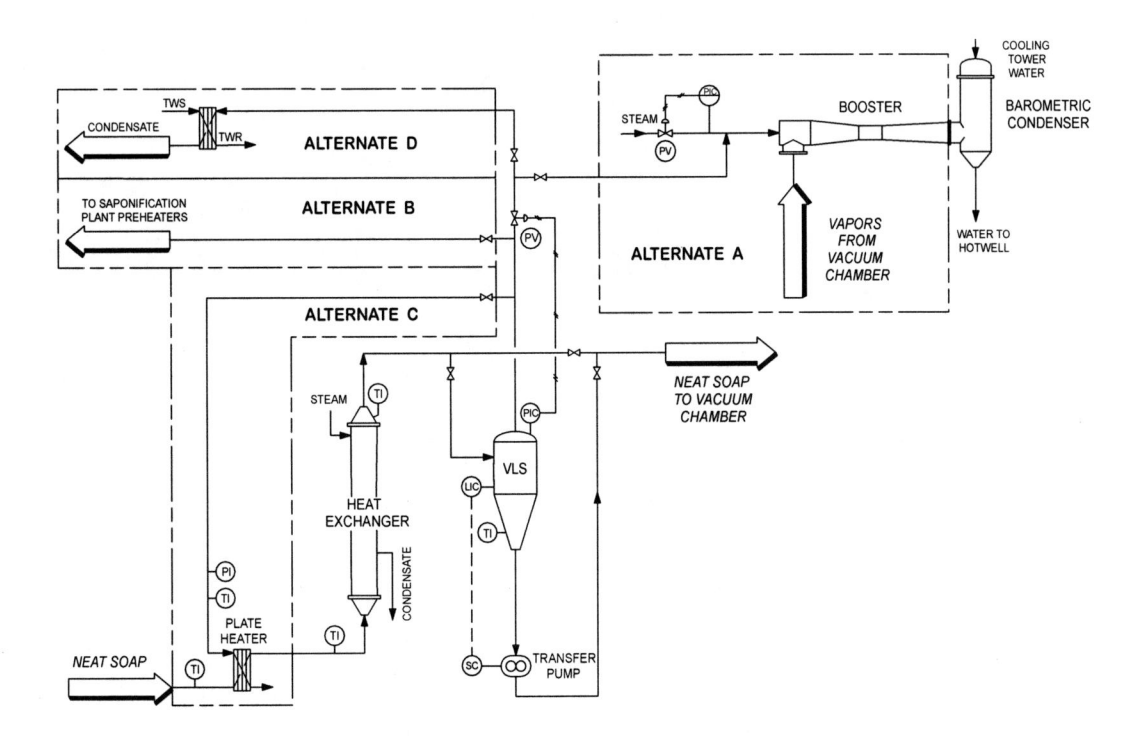

Fig. 10.11. Alternate "VLS" Heat Recovery Options.

Binacchi SDE Dryer Evaporator System

Binacchi's SDE Dryer Evaporator system consists of an evaporator with rotating inner parts and an intermediate thermocompressor unit (Fig. 10.12). The vapors are recompressed to an intermediate pressure in a thermocompressor using fresh steam, and the combined steam is then sent to the main booster. This setup allows vacuum levels lower than 30–35 torr to be obtained with limited steam consumption, which may be particularly helpful in the production of soap/synthetic combo products.

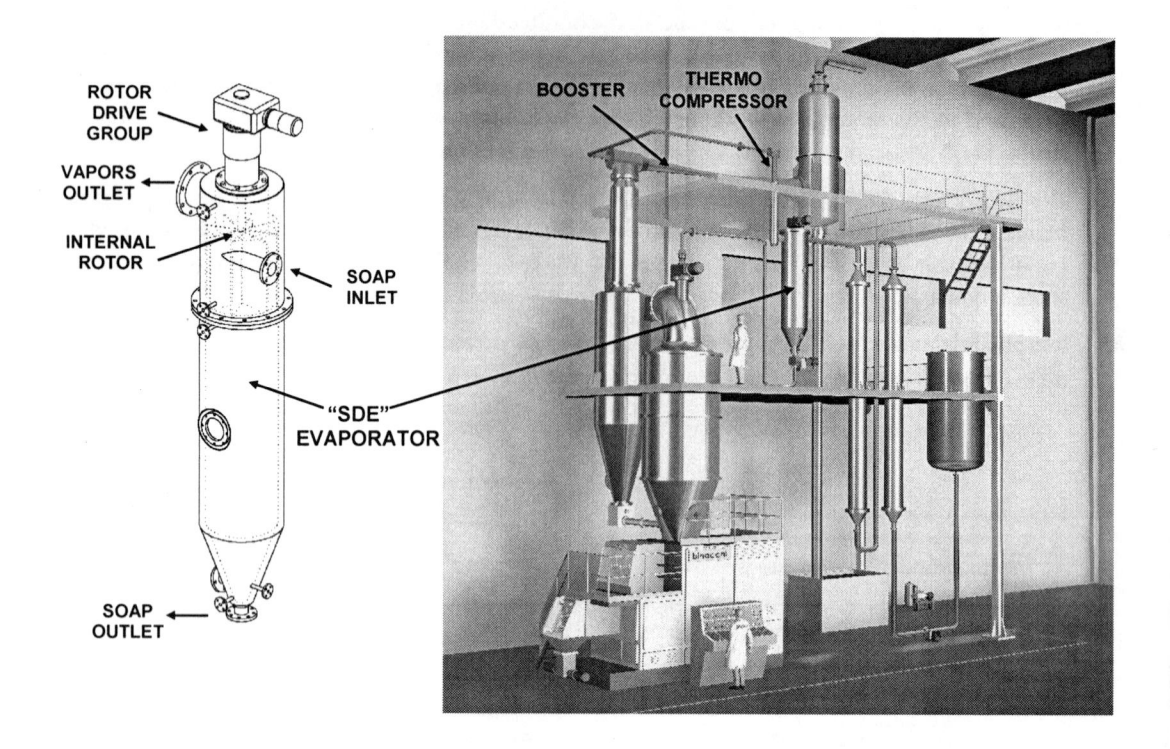

Fig. 10.12. Binacchi "SDE" Soap Dryer Evaporator Application (Courtesy of Binacchi & Co.).

Vacuum Chambers

Two types of vacuum spray chambers are used for soap drying plants (Fig. 10.13).

The most widely used rotating-nozzle type is offered by all of the manufacturers of vacuum spray drying plants, with some variation in the design (number and position of nozzles, design of scrapers, shape of vapor outlet pipes).

Mazzoni LB's current standard is the innovative "no-spray" vacuum chamber, which features a tangential injection of the liquid soap in lieu of the spraying-nozzle technique of conventional units.

The no-spray technology offers the following production advantages: significant reduction of fines formation thanks to the absence of a spraying nozzle, reduced carryover of fines due to an internal anti-entrainment baffle.

These two points result in a reduced contamination of the water flow from the barometric condenser. The new design of the no-spray vacuum chamber also demonstrates the following mechanical improvements compared to the conventional design:

- Significant reduction of the vacuum chamber's dimensions and simplification of the rotating components

- A simplified main drive shaft that doesn't require the internal soap pipe or mechanical seal on the main shaft.

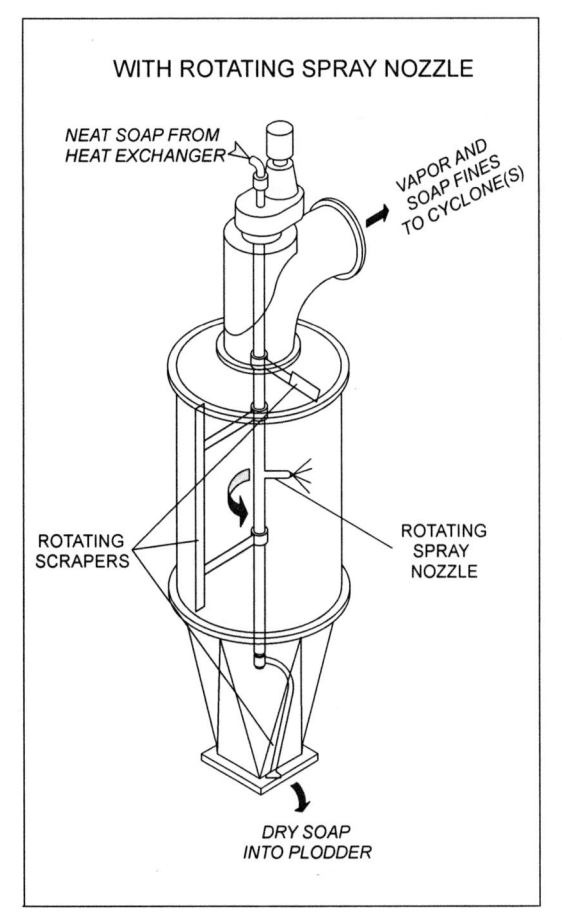

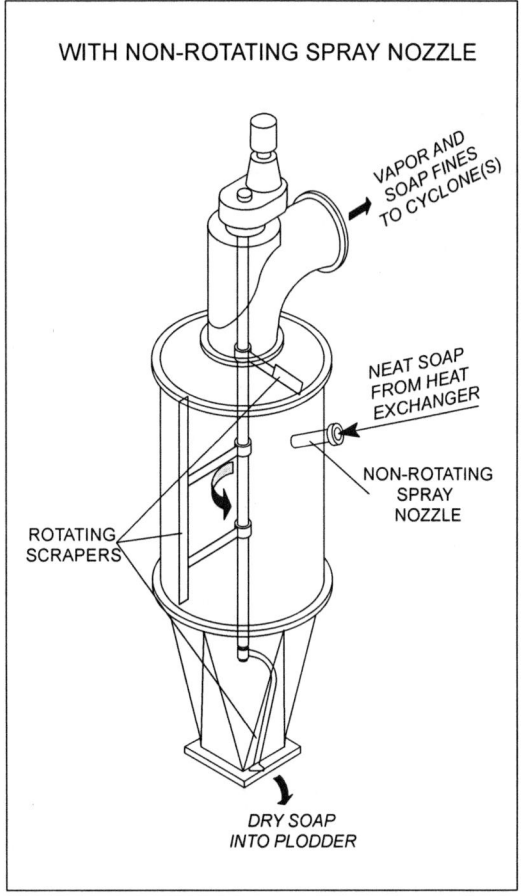

Fig. 10.13. Vacuum Chambers.

Soap Fines Recovery Systems
Various cyclone-type separators are used to recover the soap fines produced in the vacuum spray chamber.

Dual Cyclone
Two cyclones connected in series is the traditional system, commonly used today even for large dryers with capacities of 8–10 tons/h. In the past this technique was limited to smaller plants. Today the additional capacity is possible thanks to the reduction of dust carryover due to use of vapor separators and no-spray chambers. The soap fines recovered in the second cyclone are fed with a worm conveyor to the first cyclone, from which another worm conveyor feeds it to the vacuum spray chamber. Both worm conveyors are fitted with bridge breakers. To avoid recycling the fines into the vacuum chamber, a fines extruder can be installed under the first cyclone, and it is even possible to use a reversible single extruder/conveyor unit. The multicyclone design previously used as the second-stage cyclone has been abandoned in recent years, due to high investment and installation costs and more difficult cleaning.

Single Cyclone

A single large-size cyclone can be used for all laundry soap plants and for toilet soaps plants with capacities up to 3,000 kg/h. This is applicable for tallow and coconut type soaps, but it is less convenient with vegetable-based soaps, which generate more soap fines.

Soap fines are not recycled to the spray chamber, but are discharged and collected in a drum, as is regularly done for laundry soaps, but less often for toilet soaps. In this case, a second single cyclone still continuously feeds the fines to the first cyclone, from which the fines are discharged with a worm extruder conveyor. Because handling and reusing the fines poses major problems, the noncontinuous recycle system for drying toilet soaps is seldom used.

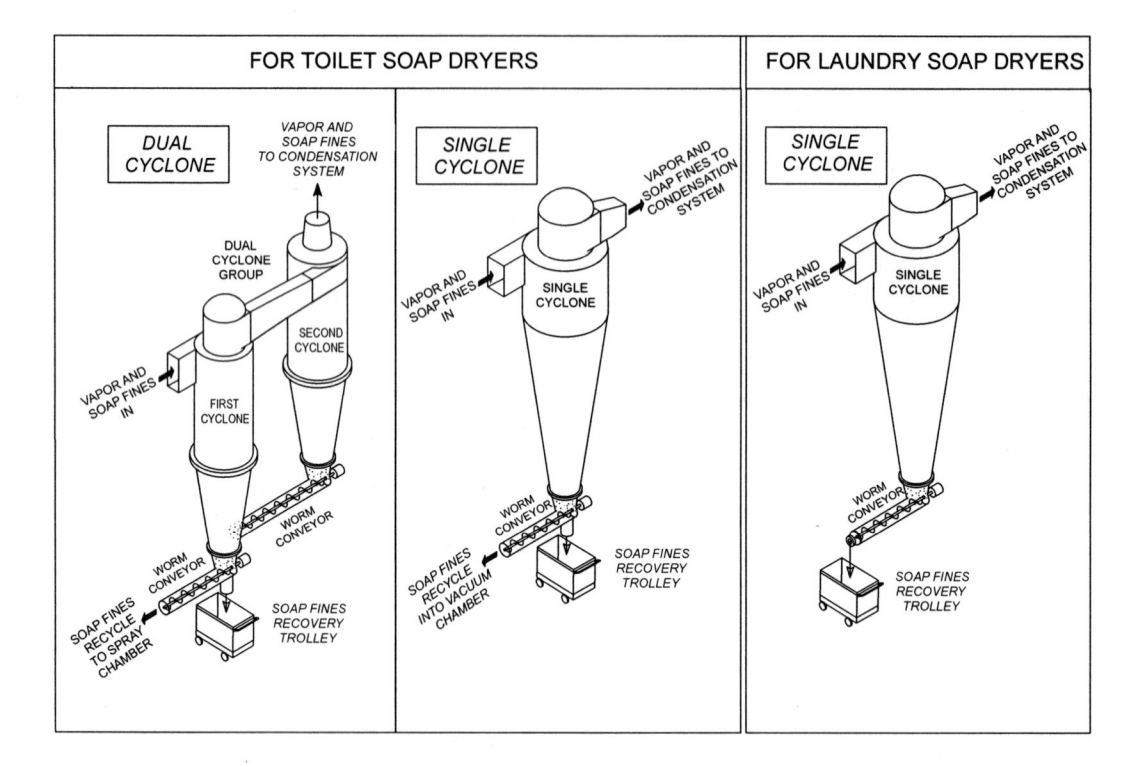

Fig. 10.14. Soap Fines Recovery Systems.

Vacuum Producing Systems

Vacuum-producing systems for soap dryers consist of vapor condensation equipment (barometric and surface condensers and cooling tower systems) and vacuum-producing equipment (vacuum pumps and steam jet ejectors).

Vapor Condensation Equipment

The sole purpose of a condenser is to continuously condense the vapors generated during the drying process back into water. Three basic types of condensers are used: barometric, surface, and indirect condensers.

Barometric Condensers

Barometric condensers are also called direct-contact condensers because they condense the vapors by direct contact with the cooling water. The condensed vapors, the cooling water used for condensation, and the soap fines are discharged together into a hot well. Contact condensers may be designed with water-spraying nozzles or with a traditional "plate" design. Carbon steel is a common construction material. When the cooling water temperature is higher than 27°C, and pellet (noodle) temperature lower than 38°C is required, then a steam booster must be used.

Surface Condensers

Surface condensers are shell-and-tube heat exchangers. The vapors flowing inside the tubes will be condensed by cooling water flowing on the shell side. The cooling water is not contaminated with the soap fines because it does not contact the process side. The condensed vapors that carry the final traces of soap fines are discharged into the hot well, and the clean water, free from soap fines, is recycled. To limit the size and cost of a surface condenser, the cooling water temperature should not exceed 20°C. The cooling water can be efficiently used in a closed circuit, using a cooling tower in conjunction with a water chiller. Surface condensers can also be mounted at a lower height than barometric condensers, using a collecting vessel under vacuum and a self-priming pump.

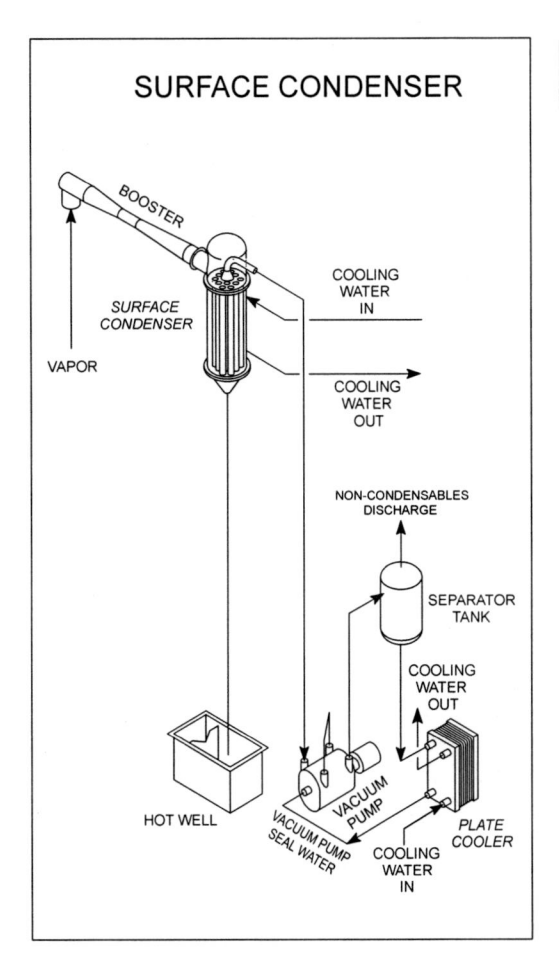

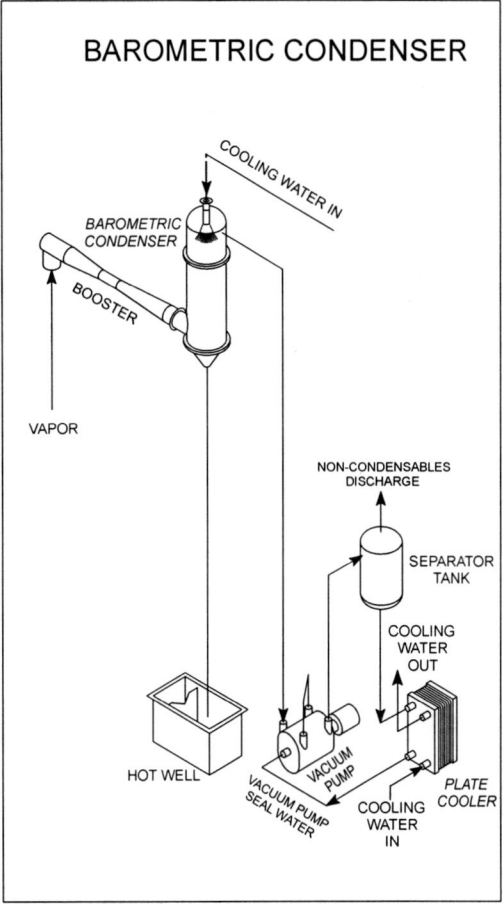

Fig. 10.15. Vapor Condensation and Vacuum Formation.

Indirect Cooling Water System (ICS)

The ICS (Fig. 10.16) is an alternative system to the previously described surface and barometric types (Fig. 10.15). The ICS system consists of a centrifugal pump to circulate water in a closed loop between the hot well and a barometric condenser, passing through a plate heat exchanger, which cools down the recycled water with the cooling tower water. Due to the high heat transfer rate obtained with the plate heat exchanger, this is an economical system which can also be operated with cooling water that exceeds 30°C temperature.

This system avoids contamination of the cooling tower like one which uses a surface condenser but offers two major advantages:
- lower investment and installation cost
- less frequent cleaning, once a month instead of every week or two weeks for the surface condenser type.

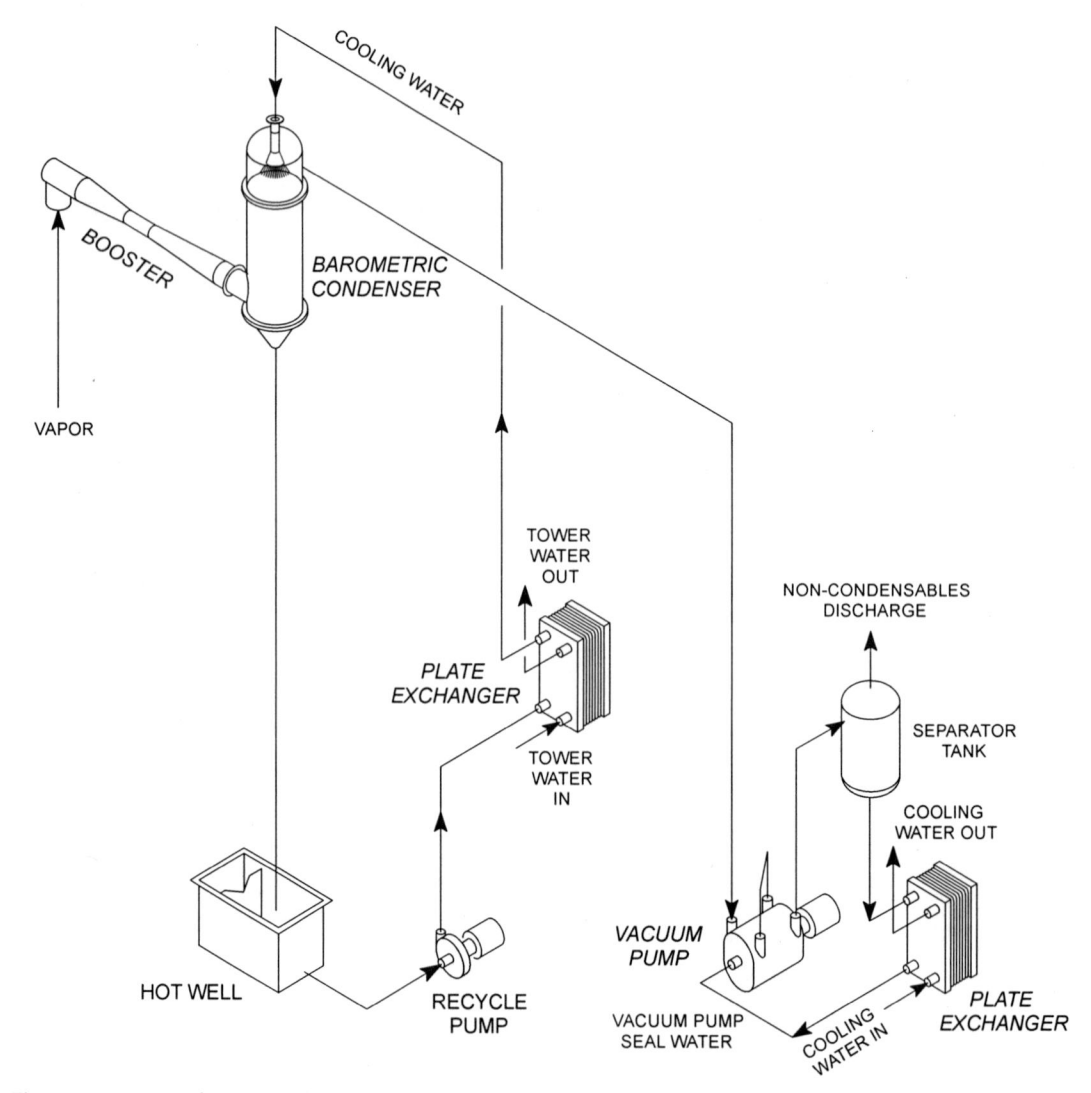

Fig. 10.16. "ICS" Indirect Condensation System.

Vacuum Formation Systems
Vacuum Pumps

Mechanical piston pumps, used extensively years ago, are practically abandoned today due to their high cost, and liquid ring pumps are now the preferred choice. They are available on the market as a skid-mounted complete group, with gas/liquid separators and plate coolers or shell-and-tube heat exchangers to maintain the seal water at the lowest possible temperature based on local conditions

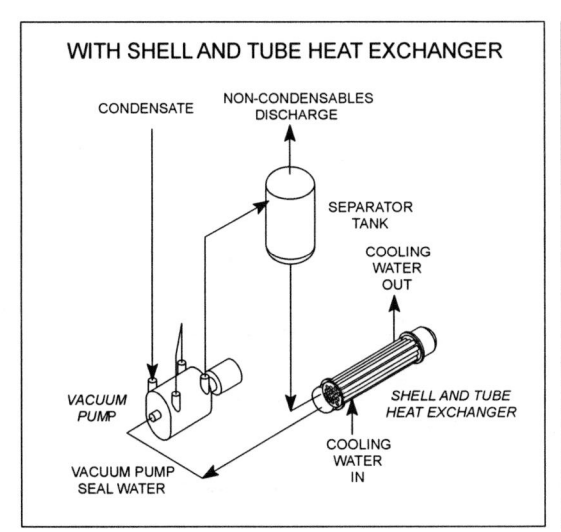

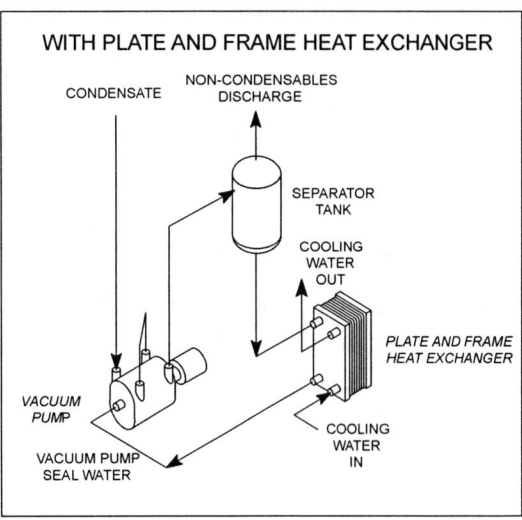

Fig. 10.17. Vacuum Formation Systems.

Steam Jet Ejectors (Boosters)

Steam jet ejectors are very simple devices consisting of a nozzle, a mixing chamber, and a diffuser. Using motive steam, according to the required operating vacuum level in the dryer, a single-stage or multistage ejector group is used.

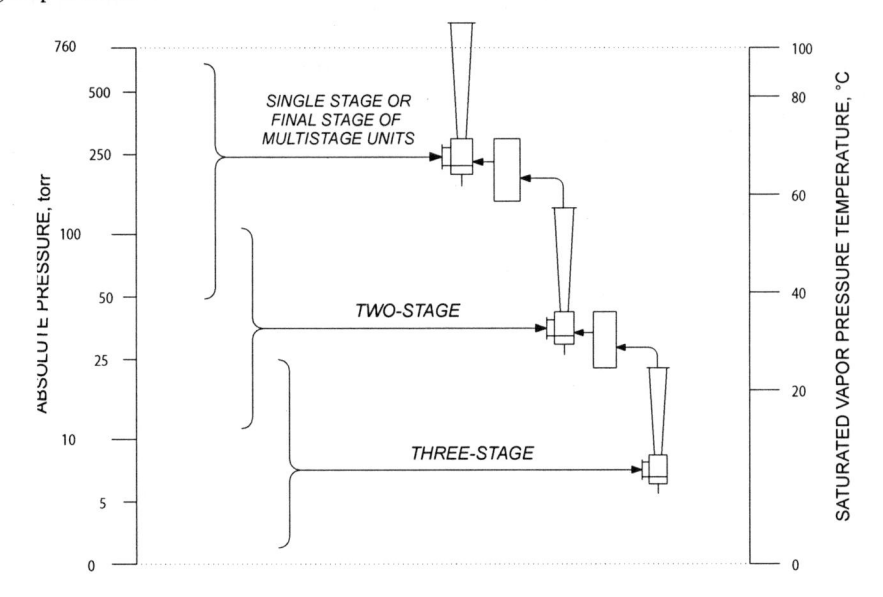

Fig. 10.18. Steam Jet Ejectors.

Typical Steam Consumption by Soap Dryers

This summary is intended only to provide typical steam requirements, for the purpose of general evaluation of the steam energy cost involved with soap drying. Data relevant to the various energy recovery systems, which may differ between plant suppliers, are not included. No discussion is presented concerning cooling water, chilled water, and electrical power as these are strongly dependent upon local climatic conditions, machine design, plant configuration (for example, the number of plodders), and other factors.

Dryer Typical Steam Requirements
Base: Production of 1,000 kg/h soap pellets or bars

	Toilet Soap 13–15% Moisture Dry Pellets			Laundry Soap 24–26% Moisture Bars
VLS use	Yes	Yes	No	Not applicable
Steam Booster use	Yes	No	Yes	Yes
Steam for preheating, kg/h	240–270	240–270	240–270	40–80
Motive steam for booster, kg/h	120–150 (1)	0	220–250 (2)	200–220
Total steam, kg/h	360–420	240–270	460–520	240–300

For toilet soaps, the VLS configuration requires the same amount of steam for the heat exchanger as the configuration without a VLS. A reduced amount of steam is needed for the booster—compare values (1) and (2)—due to the lower amount of vapors released under vacuum that must be sucked by the booster. When the VLS vapors are used to rive a properly designed steam booster, colder pellets are produced at a lower production cost as compared to a booster operated with steam from a boiler. Value (1) can be reduced to zero by proper design.

For laundry soaps, very low steam consumption is required for the heat exchanger. Steam for the booster in this case is a must, as laundry bars are normally extruded directly from the dryer plodder for outlet, so that the bar must be cold enough for efficient processing.

Cooling Tower Systems

In Fig. 10.19, two cooling tower system configurations are illustrated for cooling the hot condensing water from the hot well and returning it to the barometric condenser: a standard system, and another illustrating the ICS indirect cooling water system.

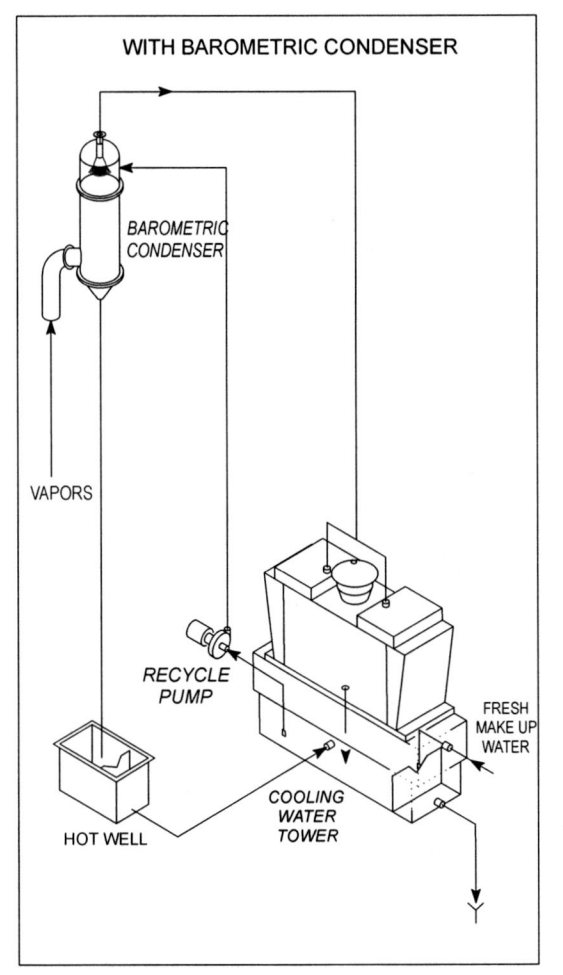

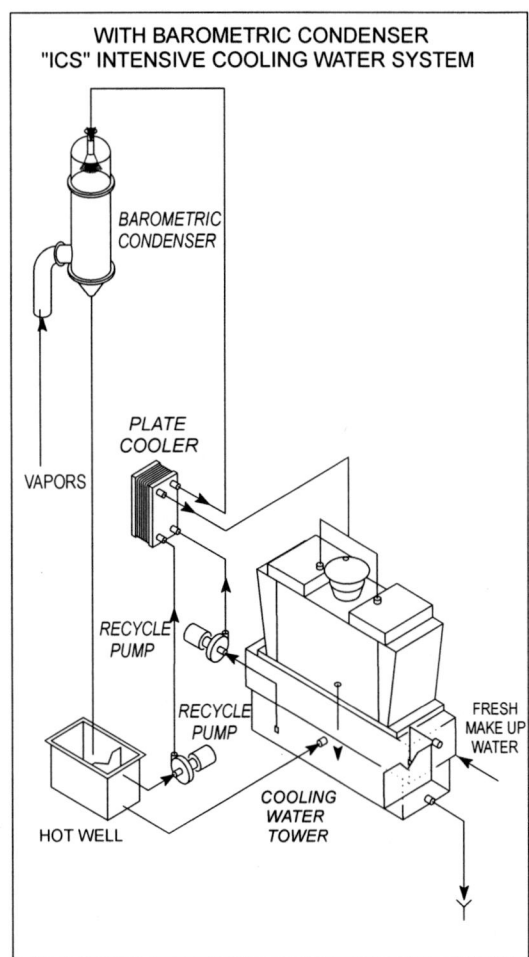

Fig. 10.19. Cooling Water Tower Systems.

Solid and Liquid Additive Systems

Soap formulations are becoming more complex, and various liquid and solid ingredients are added to the neat soap to enhance the final produce performance characteristics. Two systems are illustrated in Fig. 10.20.

The addition of citric acid, brine solution, mineral oil, and others to toilet soap soaps can be added inline after the heat exchangers. Superfatting agents are normally also added after the heat exchanger to prevent metallic stress correction.

Talc and other solid additives are added into a crutcher prior to feeding the soap into the dryer. Silicate and sometimes brine solutions are added inline prior to the vacuum chamber inlet, by-passing the heat exchanger.

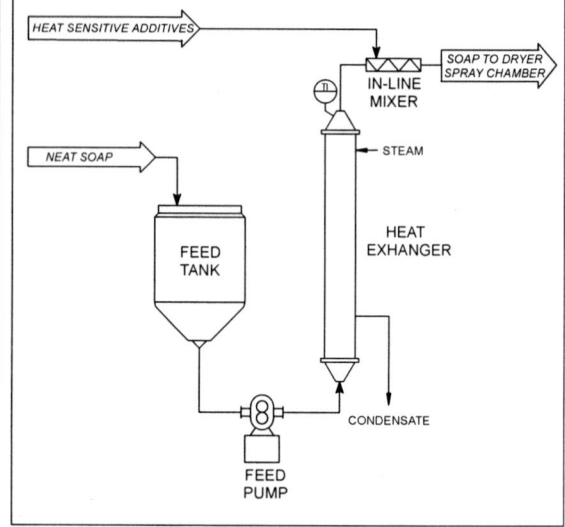

Fig. 10.20. Solids and Liquids Additive Systems.

Plant Automation

Instrumentation and Computer Control

Electronic instrumentation is now standard for soap drying installations. The majority of new plants are computerized, and existing ones are converted to computer control. Besides computer control, a modern drying plant should include the following main features:

- Feed tank level control, a level transmitter operating on the inverter of the soap transfer pump from storage, in order to keep the level in the feed tank stable
- Soap flow control loop using a magnetic or mass flow meter
- Steam pressure control loop (independent for any single heat exchanger)
- Vacuum chamber residual pressure transmitter (mm Hg or mbar)
- Soap spraying pressure control (for no-spray chambers)
- Soap level in VLS

- Pressure in VLS
- Vacuum in the spraying chamber
- Temperature at inlet/outlet of cooling water to condenser

Computer control systems (CCS) are used for soap dryer process control. CCS based on hardware consisting of PC and PLC are favored because of lower investment costs and similar capability compared to more costly distributed control systems (DCS), which are more suitable for larger plants (such as those in the petrochemical industry, for example). For small plants the PC can be replaced by an operator panel integrated into the electrical panel board.

Depending upon the provided field instrumentation, the CCS memorizes production recipes and operating conditions, summarizes consumption of neat soap, additives, and so on, and records process parameter trends. It also performs emergency procedures and alarm detection.

Safety interlocks are used for all motors, doors, high soap levels in the vacuum chamber, and so on. A soap-recycle mode is used for efficient startup and shutdown or in emergencies, using an automatic set of valves that allow the operator to pump the soap from the heat exchanger back to the feed tank. Once the proper temperature is reached, the soap is fed to the spray chamber. Steam cleaning of all the soap pipes with automatic or computerized steam injection valves is part of the overall system. Figure 10.21 shows the flow diagram of a complete, fully instrumented, automated dryer with VLS booster with bypass.

Neat Soap Flow Control

The use of mass flow meters is becoming very popular, since it offers the following important advantages (Fig. 10.22):

- Instantaneous control of neat soap feed flow rate and monitoring (per shift, day, week, etc.)
- Very easy calculation of production rate
- Easy detection of anomalous conditions (such as "no-flow") with consequent activation of proper alarms/interlocks

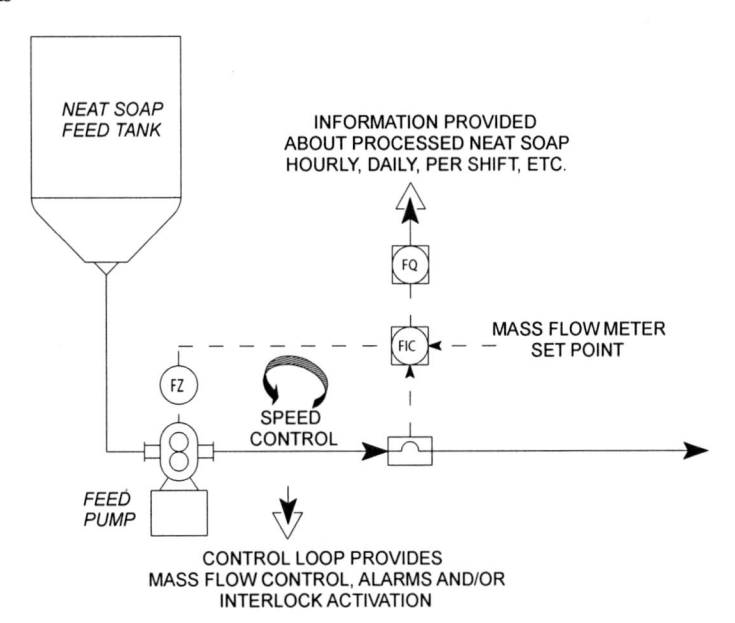

Fig. 10.22. Neat Soap Flow Control with Mass Flow Meter.

Fig. 10.21. Soap Drying Plant Automation and Control.

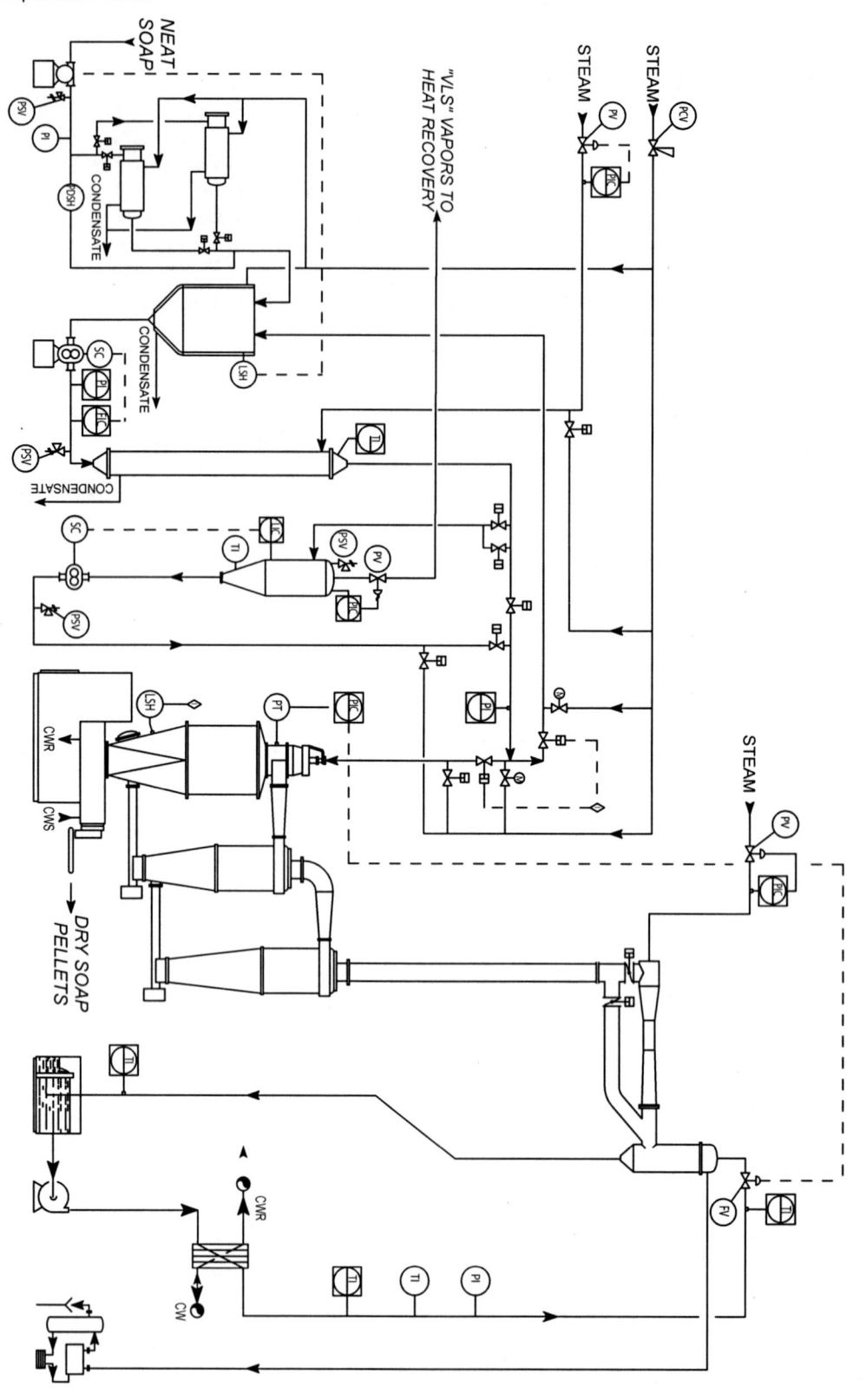

Appendix

This appendix includes the following:

- Definitions and terminology relating to drying and heat transfer
- Material balance calculations (Equation 1) and graphs for liquid neat soap to dry toilet soap pellets (Fig. 10.23) and laundry soap bar conversions (Fig. 10.24)
- Formulas for calculating the amount of water evaporated (flashed off) under vacuum and pre-evaporated in the heat exchanger (Equations 2 and 3), and a graphic illustrating toilet soap dried to different moisture levels (Fig. 10.25)
- The well-known formulas for heat exchanger duty (Equations 4 and 5) and/or heat transfer coefficient calculation for sizing a heat exchanger (Equation 6)
- Overall material balances *for* toilet soap (with and without VLS) and laundry soap drying (Fig. 10.26, 10.27, and 10.28)
- A guide to determine the actual operation vacuum (absolute pressure) and the relationship to the condensed water discharge temperature (downleg temperature) into the hot well in a drying system with a barometric or a surface/indirect condenser
- Total fatty matter (TFM) and moisture content (%) calculation table for toilet soaps (Table 10.3)

Definitions and Terminology

Density: The density or specific weight of a fluid is its weight per unit of volume.
The density of common neat soap base is 59.3–62.4 lb/ft^3 (950–1000 kg/m^3).

Specific gravity: The specific gravity of a fluid is the ratio of its density to the density of water (dimensionless).

Vapor pressure: The vapor pressure of a pure liquid is the pressure (at a given temperature) at which a liquid will change to a vapor. If this liquid is mixed with nonboiling substances (e.g., water with soap), the vapor pressure must be corrected by the mole fraction of water.

Viscosity: The viscosity of a liquid is a measure of its tendency to resist a shearing force. Neat soap behaves as a non-Newtonian fluid exhibiting a nonlinear shear stress/shear rate behavior. Its viscosity decreases with increasing shear rate (velocity).

Velocity (Feet/sec)	Velocity (Meters/sec)	Viscosity (Centipoise)
0	0	2000–3000
0.5	0.15	1000
1.0–2.0	0.3–0.6	300–500

Viscosity of 30–32% moisture content 90°C temperature neat soap at different pumping speeds.

Specific heat (Cp): The specific heat of a substance is the amount of heat necessary to raise the temperature of a unit mass by one degree. The following formula is used to calculate the amount of heat gained (lost) by a given mass due to a change in temperature:

$$Q = m \times Cp \times \Delta T$$

where Q is the amount of heat (kcal), m is mass (kg), Cp is the specific heat (kcal/kg/°C), and ΔT is the change in temperature (°C). The specific heat is calculated according to the water content in the neat soap using the following formula:

$$Cp = 0.6 \times \% \text{ anhydrous soap} + 1 \times \% \text{ H}_2\text{O}$$

Specific Heat (Btu/lb/°F) or (Kcal/kg/°C)	%TFM
0.73	63
0.68	73
0.64	84

Thermal conductivity (k): This value depends on the water content and is expressed by the formula

$$k = (0.58 \times W) \times 0.9 + (0.15 \times S) \times 0.9$$

where S = % of soap and W = % of water. k is expressed in kcal/m/°C.

Sensible heat: The sensible heat (kcal/kg) is the heat necessary to increase the temperature of a liquid from an initial to a final temperature without starting evaporation of the liquid. For example, it takes 100 kcal to bring the temperature of 1 kg of water from 0° C to 100°C, and this is calculated as follows:

$$SH = 1 \text{ kg} \times 1 \text{ kcal/kg/°C} \times (T_{final} - T_{initial}) = 1 \times 1 \times (100 - 0) = 100 \text{ kcal/kg}$$

Here 1 kcal/kg/°C is the specific heat of water.

Latent heat of vaporization: The latent heat of vaporization (kcal/kg) is the heat that produces a change of state (e.g., from liquid to vapor) without a change in temperature. For example, it requires 540 kcal to convert 1 kg of water at 100°C to 1 kg of steam at 100°C.

Material Balances

In the soap industry, the production of soap dryers refers to the plant output at a given moisture. The quantity of incoming liquid neat and the resulting quantity of finished product at various moisture levels is given by Equation (1), and it is represented for pure toilet and laundry soaps in Fig. 10.25 and 10.26.

(1) $$G_{in} = G_{out} (100 - M_{out}) / (100 - M_{in})$$

where

G_{in} = incoming neat soap quantity (kg/h)
G_{out} = outgoing dry soap quantity (kg/h)
M_{in} = incoming neat soap moisture (%)
M_{out} = outgoing dry soap moisture (%)

The quantity of water evaporated by vacuum expansion is

(2) $$W_v = G_{in} \times Cp_{he} \times (T_{he} - T_{vc}) / LH_v$$

where

W_v = quantity of water evaporated by vacuum expansion (kg/h)
LH_v = latent heat of water at vacuum chamber conditions (kcal/kg)
Cp_{he} = specific heat of soap at heat exchanger outlet (kcal/kg/°C)
T_{he} = temperature of soap at heat exchanger outlet (°C)
T_{vc} = temperature of soap in vacuum chamber (°C)

The total amount of water to evaporate is

$$G_{total} = G_{in} - G_{out}$$

The water to be pre-evaporated in the heat exchanger is

(3) $$W_{he} = G_{total} - W_v$$

When the plant includes a VLS it can be assumed that all this water is removed in the VLS before soap enters into the vacuum chamber. In most cases the water eliminated in the VLS is slightly more, due to a small expansion between the outlet of the heat exchanger and the VLS.

The heat exchanger duty is the sum of the following two partial duties:

(4) Sensible heat = $G_{in} \times Cp (T_{he} - T_{in})$
(5) Latent heat = $W_{he} \times LH_{he}$

where LH_{he} = latent heat of water in the heat exchanger (kcal/kg).

The amount of water evaporated (flashed) by expansion into the vacuum spray chamber and the quantity pre-evaporated in the heat exchanger, per Equations (2) and (3), is illustrated in Fig. 10.26 for 32.5% moisture content neat soap and dry toilet soap pellets at different moisture levels.

The overall material balance for dry toilet soap pellets produced from neat soap with a single-stage vacuum spray dryer (with and without the VLS) is summarized in Fig. 10.26 and 10.27. Balance for laundry soap bars is represented in Fig. 10.28.

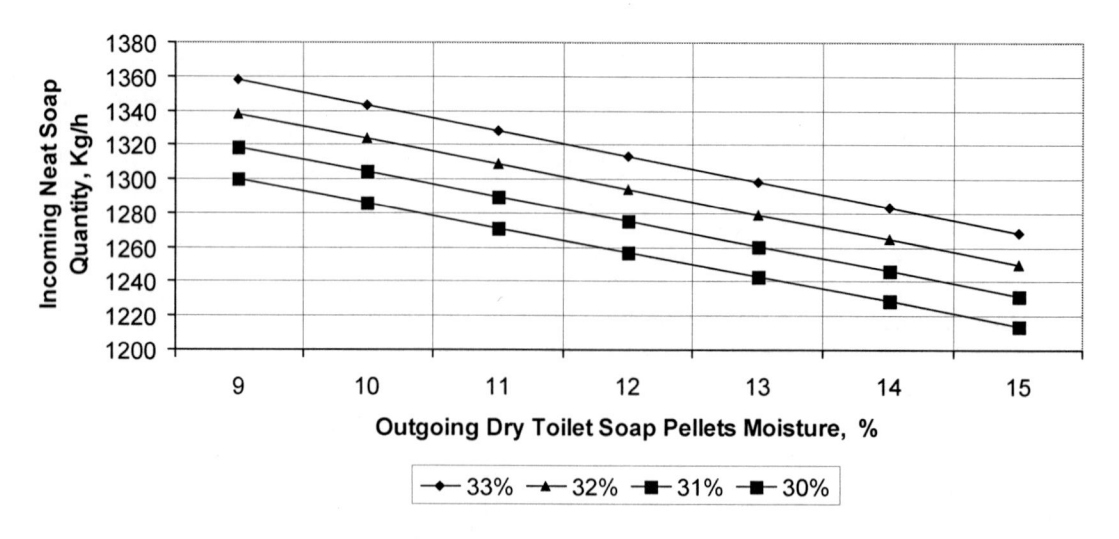

Fig. 10.23. Toilet Soap Mass Balance per 1000 Kg/h of Dried Product.

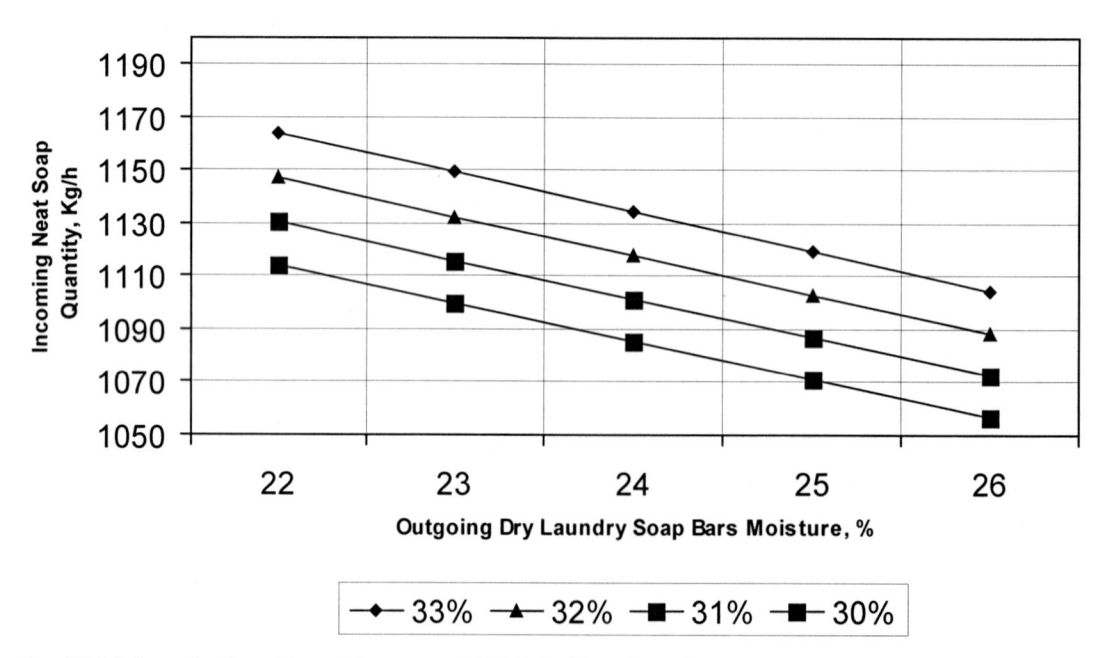

Fig. 10.24. Laundry Soap Mass Balance per 1000 Kg/ of Dried Product

Soap Drying Systems ● 297

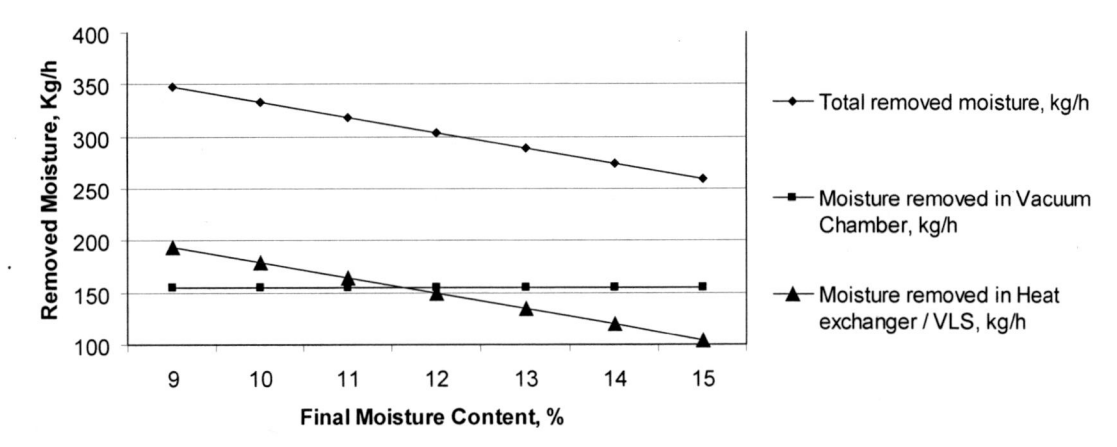

Fig. 10.25. Moisture removal distribution for Toilet Soap starting from 32% Neat Soap

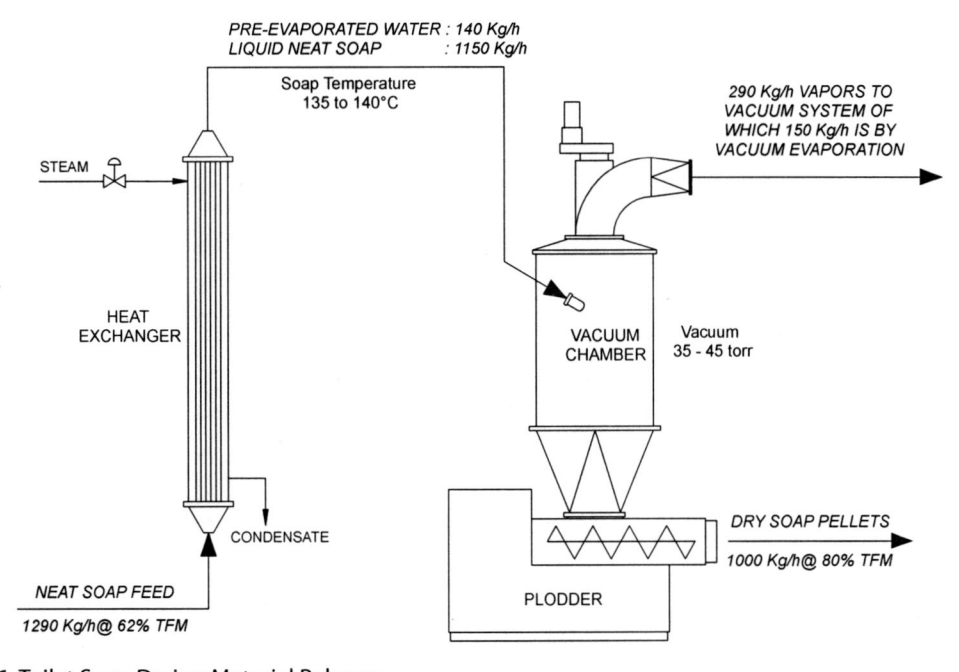

Fig. 10.26. Toilet Soap Drying Material Balance.

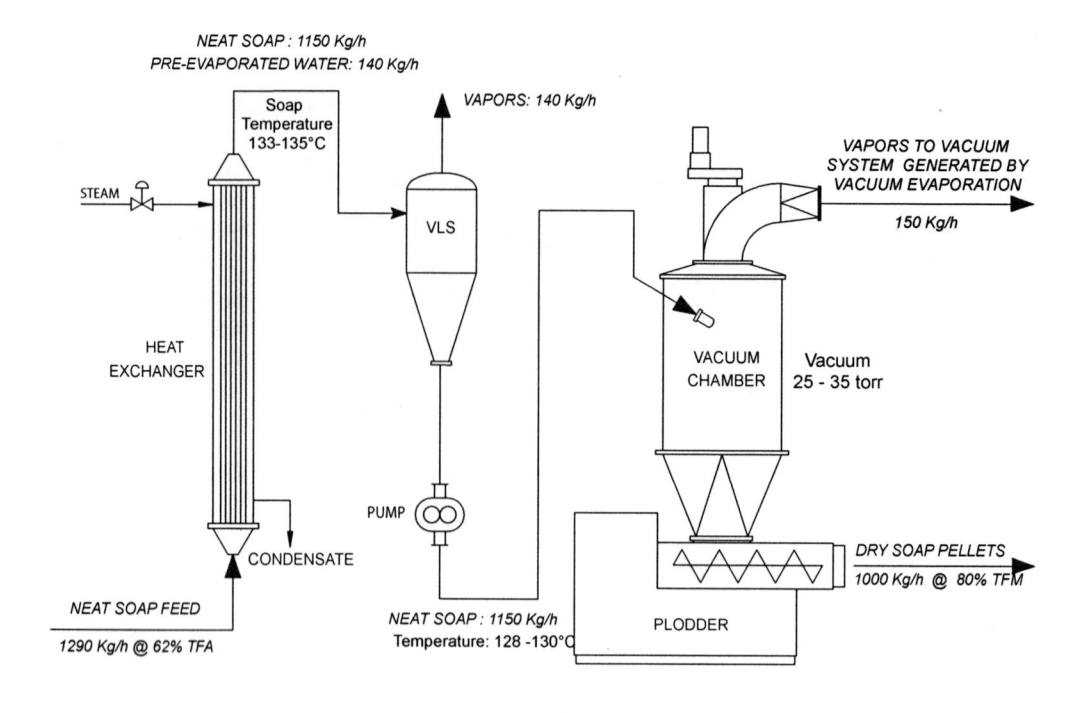

Fig. 10.27. Material Balance for Toilet Soap Drying with "VLS" System.

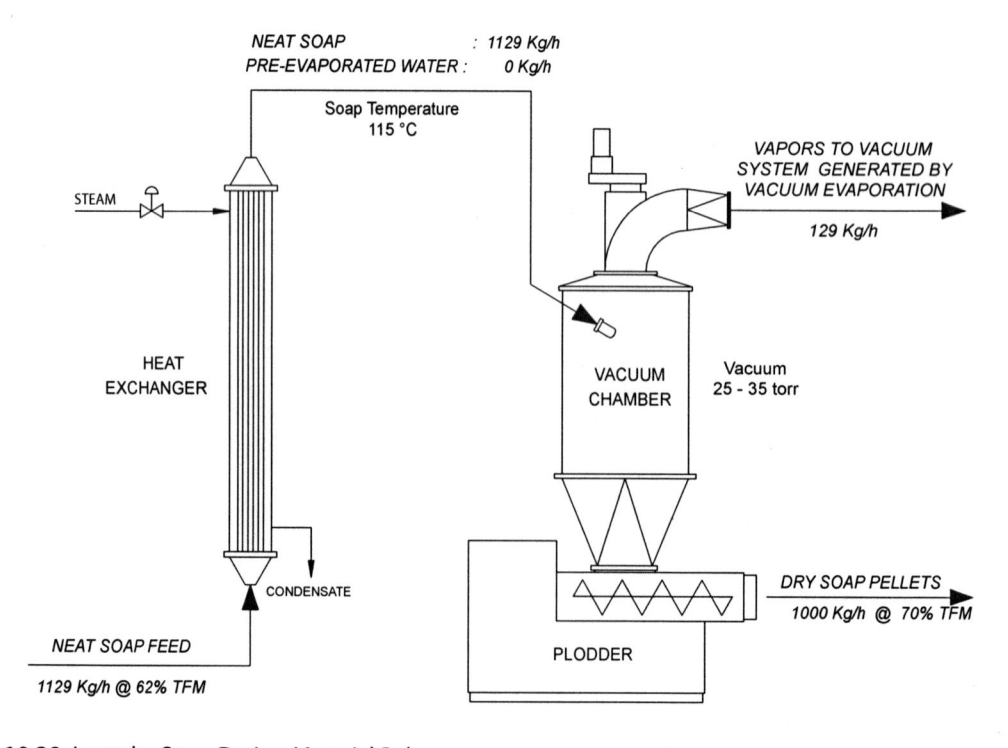

Fig. 10.28. Laundry Soap Drying Material Balance.

Heat Transfer Rate

Soap behaviour is pseudo-plastic, and its viscosity range under shear is from 50 to 100 cps at working temperatures in the heat exchanger. The global heat transfer rate, tube side, is calculated by the following quantities

Reynolds number: $Re = DVp/\mu$
Nusselt number: $Nu = hdlk$
Prandtl number: $Pr = Cp\mu/k$

using the following formula:

$$(hd / k) = 1.86 [(Re)(Pr)(d / l)]1/3 (\mu / \mu p)0.14$$

The value of h can be determined and substituted in

$$U = 1 (1/h + x/ks + 1/hv)$$

obtaining the overall heat transfer rate, where

D = tube diameter (mm)
V = velocity of soap flow (m/s)
p = density (kg/m^3)
$\mu/\mu p$ = viscosity in *cPoise* × 3.6
d = equivalent diameter (mm)
k = thermal conductivity (Kcal/m × h ×°C)
Cp = specific heat (Kcal/kg ×°C)
L = tube length (cm)
h = heat transfer coefficient (soap side) (Kcal/m^2 × h ×°C)
hv = heat transfer coefficient (steam side) (Kcal/m^2 × h ×°C)
x = tube wall thickness (mm)
ks = steel thermal conductivity (Kcal/m × h ×°C)
U = overall heat transfer coefficient (Kcal/m^2 × h ×°C)

Except for the production of a few types of laundry soaps, the water evaporated by expansion under vacuum is normally not enough for industrial drying purposes. The rest of the water must be pre-evaporated in the heat exchanger. Heat transfer areas utilized for vacuum dryer shell-and-tube heat exchangers are 20, 25, or 30 m^2/ton of dry soap; the size selected depends on the dry soap's final moisture and the product type.

Laminar flow is assumed for calculating the required heat transfer area. This allows for an adequate safety margin, since turbulent flow occurs when water starts boiling (pre-evaporation) in the heat exchanger.

Applying the equation

$$U = 1 / (1 / h + x / ks + 1 / hv)$$

we see that the heat transfer coefficient U is in the range of 80–200 Kcal/m^2 × h × °C. The average value utilized is 150 Kcal/m^2 × h × °C.

Practical experience indicates that a minimum 4–5 cm/s velocity must be considered for soap inlet into the heat exchanger tubes. This speed will avoid preferential flow of the soap to the central tubes and allow some chocking the other tubes.

Operating Vacuum (Absolute Pressure) and Condensate Temperature
With Barometric Condenser
For plants with direct-contact barometric condensers, it can be estimated that the residual pressure in the vacuum spray chamber will be that corresponding to the vapor pressure of condensate (cooling water + condensate process vapors) at the downleg temperature, increased by 3°C.

Example:
Cooling water temperature into barometric condenser: 30°C
Condensate temperature into hotwell (downleg temperature): 34°C
Vapor pressure of water at 34°C (absolute pressure): about 41 torr (from Table 10.2)
Actual vacuum in the dryer at 34°C + 3°C = 37°C: about 47 torr (from Table 10.2)

With Surface Condenser and ICS
For noncontact surface condensers, it can be estimated that the residual pressure in the vacuum spray chamber will be that corresponding to the vapor pressure of condensate process vapors at the downleg temperature, increased by 2°C. A pressure drop of 2–3 torr through the unit also must be taken into account.

Example:
Cooling water temperature into surface condenser or water temperature from ICS plate cooler: 30°C
Cooling water temperature from surface condenser (or water temperature to ICS plate cooler): 34°C
Condensate temperature into hotwell (downleg temperature): 37°C
Vapor pressure of water at 37°C (absolute pressure): 47 torr (from Table 10.2.)
Actual vacuum in the dryer at 37°C + 2°C : 39°C: about 54 torr (from Table 10.2.)
Plus 3 torr due to pressure drop, or 54 torr + 3 torr: 57 torr

Total Fatty Matter (TFM) and Moisture Content of Toilet Soaps
Assumptions:

- 80:20 tallow/coco mixture with an average 216 acid value

- Average molecular weight (MW): 56,100 / 216 = 260

- 63% TFM neat soap contains 0.40% glycerine, 0.05% NaOH, and 0.35% NaCl

Sample Calculation and Formulas:

$$\text{Fatty Acid} + \text{Caustic Soda} \rightarrow \text{Soap} + \text{Water}$$
$$RCOOH + NaOH \rightarrow RCOONa + H_2O$$
$$260 + 40 \rightarrow 282 + 18$$

The ratio between TFM (or TFA) and anhydrous soap in this case is 282 / 260 = 1.0846.
This means that neat soap with 63% TFM will in reality contain 63% × 1.0846 = 68.3% anhydrous soap.

Table 10.2. Vapor Pressure of Water

Temperature °C	Absolute Pressure torr	Temperature °C	Absolute Pressure torr	Temperature °C	Absolute Pressure torr
0	4.6	27	26.7	36	44.6
5	6.5	28	28.3	37	47.1
10	9.2	29	30	38	49.7
15	12.8	30	31.8	39	52.1
18	15.5	31	33.7	40	55.3
20	17.5	32	35.7	41	58.3
22	19.8	33	37.7	43	64.8
24	22.4	34	39.9	44	68.3
26	25.2	35	42.2	45	71.9

Table 10.3. Total Fatty Matter (TFM) and Moisture Content

TFM %	Anhydrous Soap %	Glycerine %	NaOH + NaCl %	Water %
58	63	0.4	0.4	36.3
59	64	0.4	0.4	35.2
60	65.1	0.4	0.4	34.1
61	66.2	0.4	0.4	33
62	67.3	0.4	0.4	31.9
63	68.4	0.4	0.4	30.8
64	69.5	0.4	0.4	29.7
65	70.6	0.4	0.4	28.6
66	71.6	0.4	0.4	27.5
67	72.7	0.4	0.4	26.4
68	73.8	0.4	0.4	25.3
69	74.9	0.4	0.4	24.2
70	76	0.4	0.4	23.1
72	78.2	0.5	0.5	20.9
74	80.3	0.5	0.5	18.7
76	82.5	0.5	0.5	16.5
78	84.7	0.5	0.5	14.3
80	86.8	0.5	0.5	12.1
82	89	0.5	0.5	9.9
84	91.2	0.5	0.5	7.7

References

Bassett, G. H. *Tubular Drying of Soap*. U.S. Patent 2,710,057, 1955.

Griffiths, J.J.; R.J. Wilde. British Patent 1,237.084, 1971.

Mazzoni, G. U.S. Patent 2,945,819, 1960.

Mazzoni, G. Italian Patent 386,583, 1940.

Mazzoni, G. S.p.A. Italian Patent 623,670, 1961.

Miag GmbH. British Patent 1,063,715, 1964.

Palmason, E.H. *Concentration of Foaming Materials*. U.S. Patent 3,073,380, 1963.

Perry, R.H. *Chemical Engineers' Handbook*, 8th ed.; McGraw-Hill: New York, 2007.

Spitz, L. *Soaps and Detergents, A Theoretical and Practical Review*; Spitz, L., Ed.; AOCS Press: Champaign, IL, 1996; pp. 207–242.

Weber and Seelander. European Patent 0123812, 1984.

• 11 •

Bar Soap Finishing

Luis Spitz
L. Spitz, Inc., Highland Park, Illinois, USA

Introduction

Bar soap finishing consists of six processing steps for the production of standard, superfatted, translucent, soap/synthetic (combo), and synthetic products into a solid bar (tablet) form packaged in various styles. The processing steps, the equipment used in each step, and bar soap finishing-line classification and selection are presented.

Processing Steps and Equipment

Processing Steps

- *Mixing* the main dry soap base, mostly in pellet form, with minor liquid and solid ingredients.
- *Refining* the fully formulated mixture into a uniform, homogeneous product.
- *Extruding* the finished product into a compact predetermined shape and size slug (billet).
- *Cutting* the extruded slugs into individual lengths as required by the soap press model used.
- *Stamping* the cut slug (billet) into a specified weight and shaped bar (tablet).
- *Packaging* the finished stamped bars.

Mixing

"Mixing" has no precise definition or measuring criteria in the soap industry. Macro and micro terms are used in other industries, and one can apply them to soap mixing. One can call the mixing of 1% or higher quantities of solid and liquid ingredients with the dry pelletized soap base in standard mixers macro mixing. During macro mixing, the additives only coat the outer surface of the pellets.

Intensive or micro mixing is achieved when the pellets are broken up to expose more surface area, helping the ingredients to penetrate into the pellets.

Refining

Refining is the work done on soap by the combined action of pressure and shear. The purpose of refining is threefold:

1. To produce a fully homogeneous, uniform product.
2. To improve bar feel by eliminating low solubility hard particles.
3. To enhance product lather, solubility, and firmness by affecting crystalline structure change.

Refining is performed with plodders, roll mills, or both units used in combination.

BAR SOAP FINISHING
PROCESSING STEPS AND EQUIPMENT

Dry Soap Base | **Liquid & Solid Additives**

PRE-REFINING
Simplex Refiner | Roll-Mill

MIXING
Amalgamator | Double Arm Mixer | Mixer-Kneader

REFINING
Roll-Mill | Simplex Refiner | Duplex Refiner

REFINING & EXTRUSION
Duplex Vacuum Plodder

CUTTING
Cutter

STAMPING
Press

PACKAGING
Packaging Equipment

Packaged Soaps

NOT PELLETIZED RECYCLE

PELLETIZED RECYCLE (Pelletizer)

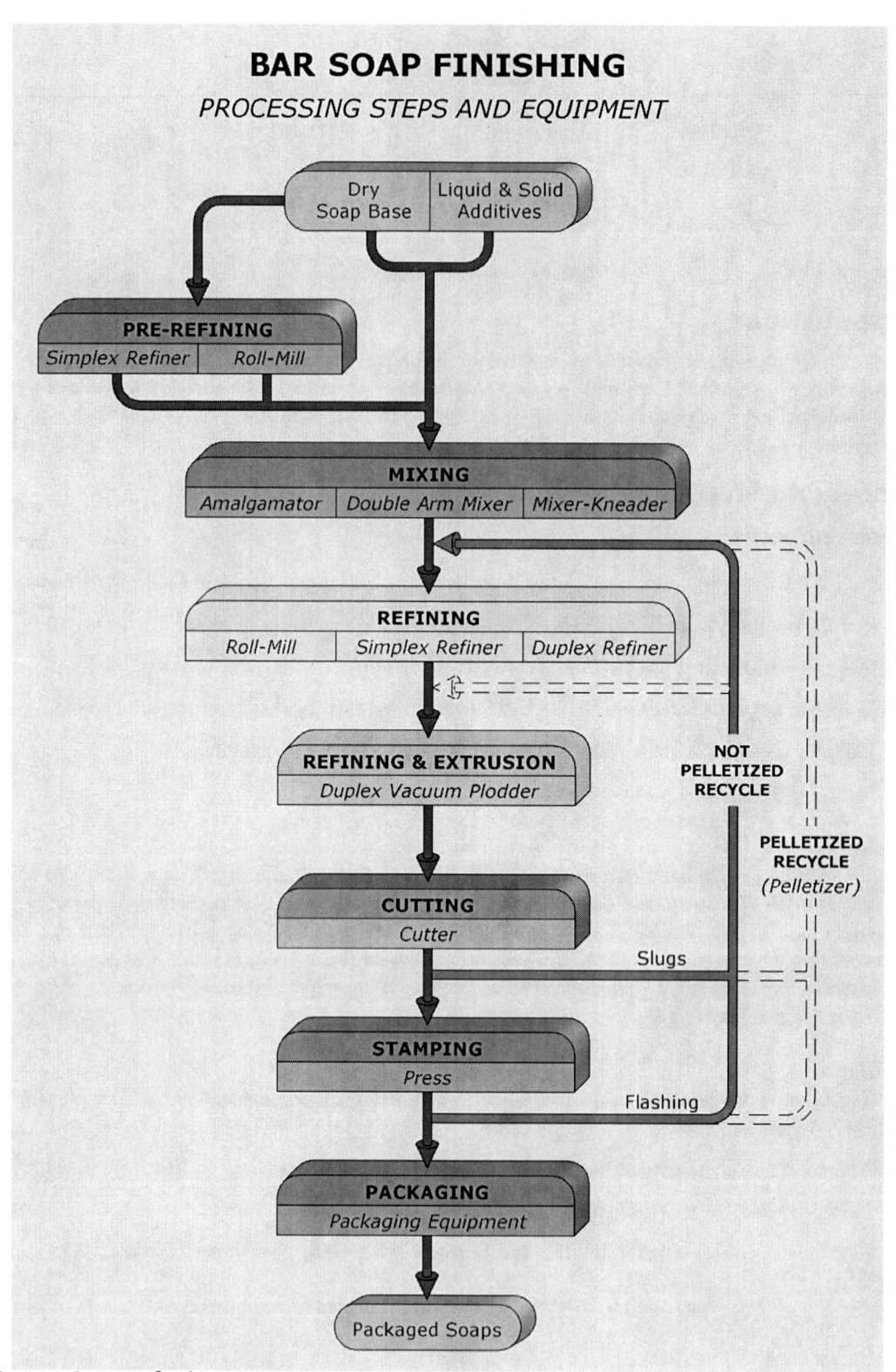

Slugs

Flashing

Fig. 11.1. Bar soap finishing: Processing steps and equipment.

Plodder Refining

One full stage of refining is achieved when a plodder is fitted with a 50-mesh sized refining screen (Fig. 11.2) When 20 or 30 US mesh size sized screens are used, the degree of refining is reduced.

The most widely used refining screen is the square mesh wire type. Screen suppliers offer the same mesh-number screens with different wire diameters, widths of opening, and percentages of open area (Table 11.1).

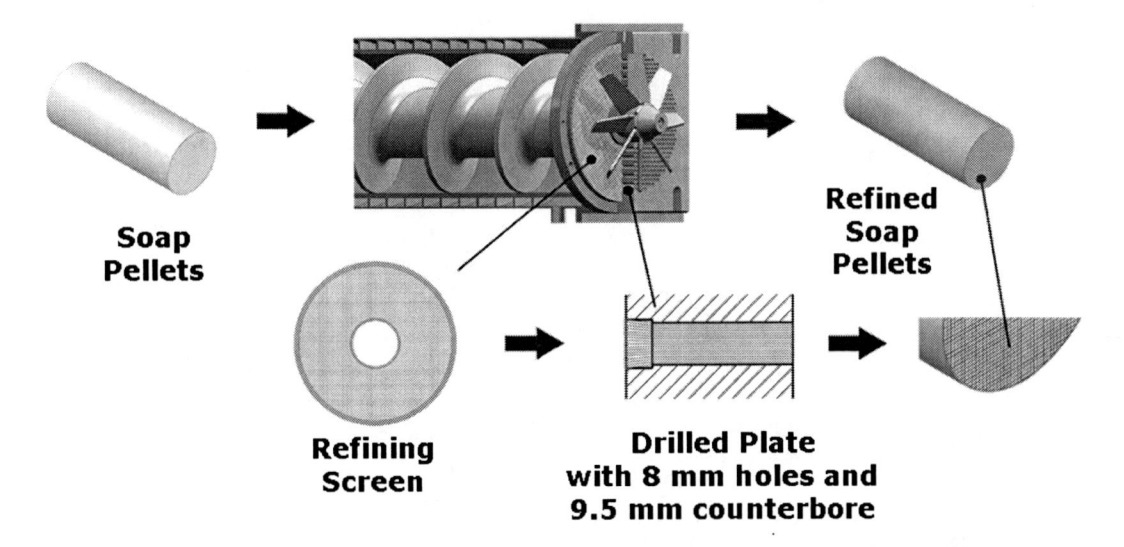

1 pass through a 50 mesh screen represents 1 Stage of Refining

Fig. 11.2. Plodder refining.

Table 11.1. Refining Screen Specifications

US Mesh Number	Wire Diamenter (in/mm)	Width of Opening (in/mm)	Open Area (%)
	0.047 / 1.19	0.059 / 1.50	34.8
10	0.032 / 0.81	0.068 / 1.73	46.2
	0.025 / 0.64	0.075 / 1.91	56.3
	0.020 / 0.508	0.030 / 0.76	36.0
20	0.018 / 0.320	0.032 / 0.81	41.0
	0.016 / 0.406	0.034 / 0.86	46.2
	0.015 / 0.381	0.018 / 0.47	30.1
30	0.013 / 0.330	0.020 / 0.52	37.1
	0.012 / 0.305	0.021 / 0.54	40.8
	0.008 / 0.203	0.012 / 0.31	36.0
50	0.009 / 0.229	0.011 / 0.28	30.3
	0.0075 / 0.191	0.0125 / 0.32	39.1

Roll-Mill Refining

The degree of refining (homogenization and dispersion) of the minor ingredients into the main soap base by a roll mill depends on two variables: the roll gap (clearance or nip) which determines the soap flake thickness and the shear generated by the roll speed differentials.

For most products, the gap between the last two rolls should be set at 0.15 to 0.20 mm for maximal performance. One must note that the actual flake (ribbon) thickness will be 0.05 to 0.06 mm more than the actual gap setting. This is due to the mechanical tolerance of the roller bearings.

The gap setting and speed differential between the last two rolls determine the degree of refining and the control of the product temperature. Unlike plodders, which always increase the product temperature during refining, roll mills are capable of maintaining and even reducing the product temperature.

Soap pellets passing through a roll mill are usually converted into thin flakes which are then usually formed into "crimped flakes" for easier subsequent conveying. The flaker knife (take-off) is a blade designed with multiple cuts and angles to produce a thin flake. The crimped, compacted flakes are formed with a scraper blade and a crimping bar (Fig. 11.3).

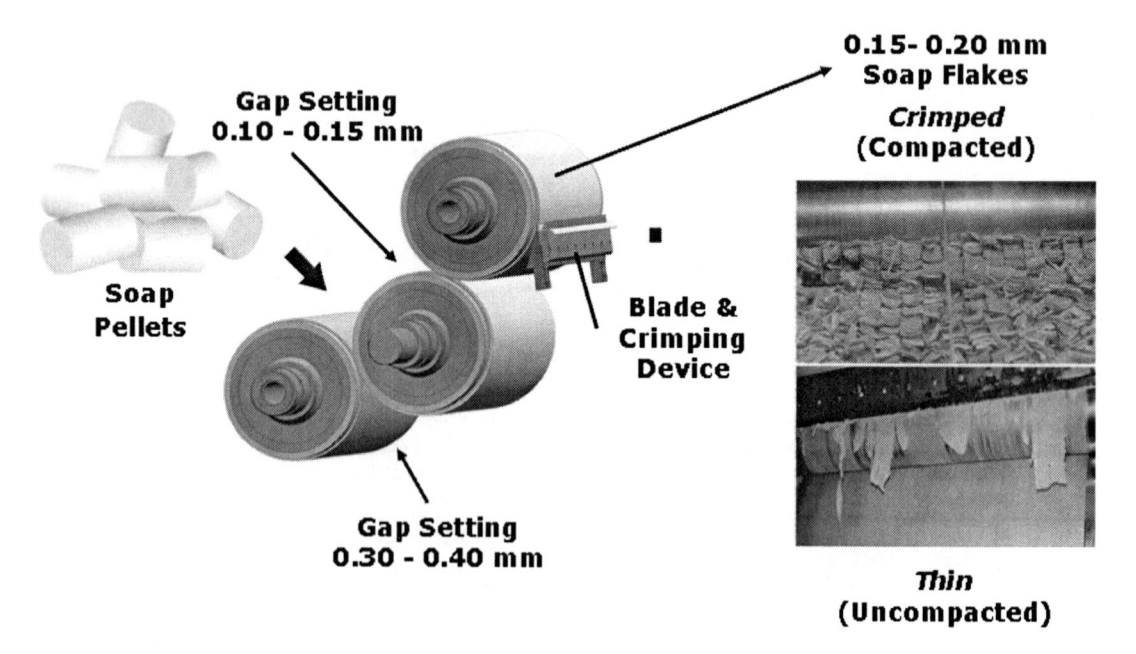

Fig. 11.3. Roll-mill refining.

Plodder and Roll-Mill-Refining Stages Comparison

A general comparison guide for the stages of refining between plodders and roll mills is presented; the classification of finishing lines is based on the number of refining stages (Table 11.2).

Table 11.2. Plodder and Roll-Mill-Refining Stages Comparison

Number of Refining Stages	Plodder Refining Screen Mesh Number	Roll Mill Flake Thickness (mm)
1½	80	0.15
1	50	0.20
¾	30	0.25
½	20	0.30
¼	10	0.40

The use of finer than 50 mesh screens (such as 80 mesh) is not recommended due to an excessive product temperature rise and a potential production rate reduction.

Methods for Measuring the Degree of Refining
Washdown Temperature

The washdown-temperature test is used to measure the degree of refining—evaluating the presence of hard particles (specks) by soap feel (grittiness, sandiness, roughness). The bar is washed with both hands for 1 minute in a sink with 30°C water. Once the bar surface is smooth and all protruding lettering and designs are washed away, the water temperature is decreased. Washdown temperature is the temperature at which one can detect the first hard specks.

- A smooth bar without any hard specks has a 22°C washdown temperature.
- A "slightly gritty" feel appears at 23 to 24°C.
- A "moderately gritty" feel appears at 25 to 26°C.
- A "gritty" feel appears at 26 to 27°C.
- A "very gritty" feel occurs when the washdown temperature reaches 28°C or higher.

Photo-Evaluation Scale

One can also use a visual method for roughness evaluation. The bar is washed for 1 minute in 20°C water and then left to dry. If the bar is held at an angle in front of a high-intensity light source and below eye level, one can easily see the dry specks. By using photographic standards, one can grade the bar as 0, 1, 2, 3, 4, or 5. Zero represents a smooth product, whereas five refers to a very gritty bar (Fig. 11.4).

Pre-refining

Pre-refining is a refining step performed before the addition of any minor liquid and solid additives to an old (fully aged) or to a new (fresh or partially aged) dry soap base. The use of a pre-refining step is especially advantageous for aged soap, low-moisture-content syndet, and high-titer soap. The main advantages of pre-refining are:

- Easier processing of hard, low-moisture-content syndet, high titer, and translucent soap.
- Better mixing of the liquid additives with the plasticized, higher temperature, partially refined base. This facilitates the refining action in the subsequent processing stages.
- Improved refining (lower washdown temperature of the finished product).

Soap Base Aging

If the soap base (with or without additives) is stored and aged before final refining, extrusion, and stamping, line efficiency increases considerably. During aging, soap crystallization is completed, and soap temperature is reduced. The optimal aging time has to be determined experimentally for each specific soap formula.

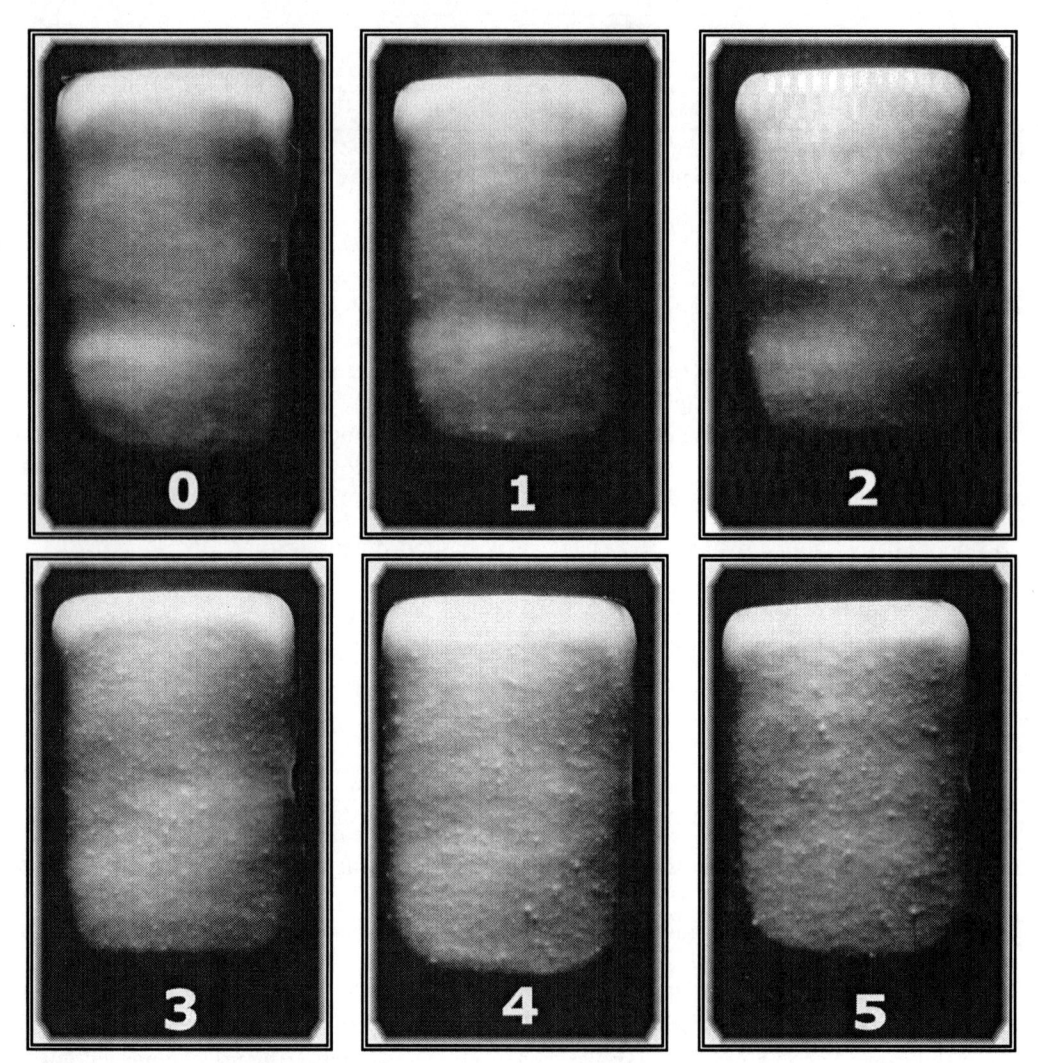

Fig. 11.4. Photo-evaluation scale.

Finishing-Line Equipment

Mixers

Amalgamator with Open-Arm Sigma Blades

The most popular and widely used mixer, called amalgamator in the soap industry, is a top-loading, bottom-discharging, non-tilting unit with "open-arm type" Sigma profile blades. This easy-to-clean, efficient blade design is derived from the Sigma blade, which is the universal mixing blade in the chemical industry. These mixers mainly coat the outer surface of the pellets with the additives. Mixers only partially break up the pellets, thereby limiting the penetration of the additives (Fig. 11.5).

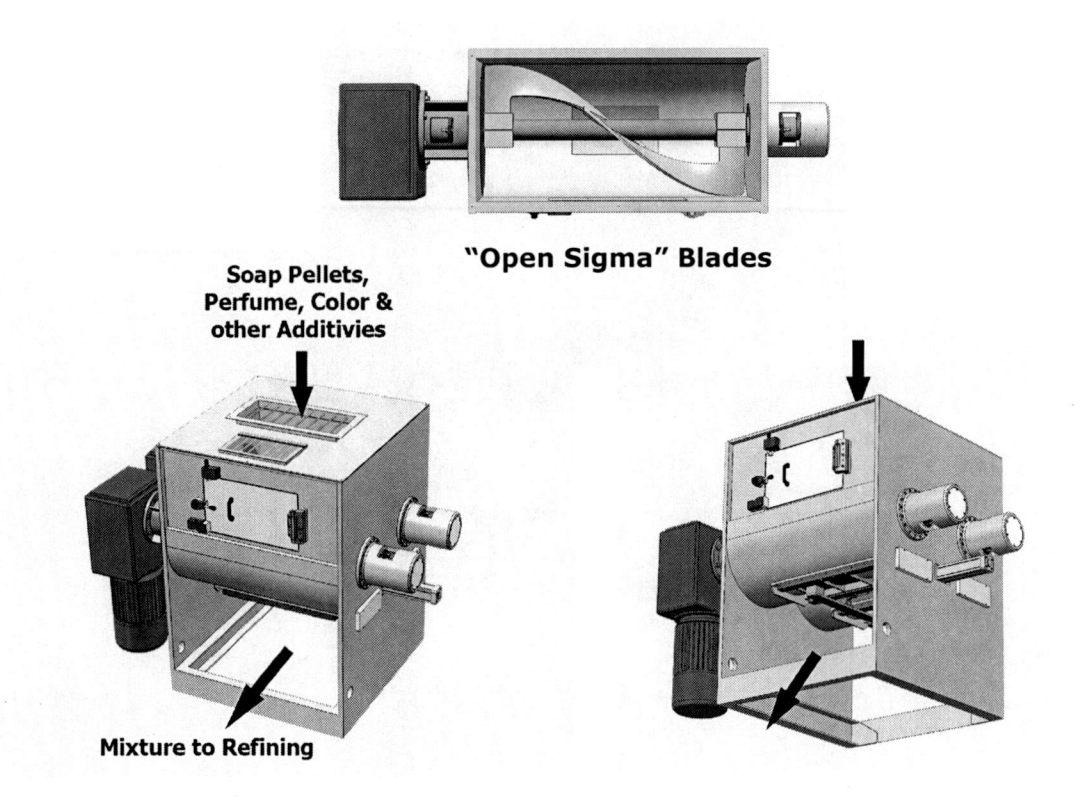

Fig. 11.5. Amalgamator.

Double-Arm Mixers with Sigma Blades
The double-arm mixers with two counter-rotating tangential or overlapping Sigma blades (Z blades) are very efficient units; in spite of their higher cost, they are gaining increased acceptance in the soap industry. They are mainly used for synthetic and combo toilet soaps, synthetic laundry bars, and translucent soap production. They are offered in fixed- and tilting-type discharge versions (Fig. 11.6).

Mixer–Kneaders
A mixer–kneader consists of a double-arm mixer with two tangential Sigma blades and a discharge extruder screw. The extruder is located at the bottom of the mixer vessel trough (bowl), and is fitted with a pelletizing head similar to a plodder. The rotation of the two Sigma blades creates intensive mixing/kneading action. Intensive mixing/kneading action is achieved by the countercurrently rotating Sigma blades (one rotating twice as fast as the other) and the screw pushing the product up into the blades. At the end of the mixing cycle, the screw rotation is reversed to facilitate product discharge and pelletizing.

Mixer–kneaders for soap applications were introduced in the 1960s by Miag from Germany. They have not gained much acceptance over the years because of their high cost and because they do not offer real advantages over the use of a batch amalgamator followed by a separate simplex refiner. Binacchi recently reintroduced these mixers to the soap industry. Mixer–kneaders are especially suited for translucent-soap production lines (Fig. 11.7).

Ribbon blender, paddle, and plow type mixers have found limited use in the soap industry.

310 • L. Spitz

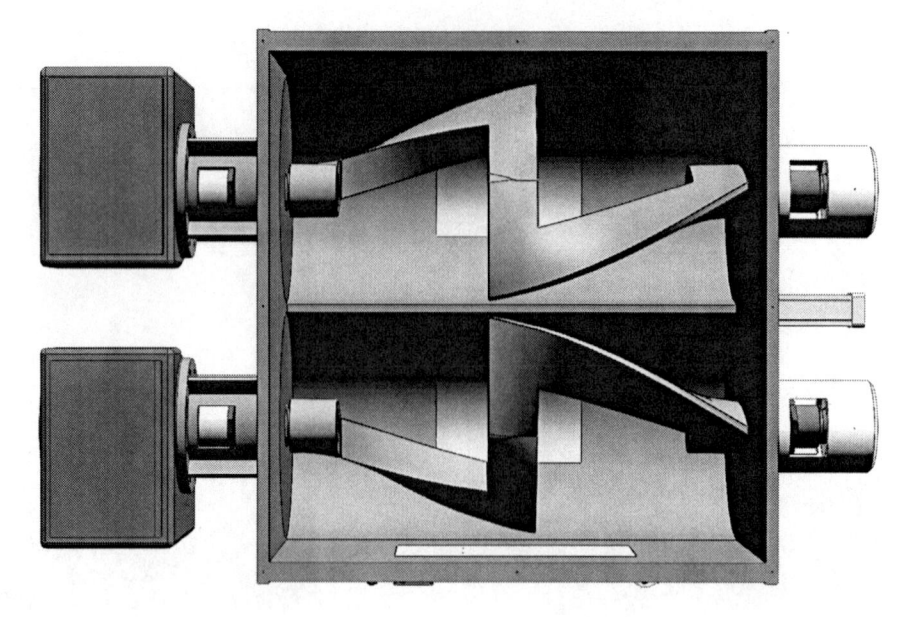

Fig. 11.6. Double-arm sigma mixer.

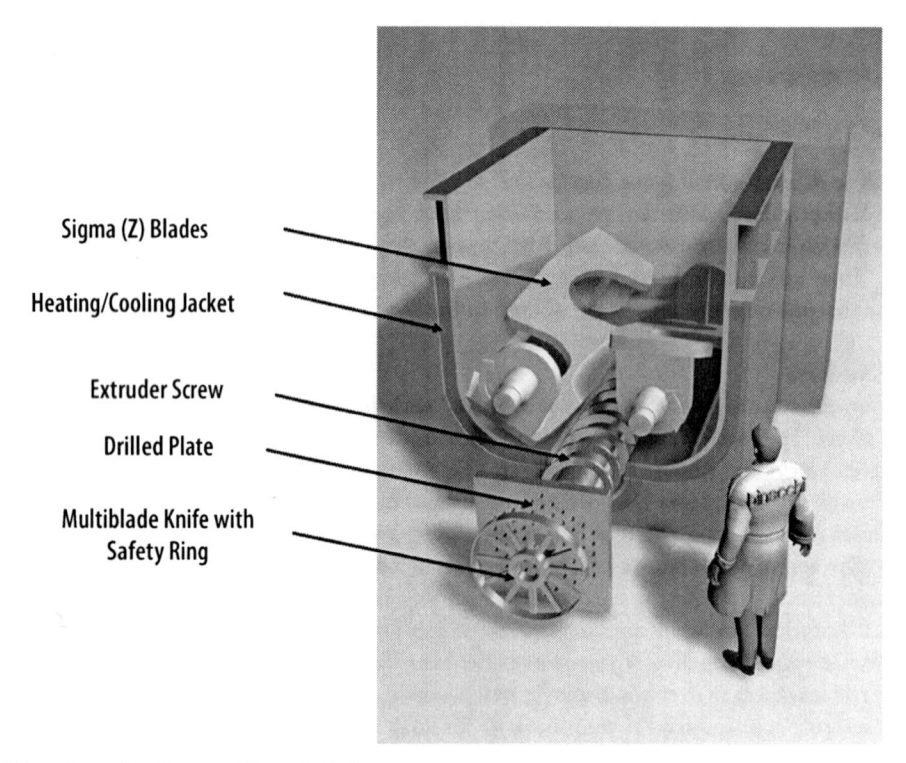

Sigma (Z) Blades
Heating/Cooling Jacket
Extruder Screw
Drilled Plate
Multiblade Knife with Safety Ring

Fig. 11.7. Mixer-kneader. Source: Binacchi & Co.

Roll Mills

Soap mills were always available in three-, four-, and five-roll variations. Over the last decade, the four- and five-roll mills practically disappeared; three-roll mills became the industry standard for the following reasons:

- Contact areas are large enough to ensure proper product refining and cooling.
- Most of the total power absorbed, that is, the heat input to the soap, takes place between the last two rolls.
- Gap-clearance-setting adjustment and control are easier.
- Three-roll mills have lower operating costs due to lower power and cooling-water requirements.

Roll-configuration geometry is very important because it determines the total contact area available for refining and cooling (Fig. 11.8).

Binacchi's model BRM-V three-roll mill was introduced this year. This novel design with a V-shaped roll positioning has a 510° total contact surface area higher than other three-roll mills. Also, this design is claimed to assure that no soap falls to the ground. No soap contamination occurs, and optimal temperature control is achieved. Each roll has an independent drive and gear box, features which allow the speed of each roll to be changed as required (Fig. 11.9).

THREE-ROLL MILL CONTACT AREAS

265°
225°
Total Contact Area = 490°

235°
180°
Total Contact Area = 415°

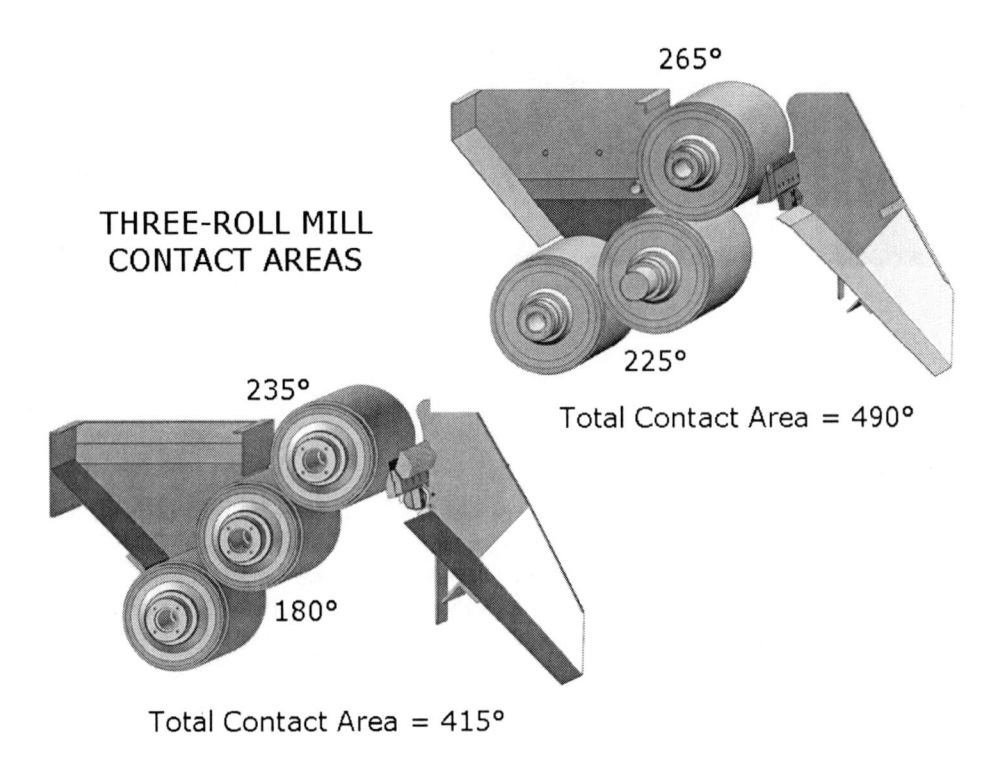

Fig. 11.8. Three-roll mill contact areas.

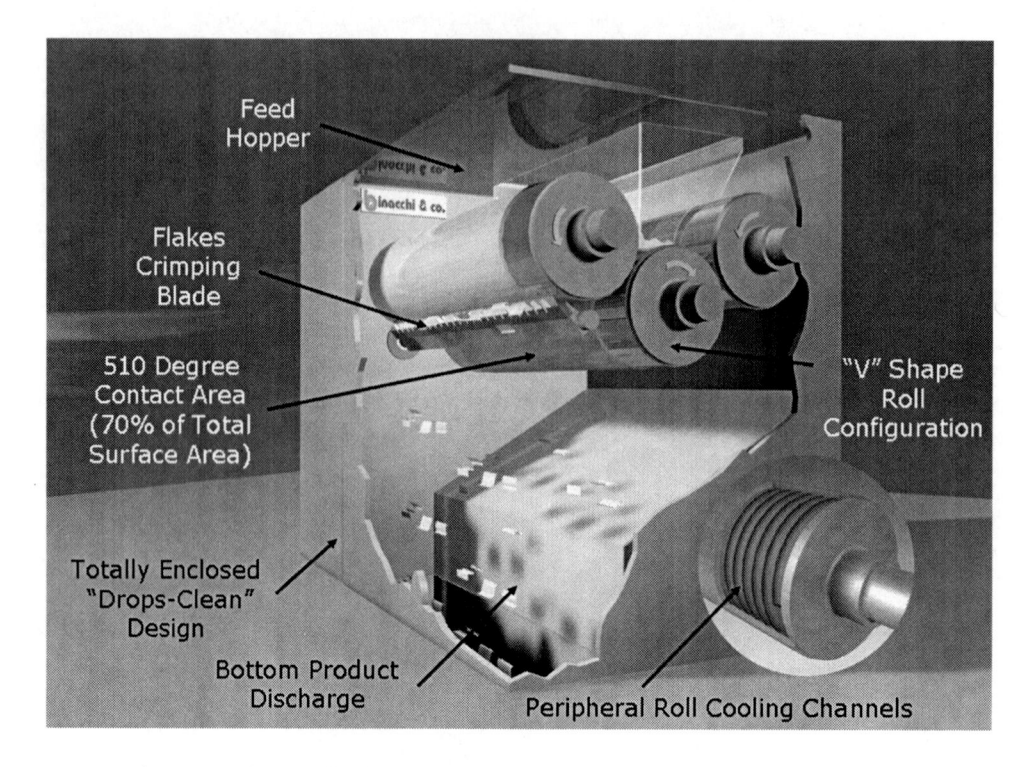

Fig. 11.9. Binacchi BRM-V Three-Roll Mill. Source: Binacchi & Co.

Roll Design Metallurgy
Standard quality rolls are made of compound, chilled cast iron with 500–550 HB (Brinell) hardness. For reduced corrosion and wear, high-chromium cast-iron rolls are indicated with 520–580 HB to a depth of chill of 15 to 20 mm. The cooling efficiency depends on the minimal thickness of the chilled white-iron outer layer of the cast-iron rolls, which has a thermal conductivity of 20 to 25 W/m°C. The thermal conductivity of the rolls' gray-iron core is higher, ranging from 45 to 60 W/m°C.

Roll Cooling Systems
Several roll water-cooling systems are illustrated. The most widely used system sprays water onto the inside wall of the rolls (Fig. 11.10).

The standard and the more expensive, seldom used, high-efficiency peripheral roll cooling systems are shown in Fig. 11.11.

The peripheral system with a temperature-control unit is shown in Fig. 11.12.

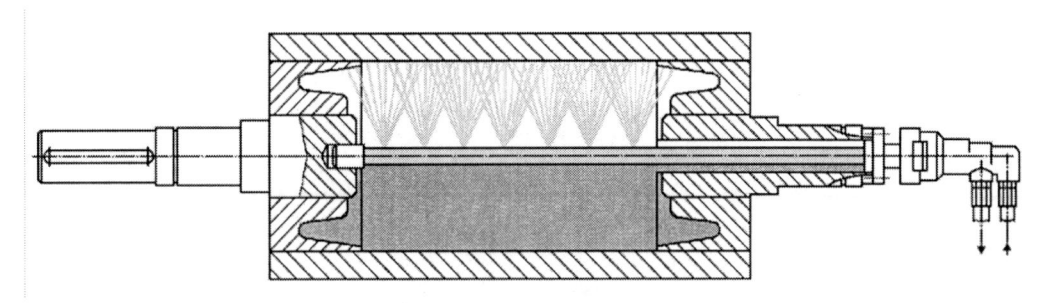

Fig. 11.10. Traditional water-spray roll-mill cooling system. Source: Sela GmbH.

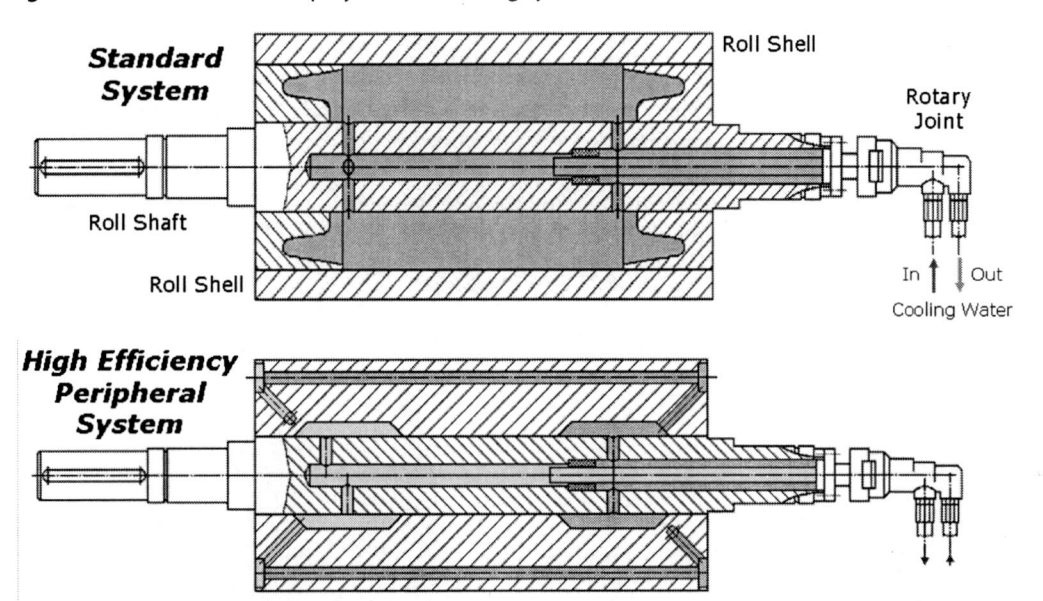

Fig. 11.11. Standard and peripheral roll-mill cooling system. Source: Sela GmbH.

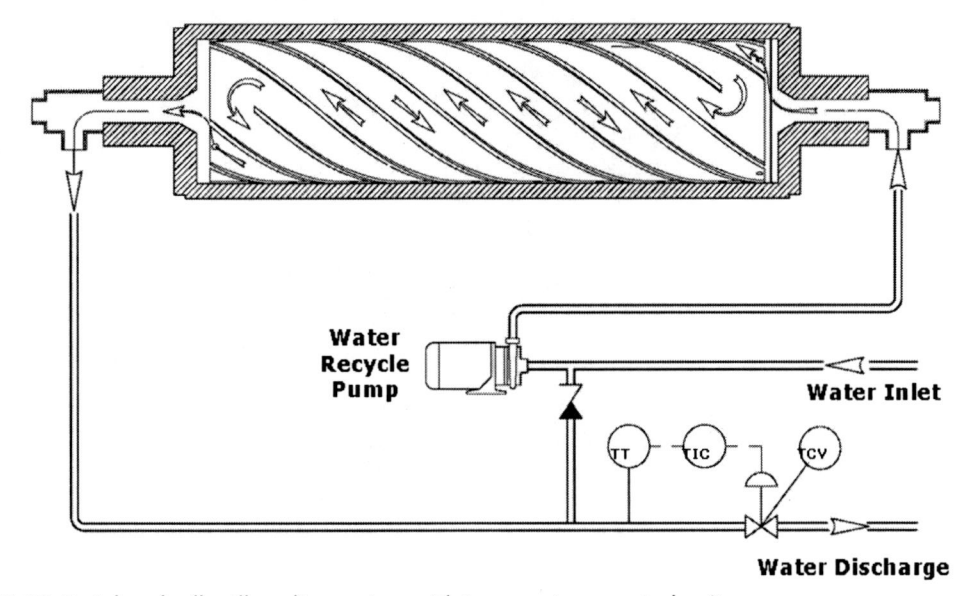

Fig. 11.12. Peripheral roll-mill cooling system with temperature-control unit.

Plodder Types

The soap industry uses three basic types of plodders, each of which is available in single-worm and twin-worm versions:

Simplex Refiners
A simplex refiner consists of one plodder designed to operate with a 50-mesh refining screen at a maximal pressure of 60 bar.

Duplex Refiners
A Duplex Refiner consists of two Simplex Refiners mounted in tandem.

Duplex Vacuum Plodders
A Duplex Vacuum Plodder consists of two plodders mounted in tandem and connected by a vacuum chamber. The preliminary stage plodder is exactly the same as a simplex refiner, and one can use it with a 50 mesh refining screen. In the final-stage plodder, the refined pellets are compacted and extruded as a continuous slug (billet), free from any entrapped air (Fig. 11.13).

One must note that a plodder functions as a refiner only when it is fitted with a 50-mesh screen. When 30 or 20 screens are used, the refining degree is reduced. The 10-mesh and coarser screens are used at times as backup, protective screens for the finer 30- or 50-mesh refining screens.

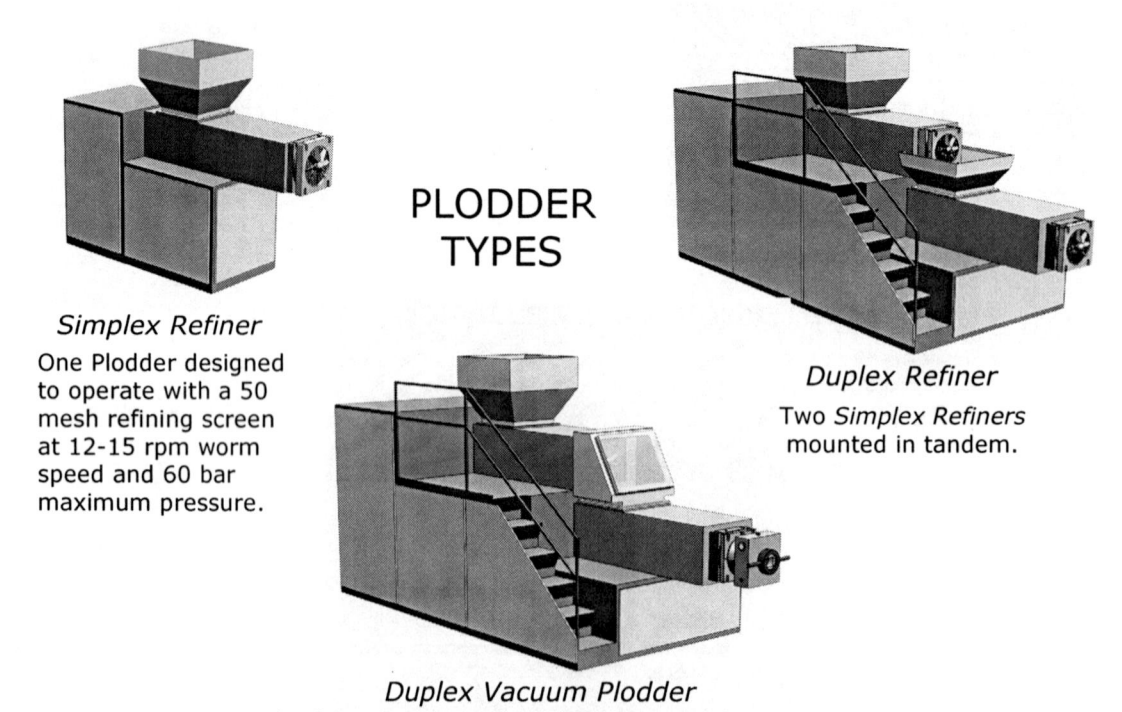

PLODDER TYPES

Simplex Refiner
One Plodder designed to operate with a 50 mesh refining screen at 12-15 rpm worm speed and 60 bar maximum pressure.

Duplex Refiner
Two *Simplex Refiners* mounted in tandem.

Duplex Vacuum Plodder
The first stage is a *Simplex Refiner* and final stage is an *Extruder*. The two plodders mounted in tandem are connected by a vacuum chamber.

Fig. 11.13. Plodder types.

Plodder Refining and Pelletizing Group

The refining stage of any plodder has a refining and pelletizing group as shown in Fig.11.14, 16 and 17.

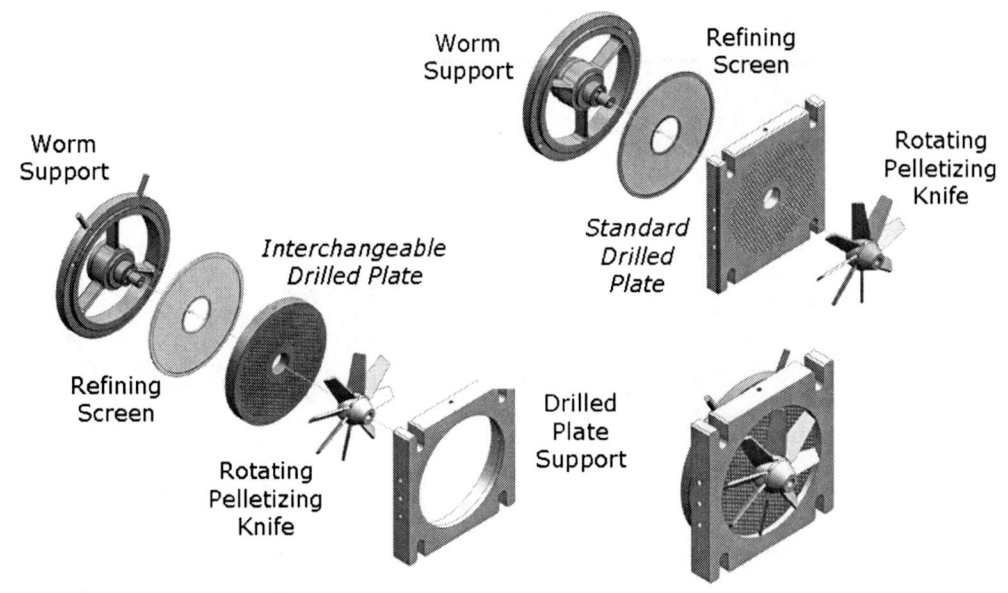

Fig. 11.14. Plodder refining and pelletizing group.

Plodder Extrusion Group

All of the components of the extrusion group of a Duplex Vacuum Plodder are illustrated in Fig. 11.15.

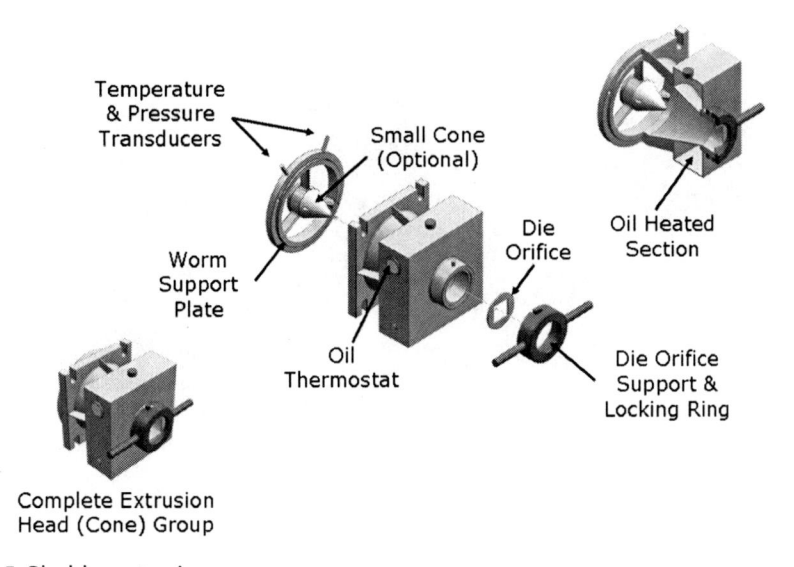

Fig. 11.15. Plodder extrusion group.

316 • L. Spitz

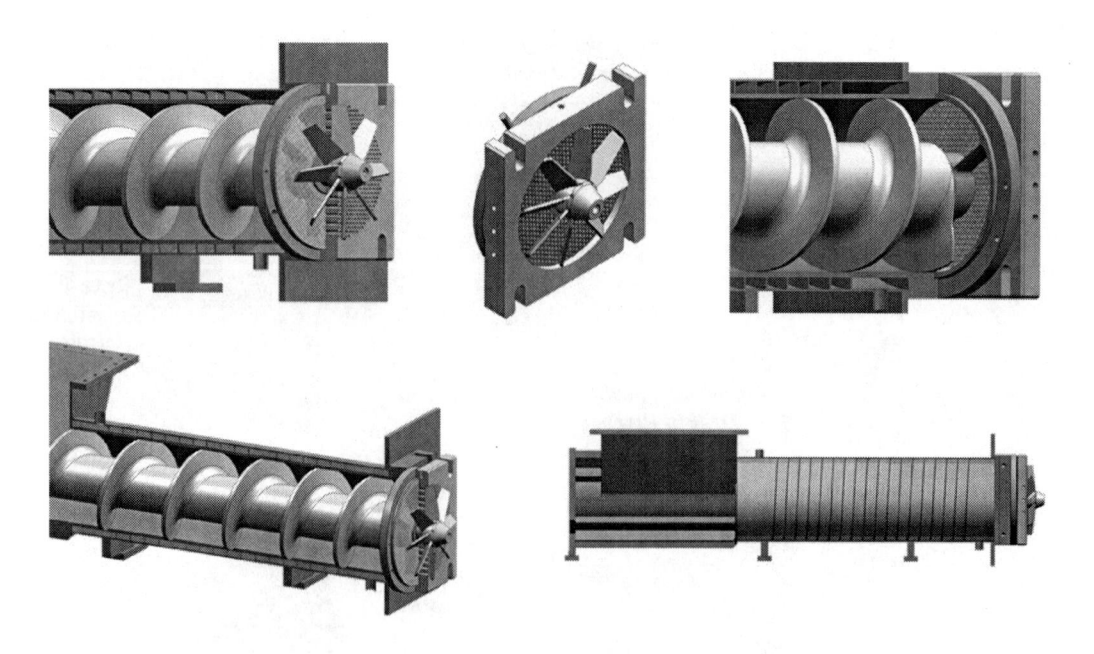

Fig. 11.16. Refining/Pelletizing Stage details.

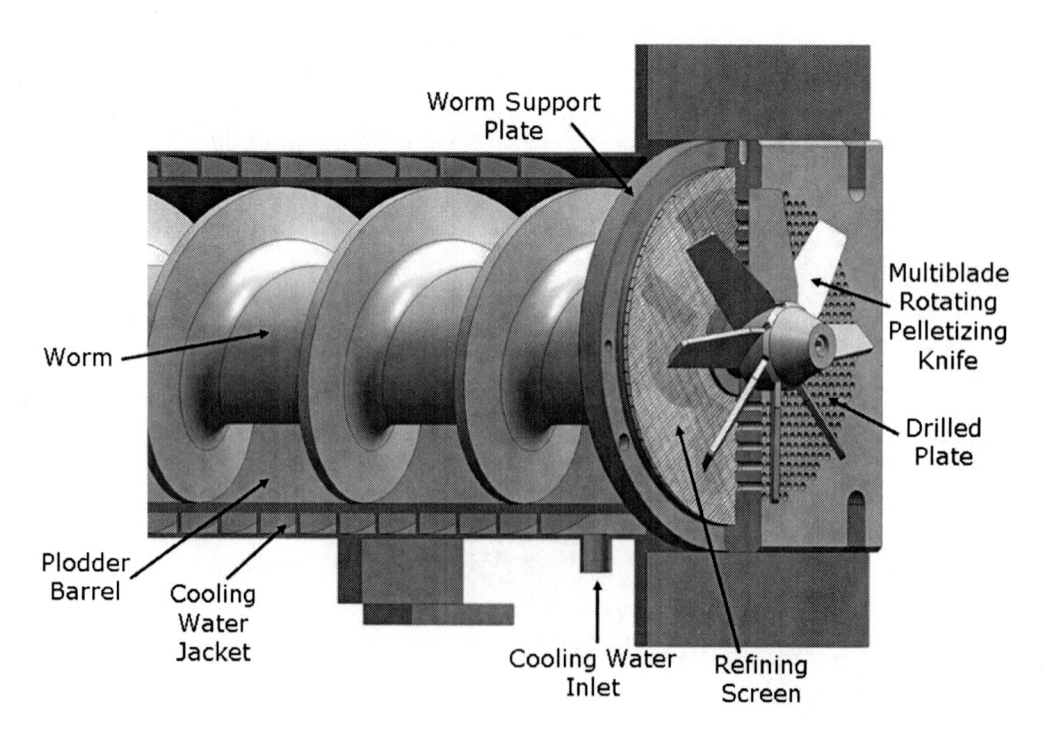

Fig. 11.17. Refining/Pelletizing Stage details.

Plodder Worms
Plodder worms are designed to perform refining, compression, and extrusion functions.

Worm Types
Three types of worms are available: (i) single-worm (ii) tangential twin-worm and (iii) nontangential twin-worm:

- Twin-worm plodders are available with tangential (touching) counter-rotating worms in a single barrel or nontangential (nontouching) counter-rotating worms in two separate barrels.
- Twin-worm plodders with tangential (touching) counter-rotating worms in a single barrel are recommended for processing sticky products and for high-capacity production lines.
- Nontangential twin-worm plodders are used for high-speed lines and multicolored soaps (Fig. 11.18).

Terminology
Lc is the closed barrel section and D is the worm diameter. Please note that the Lc/D ratio is not the same as the L/D ratio, which is the total worm length L to the D worm-diameter ratio (Fig. 11.19).

Worm Styles
Several worm styles are available with different profiles suitable for various applications. A summary of Binacchi, Mazzoni LB, and Sela worms is shown in Figures 11.20, 11.21, and 11.22.

Optimizing a plodder-worm design depends on extensive testing with different products. This is an ongoing challenge for equipment suppliers.

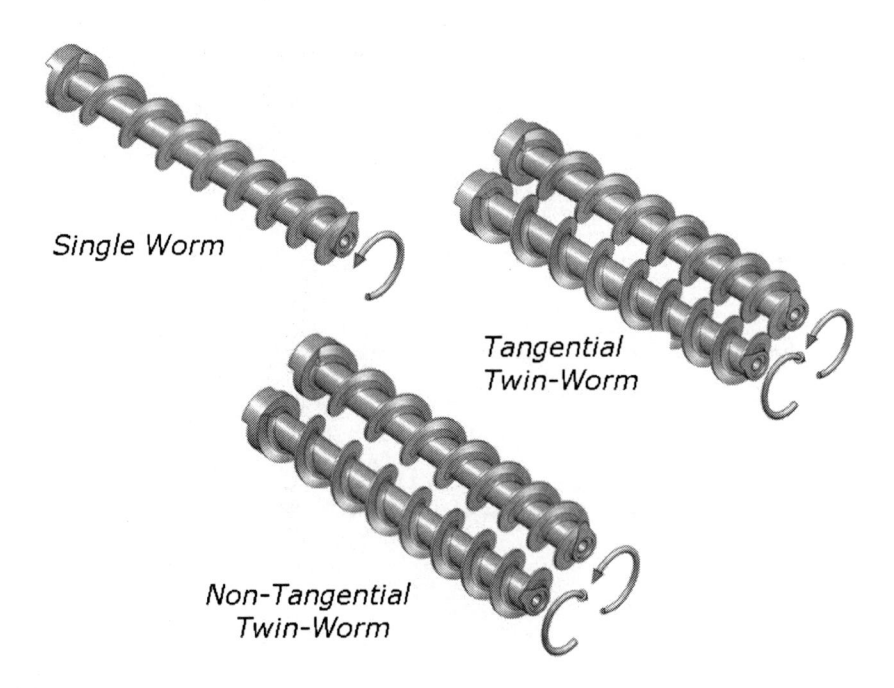

Fig. 11.18. Plodder worm types.

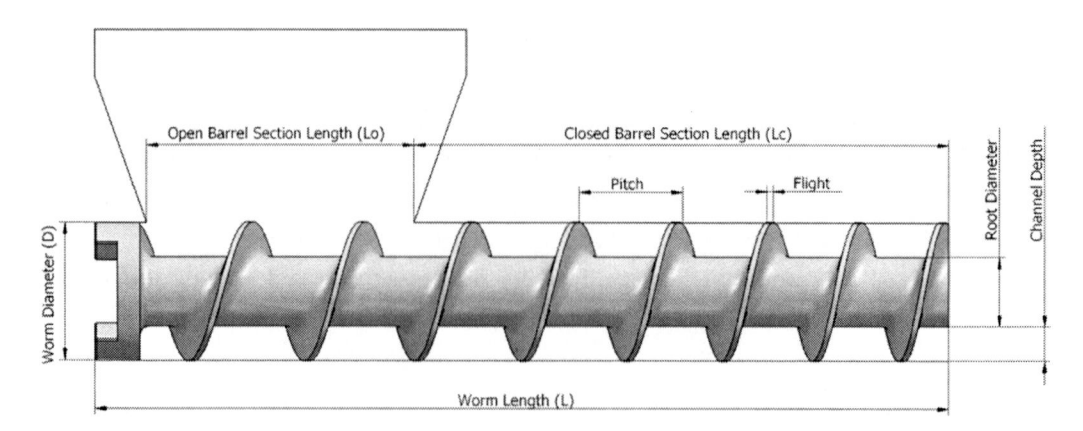

Fig. 11.19. Plodder-worm terminology.

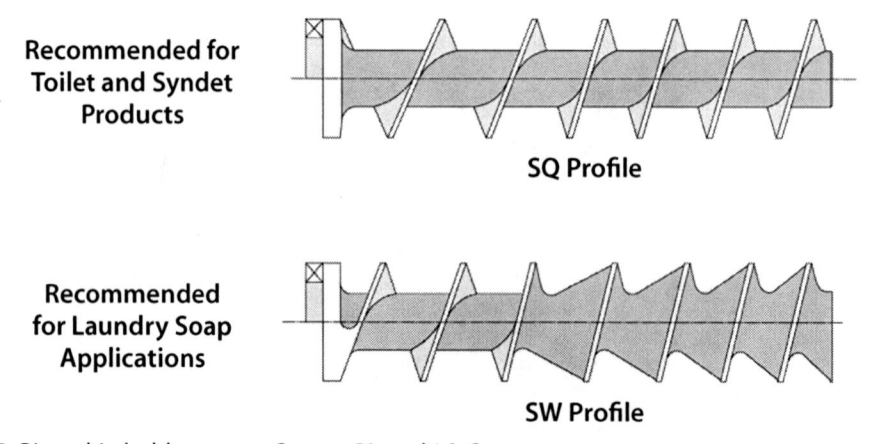

Fig. 11.20. Binacchi plodder worms. Source: Binacchi & Co.

Recommended for Toilet and Syndet Products — SQ Profile

Recommended for Laundry Soap Applications — SW Profile

Mazzoni LB Plodder Worms
4:1 Lc/D Ratio

HE - High Efficiency
For All Types of Products

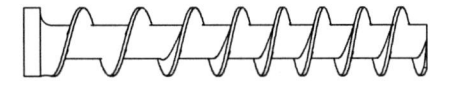

Constant Root Diameter
Decreasing Pitch

HV - High Volume
For Large Capacity Toilet and Laundry Soap Dryers

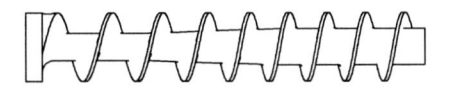

Increasing Root Diameter
Decreasing Pitch

CT – Constant
For Products Requiring Reduced Mechanical Work

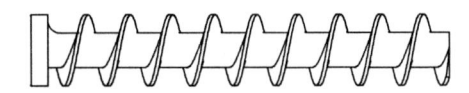

Constant Root Diameter
Constant Pitch

9:1 Lc/D Ratio for MRP plodders

HEL– High Efficiency Long
For All Types of Products

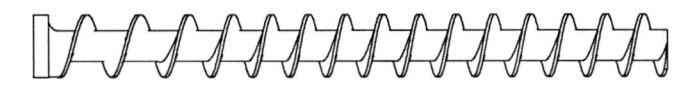

Constant Root Diameter
Decreasing Pitch

Fig. 11.21. Mazzoni LB plodder worms. Source: Mazzoni LB, SpA.

Short- and Long-Lc/D-Ratio Plodders

Conventional plodders are designed with 3:1 *Lc/D* ratios, and operate at 12–15 rpm worm speeds. The 4:1 types operate at 18–20 rpm, and the long 9:1 *Lc/D* ratio units operate at up to 50 rpm worm speeds.

The first 4:1 *Lc/D* ratio NOVA plodders were introduced in 1984 by G. Mazzoni SpA. In 1988 Binacchi & Co. launched the Extenda models. Currently all suppliers offer mostly 4:1 *Lc/D*-ratio plodders.

Special New Plodders

Two special plodders with long *Lc/D* ratios, special worm designs, and refining and milling options were introduced by Mazzoni LB and SAS.

Mazzoni LB Multirefining Plodder (MRP)

The multirefining plodder (MRP) is also a Duplex Vacuum Plodder designed to refine, homogenize, and extrude. It is claimed that the special characteristics permit its use in a finishing line without the need of any additional equipment.

The MRP's first refining stage is a 4:1 *Lc/D*-ratio twin-worm plodder with HE style worms. The final multirefining stage consists of a single-worm plodder in two separate sections. The first section has a 7:1 *Lc/D* ratio, and the second section has a 2:1 *Lc/D* ratio. Both are HE-type worms. The second section is removable to facilitate cleaning (Figs. 11.23, 24).

320 • L. Spitz

Recommended for
Combo Bars and Rim
Blocks

Recommended for
Toilet, Syndet, and
Laundry Soaps

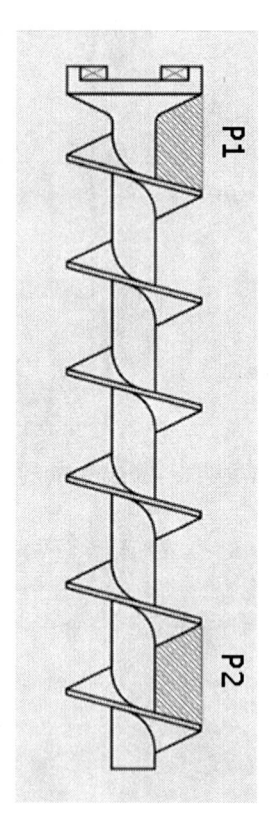

Recommended for
Translucent Soaps

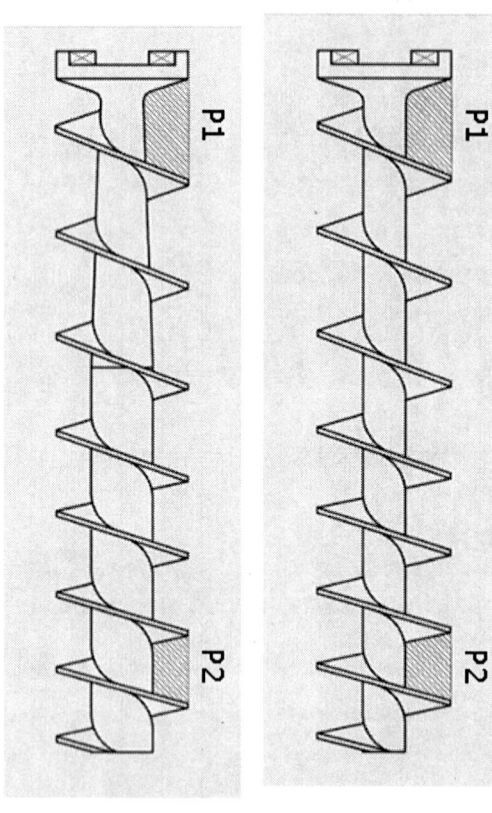

Constant Pitch
P1 = P2

Single
Decreasing Pitch
P1:P2 = 1.6:1

Double
Decreasing Pitch
P1:P2 = 2.5:1

Fig. 11.22. Sela plodder worms. Source: SELA GmbH.

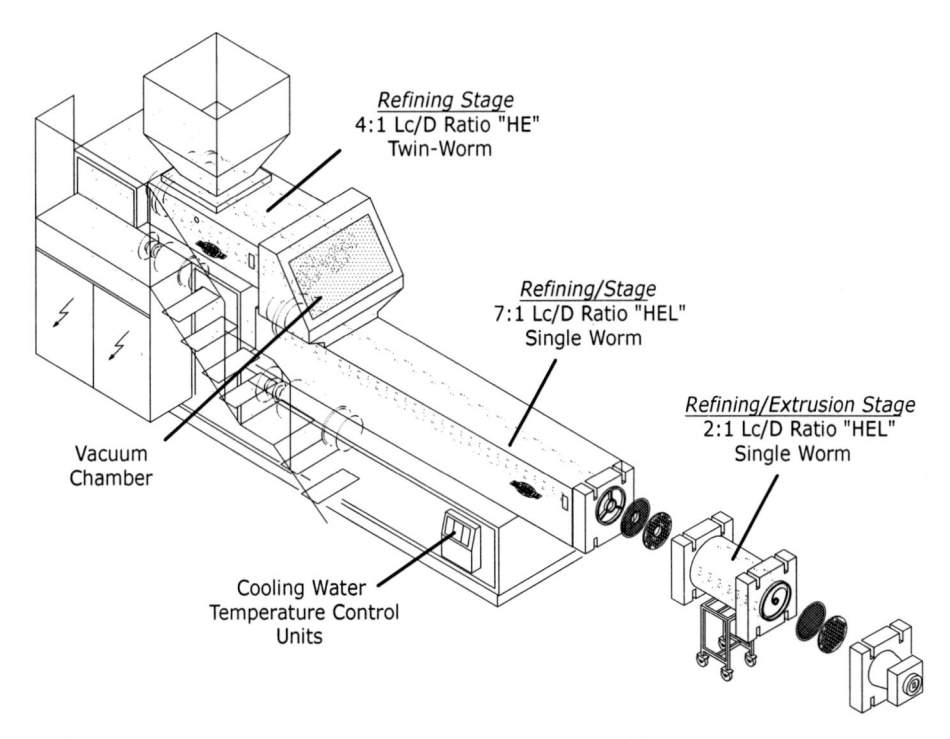

Fig. 11.23. Mazzoni LB MRP (Multirefining Plodder). Source: Mazzoni LB, SpA. For more details see Fig. 11.24.

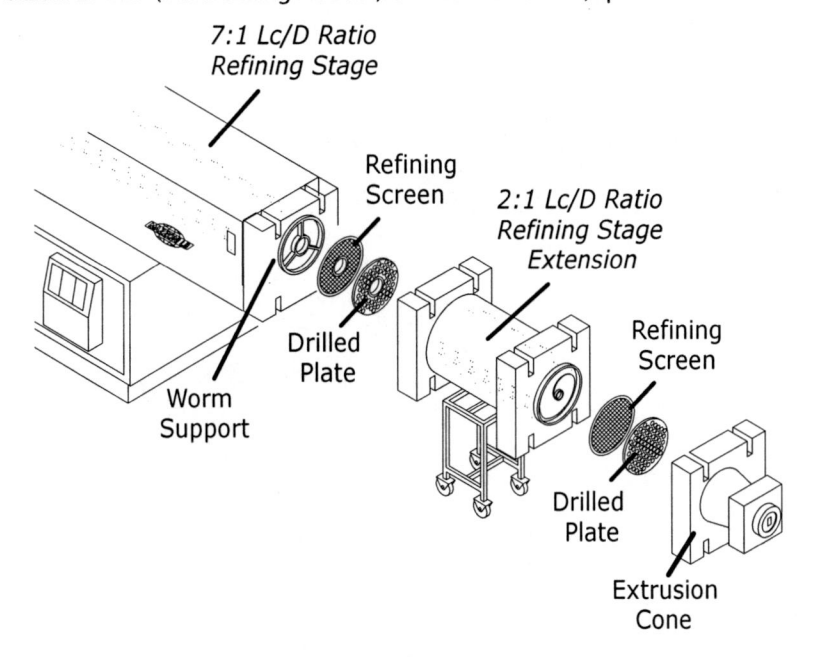

Fig. 11.24. Mazzoni LB MRP (Multirefining Plodder) final stage. Source: Mazzoni LB, SpA.

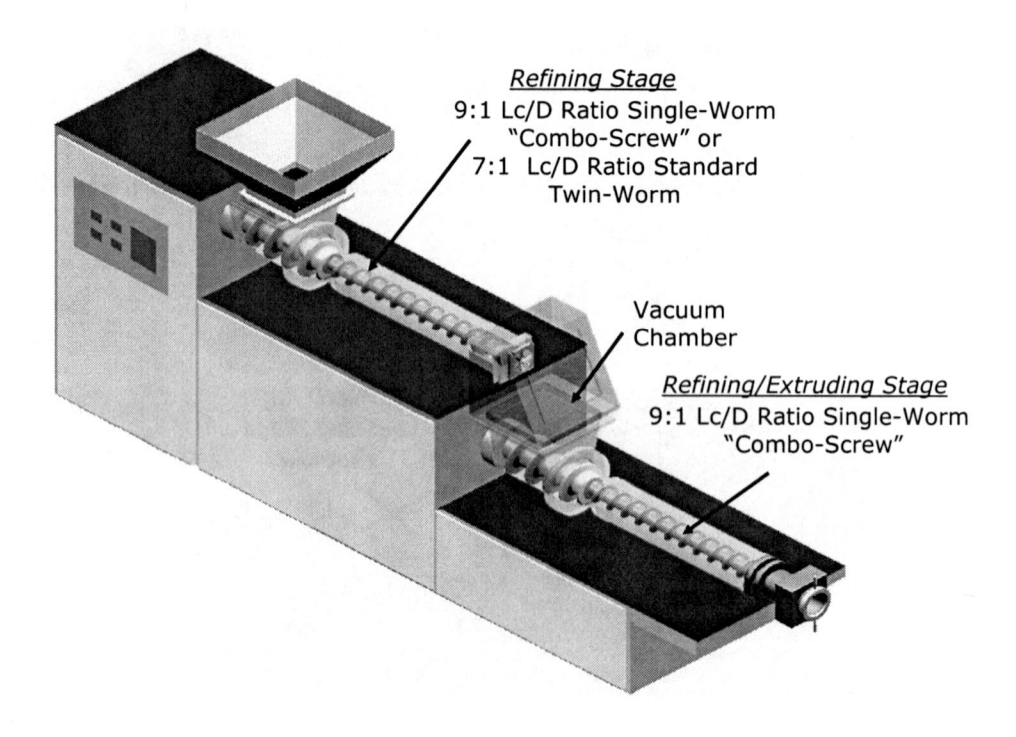

SAS Transavon Duplex Vacuum Plodder

The first Transavon units were offered for the production of translucent soap, starting with properly formulated opaque soaps. Now the Transavon Duplex Vacuum Plodder is offered for the production of any type of toilet soap.

Due to its enhanced refining action, one can use the Transavon Duplex Vacuum Plodder as the sole machine in a finishing line in the substitution for a conventional Simplex Refiner plus a Duplex Vacuum Plodder.

The SAS Transavon Duplex Vacuum Plodder consists of:

- A refining-stage plodder with either a long 9:1 Lc/D-ratio single-worm combo-screw or a 7:1 Lc/D-ratio standard twin-worm.

- A refining- and extruding-stage plodder with a long 9:1 Lc/D-ratio single-worm combo-screw.

The special profile combo-screw has a conventional diameter in the feed area, a tapered section, and a small diameter in the compression section.

The final-stage plodder has the unique adjustable milling-valve for extra refining function. The adjustable opening (gap setting) defines the achievable degree of refining. An extrusion cone completes the last stage (Figs. 11.25, 26).

Fig. 11.25. SAS Transavon Duplex Vacuum Plodder. Source: SAS.

Bar Soap Finishing ● 323

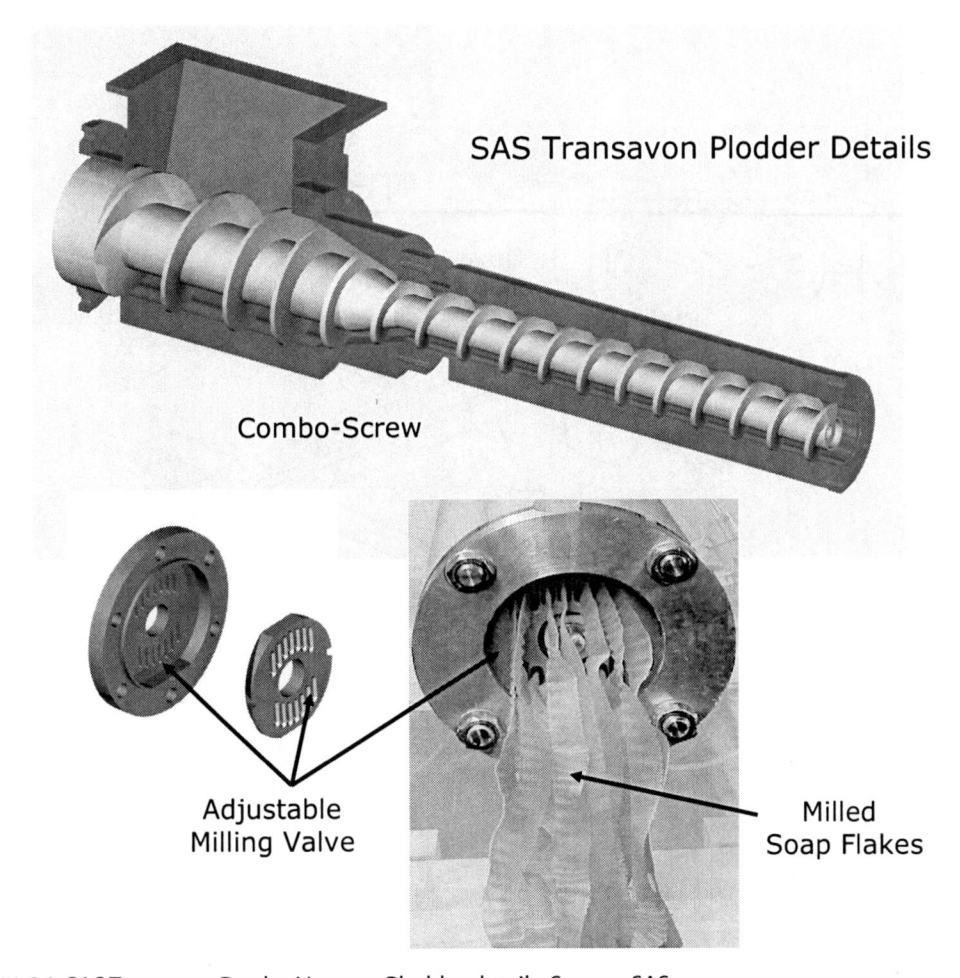

Fig. 11.26. SAS Transavon Duplex Vacuum Plodder details. Source: SAS.

Cutters

The extruded slugs (billets) from the Duplex Vacuum Plodders are cut into predetermined-length single slugs when traditional soap presses are used. The "flashstamping" presses require multiple-length slugs. During the last decade, the mechanical multiblade cutters with fixed and manually adjustable chains were replaced by the electronic/pneumatic and fully electronic cutters.

Due to constantly advancing electronic technology, today's cutters operate with ever increasing speed and cutting accuracy of: single length, multiple length, and short slugs which facilitate handling of recycle (reprocessing).

The Mazzoni LB model TE electronic cutter was introduced in 1994 followed by the TVE type in 1998 (Fig. 11.27).

The application of engraving rollers to cutters is increasing for the "natural-looking" unstamped soaps, laundry soaps, and hotel soaps. Engraving the top, bottom, and sides of the extruded product is an economical alternative to soap stamping (Fig. 11.28).

The new Binacchi ECM-2000 cutter is illustrated in Fig. 11.29.

The new SAS Easycut is shown in Fig. 11.30.

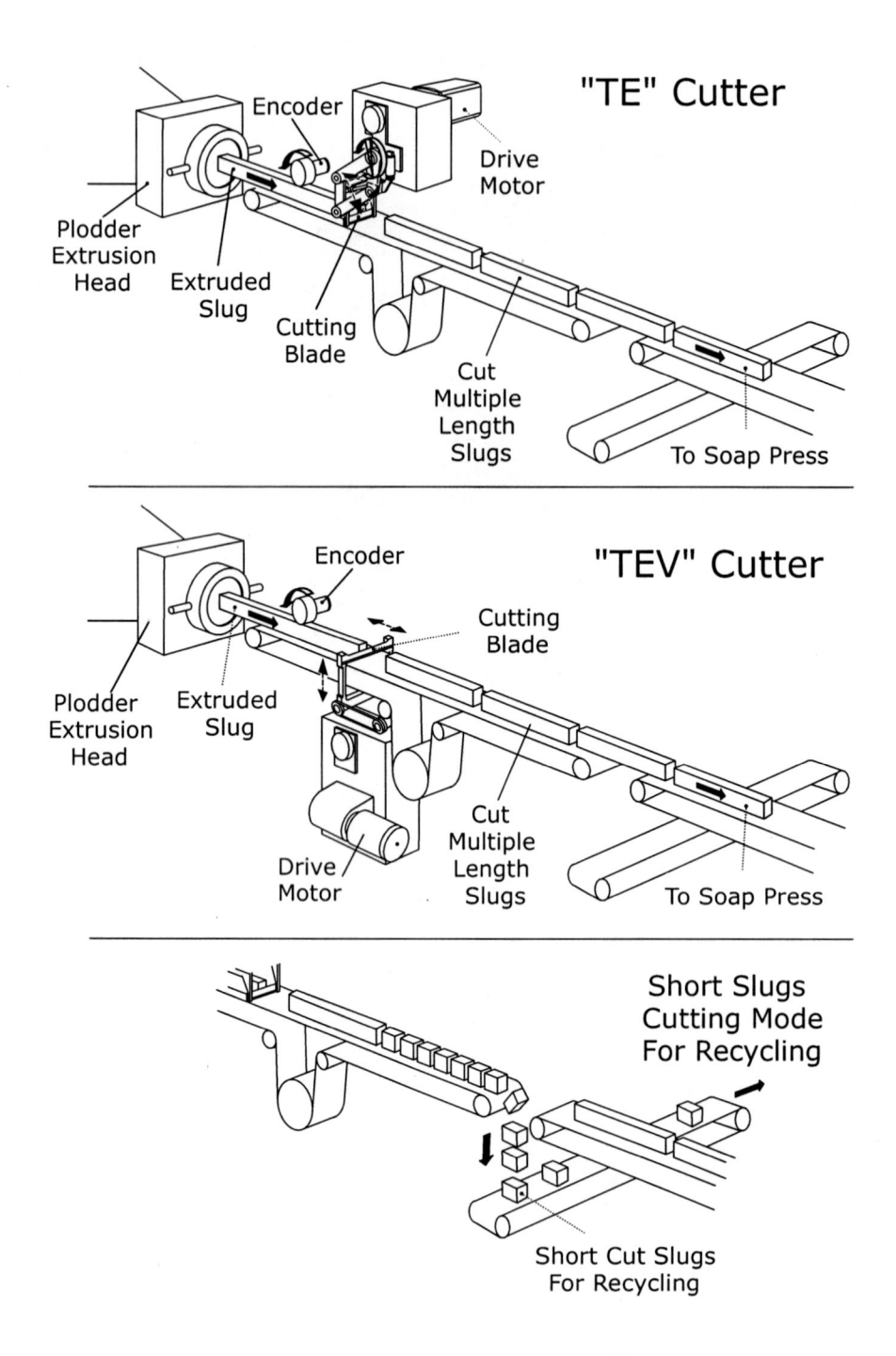

Fig.11.27. Mazzoni LB single-blade electronic cutters. Source: Mazzoni LB, SpA.

Bar Soap Finishing • 325

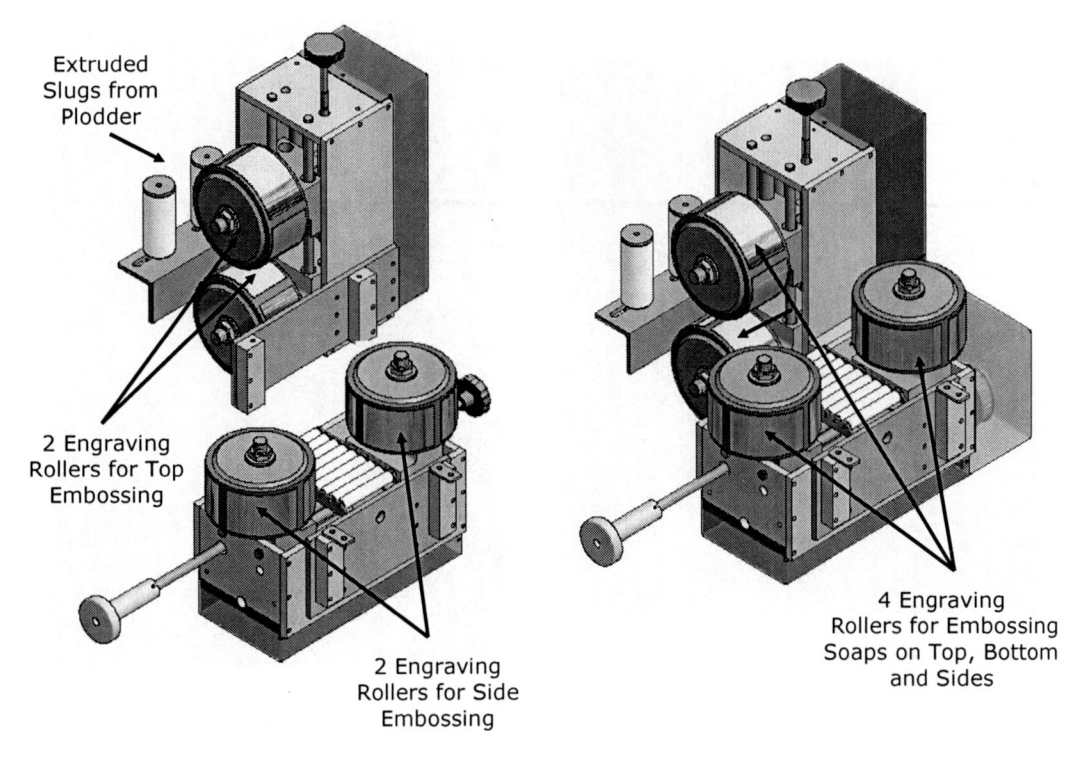

Fig. 11.28. Engraving rollers for cutters. Source: Mazzoni LB, SpA.

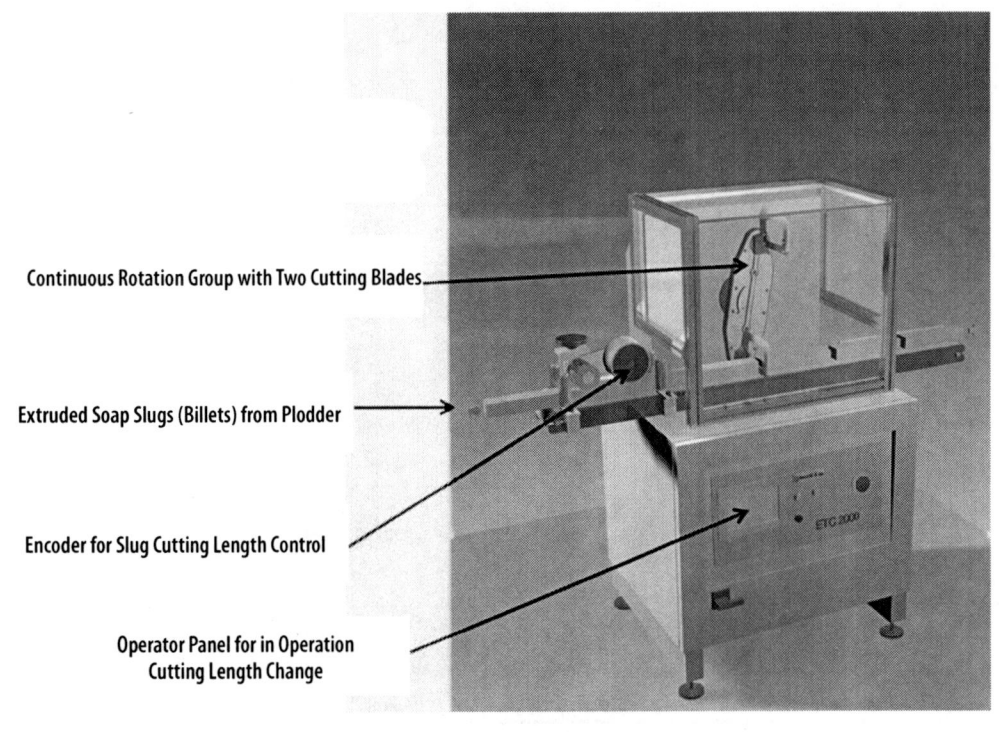

Fig. 11.29. Binacchi ETC-2000 electronic cutter. Source: Binacchi & Co.

326 ● L. Spitz

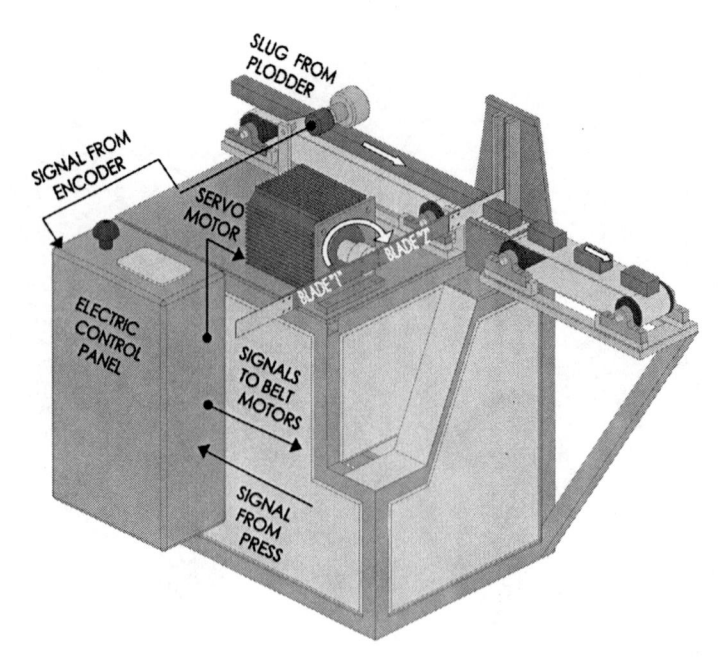

Fig. 11.30. SAS Easycut electronic cutter. Source: SAS.

Presses

Soap Shapes

Two basic shapes of soap exist:

- Banded (with a side band)—all soap shapes with vertical sides as their periphery.

- Bandless (without a side band)—all soap shapes with only one parting line, that is, without vertical sides around their circumference.

One can further classify all banded and bandless shapes into four variations: rectangular, round, oval, and irregular (Fig. 11.31).

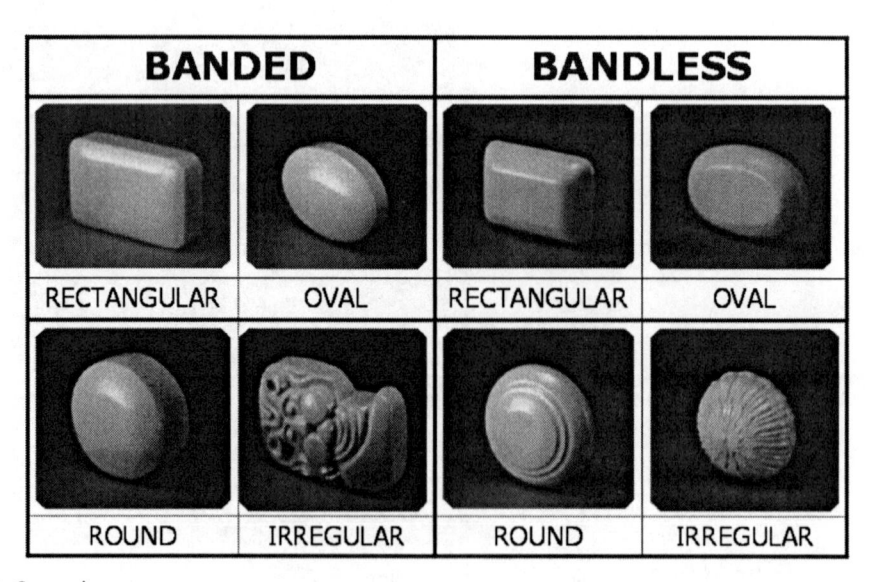

Fig. 11.31. Soap shapes.

Soap Presses

Flashstamping-type soap presses use only dies and corresponding counter-dies to stamp all banded, bandless, and specialty shaped products.

Excess soap, referred to as "flashing," is formed around the periphery of the dies as the dies come together to form the final bar shape and to set the stamped bar weight.

Flashstamping always requires a 10–30% heavier slug weight than the final stamped bar weight. Take into account this extra weight requirement to properly size the extrusion rate of the final plodder.

Soap presses utilizing the traditional die-box stamping system are in limited use due to stamping speed and soap-shape stamping limitations.

Horizontal and vertical motion flashstamping soap presses are available.

Mazzoni LB's vertical motion flashstamping STUR presses have been on the market since 1989 (Fig. 11.32).

The current STUR stamping system is illustrated in Fig.11.32.

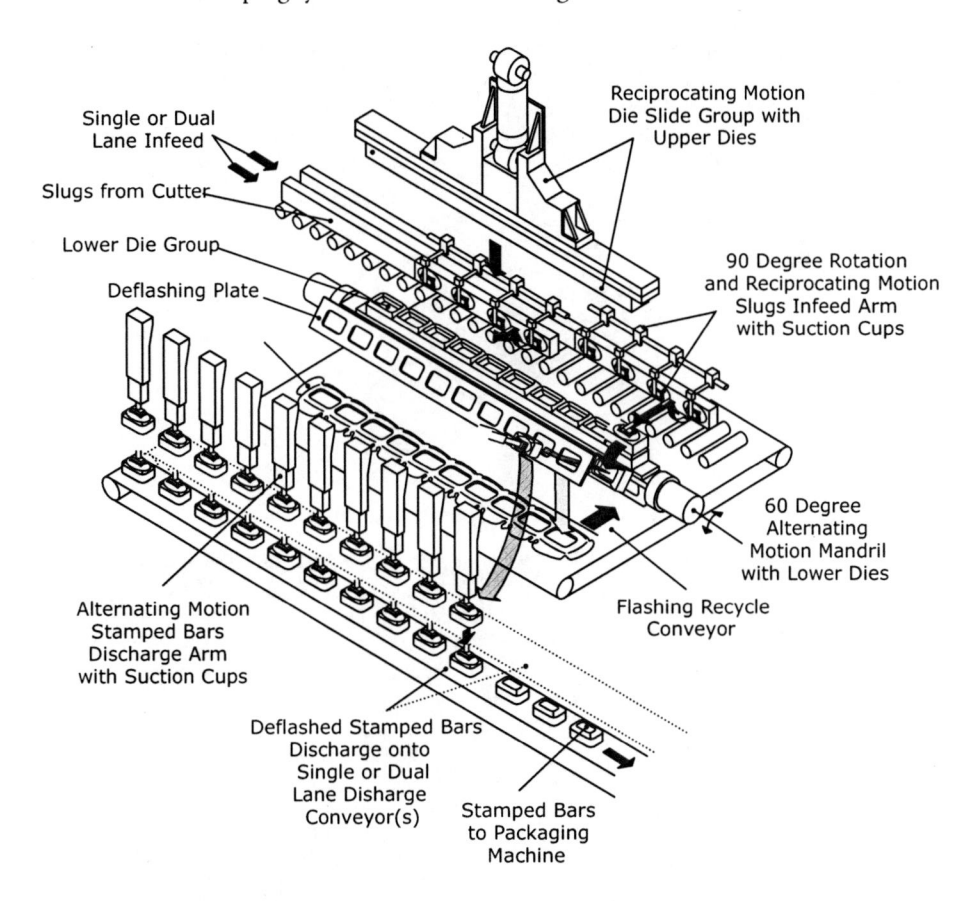

Fig. 11.32. Mazzoni LB STUR flashstamping soap press. Source: Mazzoni LB, SpA.

Binacchi's model USN-500 horizontal motion flashstamping press, introduced in 1989, was the first dual mandril press designed to accommodate up to eight dies (four on each mandril). One set was mounted on a reciprocating die slide, and the other two sets on a 180° rotating mandril.

The new USN-2000 Series presses introduced in 2006 are vertical motion units with one-third fewer dies, since they do not have rotating mandrils.

Direct Noncontact Bar soap Transfers for Cartoner and Wrapper Interface
Binacchi Direct Product Transfer (DPT)

In 1989 Binacchi was the first to offer a noncontact bar soap transfer as an integral part of their model USN-500 soap press coupled to a soap cartoner. This invention broke the then prevailing maximum 300 cartons per minute speed barrier.

In 1995 Binacchi introduced a similar transfer system for the direct interface with Binacchi soap wrappers.

The new direct product transfer (DPT) system with the Binacchi USN-2000 series presses with and without the DPT shown in Figures 11.33 and 11.34. Various optional layouts illustrate applications of these units with soap wrappers, cartoners, and flow-wrappers (Fig. 11.34).

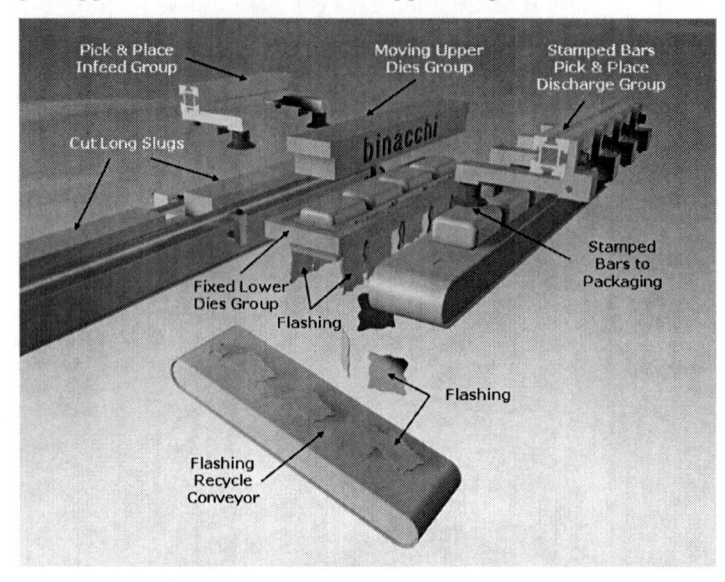

Fig. 11.33. Binacchi USN-2000 Series soap presses. Source: Binacchi & Co.

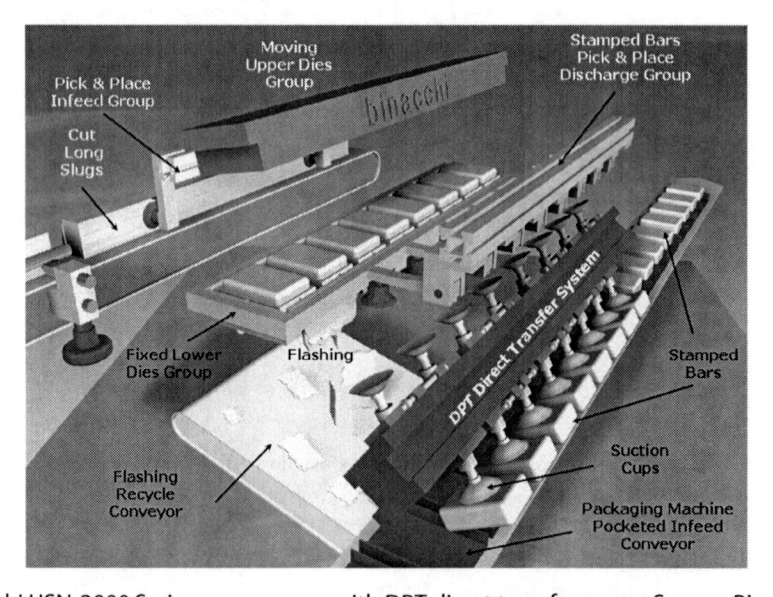

Fig. 11.34. Binacchi USN-2000 Series soap presses with DPT direct transfer group. Source: Binacchi & Co.

Mazzoni LB Direct Transfer System (DTS)

The Mazzoni LB direct transfer system (DTS) was introduced in 2006. A set of vacuum transfer cups rotates, spaces in pitch, and places the stamped bars by using a single-step rotation into the pocketed infeed conveyor of the packaging machine under the packaging section Figures 11.52, 11.53 and 11.54.

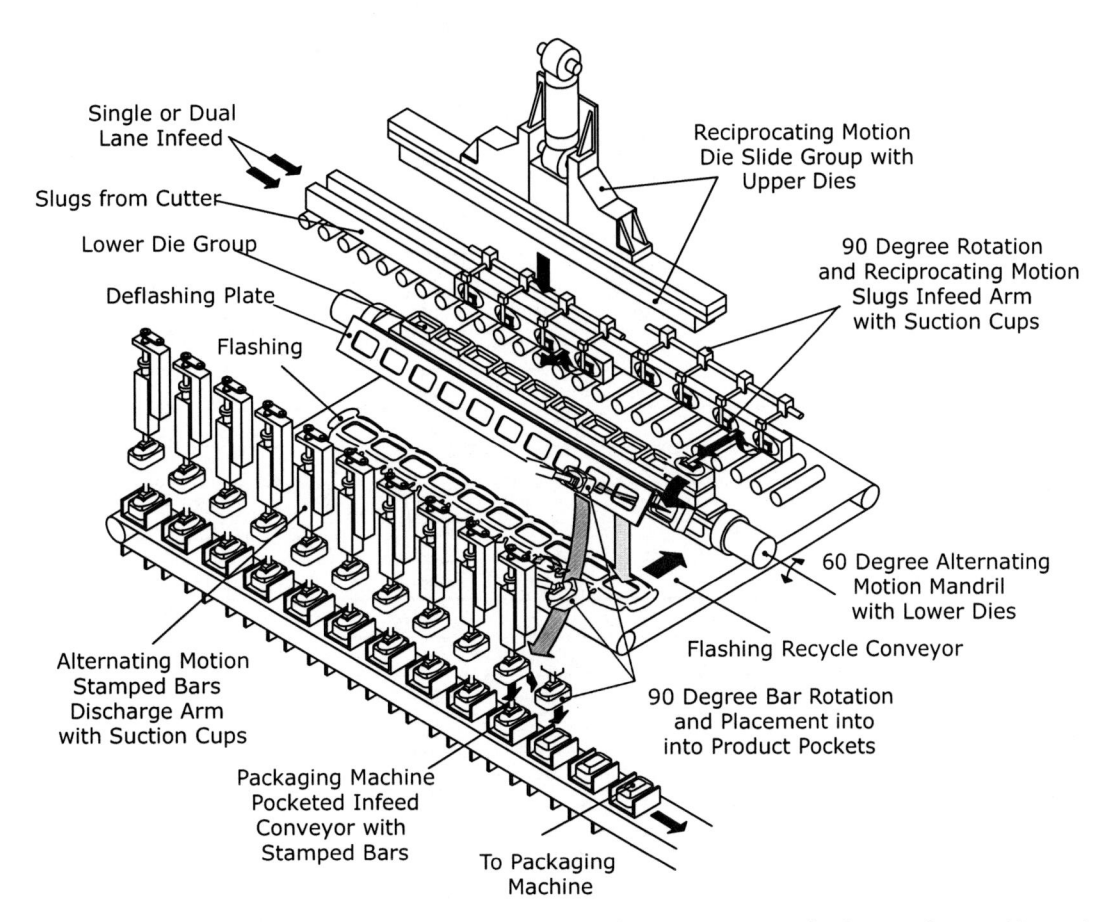

Fig. 11.35. Mazzoni LB STUR flashstamping soap press with DTS Direct Transfer System. Source: Mazzoni LB, SpA.

A summary of the most widely used soap presses offered in 2009 is listed in Table 11.3.

Table 11. 3. Flashstamping Soap Presses

Make	Model	Number of Die Support Plates	Maximum Number of Dies for Bar Weights (Grams) 100*	150**	Maximum Stamping Strokes per Minute	Stamping Speed Bars per Minute for Bar Weights (Grams) 100*	150**	Mandril Rotation (Degrees)
BINACCHI	USN-2150	2	3 4	3 4	60	180 240	180 240	Fixed
	USN-2200	2	5 7	4 6	60	300 420	240 360	Fixed
	USN-2600	2	10 12	9 11	60	600 720	540 660	Fixed
	USN-2800	2	16 20	14 18	60	960 1200	840 1080	Fixed
MAZZONI LB	STUR-2	2	2	2	80	160	160	60
	STUR-3	2	3	3	70	210	210	60
	STUR-4N	2	4	4	70	280	280	60
	STUR-7	2	7	6	65	455	390	60
	STUR-10	2	12	10	60	720	600	60
	STUR-12	2	12	12	60	720	720	60
	STUR-14	2	14	12	60	840	720	60
SAS	STAMPEX-2	2	2	2	75	150	150	60
	STAMPEX-3	2	3	3	75	225	225	60
	STAMPEX-4+	2	4	4	70	280	280	60
	STAMPEX-6+	2	6	6	70	420	420	60
	STAMPEX-8	2	8	8	65	520	520	60
SELA	SPV-3	2	3	2	80	240	160	60
	SPH	3	5	4	60	300	240	180
	SPH-S	3	6	6	60	360	360	180
	SPH-8	3	9	8	60	540	480	180
	SPH-10	3	10	10	60	600	600	180
	SPH-D	3	12	12	60	720	720	180
SOAPTEC	MFS-HY	2	4	3	25	100	75	60
	MFS-1	2	2	1	80	160	80	60
	MFS-3	2	4	3	75	300	225	60
	MFS-4	2	5	4	70	350	280	60
	MFS-6	2	7	6	65	455	390	60
	MFS-8	2	9	8	65	585	520	60
	MFS-10	2	12	10	65	780	650	60

Note:
* Binacchi USN Presses with Horizontally Positioned Dies
** Binacchi USN Presses with Vertically Positioned Dies

Packaged Water Chillers for Plodders and Roll Mills

Packaged water chillers are designed to circulate clean, chilled water to plodders and roll mills in a closed circuit. Chilled water ranging in temperature from approximately 8 to 15°C circulates in the barrels of plodders and in the roll of roll mills. The exact operating temperature depends on the product and the processing role (Fig. 11.36).

In most cases, the initial investment in and operating cost of chillers are offset by the cost of nonrecycled cooling water, varying seasonal water-temperature limitations, elimination of periodic maintenance required to remove hard-water deposits, and the ability to use the optimal chilled water temperature for the given process.

Units

Cooling units are rated in tons of refrigeration, which measure the cooling effect of 2,000 pounds (a short ton) of ice melting in 24 hours. This turns out to be 288,000 heat units in 24 hours, or 12,000 heat units per hour. For the cooling capacity of water, note:

- In the United States: 1 ton = 12,000 Btu/hour = 3,024 Kcal/hour
- In Great Britain: 1 ton = 14,256 Btu/hour = 3,592 Kcal/hour
- In Europe: 1 frigorie/hour = 756 Kcal/hour = 3,000 Btu/hour.

Chiller capacities in the United States are based on a flow rate of 2.4 gallons per minute per ton and a temperature drop of 10°F.

Chiller Sizing

Chiller sizing is calculated by the following formula:

$$\text{Tons of refrigeration} = GPM \times \Delta T/24$$

(Eq. 11.1)

where GPM is the water flow rate in gallons per minute and ΔT is the temperature differential in °F between the water leaving and entering the system.

The design capacities are also based on ambient air at 35°C for air-cooled chillers and 30°C condenser water temperature for water-cooled chillers. As a rule of thumb, the cooling capacity of a chiller is reduced by 2% for each 0.5°C below 10°C.

Low-Temperature Packaged Glycol/Water Chillers for Soap Press Dies

Low-temperature packaged glycol chillers are designed to supply a glycol/water coolant solution at −30°C (−22°F) to soap press dies. The cooling capacity of packaged chillers is expressed in tons; however, the actual capacity varies with process temperatures and ambient temperatures. For example, a unit rated for about 5,000 Kcal/hour at −30°C would provide 8,000 Kcal/hour at −20°C, 11,000 Kcal/hour at −10°C, and 17,000 Kcal/hour at 0°C. As another example, with nominal cooling capacity at 28°C ambient temperature, the cooling capacity would decrease by about 12% if the ambient temperature increased to 38°C (Fig. 11.37).

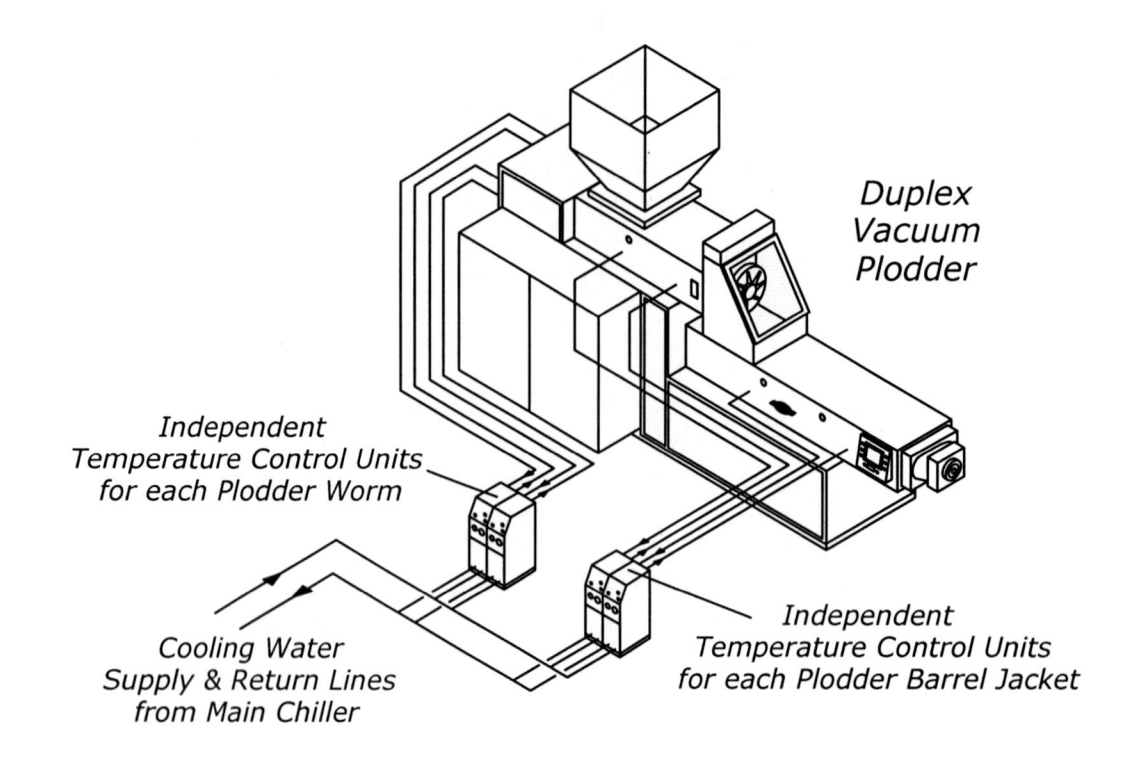

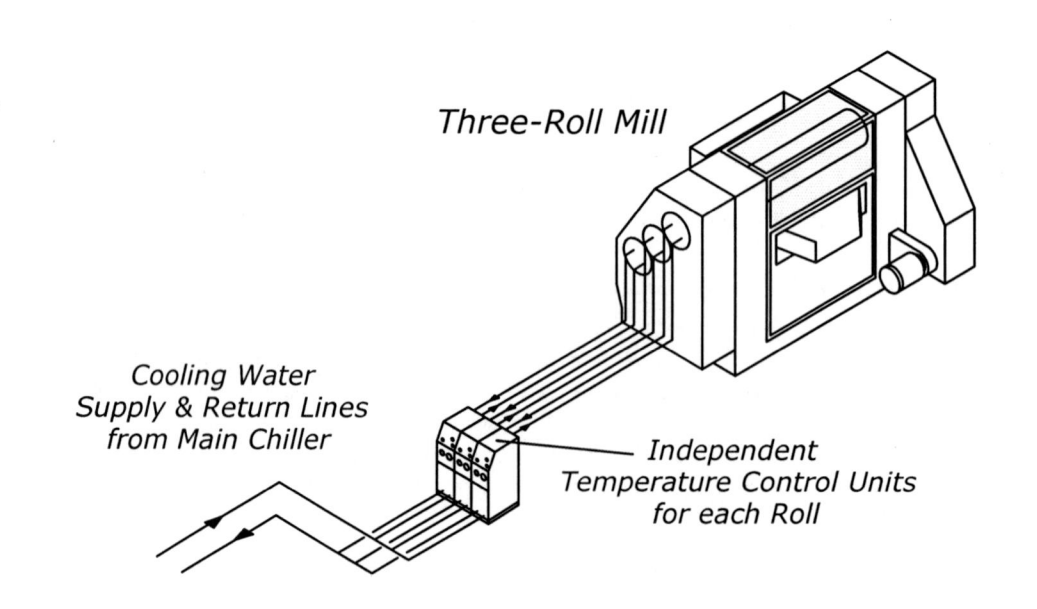

Fig. 11.36. Water-chiller system for plodders and roll mills. Source: Mazzoni LB, SpA.

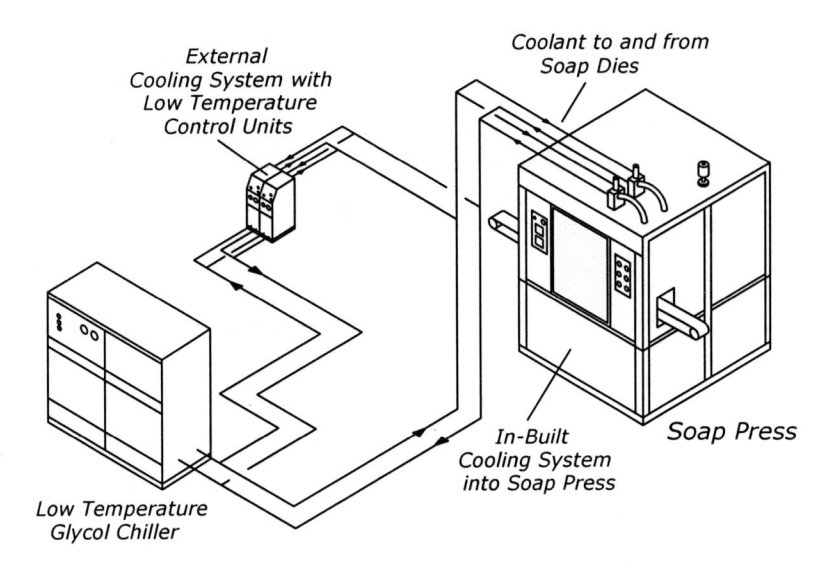

Fig. 11.37. Low-temperature gylcol/water chiller system for soap press dies. Source: Mazzoni LB, SpA.

Temperature Control Units

Temperature control units are recommended for cooling and heating the water circulating in the barrel of each plodder and in each roll of a roll mill.

Standard temperature control units are used for plodder applications.

Special low-temperature glycol control units are used for handling the glycol/water coolant solution circulating through soap press dies.

Temperature control units are free-standing and self-contained. They incorporate a high-pressure centrifugal pump, an immersion heater, a cooling (solenoid or modulating) valve, and a microprocessor-based controller.

These units are shown in Figures 11.36 and 11.37.

Soap Finishing Line Types and Selection

The selection of a bar soap finishing line depends on the following:

- Types of products to be produced
- Line-operating speed
- Number of refining stages required
- Choice of an all-plodder or a combination plodder and roll-mill line
- Rework/recycle quantity and recycle location
- Multifunction-line layout
- Pre-refining requirement

The three most widely used lines in various layout configurations are:

- Line with 3 Plodders (2 Refining Stages)
  A Simplex Refiner and a Duplex Vacuum Plodder (Fig.11.39).

- Line with 4 Plodders (4 Refining Stages)
  A Duplex Refiner and a Duplex Vacuum Plodder (Fig.11.40).

- Line with 3 Plodders and 1 Roll-Mill (3 Refining Stages)
  A Simplex Refiner a Three-Roll Mill and a Duplex Vacuum Plodder (Fig.11.41).

Pre-Refining Lines
One can make any line into a pre-refining line by placing a simplex refiner or, alternatively, a roll mill before the mixer.

Multifunction Lines
In standard lines, products have to pass from one machine to another without the option of bypassing one stage. For some products, using less mechanical work (less refining) is indicated. All layouts indicate the multifunction and recycle options.

As an example, a multifunction line with three plodders and one roll mill is shown. This is a dual function line in which one can bypass the roll mill if so desired. A similar line with four plodders would have a right-angle Duplex Refiner to allow the bypassing of the first-stage plodder when it is not required by the specific soap formulation.

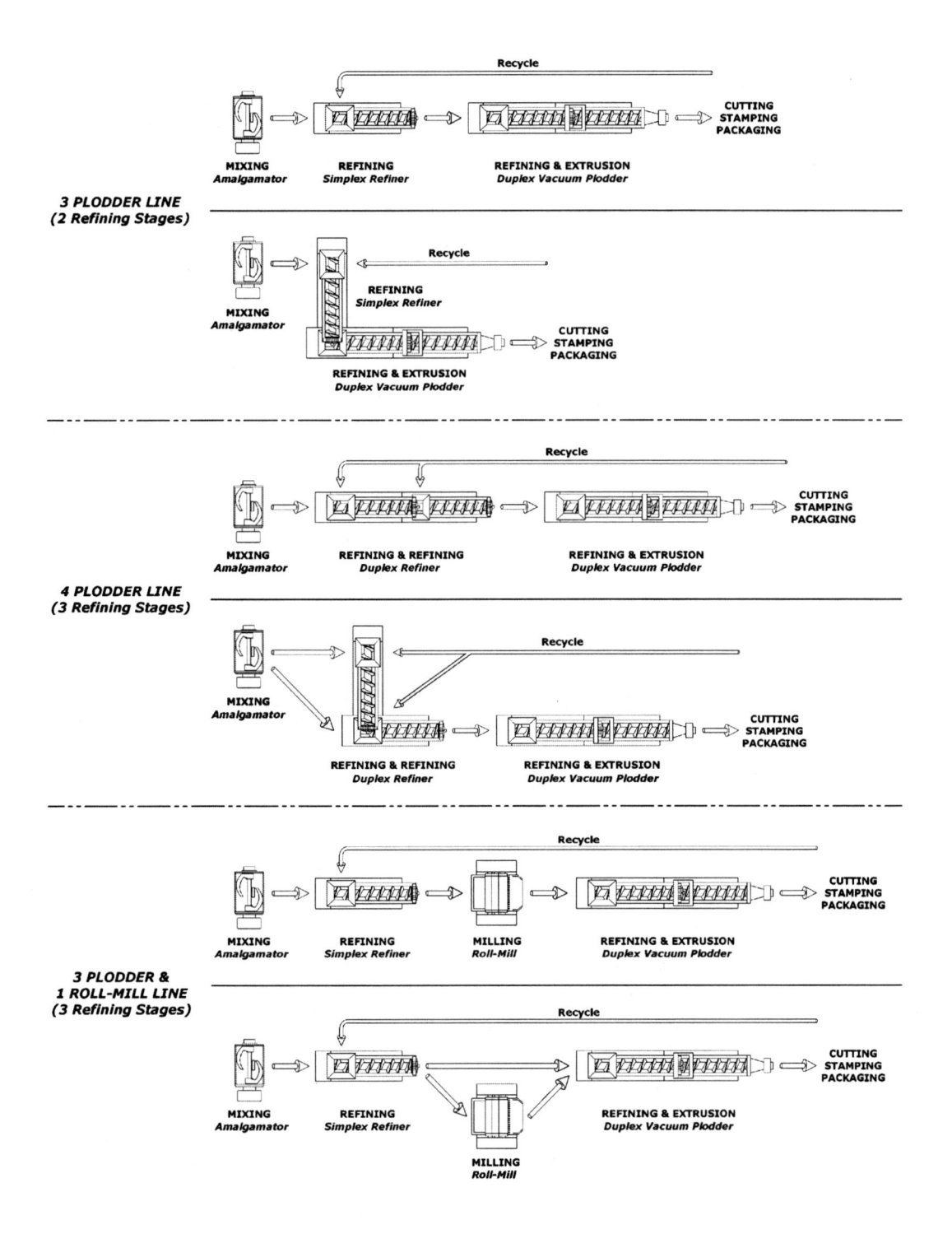

Fig. 11.38. Bar Soap Finishing Line Types Summary.

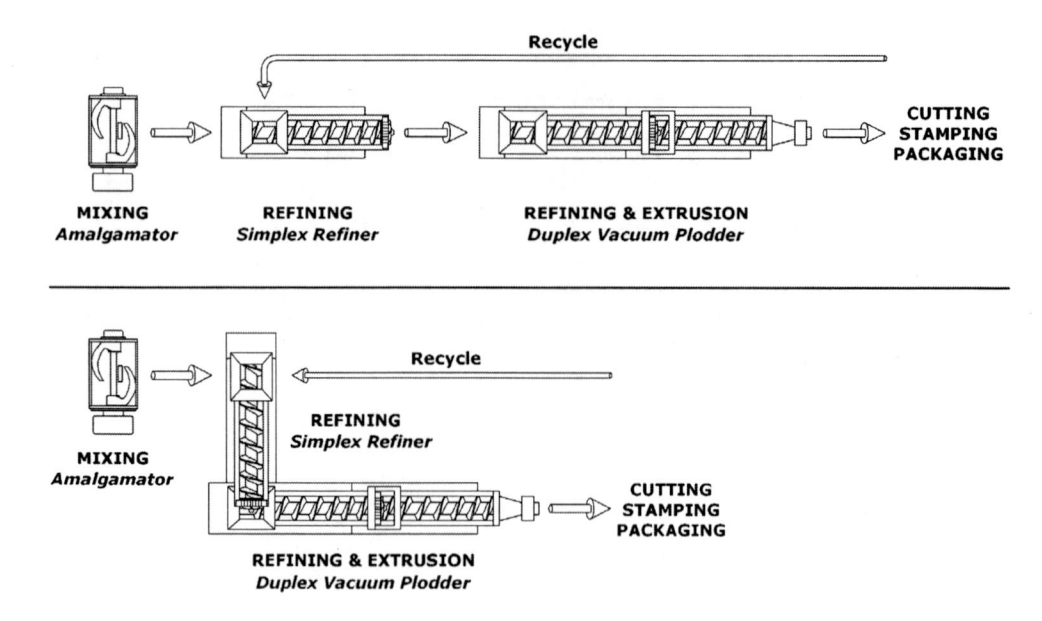

Fig. 11.39. Bar Soap Finishing Line with 3 Plodders (2 Refining Stages).

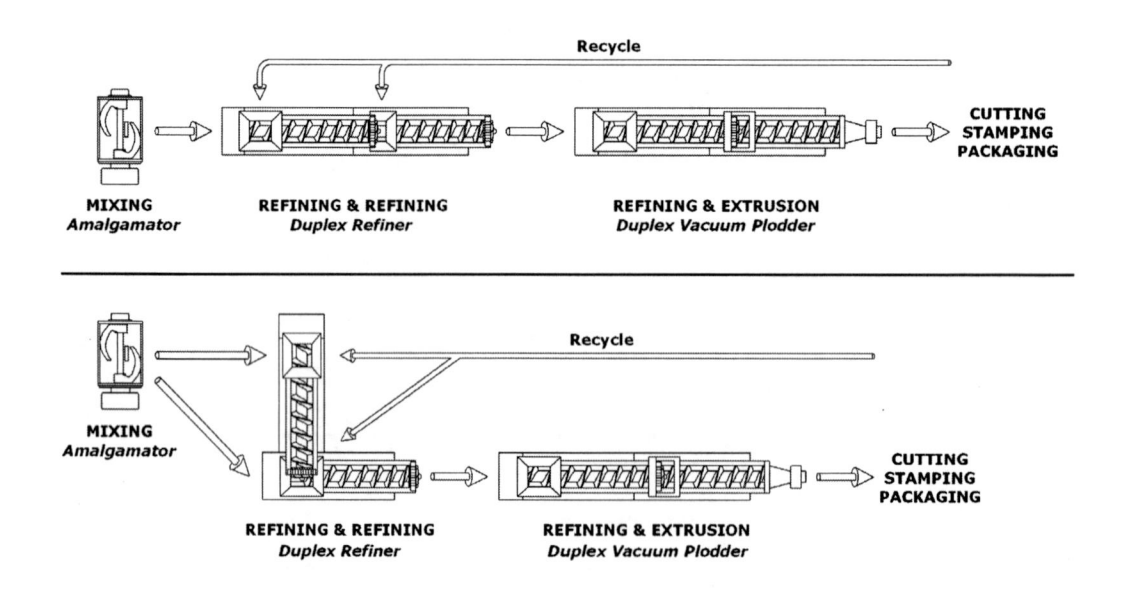

Fig. 11.40. Bar Soap Finishing Line with 4 Plodders (3 Refining Stages).

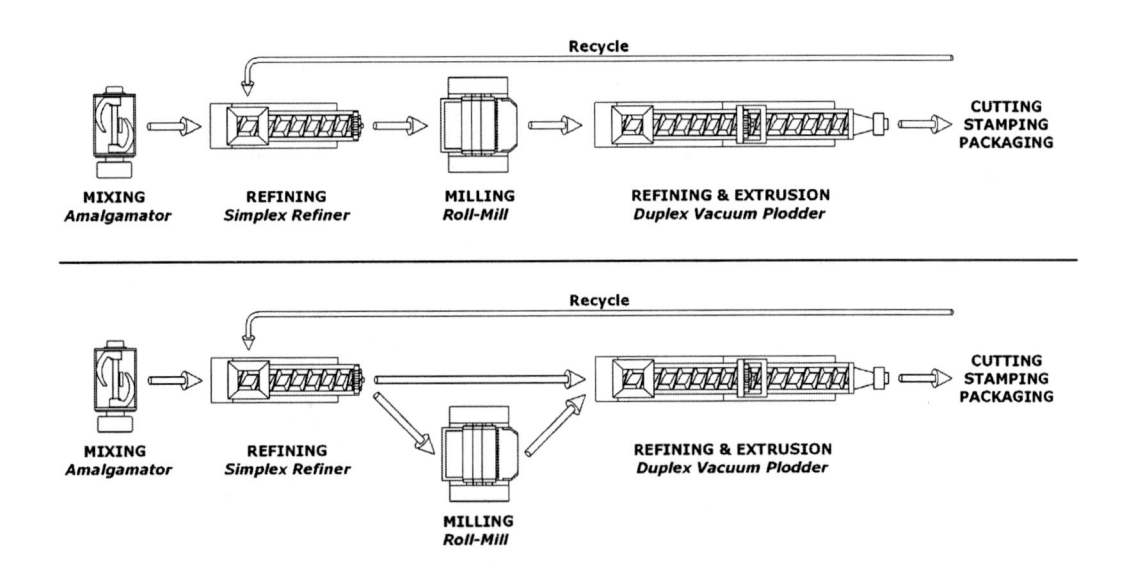

Fig. 11.41. Bar Soap Finishing Line with 1 Roll-Mill and 3 Plodders (3 Refining Stages).

Packaging

Soap Packaging Styles

Mass-market soaps are wrapped, cartoned, flow-wrapped, overwrapped, and/or bundled (U-banded). Specialty cosmetic, gift, and novelty soaps are pleat-wrapped, stretch film-wrapped, and flow-packed wrapped. High-priced soaps are also packaged in special cartons.

During the last decade, the sale of single-pack toilet soaps practically disappeared and was replaced by multipacks containing up to 20 individually wrapped or cartoned bars. The multipacks always offer a few free bars. Also, to save cost per bar, most wrappers and cartons are white and unprinted. Most multipacks are overwrapped, and some are bundled (U-banded).

Packaging Equipment

New bar soap wrappers, cartoners, bar soap infeed (transfer) systems, and an increased acceptance of existing and new direct soap press transfer systems allow one to reach up to 600-bars-per-minute packaging speeds.

Flow-packed wrapped-style (fin-seal style) packaging grew in many countries during the last few years.

Horizontal flow wrappers are offered with in-line, noncontact infeed systems for higher speed units.

Specialty cosmetic, gift, and novelty soaps are pleat- wrapped, stretch film-wrapped, and flow-packed wrapped. High-priced soaps are also packaged in special cartons.

This portion of the presentation consists of an updated listing and illustrations of the most widely used soap wrappers and cartoners.

The most widely used soap wrappers and cartoners are summarized in Tables 11.4 and 11.5, and illustrated in Figures 11.42 through 11.54.

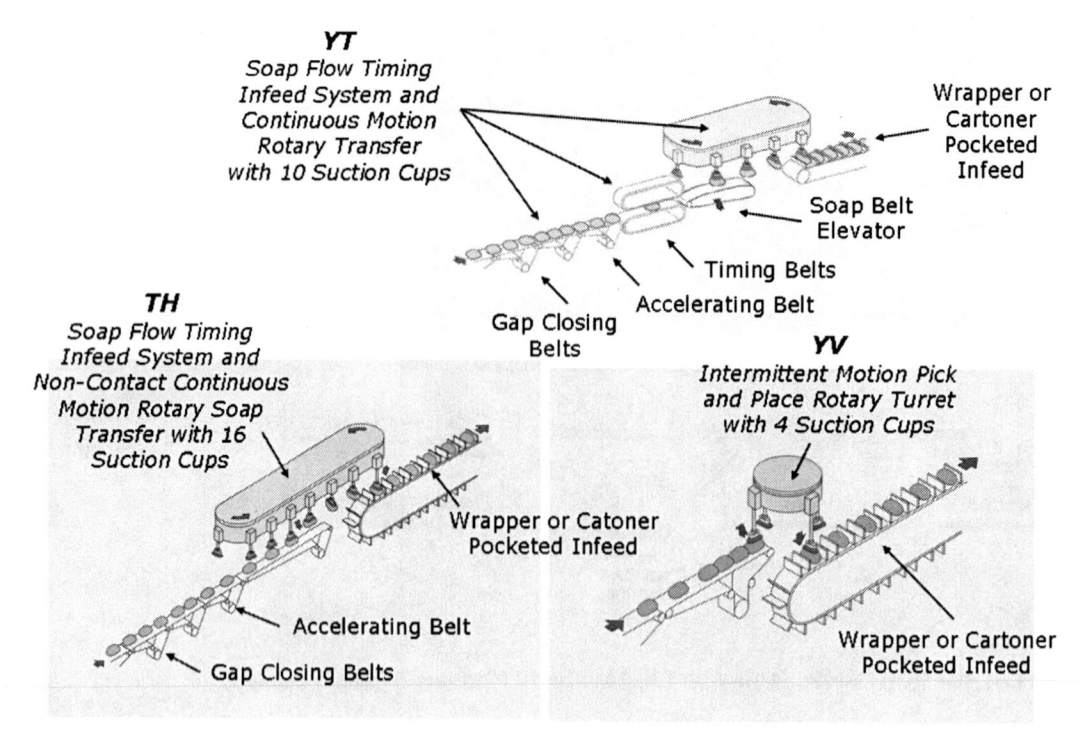

Fig. 11.42. ACMA Rotary Infeed/Transfer Systems for Soap Wrappers and Cartoners.

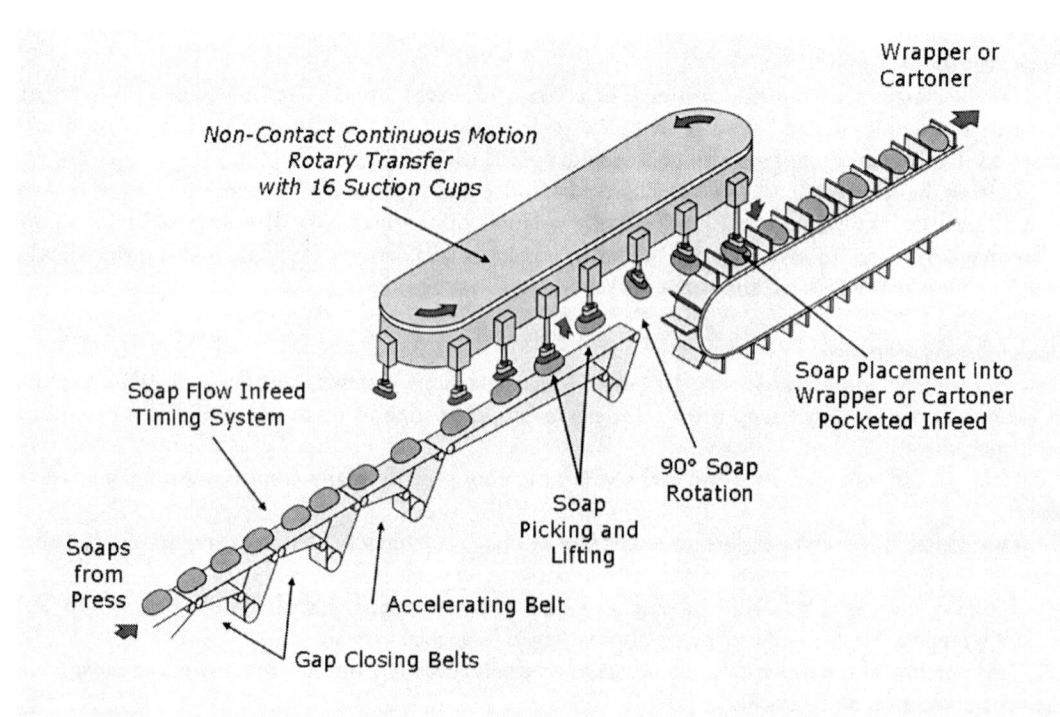

Fig. 11.43. ACMA TH-Non-Contact Soap Infeed/Transfer with 16 Suction Cups for Wrappers and Cartoners.

Bar Soap Finishing • 339

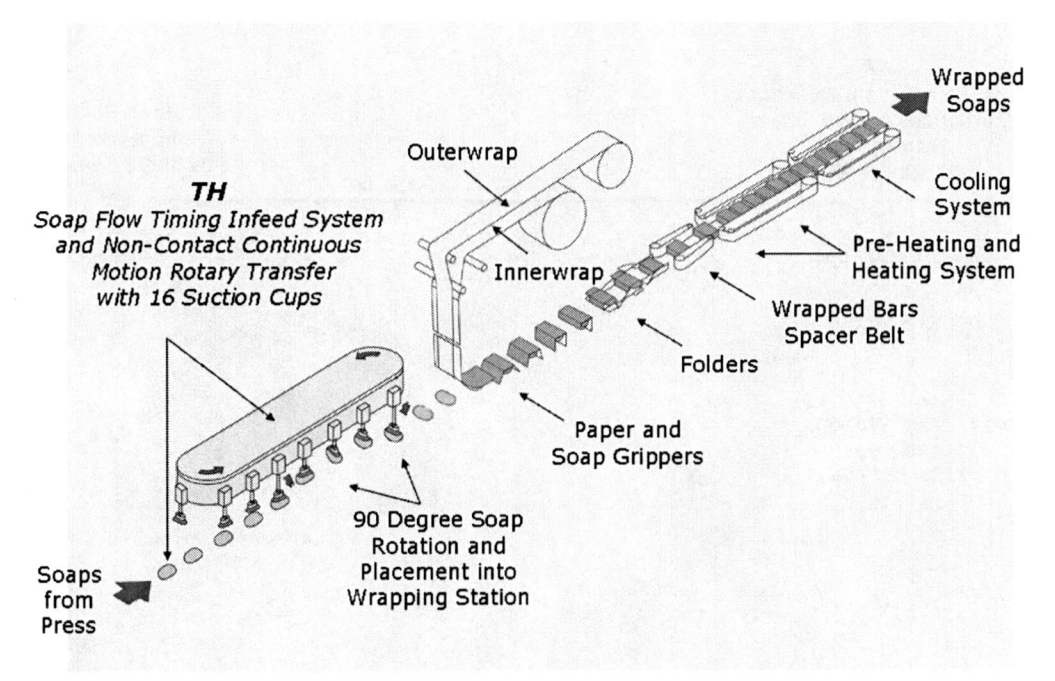

Fig. 11.44. ACMA 771 Soap Wrapper with TH Non-Contact Infeed/Transfer.

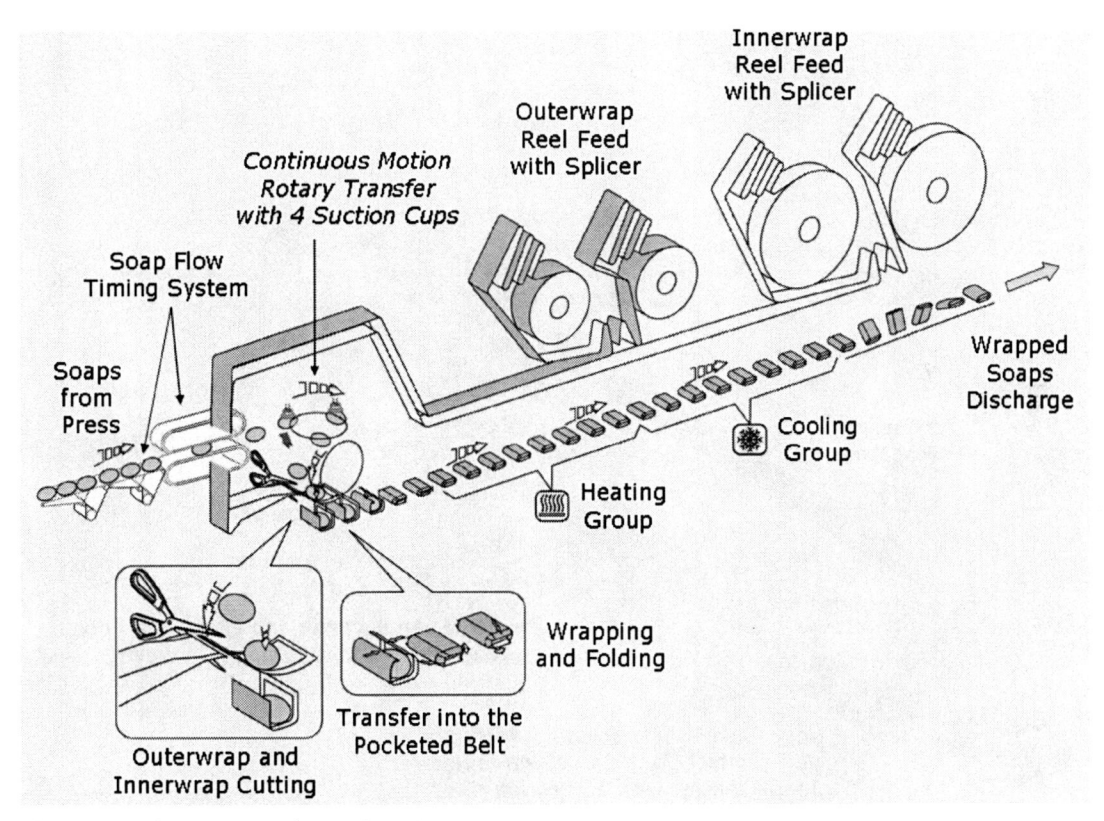

Fig. 11.45. ACMA 7250 and 7350 Soap Wrappers.

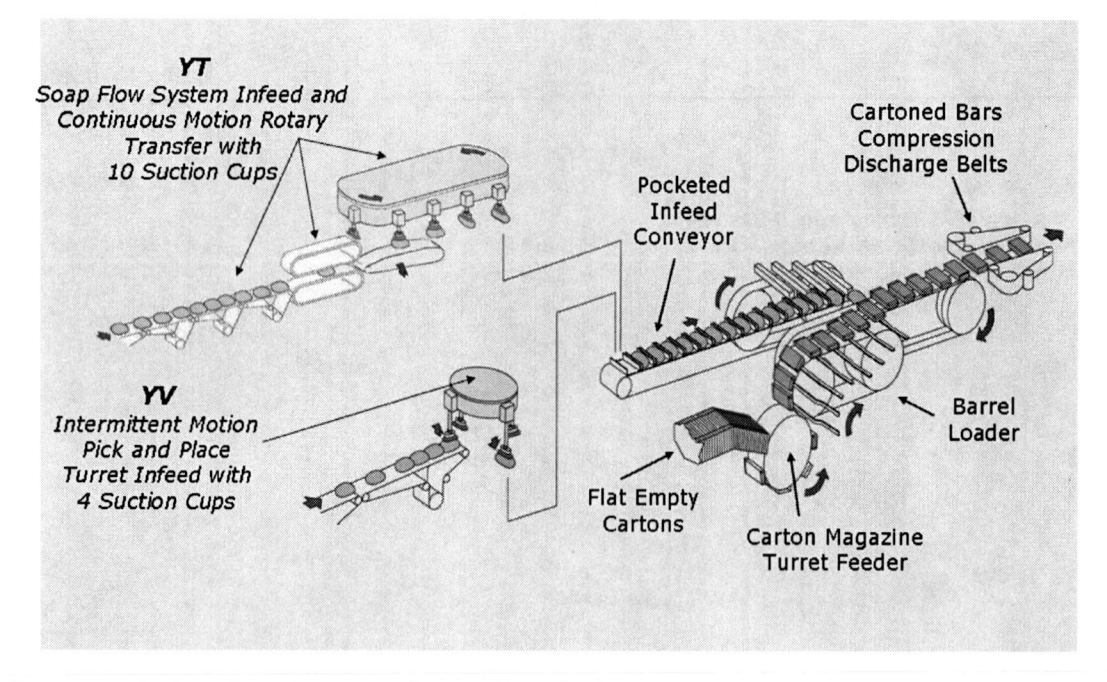

Fig. 11.46. ACMA 330 Soap Cartoner with YT and YV Infeeds.

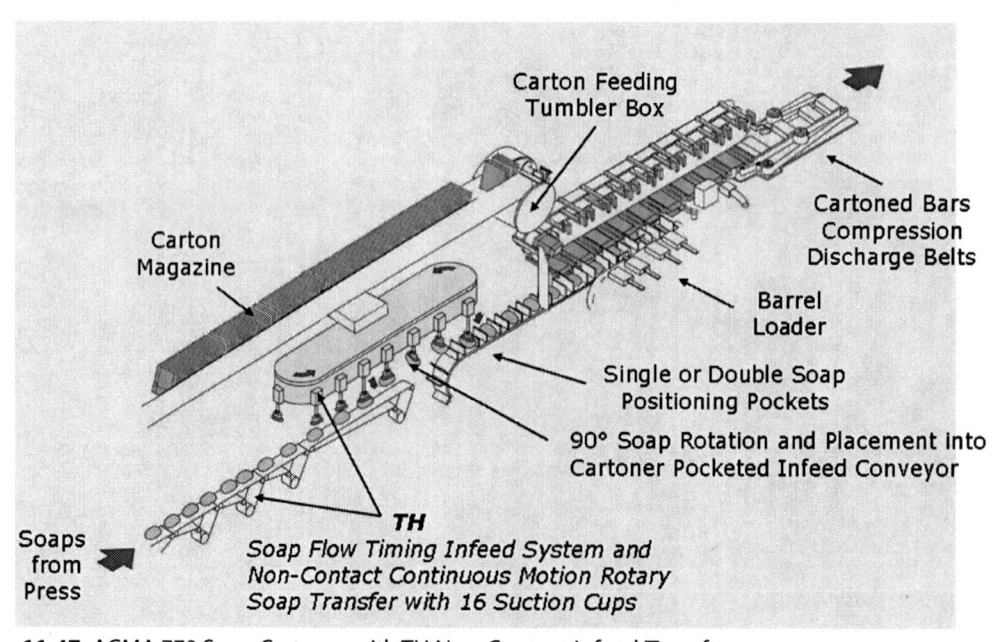

Fig. 11.47. ACMA 770 Soap Cartoner with TH Non-Contact Infeed/Transfer.

Bar Soap Finishing • 341

V512

Continuous Motion
Non-Contact
Rotary Infeed with
6 Grippers

V520

Continuous Motion
Non-Contact
Rotary Infeed with
18 Suction Cups

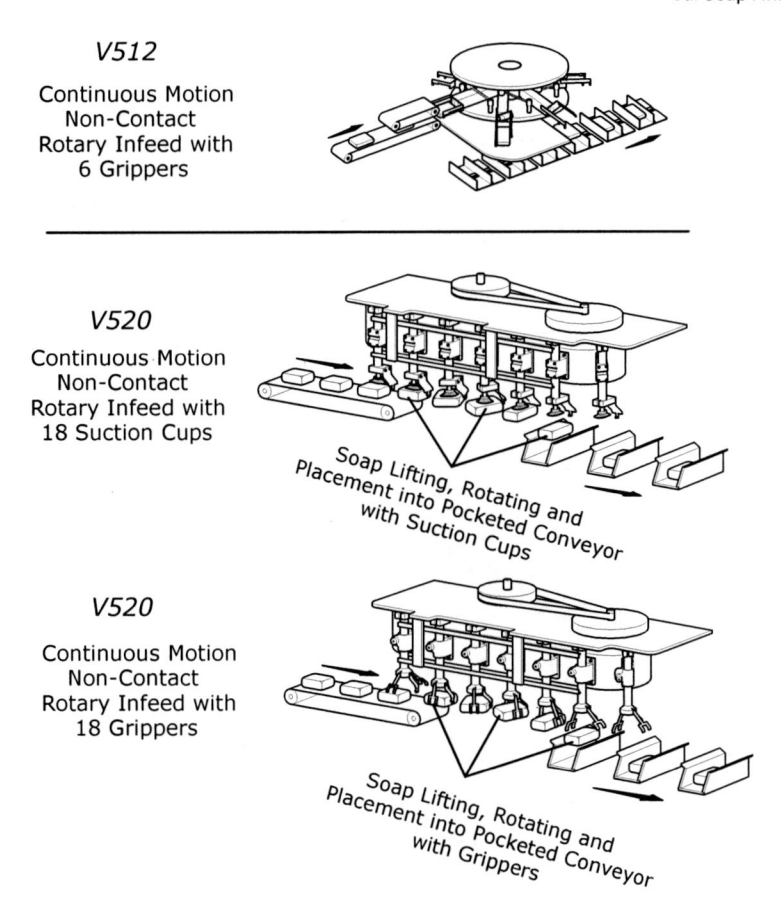

Soap Lifting, Rotating and Placement into Pocketed Conveyor with Suction Cups

V520

Continuous Motion
Non-Contact
Rotary Infeed with
18 Grippers

Soap Lifting, Rotating and Placement into Pocketed Conveyor with Grippers

Fig. 11.48. CAM Soap Cartoner Infeed/Transfer Systems. Source: Tecnicam Srl.

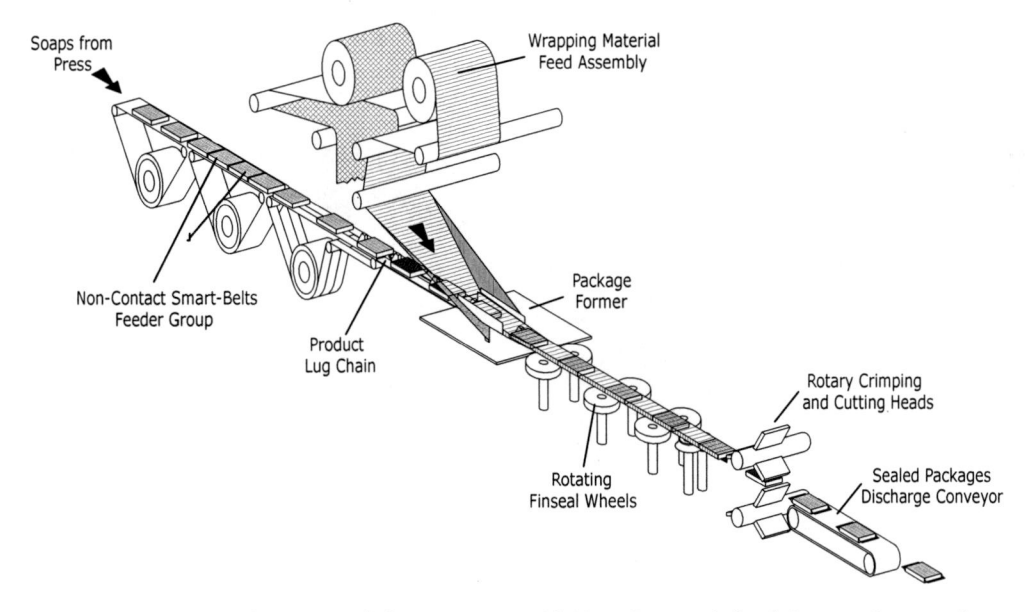

Fig. 11.49. Guerze 1000HS Horizontal Flow Wrapper with Non-Contact Infeed. Source: Guerze srl.

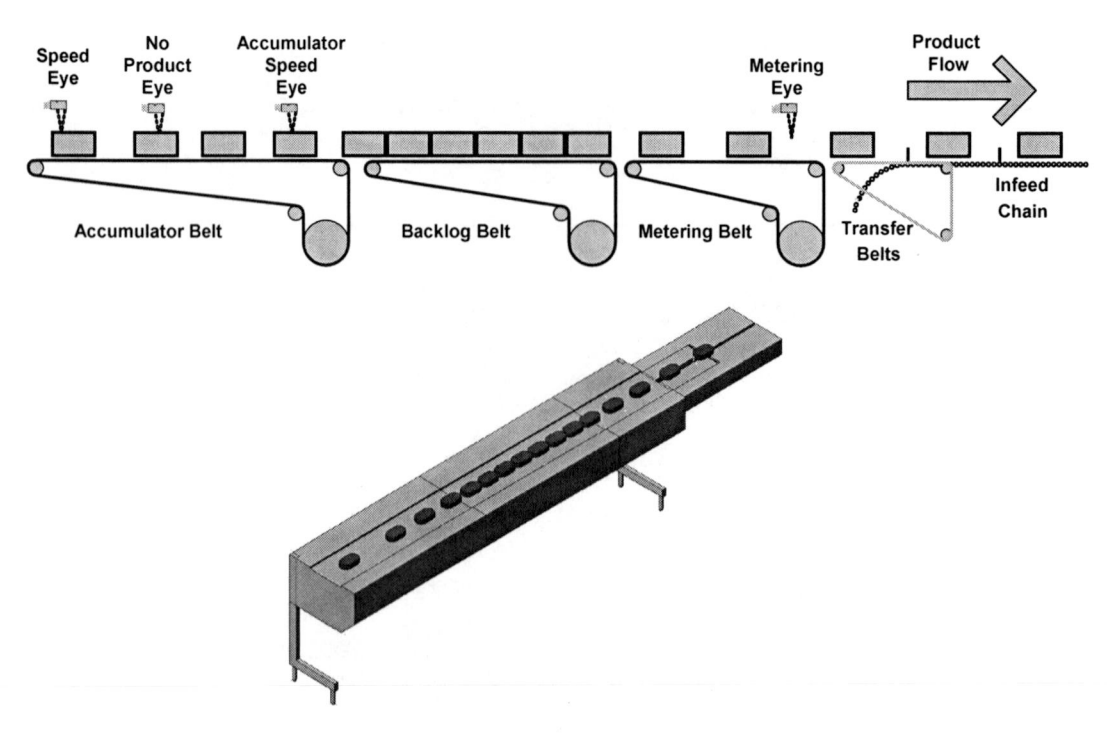

Fig. 11.50. Doboy IL3 Three Belt Contact Feeder for Horizontal Flow Wrappers. Source: Doboy, Inc.

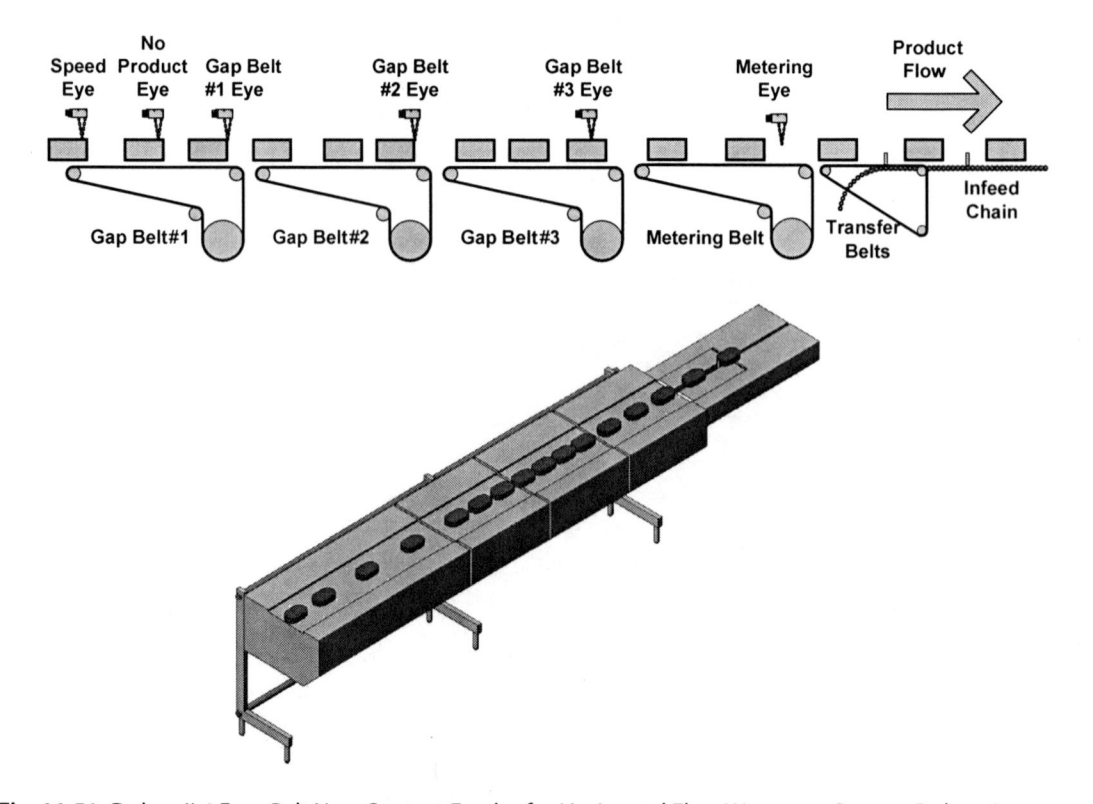

Fig. 11.51. Doboy IL4 Four Belt Non-Contact Feeder for Horizontal Flow Wrappers. Source: Doboy, Inc.

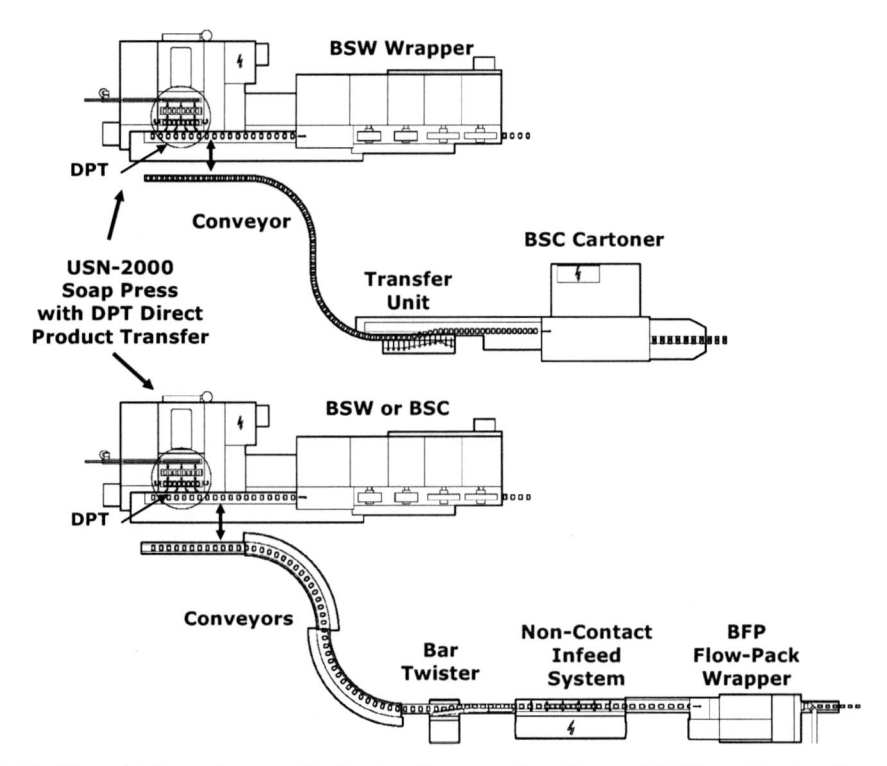

Fig. 11.52. Binacchi Stamping and Packaging Systems. One Press with Direct Product Transfer and Two Alternate Packaging Options. Source: Binacchi & Co.

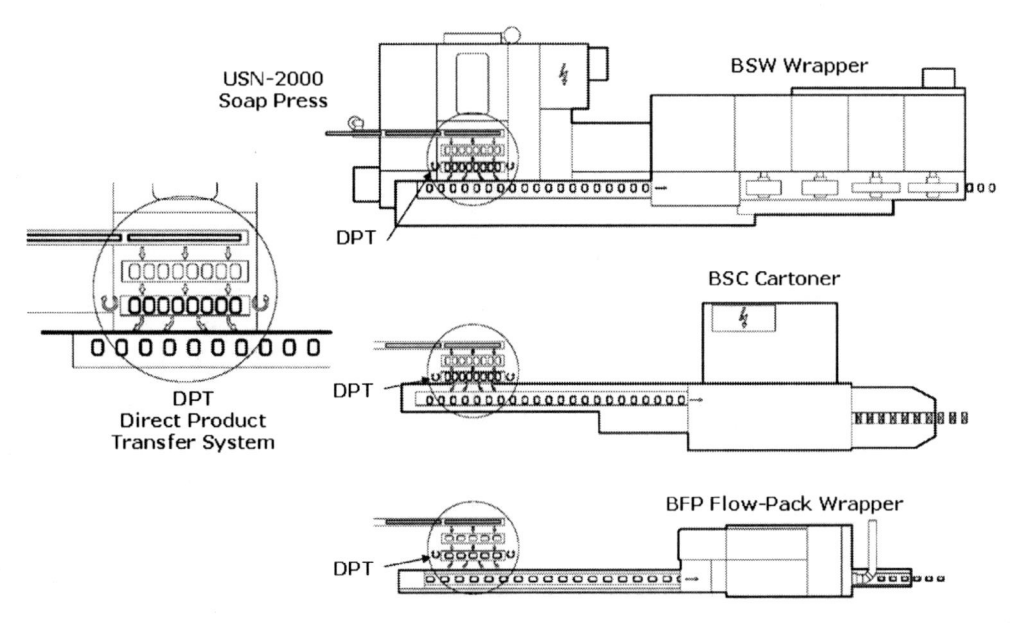

Fig. 11.53. Binacchi Stamping and Packaging Systems. One Press with Direct Product Transfer and Three Alternate Packaging Options. Source: Binacchi & Co.

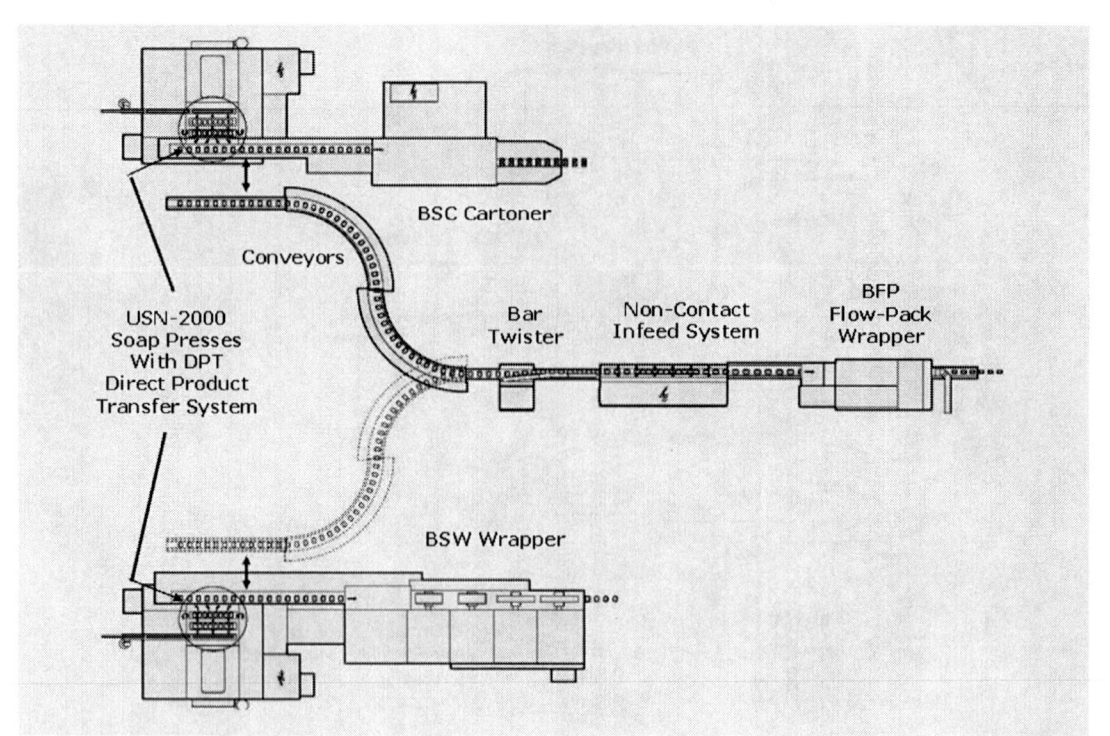

Fig. 11.54. Binacchi Stamping and Packaging Systems. Two Presses with Direct Product Transfer and Three Alternate Packaging Options. Source: Binacchi & Co.

Table 11.4. Soap Wrappers.

Make	Model	Speed (Wrapped Bars per Minute)	Soap Infeed/Transfer System
ACMA	7250	250	Timing belts and suction cups.
	7350	350	Timing belts and suction cups.
	7350/DL	380	Direct linkage with press.
	C701/HS	300	Flat belt.
	771/TH	500	Pick and place turret.
	771/DL	550	Direct linkage with press.
BINACCHI	BSW-220	220	Direct linkage with press.
	BSW-330	330	Direct linkage with press.
	BSW-550	550	Direct linkage with press.
GUERZE	Packsavon	250	Pick and place turret.

Bar Soap Finishing • 345

Table 11.5. Soap Cartoners.

Make	Model	Speed (Cartoned Bars per Minute)	Pitch	Soap Infeed/Transfer System
ACMA	330/YV	240	95 mm	Pick and Place with 4 cups
	330/YT	300	95 mm	Pick and Place with 10 cups
	770/TH	500	120 mm	Pick and Place with 16 cups
	770/DL	600	120 mm	Direct linkage with press
JONES	Legacy	300	6 inches	2 Pick and Place Groups with 4 cups each
	Legacy	400	4 inches	2 Pick and Place Groups with 4 cups each
	Criterion	500	4.5 inches	Direct linkage with press
	Criterion	600	4 inches	Direct linkage with press
GUERZE	Boxsavon	250	127 mm	Pick and Place Group with 4 cups

Refer to Fig. 11.55 for carton blank terminology carton styles, Fig. 11.56 for carton styles, and Table 11.6 for wrapping material specifications and conversion factors data.

The description of banders, bundlers, stretch film, pleat wrappers, and end-packaging machinery is beyond the scope of this chapter.

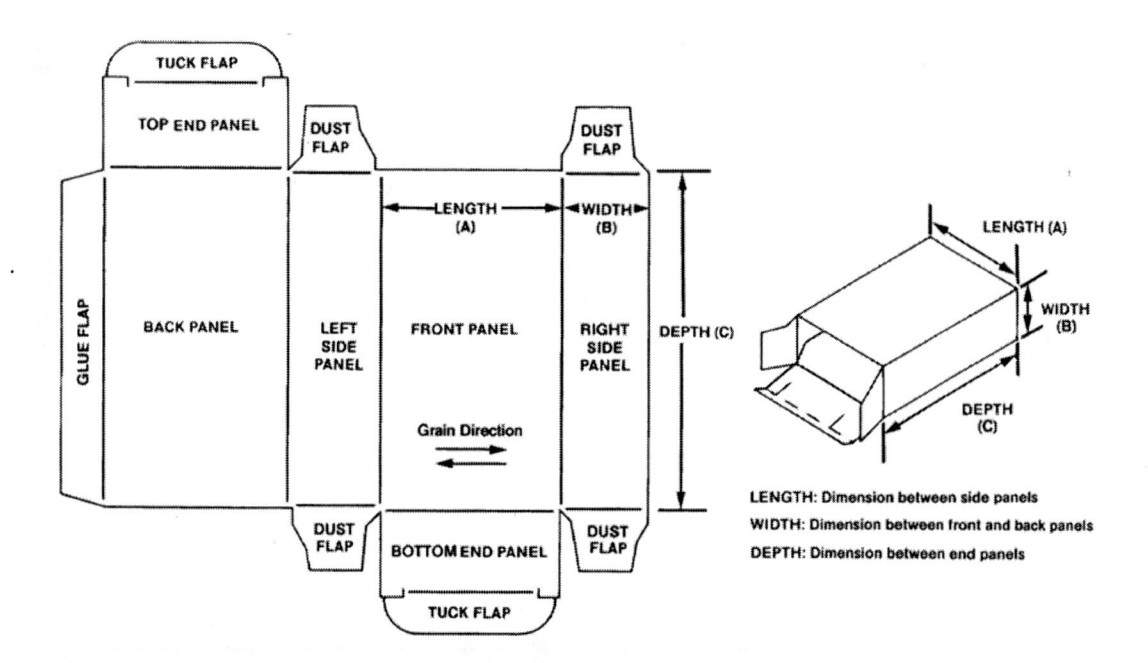

Fig. 11.55. Carton blank terminology.

346 ● L. Spitz

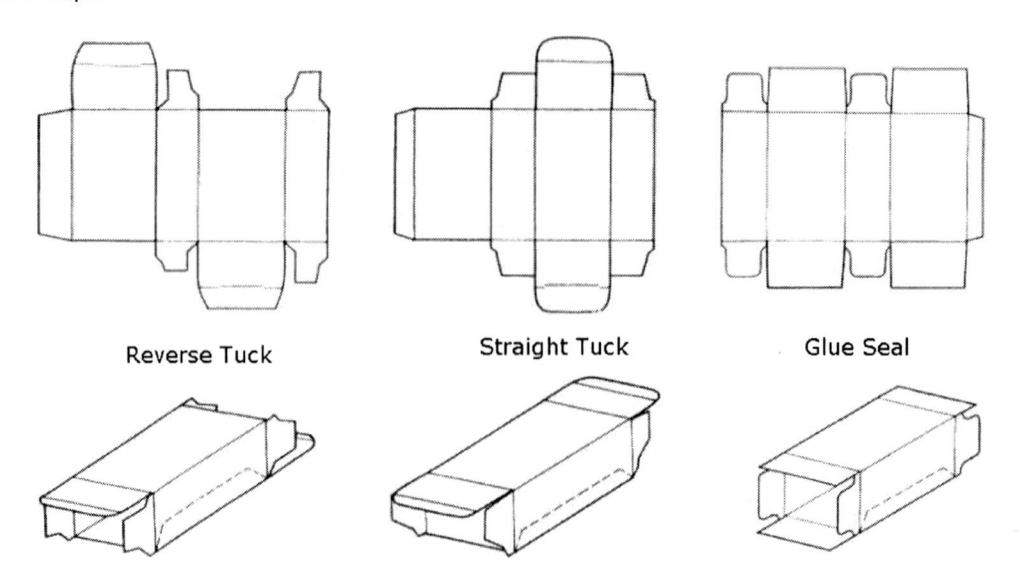

Fig. 11.56. Carton styles.

Table 11.6. Packaging Material Definitions and Conversion Factors

Basis Weight

In the USA basis weight refers to the weight expressed in pounds of 500 sheets of 24 x 36 inches size paper (24" x 36" = 3,000 ft^2 = 270 m^2)

Grammage

Grammage is the basis weight expressed in g/m^2 (grams per square meter)

Caliper/Point

Caliper or Point is the thickness of paper or board expressed in thousands of an inch
1 caliper or 1 point = 0.001 inch
Example: a 15 point board is a 0.015 inch thick board

Gauge

Film Thickness is expressed in terms of mils.
1 mil = 0.001 inch
For thin films the term gauge is used. 100 gauge = 1 mil
Example: a 90 gauge film is a 0.00090 inch thickness film

Micron

1 micron = 0.001 mm
1 millimeter = 1000 microns

To Convert	Multiply by
lbs/ream to g/m^2	1.627
in^2/lb to m^2/kg	0.0014
reams to m^2	278.7
mils to microns	25.4
gauge to micron	0.254
micron to gauge	3.937

Acknowledgments

Information and illustrations provided by Acma SpA, Binacchi & Co., Doboy, Inc. Guerze Srl, Mazzoni LB SpA, SAS, Sela GmbH, Soaptec Srl, and Tecnicam Srl. are very much appreciated.

I wish to extend a special note of thanks to Mr. Andrea del Corno from Soaptec srl who prepared many Solid-Edge software based drawings.

References

Spitz, L. (ed.) *Soap Technology for the 1990s;* AOCS Press: Champaign, Illinois, 1990; pp.173–208.

Spitz, L. (ed.) *Soaps and Detergents, A Theoretical and Practical Review;* AOCS Press: Champaign, Illinois, 1996; pp. 243–287, 474–492.

Spitz, L. Bar Soap Finishing Lecture, SODEOPEC2006 Conference, Hollywood, Florida, April 5, 2006.

Manufacture of Multicolored and Multicomponent Soaps

Luis Spitz
L. Spitz, Inc., Highland Park, Illinois, USA

Introduction

The first multicolored soaps were the "mottled" laundry soaps. Introduced over a century ago in Germany and later in France, Spain and Italy, they were made from bleached palm and coconut oils, and became accepted by the public as high-quality soaps. Today, most laundry soaps are blue, and the mottling effect is obtained by using ultramarine blue dye.

This chapter introduces the subject with a brief history of multicolored and multicomponent toilet soaps. More soap history details are found in Chapter 1.

Multicolored/multicomponent soaps are classified into marbleized, striped, speckled, and two-tone types. The manufacturing system for each type is described and illustrated.

These soaps offer potential marketing advantages over single-color soaps with or without additives. The visual differentiation over single- color soaps provides aesthetic advantages for the multicolored types, and for the multicomponent types can show the ingredient(s) which claim to enhance product performance.

The "freshness" multicolored soap category was introduced in 1968 with Henkel's Fa bar which rejuvenated the bar-soap market. Shortly thereafter, many multicolored "fresh" soap brands appeared worldwide.

Line extensions of existing and new multicomponent/multicolored bars by Henkel (Dial), Colgate-Palmolive, Unilever, Evyap, Dalan, and other companies are launched periodically, confirming the longevity and growth potential of this successful forty-year-old soap category.

Fa (1968): The Marbleized "Freshness" Soap

The original 1954 "Die Seife Fa" "Fa"ntastic was a skin-care bar. In 1968 "Die Frische Fa" (The Fresh Fa), a marbleized green-and-yellow soap with the "wild freshness of limes," started the new "freshness" soap category . Fa's success prompted many competitors to enter the market. "Freshness," combined with deodorant and antibacterial claims, is an active soap category today.

Atlantik and Pacific (1969–1970)

In 1969 Unilever introduced Atlantik soap in Germany, highlighting its "Seaweed Extract" ingredient. The soap's name, shape, ingredients, color, and package design are the best examples of an "integrated" consistent bar-soap product concept and execution. The Pacific companion soap followed in 1970.

Irish Spring (1972)

In 1972, Irish Spring, "The Double Deodorant Soap" bar with "The Freshness of an Irish Morning," was introduced. It was the first freshness category green-and-white marbleized soap in the United States. Later, Irish Spring bars with different names were launched in many countries.

In 1986 the product was changed into a deodorant soap with skin conditioners and packaged in a green carton.

In 2008 six Irish Spring variants were offered in the U.S. market.

Coast (1974)

Procter & Gamble (P&G) entered the multicolored category in 1974 with the blue-and-white marbleized Coast bar. The original "Eye-Opener Refreshing Deodorant Soap" was produced with a patented solid–solid manufacturing system. White-and-blue pellets of different diameters pelletized to different lengths and in different ratios were fed into the final stage of the Duplex Vacuum Plodder (Fig. 12.5 later in chapter).

To enhance the marbleizing effect, one can shave off a thin surface of the extruded slugs (billets), and recycle the shavings. Stamping the slugs at an angle (on a bias) will also improve their appearance. Irish Spring and Coast are stamped at an angle (Fig. 12-19 later in this chapter).

P&G sold the Coast brand to The Dial Corporation in 2000.

In 2008 two marbleized bars were offered: Coast Arctic and Coast Pacific Force.

Dove Nutrium (2000)

The Dove Nutrium bar was introduced in late 2000 after the success of the Nutrium Moisturizing Body Wash. This was the first time that a liquid product was later introduced in a solid-bar version.

Unilever's Dove Nutrium is a multicomponent, striped, dual-formula bar with a white moisturizing cleanser and a pink nutrient-enriched lotion with vitamin E.

Dove Nutrium was the only multicomponent/multicolored bar on the market which showed two distinct soaps.

The 2008 version called "Nutrium Cream Oil Beauty Bar" was a solid pink bar. The striped version has been discontinued.

Multicolored and Multicomponent Soap Types

Four distinct types of multicolored and multicomponent soaps are available. In the soap type and manufacturing method descriptions, the *primary base* refers to the larger quantity, predominant color soap base. The *secondary base* is the lesser quantity second color base.

Marbleized

Marbleized (also called marbled, variegated, and mottled) soaps are produced by dosing or injecting an additional color into the primary base soap which can be white or colored. Detail of a Dial bar is shown.

Striped

Striped soaps with well-defined linear designs are produced by the controlled addition (injection) of a secondary base of one color into a primary base of another color, such as this sample from Dalan (Turkey).

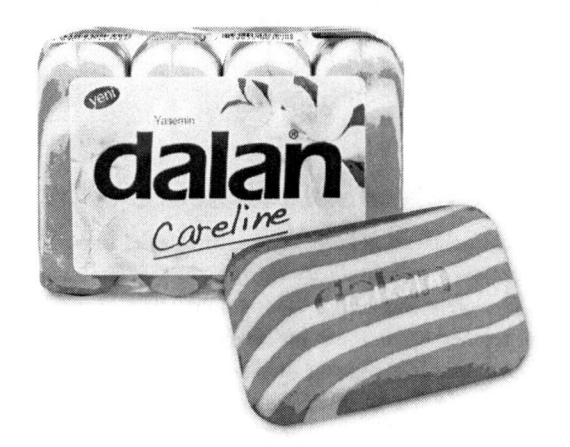

Speckled

Speckled soaps are formed by the proportioned addition of small speckles (granules) or larger chunks of different colors and/or different colors and types of product added into the primary base, such as this sample from Pre de Provence (France)

Two-Tone

Two-tone toilet and laundry soaps are formed when the primary and secondary bases are fed into non-tangential twin-worm plodders. These plodders are side-by-side with separate worm barrels and Individual Worms. The two different bases move through the plodder separately until extrusion. Side-by-side, vertical, horizontal, diagonal, radial, and multiple patterns can be produced. See the samples below for (A) Two-Tone Toilet Soaps [Evyap, Turkey] and (B) Two-Tone Laundry Soaps

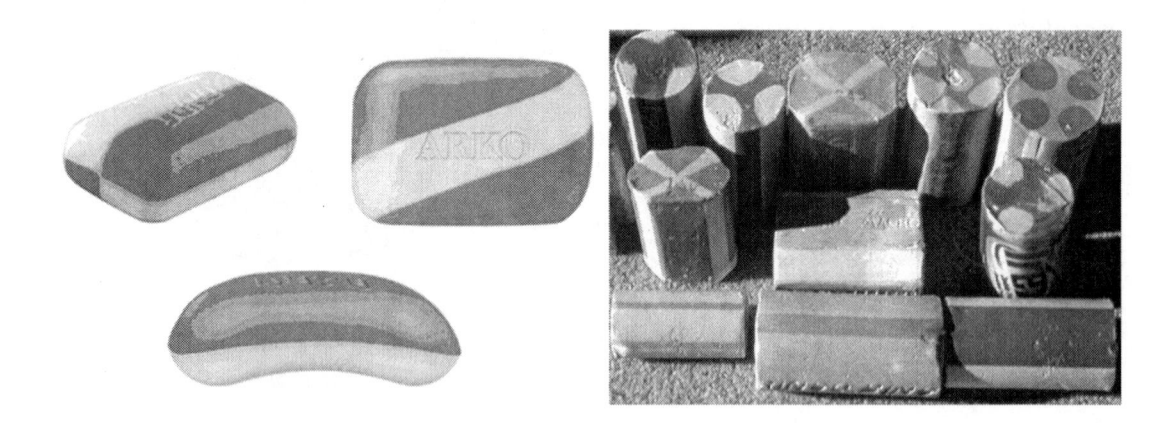

Handcrafted Artisan Soaps—An Old/New Niche Market

The handcrafted, artisan soaps with special ingredients, performance claims, and very interesting designs have grown in popularity during the last two decades. What started as a hobby for many home soap makers grew into small-business enterprises. In 1998 The Handcrafted Soap Makers Guild (HSMG) was formed with these objectives: "to promote the handcrafted soap industry, to act as a center of communications among soap makers, and to circulate beneficial information to them."

HSMG is a nonprofit international organization with 800 members. An annual conference is open to members and also to nonmembers. Comprehensive information for this growing niche market for the various homemade soap manufacturers, production methods, formulations, reference books, and raw material and equipment suppliers is found on the HSMG Web site www.soapguild.org.

Manufacture of Multicolored and Multicomponent Soaps ● 353

Manufacturing Systems, Methods, and Product Types

Table 12.1. All the Manufacturing Methods presented are based on Mazzoni LB technology.

Manufacturing System	Manufacturing Methods	Product Types
Solid–Liquid	• Color Dosing into the Extrusion Stage Plodder • Color Injection into the Extrusion Stage Plodder • Color Injection into the Extrusion Stage Plodder Barrel	Marbleized
Solid–Solid	• Utilizing two separate single-worm plodders for feeding two different-size soap pellets • Utilizing non-tangential twin-worm plodder • Using speckles, granules and "chunks" feeder group	Marbleized Two-tone Speckled
Solid–Solid Co-Extrusion	• Co-extruder with striping group • Co-extruder with striping and marbleizing group	Striped Striped and Marbleized
Solid–Solid–Liquid	• Combination solid–solid and solid–liquid	Striped and Marbleized Two-Tone Striped

Solid–Liquid Systems for Marbleized Soaps

Solid-Liquid Systems for marbleized soap production consist of:

- Duplex Vacuum Plodder
- Color Dosing/Injecting Group

or a

- Duplex Vacuum Plodder
- Color-Dosing Group, and a
- Marbleizing group.

Color Dosing into the Extrusion-Stage Plodder

The simplest and most economical method is dosing a color solution into the Extrusion Stage of the Duplex Vacuum plodder. The random color distribution (the marbleized effect) is difficult to control (Fig.12.1).

Color Injection into the Extrusion-Stage Plodder Barrel

A color solution is injected into the Duplex Vacuum Plodder's Extrusion Stage plodder barrel (Fig. 12.2).

Color Injection into the Extrusion-Stage Plodder

The color solution is injected through a drilled plate with injection nozzles in the Extrusion Stage of the Duplex Vacuum Plodder (Fig. 12.3).

A Rotor Drive Group can be added for independent speed variation of the rotor, allowing one to obtain more marbleizing effects (Fig. 12.4).

The striping and marbleizing group assembly is shown in Fig. 12.5.

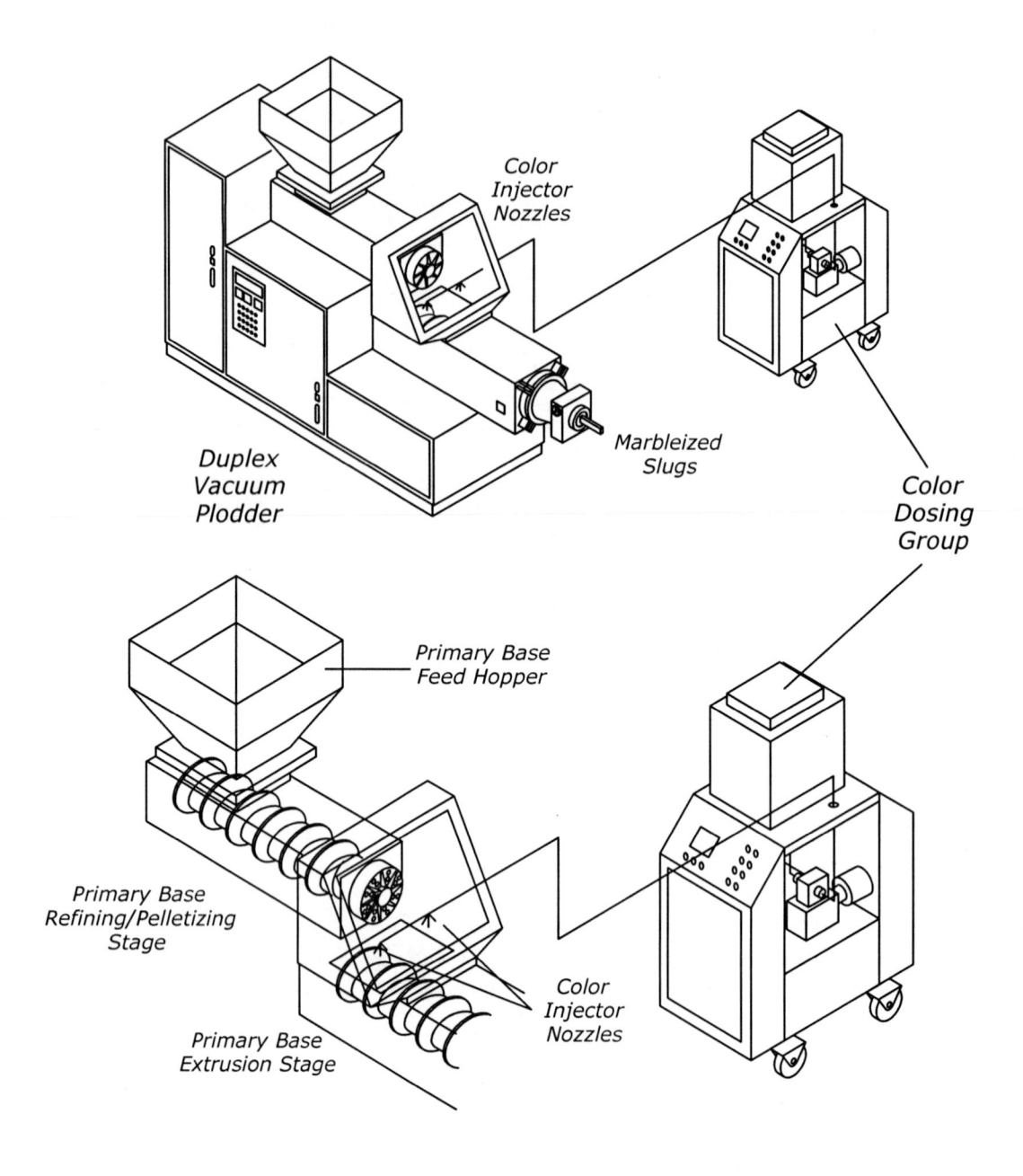

Fig. 12.1. Solid-Liquid System for Marbleized Soaps—Color Dosing into the Extrusion Stage Plodder.

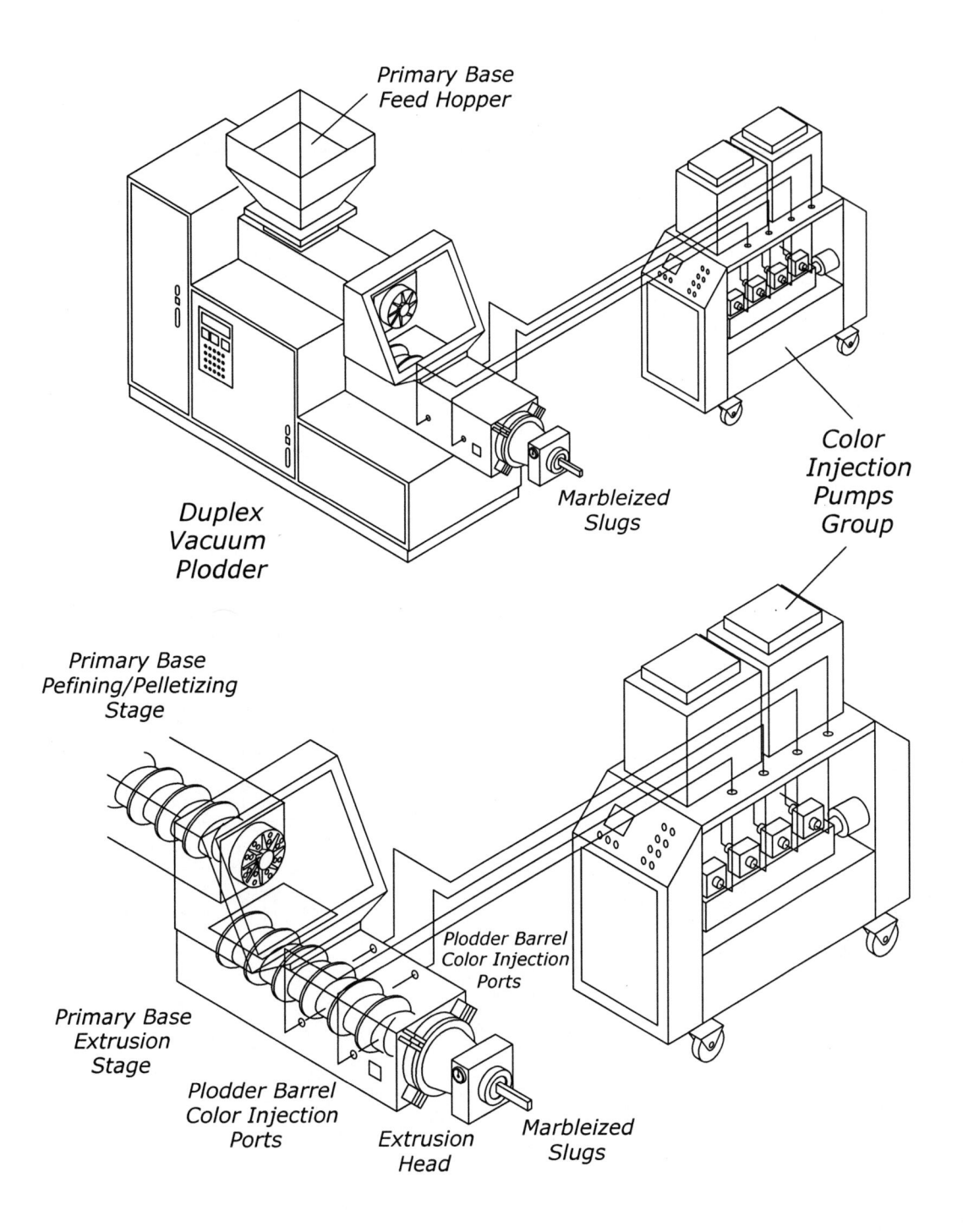

Manufacture of Multicolored and Multicomponent Soaps • 355

Fig. 12.2. Solid-Liquid System for Marbleized Soap—Color Injection into the Extrusion Stage Plodder Barrel.

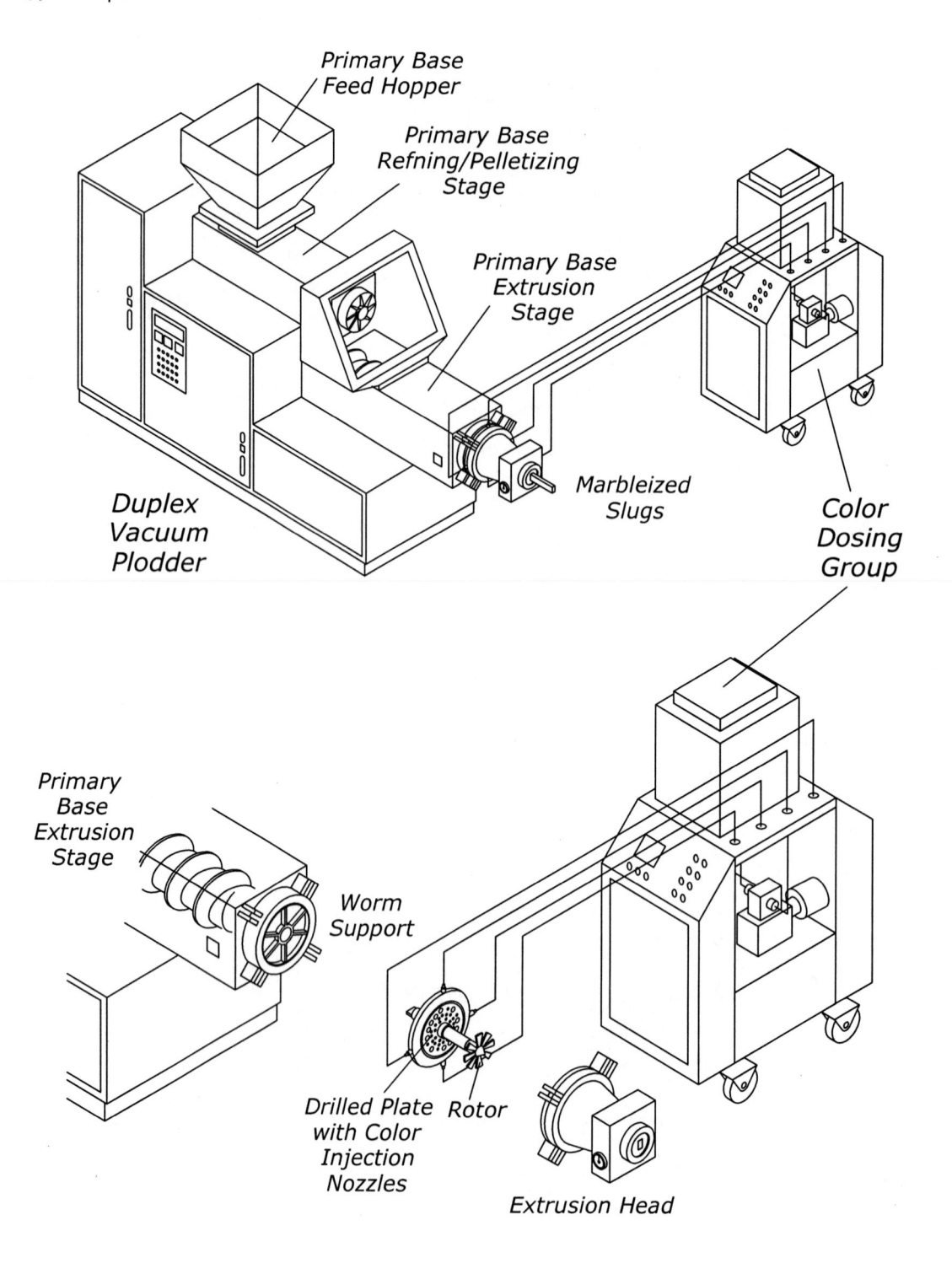

Fig. 12.3. Solid-Liquid System for Marbleized Soaps—Color Injection into the Extrusion Stage Plodder.

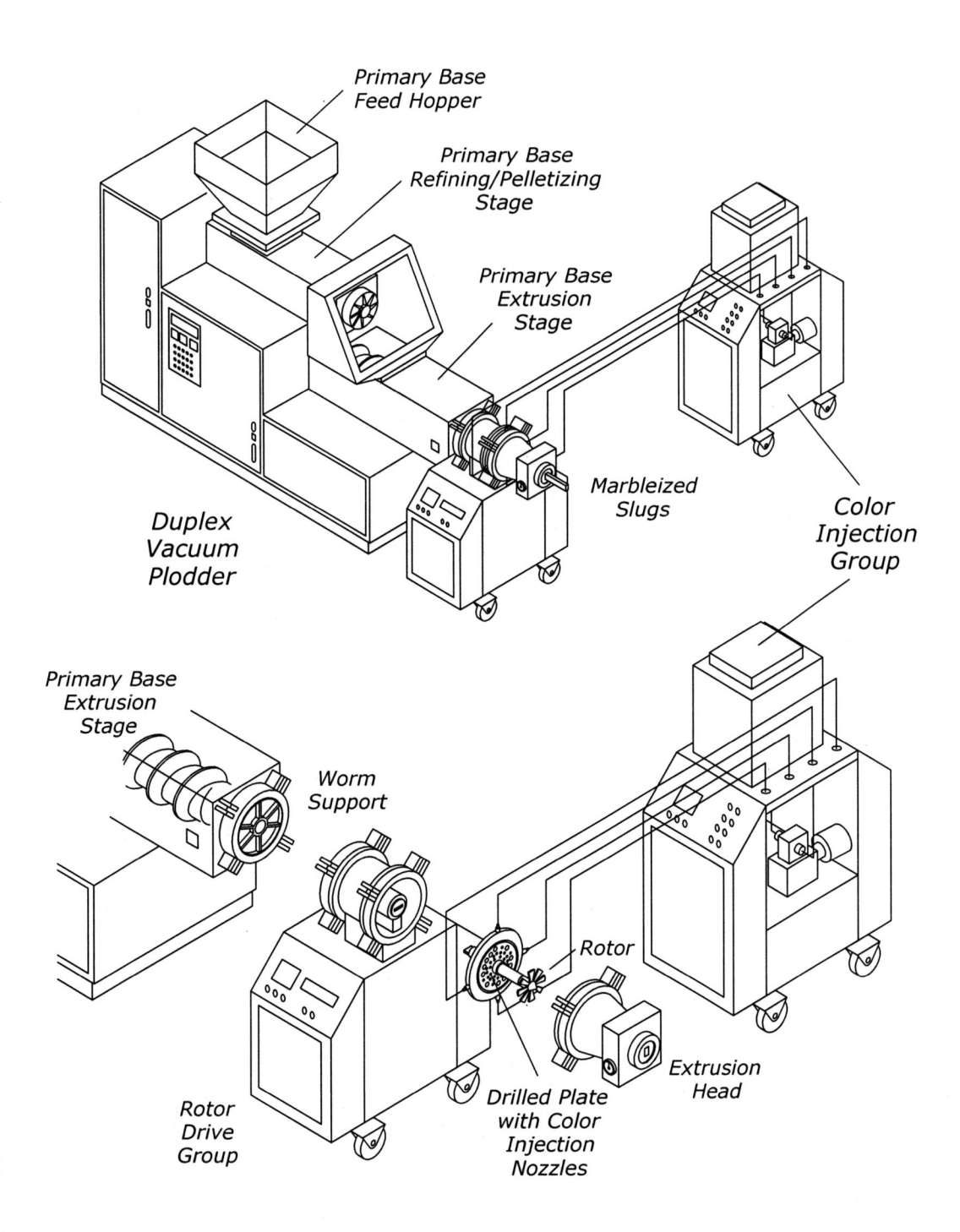

Fig. 12.4. Solid–Liquid System for Marbleized Soaps—Color Injection into the Extrusion Stage Plodder with Rotor Drive Group.

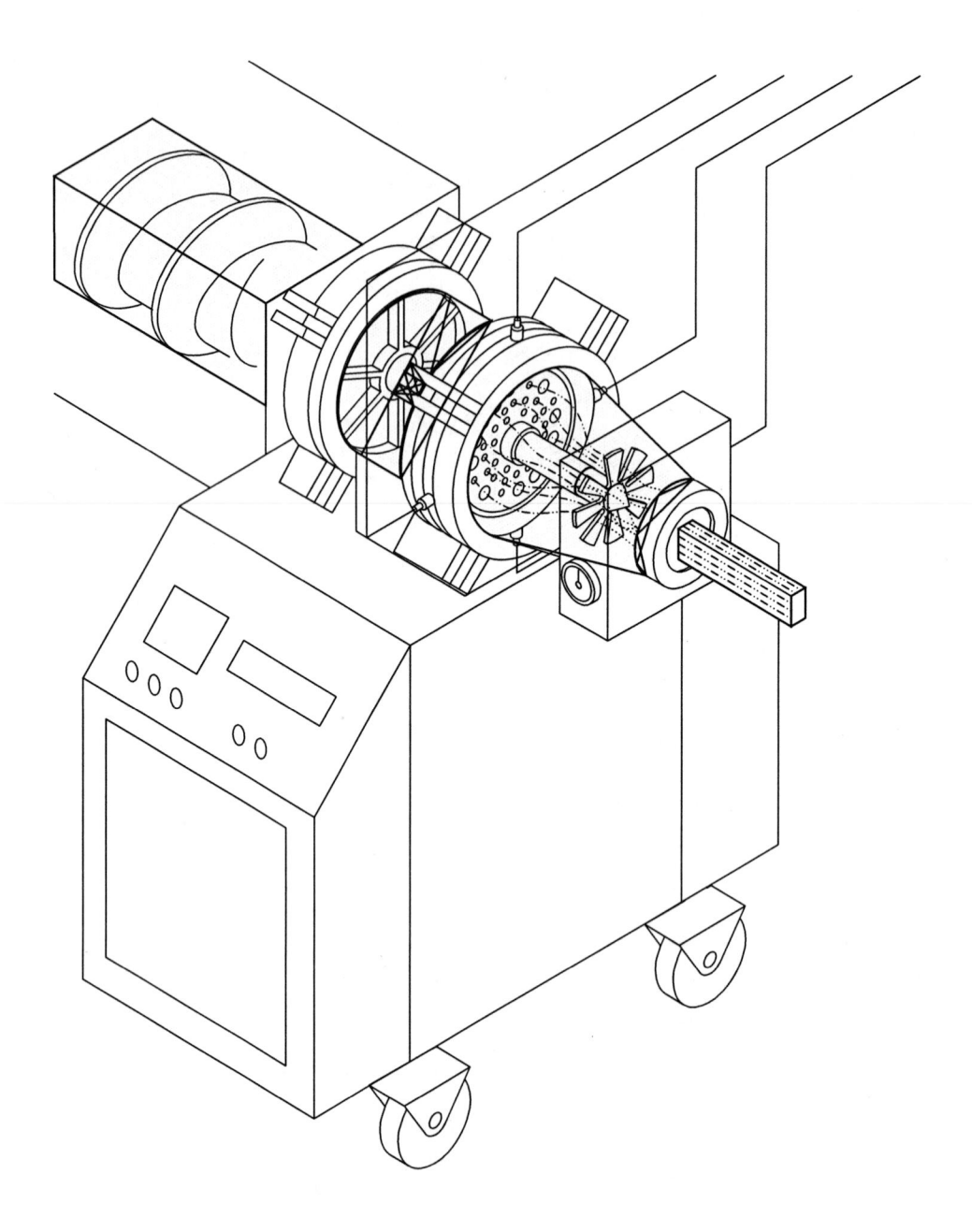

Fig. 12.5. Solid-Liquid System Marbleizing and Striping Group Assembly.

Solid–Solid System for Marbleized Soaps
Utilizing Two Separate Single-Worm Plodders
Two separate single-worm plodders are used to feed two different diameters and lengths of pellets in a predetermined weight ratio into the Extrusion Stage of a single-worm plodder. The random or partially controlled mixing of the two pellets produces the marbleized effects (Fig. 12.6).

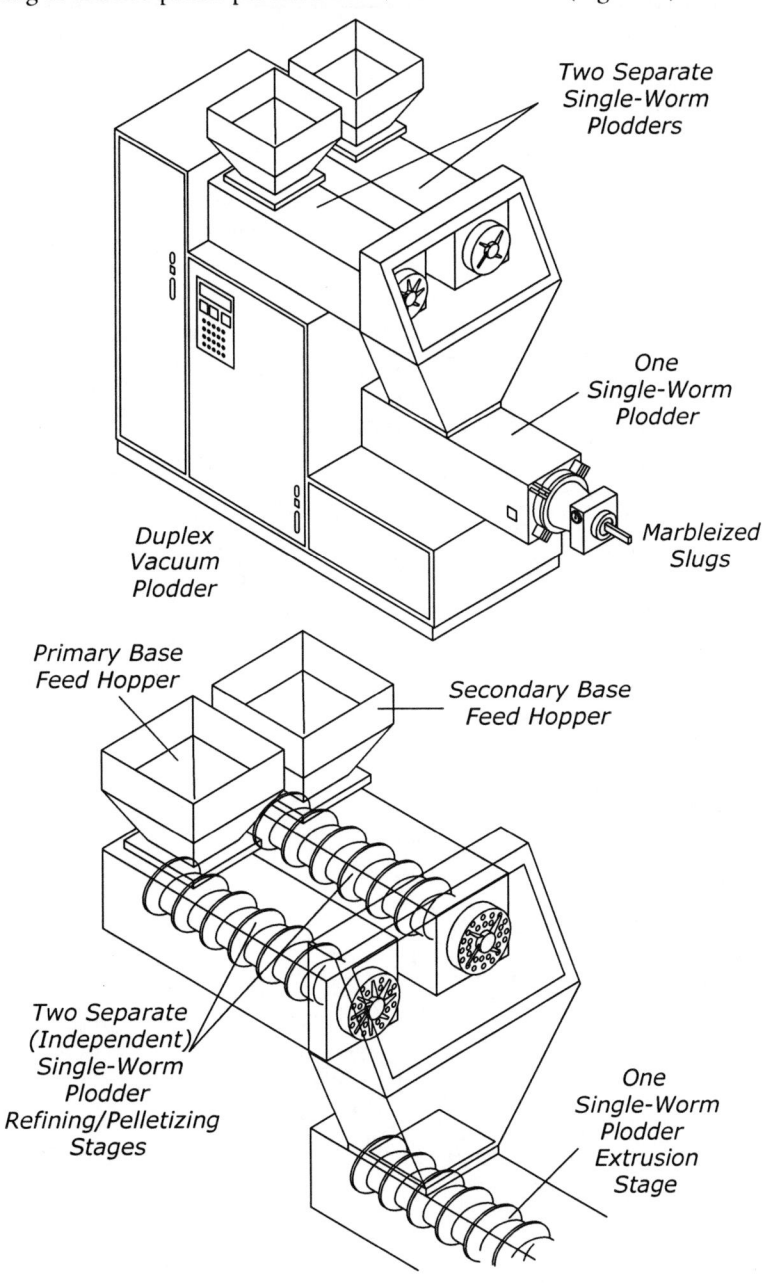

Fig. 12.6. Solid-Solid System for Marbleized Bars—Utilizing Two Separate Single-Worm 1st Stage Plodders and One Single-Worm Extrusion Stage Plodder.

Solid–Solid System for Two-Tone Soaps
Utilizing Non-tangential Twin-Worm Plodders

This system requires the use of a Non-Tangential Twin-Worm Duplex Vacuum Plodder. Two different soap bases are fed into each first stage plodder and proceed separately until extrusion. The two bases meet when they reach the extrusion head, which is provided with a baffle. This system can produce various two-tone designs (Fig. 12.7).

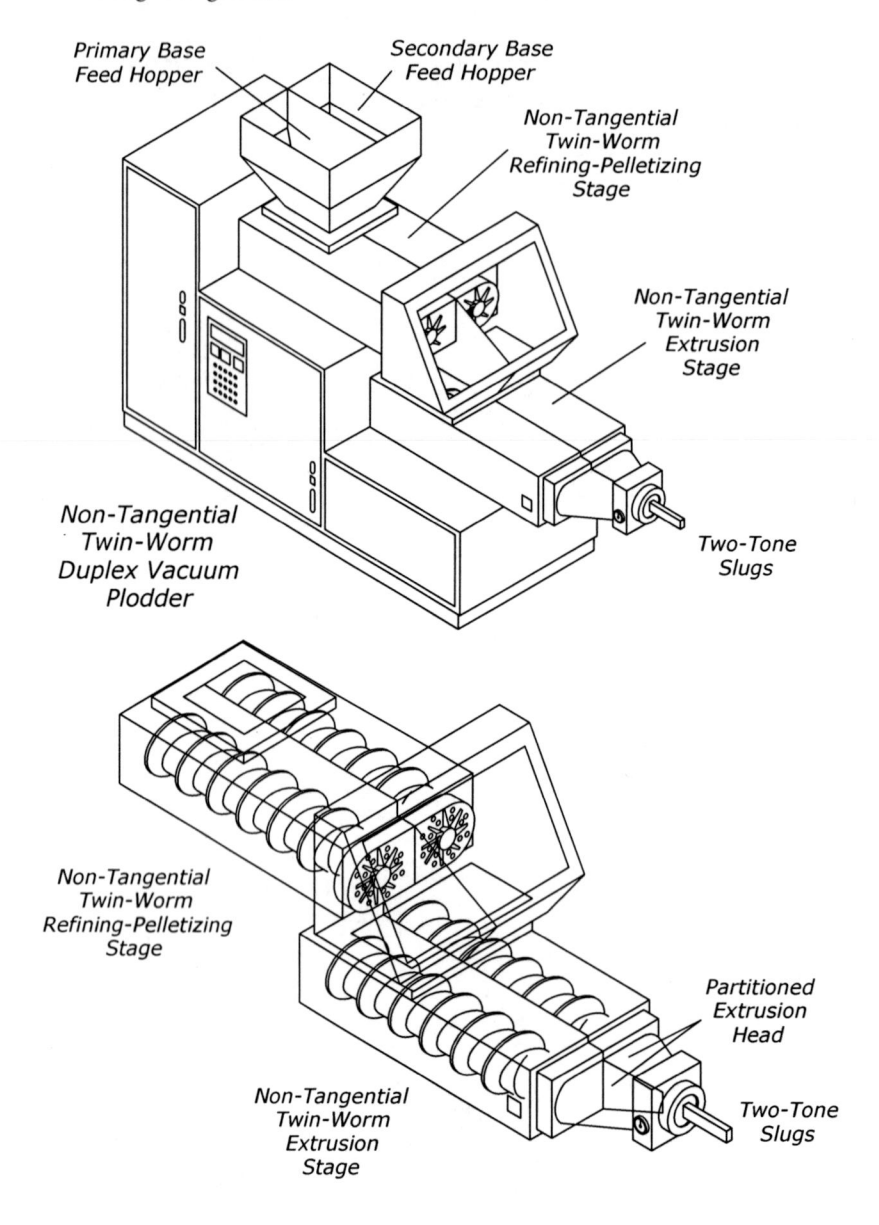

Fig. 12.7. Solid-Solid System for Two-Tone Soaps— Utilizing Non-Tangential Twin-Worm Plodders.

Manufacture of Multicolored and Multicomponent Soaps

Solid–Solid System for Speckled Soaps

Additives Proportioned into the First-Stage or Extrusion Stage Plodder

Randomly distributed speckles, granules and small colored soap "chunks" can be added with a feeder group into either the first stage or the extrusion stage plodder (Fig. 12.8. and Fig. 12.9).

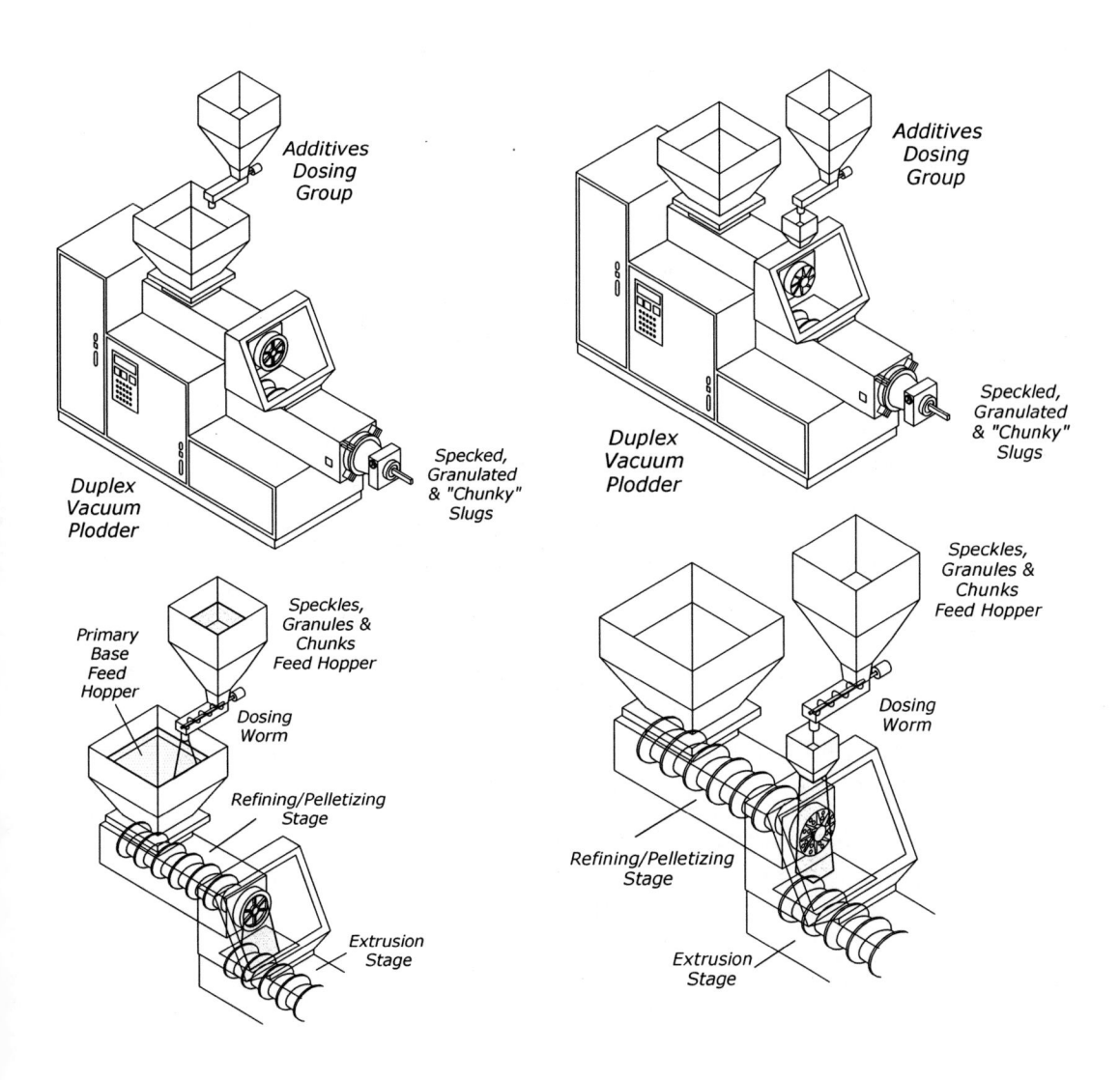

Fig. 12.8 & 9. Solid-Solid System for Soaps with Speckles, Granules, and "Chunks."

Solid–Solid Co-Extrusion Systems for Striped, Marbleized, and Two-Tone Soaps

To obtain well-defined striped soaps, a Co-Extrusion System consisting of:

- A standard Duplex Vacuum Plodder for the primary base
- A Simplex Co-extruder Plodder for the injection of the secondary base
- An interconnecting "striping/marbleizing group" with a tube-bundle cylinder: tube bundle; drilled plate; rotor-drive group—this optional group can run the rotor at different speeds than the fixed plodder worm speed, allowing one to obtain a wider range of marbleized effects. The tube bundle has different diameter tubes. The primary base soap is fed through the larger diameter tubes and co-extruded secondary base, which is always of lesser quantity, through the smaller tubes.

Solid–Solid Co-Extrusion System for Marbleized Soaps

This system includes a rotor device for the production of marbleized soaps (Fig. 12.10). If the rotor is removed, striped soaps can be made.

Solid–Solid Co-Extrusion System for Striped Soaps

The refined/pelletized primary base from the Duplex Vacuum Plodder is extruded into the tube bundle of the "striping group" assembly. A secondary base co-extruder (Simplex Plodder) feeds the secondary base also into the tube bundle. To achieve the striped patterns, the two bases remain separate until they exit from the extrusion head (Fig. 12.11).

Solid-Solid Co-Extrusion Group Assembly without Rotor for Striping

This is a complete system including a rotor-drive group (Fig. 12.12 and 13).

Solid-Solid Co-Extrusion Assemby with Rotor Drive Group for Striping and Marbleizing (Fig. 12.14.)

Solid-Solid-Liquid Co-Extrusion Multipurpose System for Striped and Marbleized Soaps

This multipurpose combination system includes all the components of the Solid-Solid and the Solid-Liquid systems. (Fig.12.15.)

Solid-Solid Co-Extrusion System for Two-Tone Striped Soaps

This system utilizes a Tangential Twin-Word Plodder and a Single-Worm Co-Extruder for the production of interesting two-tone striped designs (Fig. 12.16).

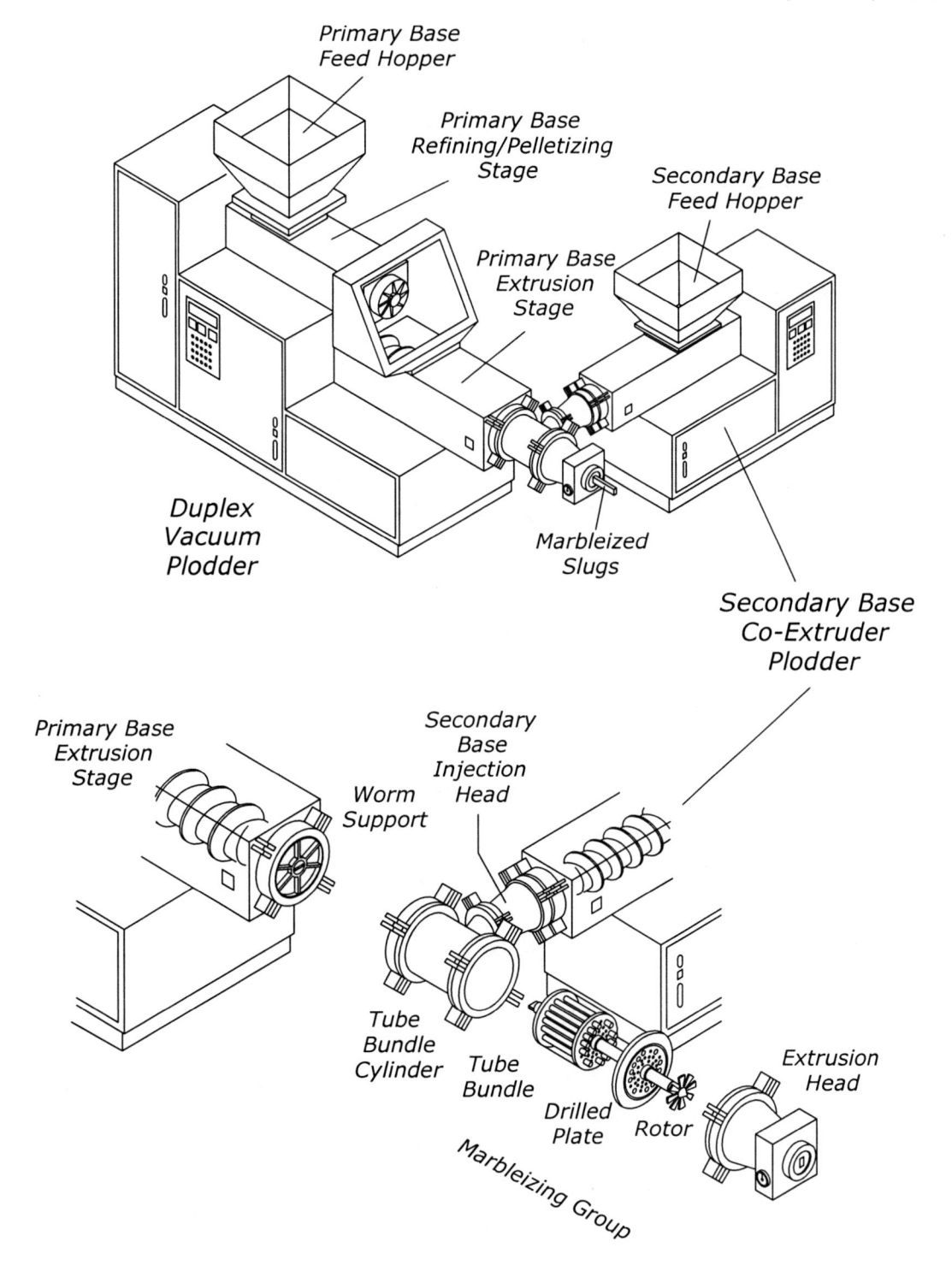

Manufacture of Multicolored and Multicomponent Soaps • 363

Fig. 12.10. Solid-Solid Co-Extrusion System for Marbleized Soaps.

364 ● L. Spitz

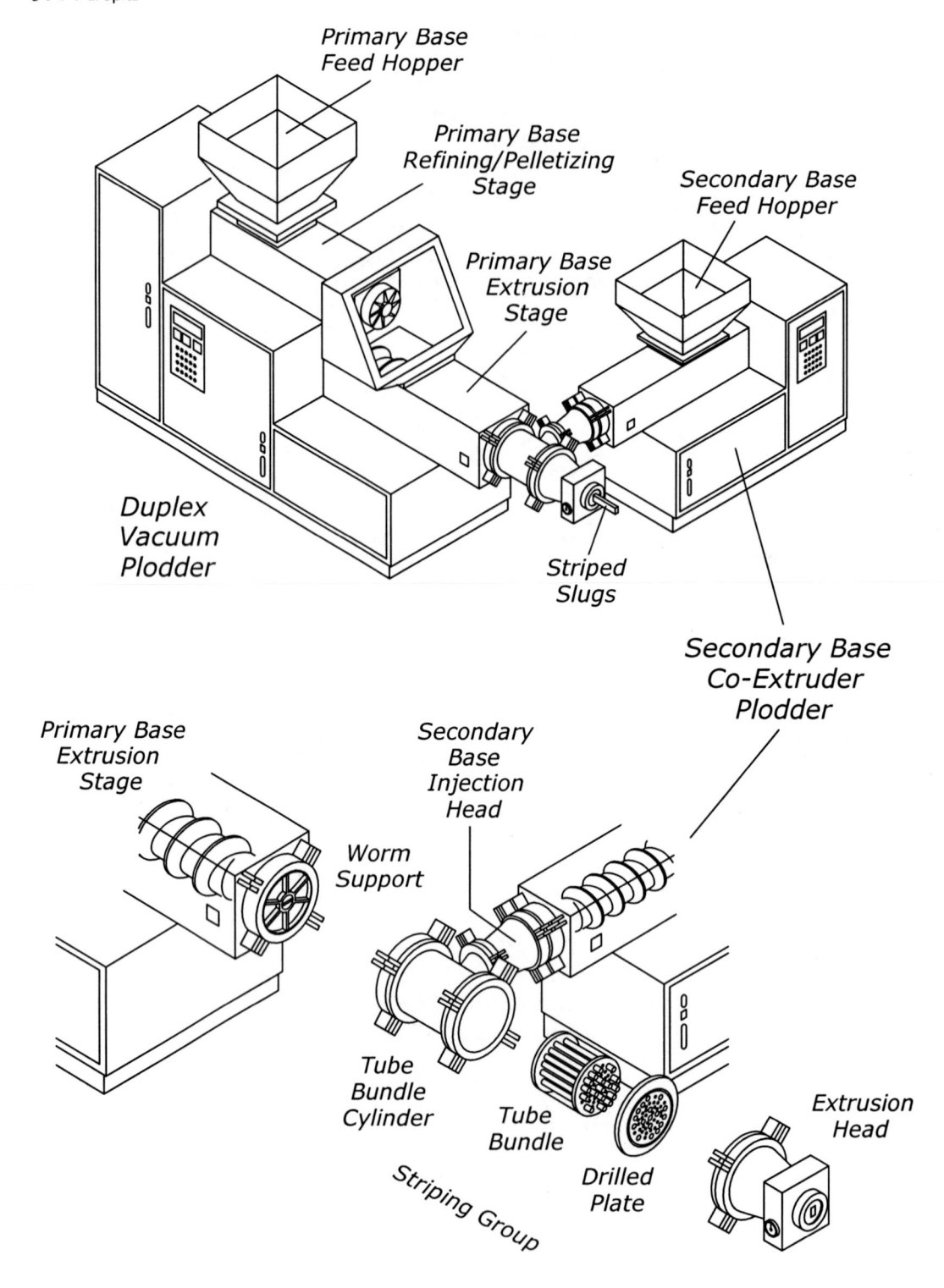

Fig. 12.11. Solid-Solid Co-Extrusion System for Striped Soaps.

Manufacture of Multicolored and Multicomponent Soaps • 365

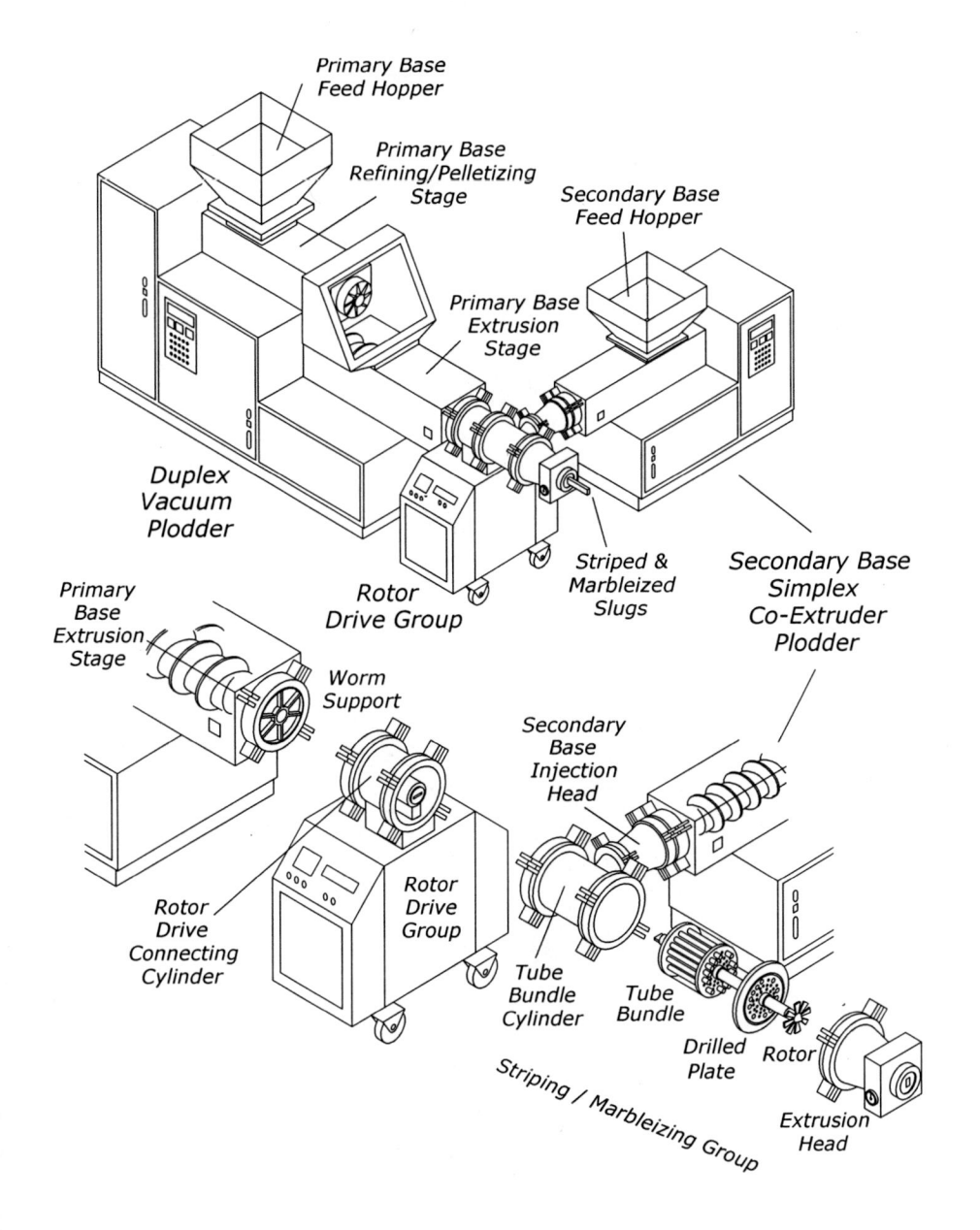

Fig. 12.12. Solid-Solid Co-Extrusion System with Rotor Drive Group for Striped and Marbleized Soaps.

366 ● L. Spitz

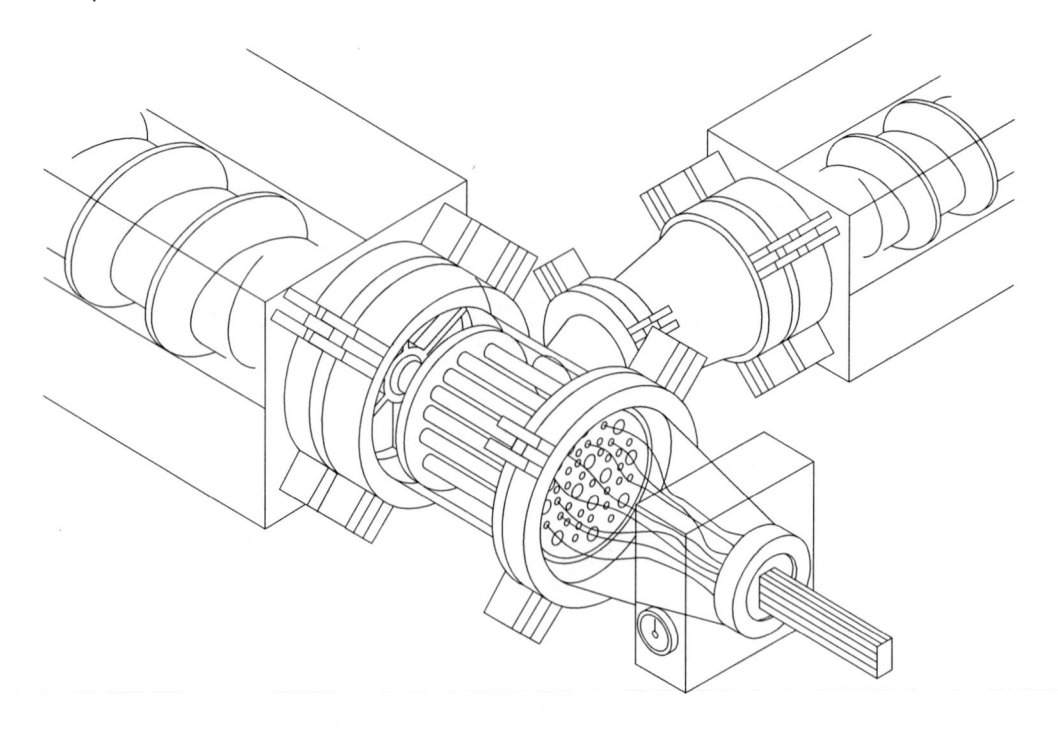

Fig. 12.13. Solid-Solid Co-Extrusion Assembly for Striping.

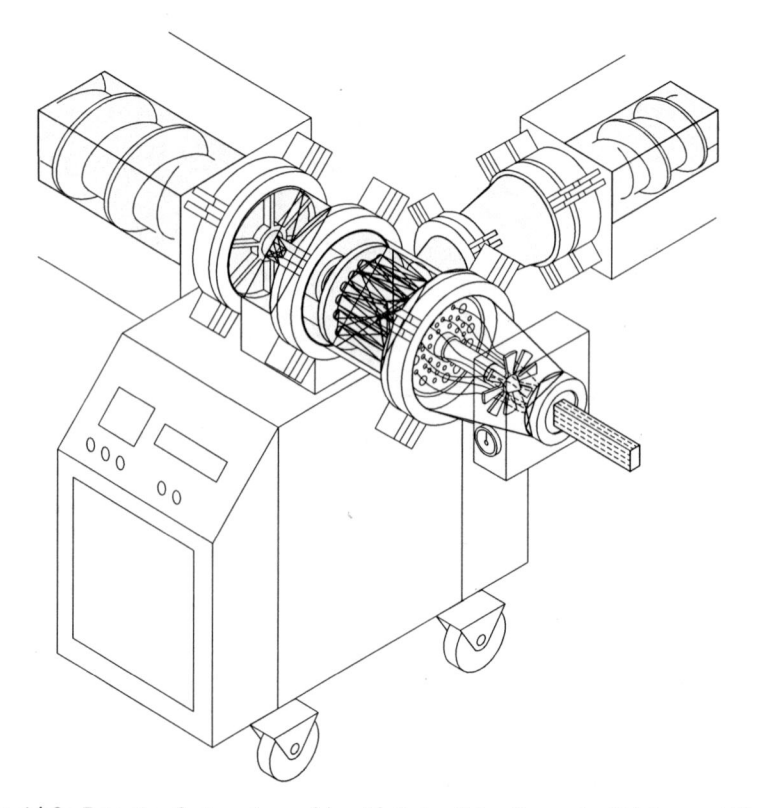

Fig. 12.14. Solid-Liquid Co-Extrusion System Assembly with Rotor Drive Group for Striping and Marbleizing.

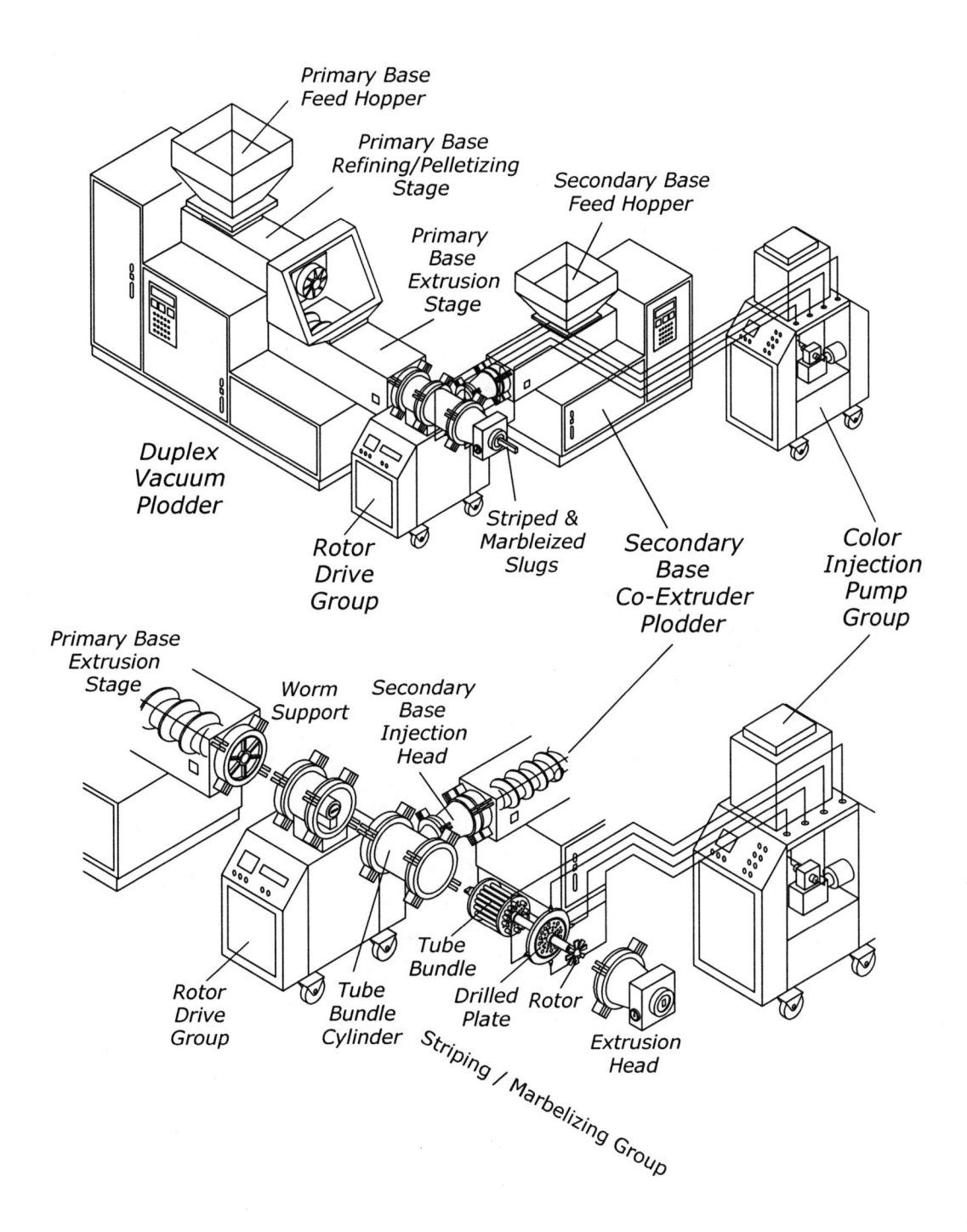

Fig. 12.15. Solid-Solid-Liquid Combination Multipurpose System for Striped and Marbleized Soaps.

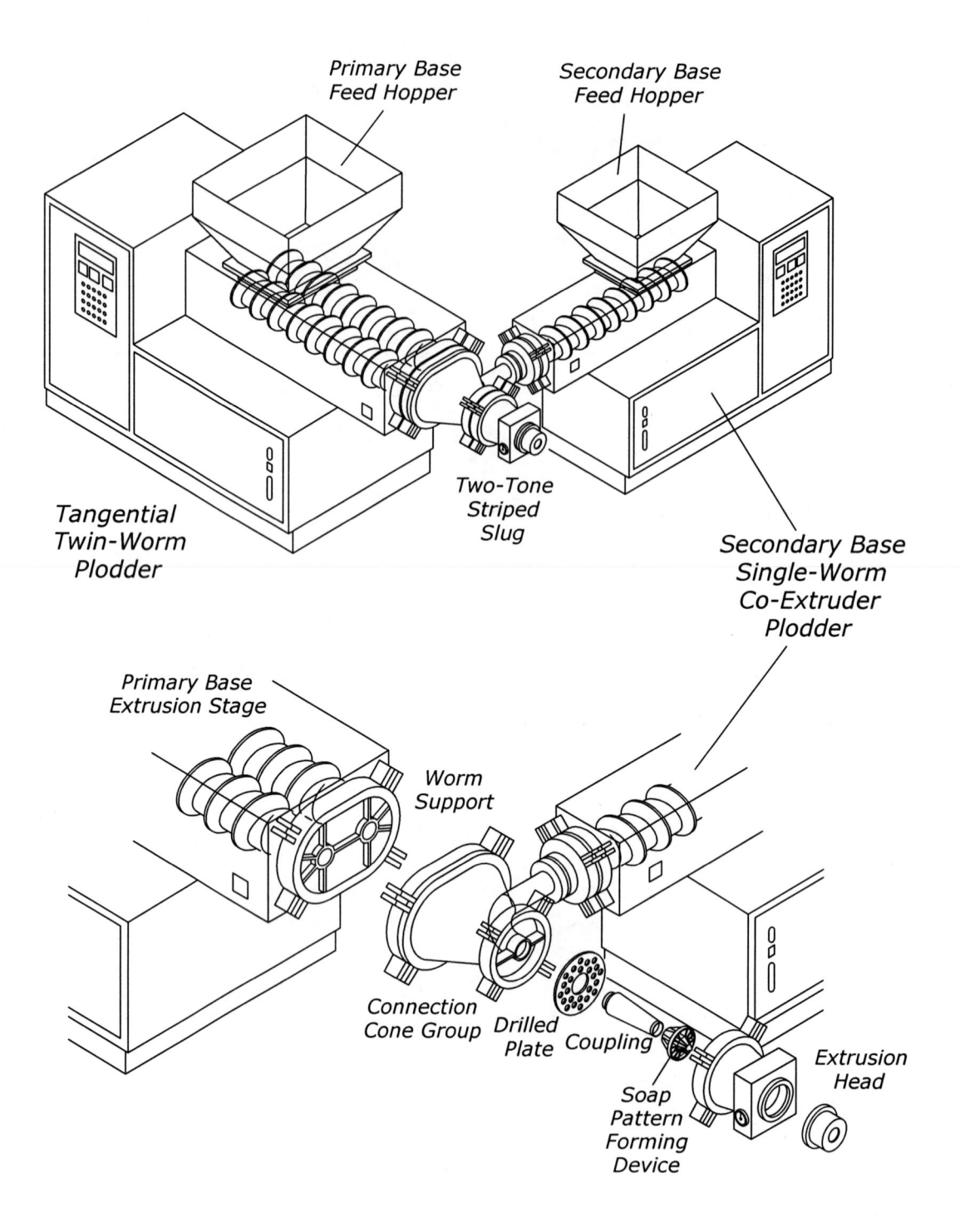

Fig. 12.16. Solid-Solid Co-Extrusion Systen for Two-Tone Striped Soaps with a Tangential Twin-Worm Plodder and a Single-Worm Co-Extruder.

Solid-Solid Co-Extrusion Assembly for Two-Tone Striped Soaps (Fig. 12.17.)

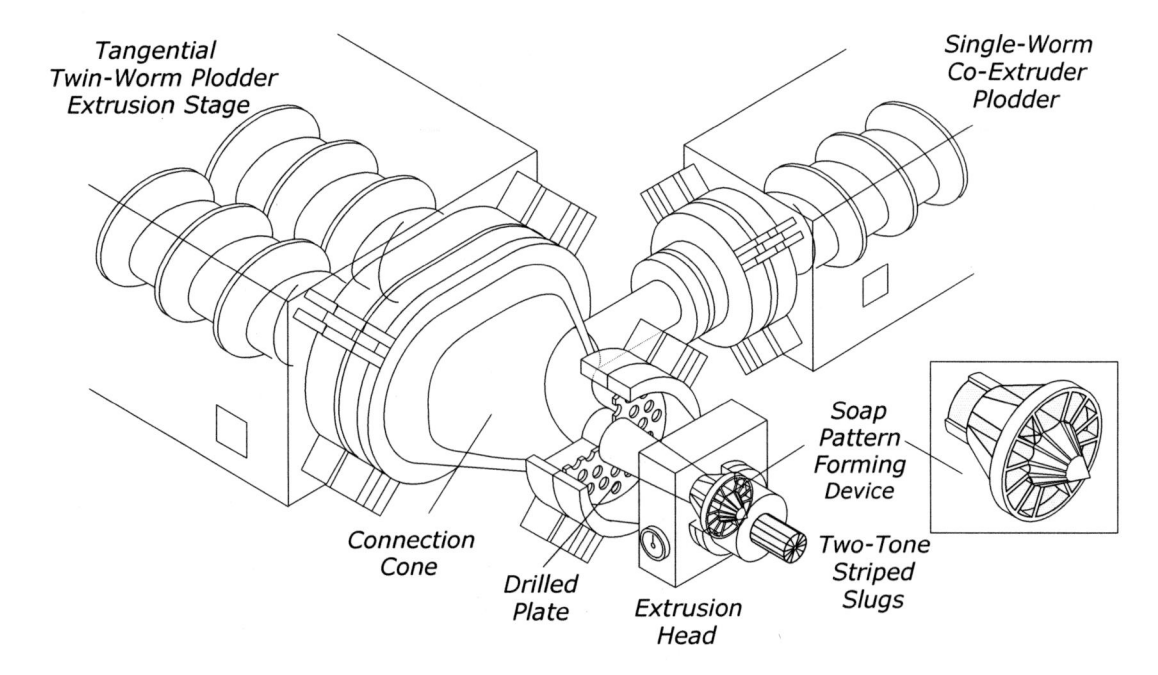

Fig. 12.17. Solid-Solid Co-extruder System Assembly for Two-Tone Striped Soaps.

Recycling Methods

To maintain the constancy of multicolored and multicomponent effects, an essential action is to control the recycling of these key variables:

- Flashing: the excess soap formed as the dies come together when the flashstamping system is used for stamping bandless shaped bars;

- Excess slugs: the result from extrusion, cutting, and stamping rate synchronization;

- Rejected stamped bars: the result from the press and packaging unit speed synchronization operation and the stamped bar inspection; and

- Shavings: (in case) a thin surface removed from the surface of the extruded slugs to expose a better defined inner pattern.

Recycling Method for Solid–Liquid Systems

The simplest method is to recycle everything into the first stage of the Duplex Vacuum Plodder. But in this case, no control exists over the total amount recycled, and therefore no control is possible over the final marbleized pattern of the extruded soaps (Fig. 12.18).

Recycling Methods for Solid–Solid Co-Extrusion Systems

All items to be recycled are recolored and repelletized in an additional Simplex Pelletizing Plodder and fed into the Secondary Base Co-extruder. Two Versions are illustrated in Fig. 12.19 and 12.20.

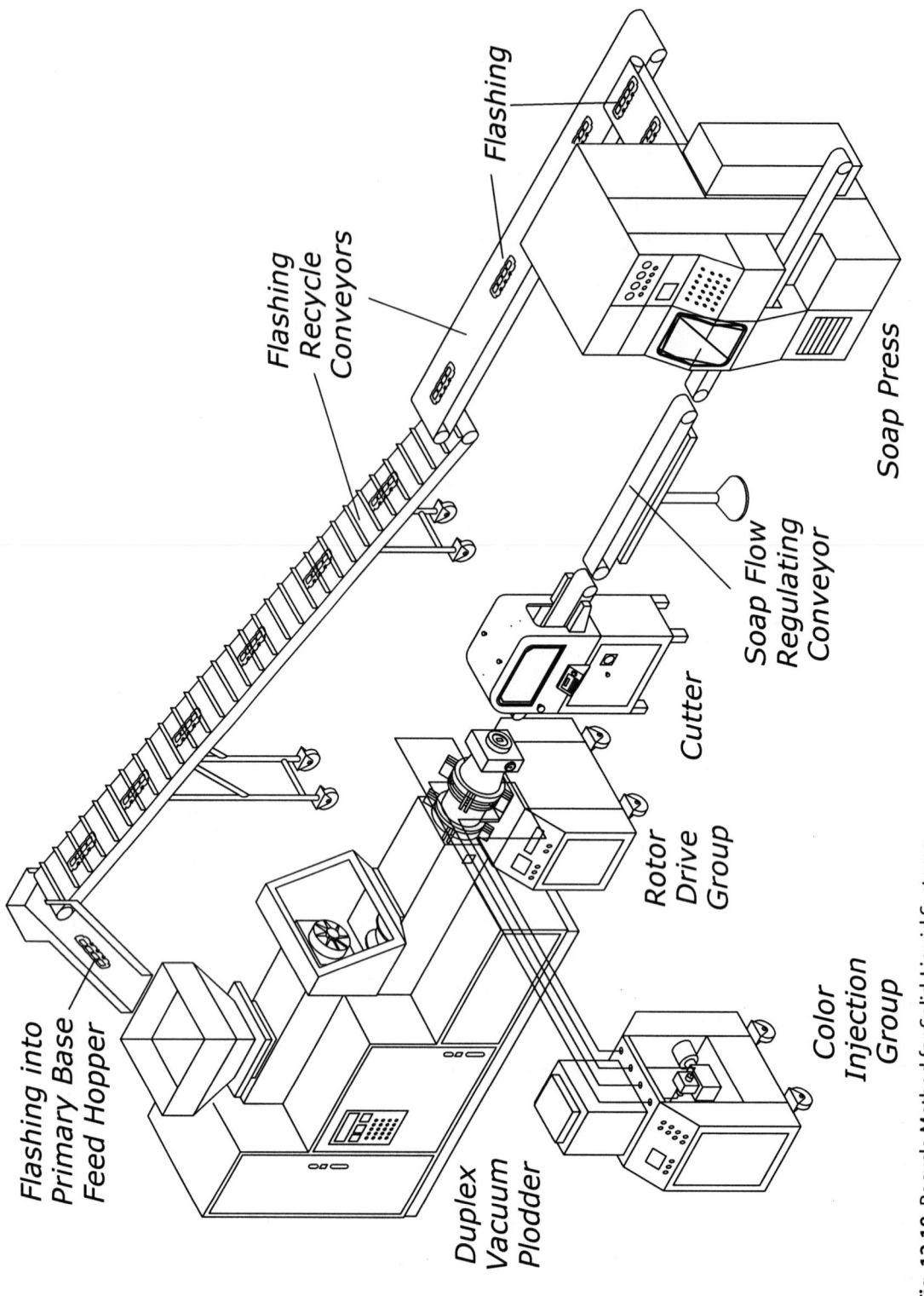

Fig. 12.18. Recycle Method for Solid-Liquid Systems.

Manufacture of Multicolored and Multicomponent Soaps ● 371

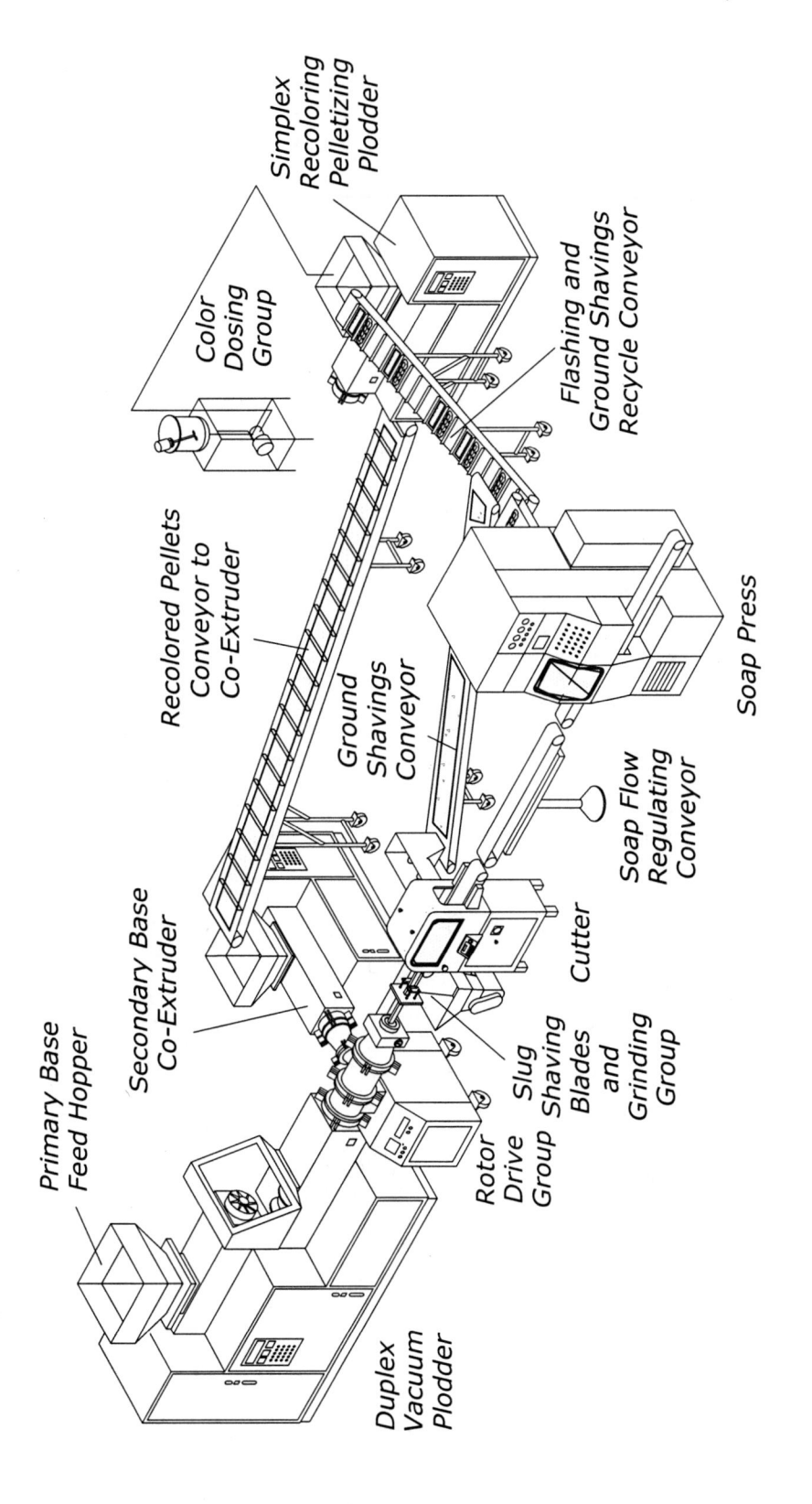

Fig. 12.19. Recycle Method for Solid-Solid Co-Extruding Systems - Version 1.

372 ● L. Spitz

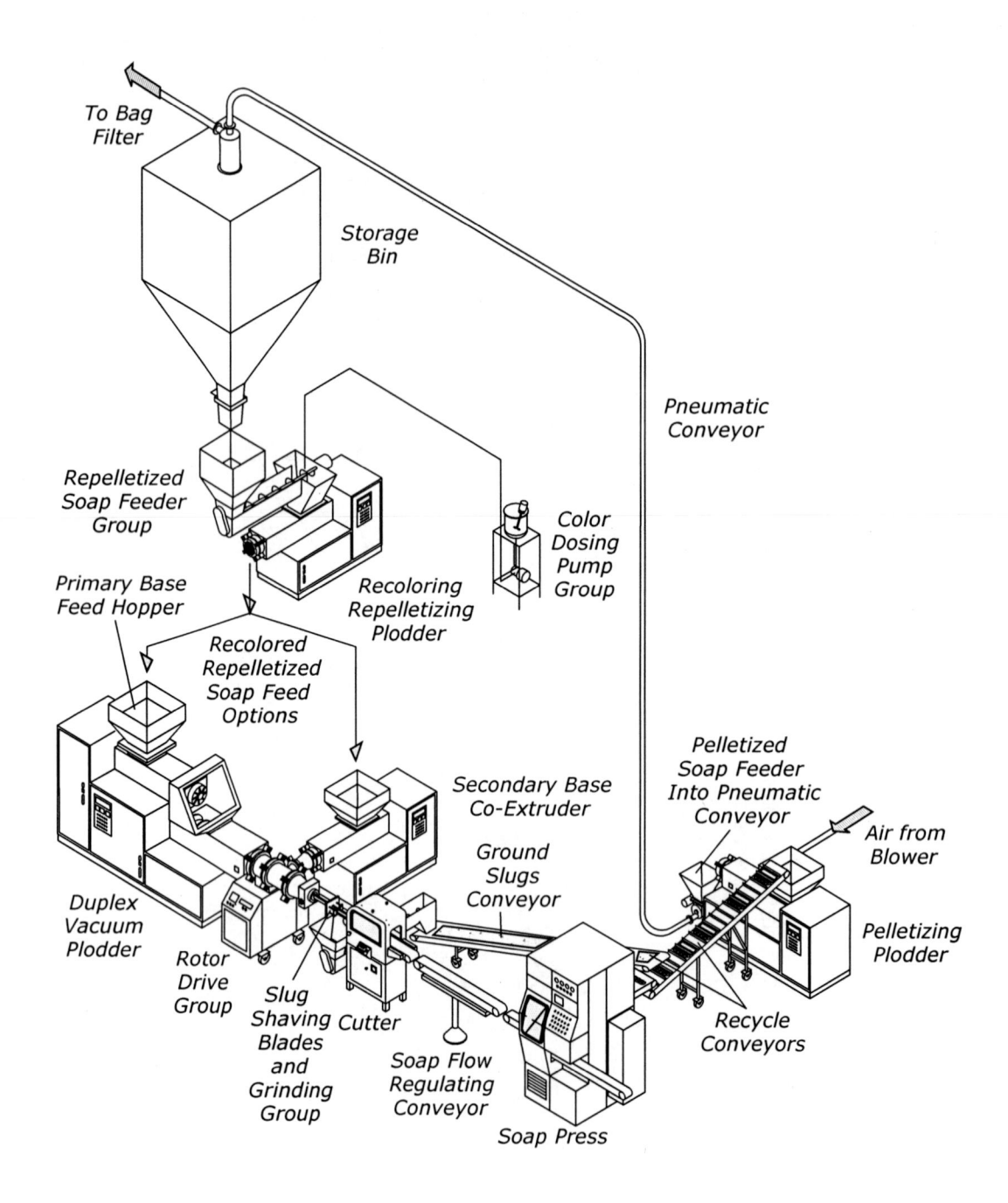

Fig. 12.20. Recycle Method for Solid-Solid Co-Extrusion System - Version 2.

Stamping Options

Standard and Angled Bar Soap Stamping Modes

Fig. 12.21 illustrates the difference between the straight (standard) and the special angled (bias) stamping modes. Stamping with the dies positioned from 20 to 30 degree angle, enhances the appearance of multicolored soaps.

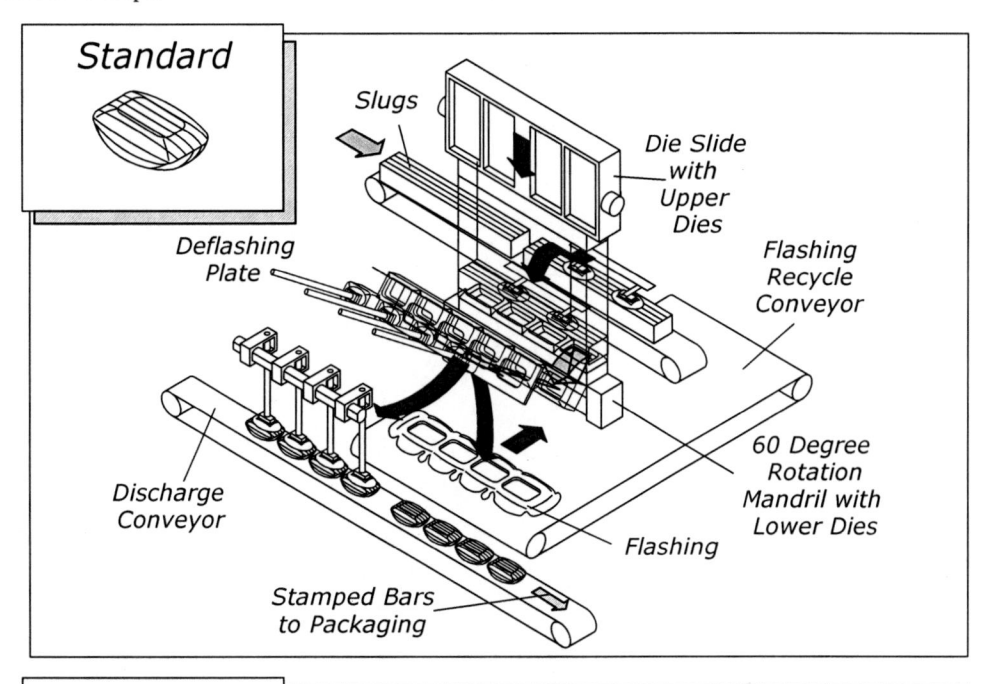

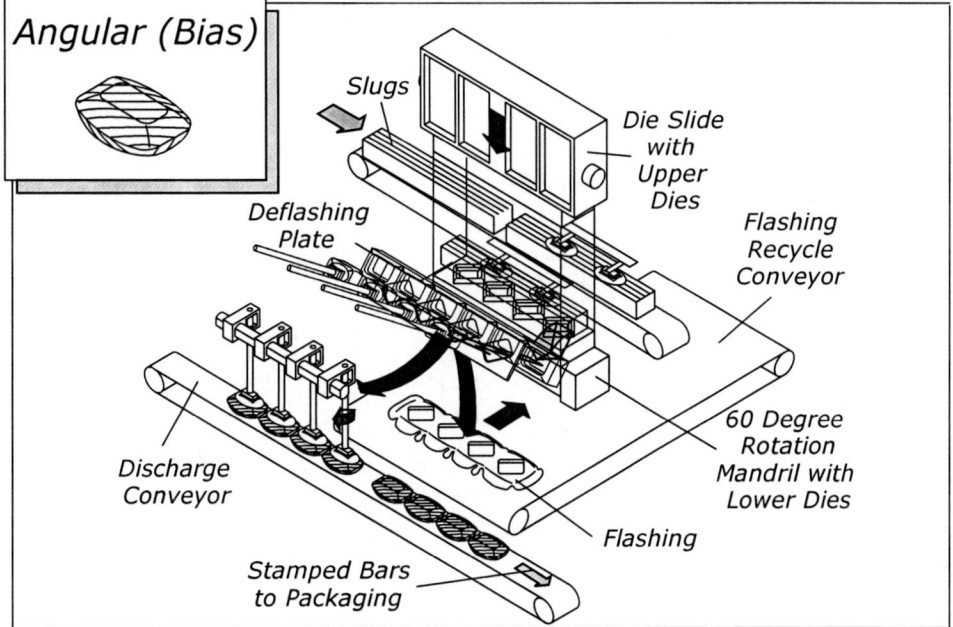

Fig. 12.21. Standard and Angular Bar Soap Stamping Methods.

Mottled Laundry Soap Manufacturing System

Mottled laundry bar soaps for laundering purposes are still widely used in many developing countries. Fig.12.22. illustrates a complete Mazzoni LB plant for the continuous production of blue mottled laundry soaps. The finished laundry bars are extruded from dryer's last stage plodder and are cut with a cutter provided with engraving rolls (see Chapter 11).

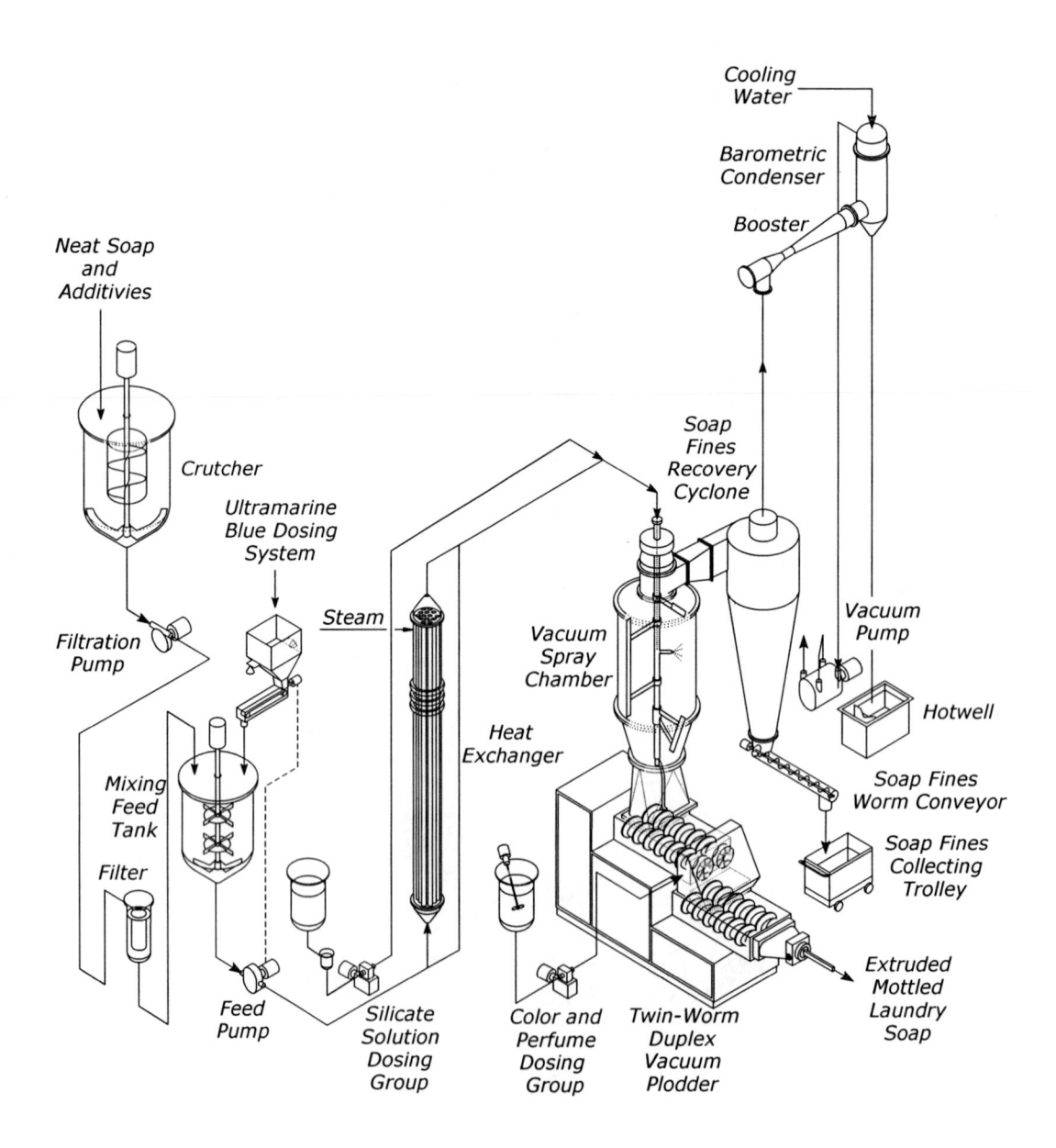

Fig. 12.22. Mottled Laundry Soap Manufacturing System.

Acknowledgment

The assistance of Mazzoni LB, SpA for the preparation of the manufacturing system diagrams is very much appreciated.

Reference

Spitz, L. (Ed.) *SODEOPEC;* AOCS Press: Champaign, Illinois, 2004; pp. 212–237.

Patents

The following is a selected list of the most important and interesting patents relating to multicolored and multicomponent soap formulation and processing.

Miscellaneous Patents

Bernard, A. Apparatus for Making Composite Product. Savonnerie Clair Bernard. U.S. Patent 3,779,676 (1973).

Hörning, H. Soap Bar and Process for Its Manufacture. Blendax-Werke. U.S. Patent 4,311,604 (1982).

Kaniecki, T. Method and Apparatus for Producing Striped Soap Bar. Armour-Dial Inc. U.S. Patent 3,890,419 (1975).

Meye, R.W.; G. Thor. Process for the Continuous Manufacture of Marbleized Soap Bars. Henkel GmbH. U.S. Patent 3,663,671 (1972).

Patterson, C. Method for Soap Bars Having Marble-like Decoration. Purex Corp. U.S. Patent 3,676,538 (1972).

Tanaka, Y. Marbleized Soap Plodder. Ideal Soap Co. U.S. Patent 4,077,753 (1978).

Colgate-Palmolive Patents

Compa, R.E. Process for Making Variegated Soap. U.S. Patent 3,485,905 (1967).

D'Arcangeli, A. Method and Equipment for the Manufacture of Variegated Detergent Bars. U.S. Patent 3,940,220 (1976).

Fischer, C.R. Soap Plodder Nozzle Plate. U.S. Patent 3,868,208 (1975).

Fischer, C.R. Apparatus for Making a Striated Soap Bar. U.S. Patent 3,891,365 (1975).

Fischer, F.; D.P. Joshi. Continuous Process for Making Variegated Soap. U.S. Patent 4,141,947 (1979).

Joshi, D.P. Method for Producing Multicolored Variegated Soap. U.S. Patent 4,156,707 (1979).

Joshi, H.H. Process of Making Variegated Soap. U.S. Patent 4,017,573 (1979).

Marchesani, C. Apparatus for Making Soap with Orifice Plate and Trimmer Plate. U.S. Patent 4,738,609 (1988).

Perla, G. Apparatus for Making a Variegated Soap Base. U.S. Patent 3,923,438 (1975).

Perla, G.; A. D'Arcangeli. Apparatus for Manufacturing Marbled and Striped Soaps. U.S. Patent 4,127,372 (1978).

Perla, G.; A. D'Arcangeli. Method for Making Soap Bars. U.S. Patent 4,201,743 (1989).

Pickin, J.; H.H. Joshi. Marbled Detergent Bar. U.S. Patent 4,011,170 (1977).

Ratz, R.M. Variegated Soap Apparatus. U.S. Patent 3,857,662 (1974).

Sanabria, J.A. Plodder Outlet Assembly. U.S. Patent 4,459,094 (1984).

Lever Patents

Alderson, D.A.; R.C. Scott. Manufacture of Multicolored Detergent Bars. U.S. Patent 4,304,745 (1981).

Aronson, M.P. and six others. Process for Making Extruded Multiphase Bars Exhibiting Artisan Crafted Appearance. U.S. Patent 6,723,690 (2004).

Coyle, L.A. and five others. Personal Washing Bar Having Adjacent Emollient Rich and Emollient Poor Phases. U.S. Patent 6,383,999 (2002).

Kelley, W.A.; P.J. Petix. Method for Making Variegated Soap. U.S. Patent 3,398,219 (1968).

Marchesani, C. Apparatus for Striated Soap Bars of Comparable Aesthetic Quality on Both Inner and Outer Log Faces for Soap Bars Produced in a Dual Extrusion Process. U.S. Patent 5,246,361 (1993).

Matthei, R.G. Method for the Manufacture of Marbleized Soap Bars. U.S. Patent 3,673,294 (1972).

Schönig, E.; H. Brückel. Manufacture of Detergent Bars. U.S. Patent 4,720,365 (1988).

Procter & Gamble Patents

Borcher, T.A.; J.R. Knochell. Apparatus for Making Variegated Soap Bars or Cakes. U.S. Patent 4,077,754 (1978).

Lewis, W.P. Apparatus and Process for Manufacture of Variegated Soap Bars. U.S. Patent 4,092.388 (1982).

Murray, G.D. Process for Manufacturing Color-Striped Stamped Detergent Bars. U.S. Patent 3,899.566 (1975).

13

Soap Making Raw Materials: Their Sources, Specifications, Markets, and Handling

Michael A. Briggs (ex Unilever)
51, Raby Drive, Raby Mere, Wirral, UK, CH63 0NQ

Introduction

Two major materials plus one by-product dominate soap-making operations, the final-product user performance, and the associated costs. These are triglyceride (TG) fatty matter, caustic soda, and glycerine, respectively. They dominate because two major processes for making soap are available—the TG process and the distilled fatty acid (DFA) process. The details of each process are different. However, both processes convert TG fatty matter and caustic soda into soap (i.e., the sodium salt of fatty acids), with the by-product glycerine released during the processing.

In the TG process, the TG is first cleaned of impurities and then reacted directly with caustic soda to produce an aqueous solution of soap and glycerine. This mixture is then separated by extracting the glycerine into a brine solution for recovery. For glycerine soap, the separation step is dispensed with, and all the glycerine is left in the final soap-bar product.

In the DFA process, the TG is hydrolyzed at high temperature with an excess of water which releases the fatty acids. The glycerine passes from the process in the excess water stream for recovery. The crude fatty acid is cleaned of impurities by distillation, and then reacted with caustic soda to form the soap. Should glycerine soap be required to be made by this process, then the recovered glycerine is added back.

These three materials—TG fatty matter, caustic soda, and glycerine—are the subjects of this chapter. All are commodities which are traded under common specifications and whose prices are set by their global-trading activities.

In this chapter, each material is covered in turn under the following subtitles:

- Production, Sources, Grades, and Users
- Characteristics, Specifications, and Analysis
- Markets and Price Trends
- Transport, Storage, and Handling (Including Oil Bleaching)

Minor soap-product ingredients, process materials, or packaging components will not be covered, except as passing references.

Triglyceride Fatty Matter

Fatty matter is by far the largest constituent of a soap bar, typically occupying about 70–80% of its total weight and over 80% of its total ex works cost. Therefore, a considerable part of this chapter is devoted to fatty matter.

Fatty Matter Production, Sources, Grades, and Users

Soap has been made since ancient times from a large number of fatty matter sources. The vast majority are TGs from vegetable oils or from animal fats. The distinction between a TG being called an oil or a fat is essentially arbitrary, but arose in history depending on whether the material was a liquid or a solid at ambient temperatures in its place of origin. So, animal fats are solids throughout the world, and seed oils are liquids. However, coconut, palm, and palm-kernel oils are liquid in the tropics, where they are produced, but solids at ambient temperatures in North America and Europe, where they are mostly used. In the rest of this chapter, traditional arbitrary descriptions are used, and often all are referred to as oils.

Many TG oils and fats are produced worldwide, and Table 13.1 covers the last 10 years of the global production of the main types and their major sources.

The materials specifically shown in Table 13.1 are the important ones for soap making because they are used directly or set the global prices for the ones that are. The "Other TG Oils" category is a collection of materials which are not normally used in soap making, and the category is included to show total global production.

Things to note are that:

- Soy, palm, sunflower, and palm kernel increased significantly as more land was allocated to their production, particularly in Brazil, Argentina, Malaysia, and Indonesia.

- Total global production over the last 10 years increased significantly under the pressure of an increased consumption as populations grew and prosperity increased, particularly in Asia.

- Consumption of TG oils is nearly all for human-food purposes, usually called edible oils, with only about 8 million tonnes used for soaps, 7 million for oleochemicals, and 5 million for animal feeds. Of the oleochemical consumption, about 5 million tonnes is used in soaps, toiletries, and detergents, with most of the rest used for rubber and plastics additives and for candles.

Table 13.1 includes all grades and derivatives of TG oils and fats of which a considerable number exist. However, for soaps, the important ones are those that one can buy in the crude form or in several cleaned forms. The grade of a crude oil is determined by its degradation level, which, in turn, is caused

Table 13.1. Global Fatty Matter Production and Sources

Oil	1997/98	2001/02	2006/07	Major Sources
Soybean Oil	26.5	31.2	36.4	USA, Brazil, Argentina, (China[1])
Palm Oil[2]	17.1	24.7	39.0	Malaysia, Indonesia
Rapeseed Oil[3]	11.4	12.3	12.1	Europe, Canada, (China[1], India[1])
Sunflower Seed Oil	8.5	7.8	10.6	Europe, Argentina
Tallow	7.6	8.0	8.0	USA, Europe, Brazil, Argentina, Australia
Coconut Oil	3.4	3.4	3.4	Philippines, Indonesia
Palm Kernel Oil	2.2	3.1	4.6	Malaysia, Indonesia
Other TG Oils[4]	27.6	30.0	28.5	
World Total	104.3	120.5	142.6	

Notes: [1] Production within China and India is all consumed domestically
[2] Includes Palm Oil derivatives e.g. Olein, Stearin, etc.
[3] Includes Colza Oil, Canola (and Indian Mustard Seed Oil)
[4] Includes Lard 6, Butter 5, Groundnut 5, Cotton Seed 4, Olive 2.5, Maize 2, Sesame Seed 1, Fish 1

by microbiological action giving free fatty acid, color, and odor. In general, the degradation is worse when the time, temperature, and moisture conditions are greater between the time the crop is ripe and the time the oil is extracted. Palm oil and tallow are particularly susceptible to this type of degradation, and so have more grades than the other oils.

When used for human food, the vast majority of oil is refined, bleached, and deodorized to remove free fatty acids, poor color, malodor, and undesired tastes. One can carry out these cleaning processes in a number of ways, but the final product grade is usually referred to as refined, bleached, and deodorized (RBD). For soap use, the cleaning requirements are less stringent. This is because the free fatty acid is not a disadvantage, the final bar's added color and perfume usually mask some of the oil's color and odor, and humans rarely worry about the taste of their soap. So, for TGs, bleaching is the only cleaning process normally required prior to TG soap making. This topic is dealt with in more detail under the Oil Handling and Bleaching section below.

Edible-oil RBD processing gives by-products like fatty-acid distillates which are suitable for DFA soap-making feedstocks and in small proportions for TG-process use.

Palm oil is used as such for human food and other applications, but it is also fractionated into two streams. One is an olein stream, which is the main commercial product for food use; the other is the by-product stearin stream. The olein contains a high oleate fraction, and mimics seed oils like soybean and rapeseed for edible liquid-oil applications; the stearin contains a high-palmitate fraction which is solid and is useful for soap making.

Tallow can be a mixture of many animal fats from various sources. It is a by-product from meat processing, and usually constitutes less than 5% of the value derived from the carcass. Tallow can also include fat recovered from meat cooking in food-processing plants and in restaurants. The highest grades of tallow are from the meat-packing operations of a single animal, particularly a beef source. The lowest grades are mixtures from many animal types and from various sources.

One can possibly use other organic acids in soap making, particularly rosins and tall oils from wood processing, but usually these are added, if at all, in relatively small amounts to TG oils. As a consequence, these materials are not covered in this chapter.

Fatty Matter Characteristics, Specifications, and Analysis

TG oils are esters of three fatty acids and glycerine with generic structures as shown in the diagram below. R', R", and R'" are carbon chains with lengths mostly ranging from 7–19 which can be fully saturated or mono- or polyunsaturated.

$$\begin{array}{c} R'CO\text{-}OCH_2 \\ | \\ R''CO\text{-}OCH \\ | \\ R'''CO\text{-}OCH_2 \end{array}$$

Diagram 13.1. Generic Structure of Triglyceride Oils.

For food use, where the different fatty acids are attached to the glycerine matters because this gives significantly different physical properties. For example, cocoa butter and palm stearin have very similar chain-length distributions, but give a completely different mouth feel. However, for soaps, the arrangement is unimportant because, in the soap mix, the TG molecule was broken, and so each fatty-acid chain contributes to the aggregate properties individually.

The chain-length distributions and properties of fatty matter differ quite considerably, and this is important when choosing them for soap making. Table 13.2 gives typical values.

Looking at the table above, one can see that the predominant components of coconut and palm-kernel oils are all short saturated chains centered on C12, lauric acid. This feature is unique, and only

Table 13.2. Typical Chain-Length Distributions and Properties of Triglyceride Fatty Matter

Oil Data Item	Coconut Oil	Palm Kernel Oil	Tallow	Plam Oil	PO Stearin	Soya Bean Oil	Rapeseed Oil	Sunflower Seed Oil
Short Chains <=C8 (wt%)	9.5	3.4	0.0	0.0	0.0	0.0	0.0	0.0
Capric C10 (+ C9) (wt%)	6.0	3.6	0.0	0.0	0.0	0.0	0.0	0.0
Lauric C12 (+C11) (wt%)	46.8	47.0	0.0	0.0	0.0	0.1	0.0	0.0
Myristic C14 (+C13) (wt%)	18.0	16.2	3.0	1.3	1.8	0.2	0.0	0.0
Palmitic C16 (+C15) (wt%)	9.5	8.6	25.3	42.9	62.5	11.0	4.5	3.5
Stearic C18 (+C17) (wt%)	1.9	2.2	21.0	4.8	5.3	4.4	1.5	2.9
Long Saturates >=C19 (wt%)	0.0	0.0	1.8	0.4	0.1	1.2	3.0	0.6
Palmitoleic C16:1 + lower unsaturates (wt%)	0.1	0.2	4.6	0.5	0.3	0.5	0.5	0.5
Oleic C18:1 cis (+C17:1) (wt%)	5.6	15.9	39.0	38.9	23.7	23.8	57.0	33.5
Linoleic C18:2 (+C17:2) (wt%)	2.6	2.9	4.7	10.6	6.3	51.0	19.5	58.6
Linolenic C18:3 (+C17:3) (wt%)	0.0	0.0	0.3	0.1	0.0	7.8	9.0	0.0
Long Unsaturates >=C19 (wt%)	0.0	0.0	0.3	0.5	0.0	0.0	5.0	0.4
Mean Molecular Weight (MMW) of Fatty Acids	205	218	273	269	264	278	282	280
Saponification Value (SV) (g KOH/kg Oil)	257	243	196	199	202	193	190	191
Iodine Value (IV) (g I$_2$/100g Oil)	10	20	49	55	33	136	116	137

Note: to give the mean molecular weights of the fatty acids present, and not those of the parent TGs, is conventional.

occurs in palm-nut oils. Other palm-nut oils do exist, but they are not in significant commercial use globally. All of these oils originate in the tropics, and as a result, are often called tropical-nut oils.

Detailed descriptions of soap formulations are given in separate chapters. However, to produce good lather in the final soap bar, it is typically necessary for the formulation to contain about 20 wt% of tropical nut oils. Both coconut and palm-kernel oils have very similar chain-length distributions, with coconut having slightly more caprylic and capric acids, C8 and C10, but less oleic acid, C16:1. Nevertheless, they are essentially interchangeable for soapmaking purposes.

None of the other fatty matter has significant contents of short chains, but they do split into two groups, tallow together with palm oil and the seed oils, like soy, rape, and sunflower. The seed oils have a large majority of unsaturated chains, while in tallow and palm the ratio is about even. Palm-oil stearin is a by-product of extracting the olein fraction from palm oil, and so has a lower oleic content than its parent palm oil.

For soap making, to enhance the lather produced by the shorter chains of the nut oils, the formulation best contains about double the amount of unsaturated to short chains (i.e., about a 40 wt%). Also, at least about a 40 wt% of insoluble long saturated chains is needed to make the bar firm enough to handle in use and to not be damaged in bar packing and during distribution. Tallow meets these criteria, as does approximately 50:50 of palm and palm stearin or palm stearin plus minor amounts of seed oils.

Overall then, one can not possibly make good soap bars using a single oil; one must use a blend of oils. Most major soap makers have several proven blends from which they can choose to give the cheapest combination over a range of oil prices.

Because TG oils are commodities which are traded globally, they have to conform to a set of common specifications. Essentially this means assuring the buyer that the oil is what the seller says it is (i.e., say, a coconut oil actually is made from coconuts and not palm kernels, and has properties which conform to the grade to which the seller claims it belongs). In general, the various markets set standards for each oil traded, and provide a "court of peers" to settle any disputes. With these in place, one must only specify the following:

- Acid value (AV)—giving the amount of free fatty acid, and thus partially confirming the amount of degradation and so the grade.

- Saponification value (SV)—giving the mean molecular weight of the fatty-acid content and thus partially confirming the likely type of oil (see Table 13.2 above).

- Iodine value (IV)—giving the degree of unsaturation, and thus partially confirming the likely type of oil (see Table 13.2 above).

- Color—partially confirming the amount of degradation and so the grade. For "clean" materials, a 5.25-inch cell is used in the tintometer, while for lower grade materials, a 1-inch cell may be necessary.

- Odor—partially confirming the amount of degradation and so the grade. No standard method exists, and odor is often only invoked in cases of a dispute about the general quality of the material in question not being up to the "usual" commercial standard. An agreement between the buyer and seller (using experienced practitioners and arbitrated by the "court of peers" in cases of unresolved dispute between the buyer and seller) determines this.

- Moisture, impurities, and unsaponifiables (MIU)—giving the amount of nontriglycerides and thus partially confirming the grade. MIU is usually recorded as the sum of the separate determinations because together they determine the total amount of fatty matter being bought.

- Bleachability—serving as an additional item only included in contracts for soap making, particularly when buying tallow and palm oil. This is because "locking in" a poor color by inappropriate treatment is possible, thus making any subsequent bleaching ineffective. It is determined by adsorption onto activated fuller's earth. Numerous standard methods exist, but no good correlation between them exists because they use different standard bleaching earths and test conditions. So, the one chosen by a buyer should be that which most closely matches the soap maker's own bleaching facilities. Results obtained under other test conditions can give very different values, and so are irrelevant to a specific factory's operations, thus leading to the purchase of unsatisfactory oils.

The analytical methods used for these specifications are relatively simple, and the methods are essentially the same whether set out by the AOCS, national standards bodies like ASTM and ISO, or trade bodies. A more detailed coverage is given in Chapter 14.

Fatty Matter Markets and Prices

Agricultural products are commodities and are traded, usually in about 1000-tonne lots, in various market exchanges. Their prices are set depending on the balance in supply and demand for each crop. One market tends to set the marker price for each commodity, and this then leads to the determination of prices globally via setting premiums or discounts according to the marker price.

TG oils and fats are no different, with the Chicago price of soybean oil as the key-marker material. This is because the soy grown in the United States is a major part of the world total, and Chicago has a well-established market system. Also, soy is an annual crop, and so its harvest volume is more uncertain than those of the perennial crops, like palm oil, whose trees continue to produce throughout the year for up to 30 years. With soy, farmers can choose at the start of each growing season whether to plant this or an alternative, usually corn. They make this choice depending on how they feel the supply-and-demand situation will be at harvest, and thus how prices are likely to be. However, once the crop is planted, the amount that will appear on the market at harvest time depends on the weather occurring since planting. So the actual price the farmer can sell for is not perfectly predictable when making the planting decision. However, one can store soy beans for a year or more after harvesting, and so the farmer has an extra dimension in the decision on when to sell. To remove some of this uncertainty, futures markets exist so the farmer and the ultimate user can trade the crop at any time of the year (i.e., before and after harvest occurs). This market helps both the farmer and the user because it enables them to "fix" a major element of their prices/costs and thus make their decisions easier. In fact, the total trades in agricultural commodities are about three to ten times the net throughput of material, and this occurs to make transfers between individual producers and users more efficient and to incorporate the effects of weather etcetera on the supply/demand balance throughout the year.

Obviously, nobody knows the real supply/demand balance at any time, and so the price is set by the aggregate of individual trader's sentiments on how the weather forecasts will impact the crop yield. It also takes into account the unfolding of the general state of the economy and how this affects user demands. Major daily-price fluctuations occur as a consequence of the different interpretations of the various events which over- and undershoot the mean. Even so, when averaged over a quarter, still significant volatility exists as one can see for the last 20 years in Fig. 13.1 below.

Up until mid-2006, market prices showed their typical behavior. In general, a 1% out-of-balance in supply and demand changes the price by about 10 to 20% depending on the level of stocks to draw on. Between 1988 to about 1993, a period of harvests occurred exceeding demand, then an overall deficit until 1999, followed by a reduction in surplus until about 2006. At this point, biofuel appeared as a significant factor, and totally disrupted the market as it created a major additional demand for many food crops, including natural oils. Prices did overshoot and are now falling back. However, nobody knows whether natural-oil prices will stabilize at a new higher level or fall back to their long-term average.

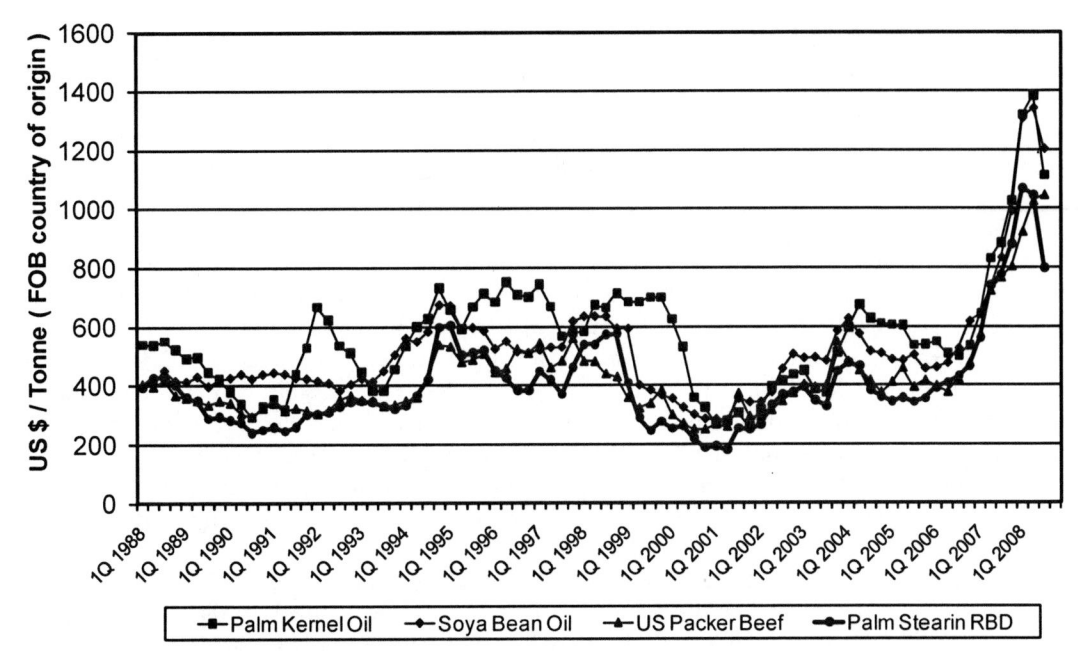

Fig. 13.1. Fatty Matter Quarterly Price Trends.

The move to biofuels, particularly bioethanol and biodiesel, has had an ever-increasing effect on agricultural-product demands. Further, because crop production has lagged behind demand, prices have increased dramatically. Again, because biofuel production is driven by government subsidies and mandates, the price could fall equally quickly if these are removed in response to the current "Food versus Fuel" debate. However, should biofuel demand continue, prices will be driven even higher because only about 3% of a biodiesel substitution needs to occur to use all the TG oils currently produced globally. Further, because nearly all the land in the world suitable for agriculture is in production already, little chance is likely of a total substitution of petrodiesel by biodiesel.

Since the expansion of biofuels is based on government subsidies and mandates to off-set their lack of economic viability, the outcome is determined entirely by political decisions, and so is unpredictable in the long term. All we can do is form business strategies and have buying tactics which accommodate both sides of these political decisions.

Looking at Fig. 13.1 again, one can see that all the key TG oils for soap making essentially follow each other's prices. The following paragraphs explain the differentials between them, mostly due to the ease of substitution with other oils and the absolute volume of production. So, groups of oils which are easy to intersubstitute closely follow each other in price because a bigger pool of volume is present across which to spread any deficit or surplus in any one of them. The opposite applies to oils with special properties which few other oils can mimic.

Soybean oil, which is generally taken as the global-marker commodity for TG oils, is mostly used by the food industry for premium uses. This has driven up its price into the higher range because its highly unsaturated carbon chains mean that it is liquid at room temperature, whereas the other stearics and laurics are solid.

Palm kernel, being a lauric oil, is also high in price because its short highly saturated carbon gives properties which can only be generated by coconut oil and not by the more abundant stearic oils. Its price is also more volatile because it represents only about 3% of total world TG production. Coconut has a price similar to palm kernel, but is usually slightly higher because it is in higher demand as a primary food owing to its more desired taste.

Beef tallow and palm-oil stearin are cheaper than the others because they are by-products of operations which generate the majority of the income for their operators from other products, namely meat and edible liquid oils, respectively. Also, these two have similar prices because they are easily substituted one for the other in products where animal-origin ingredients are acceptable. For markets where animal products are unacceptable culturally, palm oil is more expensive because it is unchallenged by tallow.

Many other materials and grades are not shown in Fig. 13.1. In general, these are of lower value to the food industry, and/or may be low-quality by-products. The prices of these are set to quit their volumes, and they end up being $50 to 100/tonne cheaper than their marker materials, depending on the balance of supply and demand and their price elasticity among users.

The prices of fatty acids are generally set by negotiations between the producer and user because few of each exist worldwide, and so an efficient commodity market is not possible. However, traders do exist who buy large lots and sell on to smaller users who would otherwise have less heft with the producers. These traders often have large storage-tank facilities at ports or other transport nodes.

The prices of fatty acids are often based on a formula which includes the market price of the parent oil multiplied by a consumption factor, then corrected for the market price of the by-product glycerine, and finally including a conversion and delivery fee.

No matter whether the TG or DFA process is used for soap making, TG oils are the major cost. However, soap makers are able to use the full range of fatty matter, particularly the cheaper and lower quality materials. Therefore, most soap makers use skillful blending and flexible soap-making techniques to yield high-quality products at cheap prices.

Fatty Matter Transport, Storage, and Handling

Oils, fats, and fatty acids for soap making are normally handled in bulk because of the amounts involved. These materials are delivered to the soap-making plant in road tankers, rail cars, barges, or ships, and then transferred into storage tanks on site. The equipment for crude TG handling is usually constructed from carbon steel, whereas that for bleached oils and fatty acids is from ASTM 304 austenitic stainless steel. Ocean-going ships carrying fatty matter must conform to the International Maritime Organization (IMO) regulations which require double hulls to prevent leakage to the environment in the case of damage to the ship.

When the fatty matter arrives at the plant, any water is run off the bottom of the transport vessel, and then the contents are weighed and tested for AV, SV, IV, color, odor, and bleachability before being transferred to the selected storage tank.

The storage tanks are usually vertical cylinders with a heating coil in the base and a stirrer in the side. The tank base slopes to allow water to be run off. The heating coil is sufficiently above the base to allow water and sludge to collect below it. Having the heating coil in the water and/or sludge accelerates degradation in the rest of the stored oil. Similarly, water should be drawn off every day, and sludge as frequently as possible depending on the oil grade and the turnaround of the tank for different grades. One always exhibits bad practice by putting higher grade oil into a tank which has held a lower grade without first cleaning it.

The fatty matter is usually transported and stored at ambient temperature, if possible, or at about 5–10°C above its melting point, to ensure that it is mobile for pumping but not too hot to cause degradation. Preferably, the heating medium is temperature-controlled hot water or electrical tracing to ensure no hot spots which again cause degradation.

If storage is to be for more than about two weeks, then the heating should be turned off and the material allowed to solidify, again to minimize degradation. Note that if solidification is practiced, then the heating coil must have a vertical leg to the top of the tank so that during re-melting a column of liquid is produced which allows for the expansion of the newly melted oil at the base of the tank.

Note that high-IV fatty acids, like those made from tallow, are much more prone to degradation than the parent oil, and they have a maximal life of only a few days between their production and use.

Oils, fats, and fatty acids are flammable, but their flash points and auto-ignition temperatures are high, in the range of about 250 and 290°C, respectively. So, no particular precautions need to be taken in the plant storage areas except to bund the tanks and provide mobile fire-fighting equipment for use in emergencies.

As well as protecting the general environment against leaks spreading, the bunds also collect rainwater. Discharging this water from the bund should be via a fat trap before running to the plant's effluent system.

Soap plants usually have many fatty matter storage tanks. Usually they are allocated in pairs to a particular grade, and are used alternately, the one being used for receiving deliveries, storage, and/or cleaning, while the other is in use supplying soap making.

Traditional soap plants often had much more storage tankage than was necessary to smooth out fluctuations in deliveries and production scheduling. The extra storage was used to hold material which was bought when it was cheap, and was also used to commercial advantage when prices increased. Whether this is a good commercial proposition must be evaluated from the working capital and operating costs of holding the material against the prices that would be paid on the spot or futures markets when much less storage would be required. Each plant location has its own commercial context, and the optimum must be evaluated separately for each.

Oil Bleaching

DFAs are cleaned as part of their production process, and so do not need any further bleaching to produce all the desired qualities of a finished soap product.

Oils come in a wide range of colors—from Lovibond 10 to 300 red—but to make appropriate quality soap requires a color of less than 1 for white toilet soap and less than 3 for colored toilet soap or laundry soap. As a consequence, all oils for soap manufacture need to be bleached either in the soap factory or bought pre-bleached.

Although a target color should be set for bleached oil, this is not directly related to the soap color. In fact, each soap-making plant has to set its own target for each soap type that it makes. This is needed because the various soap-making routes and the details of their operating practice can affect the final soap color even with the same bleached-oil color. Thus, a TG-process soap contains only a few of the water-soluble color bodies because these are removed into the recovered glycerine stream during the soap washing and fitting operations. Obviously, the better the extraction, the lower the final soap color. On the other hand, glycerine soap, in which most, or all, of the glycerine is retained within the soap stream, will have a worse color unless a more stringent oil-color target is set.

Adsorption bleaching onto fuller's earth is by far the most common process and the subject of most of this section. However, one can use air bleaching when dark-colored toilet and laundry soaps are made from palm oil. This will be dealt with briefly. One can also use numerous chemical-bleaching agents, particularly on the lowest grades of tallow or the soap made from them. The most common reagent here is sodium hydrosulfite ($Na_2S_2O_4$) also called dithionite, bisulfite, and blankite. However, this process is not common, and must be developed on a case-by-case basis. So, it is not covered in this chapter.

One can carry out adsorption bleaching in a number of ways and with a wide range of equipment types. Both batch and continuous processes are used. Which is chosen depends on the economics at the factory involved with different local circumstances giving different results. The batch process plus its variants covered below is the one most commonly used. The schematic diagram of the process is shown in Fig. 13.2 below, and described in the subsequent paragraphs.

Bleaching is carried out in a vessel typically with a capacity of 5 to 50 Tonnes per batch. This vessel is equipped with a stirrer, a vacuum system, and dual use heat-transfer coils which can take either team or cooling water. The bleaching vessel also has feed points for crude oil and bleaching earth, an outlet

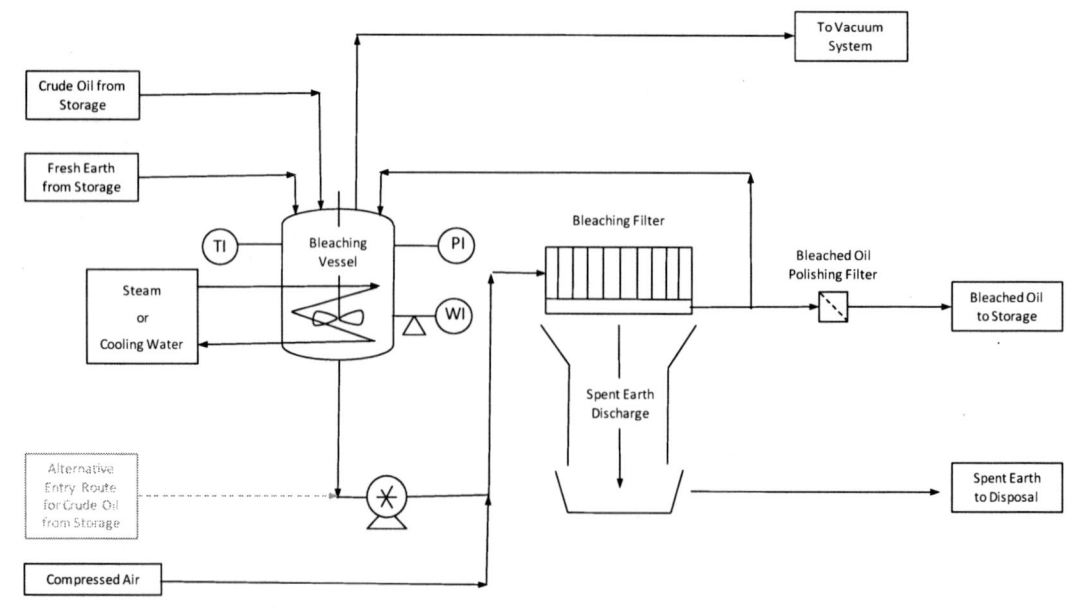

Fig. 13.2. Oil Bleaching – Earth Adsorption – Schematic Diagram.

via a pump to the bleaching filter, plus a return line from this filter. Note that the vacuum system will receive some fatty acids, and so its condensate overflow must include a fatty matter trap before running to effluent.

The bleaching filter removes the vast majority of the earth from the bleached oil, but also a polishing filter is present which removes any residual earth and acts as a long stop should the bleaching filter fail and pass an undue amount of earth with the oil. The bleaching filter can be of the plate-and-frame type or of the leaf type. These filters are equipped with filter cloths, but the majority of the filtration is carried out by the "precoat" of bleaching earth which is laid down when the batch is first pumped into the filter. The polishing filter is usually of the cartridge type.

All the equipment is typically made from ASTM 316L, although the bleaching filter may be made from high-temperature plastics, like polypropylene.

Spent earth from the bleaching filter is discharged into a carbon-steel container which in turn is sent for disposal.

The typical operation of the bleaching plant is described below.

Crude oil from storage is measured into the bleaching vessel. It can pass there directly, or one can pass it through the spent earth from the previous batch which is held in the bleaching filter for this purpose. This operation is designed to use any residual bleaching capacity in the spent earth because the feed oil has a higher color-body content than the previous batch's oil had at the end of the filtration cycle. However, whether this operation is effective in bleached-oil color, economics, and plant capacity depends on many factors, and thus must be worked out for each grade of raw oil and bleaching earth, taking into consideration the plant's equipment configuration and loading.

Once the bleaching vessel is filled, steam is admitted to the vessel's coils, the temperature set-point is raised to a pre-decided value (usually 90–110°C), and an evacuation to about 50 mm of Hg absolute is started. As the temperature rises and the pressure falls, water and other volatiles, like short-chain fatty acids, are boiled off and pass to the vacuum system. From the trace of the temperature, one can discern the point at which the oil is fully dried. Note that the water must be removed before the earth is added because it binds strongly to the adsorption sites, thus, preventing the color bodies from being captured there. If the water were not removed first, then a greater amount of earth would be required to achieve the same color-target value.

A pre-determined amount of bleaching earth is sucked into the bleaching vessel. Note that the earth can contain free silica; thus, any operating staff must be protected against breathing any of the dust by handling the powder in downdraft booths.

Different oils have different combinations of color bodies, and so require different amounts of bleaching earth, holding times, and temperatures with different grades of bleaching earth. The only way to find the optimum for bleached color and cost is by practical experimenting in the plant laboratory combined with operating experience in the plant on each type of oil. If the types and grades of crude oil vary, then one may have to do these tests on each delivery batch of oil. However, experience suggests that typical requirements are as follows:

- Coconut and palm-kernel oils 2 – 3% of earth at 90 – 100°C for 30 minutes
- Higher grade tallows 2 – 3% of earth at 100 – 110°C for 30 minutes
- Lower grade tallows 3 – 5% of earth at 100 – 110°C for 30 – 60 minutes
- Palm-oil 3 – 5% of earth at 100 – 110°C for 30 minutes then 160°C for 30 minutes.

All bleaching earths are the mineral montmorillonite, a naturally porous calcium-rich alumina–silicate clay, commonly called fuller's earth. This is mined all over the world. Numerous types of bleaching earth exist ranging from low activity (i.e., as mined) to high activity. The particle size is important—small enough to expose a large surface area, while large enough to be easily filterable and providing good oil flow through the cake. One can increase the activity by acid treatment of the native earth which increases the number, size, and depth of the pores.

The earth-usage values given above are based on moderately high activities: lower grades require as much as double or triple the amounts, and superactive ones require only one-half. The one to choose is based on the economic evaluation at each factory location which includes not only the cost of the earth itself, but also the loss of oil in the spent earth.

To bleach oils with activated carbon is possible, and this is practiced for some food applications. However, rarely is it economical for soap-making use.

The reason for the extra time at a higher temperature for palm-oil bleaching is that this oil has a very significant content of carotene and its precursors. If these are not destroyed thermally and adsorbed, they will pass through bleaching essentially uncolored but develop their color later, thus giving poor color to the final bar after leaving the soap factory.

Once the bleaching time is completed, a sample is taken and its color checked. Note that the target color at this point will be higher than the target for the final bleached oil because further bleaching will occur during the filtration operation.

If the sample is not good, then the bleaching cycle is continued for a further 15 minutes, and if necessary a further 1% of earth is added. When the color target is reached, the oil is cooled by passing cooling water through the bleaching vessel's coils. The target temperature is usually less than 85°C. However, it can be 75°C if, as in the case of low-grade tallows, the oil contains dissolved polyethylene which entered the rendering process via butchers' trimmings, etcetera. At this lower temperature, the majority of the polyethylene precipitates, and is caught by the bleaching filter. If this is not done, then the polyethylene often deposits on the soap dryer's tubes, degrades, and breaks off as black specks which give off-quality soap bars.

The cooled oil plus earth is now passed into the bleaching filter and returned to the bleaching vessel. The turbidity of the returning liquid is checked, and when clear, the filtrate is directed via the polishing filter to bleached-oil storage. Note that one must make a close monitoring of the performance of the polishing filter because at the relatively high temperatures involved, bleaching earth catalyzes the degradation reactions which lead to poor oil color and odor. Even with ideal conditions, one should not store bleached oil for more than about 8 to 12 hours before using it in soap making. What is more, fuller's earth in the final soap bar also catalyzes degradation reactions and rancidity development so seriously that poor color and malodor can develop in the product during the distribution operations.

Once the bleaching vessel is emptied, compressed air is blown down the lines and through the bleaching filter to remove any residual oil to the bleached-oil storage tank. This air-blowing must be stopped as soon as all the readily available oil has passed through because starting a fire in the hot-spent bleaching earth within the filter is easily possible.

When the filter is empty of bleached oil, one can use it to pre-bleach the feed oil for the next batch by directing the charge to the bleaching vessel via this filter. When charging is finished, the remaining feed oil is chased from the filter into the bleaching vessel by using compressed air.

The spent earth is now discharged from the filter into a disposal container. Again, this spent earth is a serious fire hazard, and so one should watch closely this container and remove it from the process building every 8 hours, at least. This is because the self-heating of the oil on the hot earth can result in the auto-ignition temperature being exceeded within a few hours, leading to a serious danger of fire for the whole building.

Traditionally, the spent earth was disposed of to landfill, but modern practice uses its calorific value and mineral content in many applications, for example in brick- and tile making and animal feeds.

Air bleaching is practiced only on palm oil because of the high content of carotene and its precursors. One can destroy them by passing air through the oil held at between 100 and 140°C. The exact temperature and air-blowing time are highly variable, and they depend on the state of degradation of the crude oil. At first, the color level reduces, but then starts to rise again due to the carotenes and the oil itself being oxidized. From this point on, the color level cannot be recovered. So, great care is needed when operating this process. The ultimate level of bleaching is not as good as two-stage adsorption onto fuller's earth. However, the process is much cheaper, and does give adequate oil for making deeply colored laundry soaps.

Caustic Soda

Caustic Soda Production, Sources, and Users

Caustic soda is a co-product with the chlorine of electrolyzing sodium chloride brine. The two coproducts are locked by chemistry into being produced in stoichiometric proportions, that is, 40 caustic soda to 35.5 of chlorine or 1.13:1. Production is carried out all over the world, and most of the two products are used locally. However, local production and use for both are seldom in balance, and so caustic soda, being the most easily transported, is the one traded globally.

Three main electrolysis processes are available: mercury, diaphragm, and membrane cells. All three produce to the same basic product specifications, but their trace impurities differ. Mercury- and diaphragm-cell plants are generally old, and at the end of their economic lives, are being replaced by membrane plants.

Caustic soda and chlorine are basic chemicals for many industries, and their production has increased in North America and Western Europe by about 2–3% per year for the last 10 years but by 50% in China in the last 5 years. World caustic-soda production was at about 50 million dry tonnes in 2007, with about 27% in the United States; 21% in Europe; and 20% in China, 7% in Japan, and 12% in other Asian countries. Expectedly, China and other developing Asian countries will continue to grow relatively quickly, while demand in the rest of the world remains essentially constant.

The proportions of the users differ in parts of the world, but Fig. 13.3(a and b) gives the overall picture for the two coproducts, chlorine and caustic soda.

Soap Making Raw Materials • 389

Fig. 13.3a. Chlorine User Products (%).

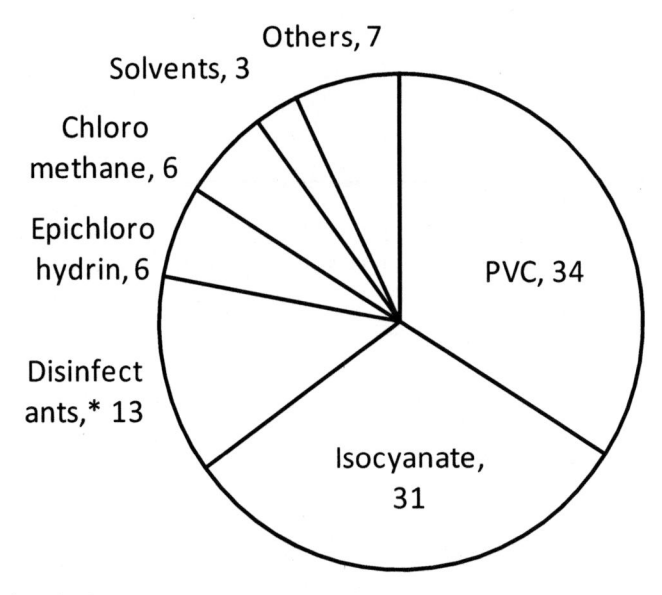

Pie chart values:
- PVC, 34
- Isocyanate, 31
- Disinfectants,* 13
- Epichlorohydrin, 6
- Chloromethane, 6
- Solvents, 3
- Others, 7

*includes water treatment

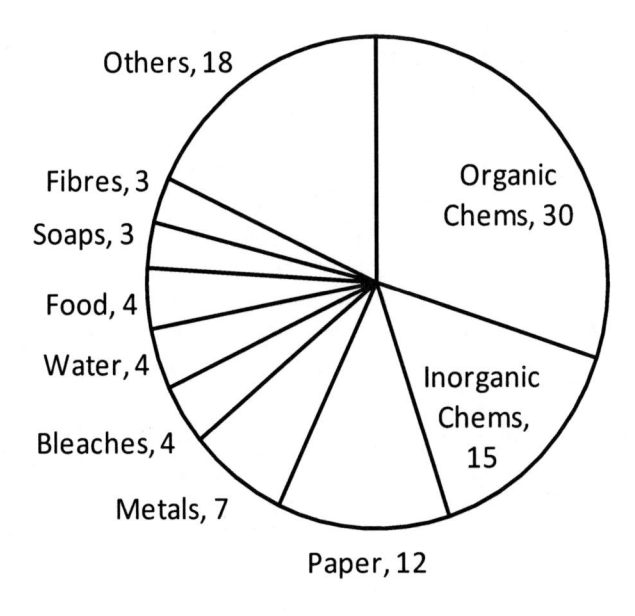

Fig. 13.3b. Caustic Soda User Industries (%).

Pie chart values:
- Organic Chems, 30
- Inorganic Chems, 15
- Paper, 12
- Metals, 7
- Bleaches, 4
- Water, 4
- Food, 4
- Soaps, 3
- Fibres, 3
- Others, 18

Obviously, polyvinyl chloride (PVC), mostly used in plastics for the construction industry, is the biggest single consumer of chlorine, but the soaps industry is a relatively small user of caustic soda.

Because chlorine is dangerous and difficult to transport, its demand sets the production schedule of the electrolysis plants, with caustic soda acting essentially as a by-product. Chlorine is used in many chemical applications and for treating water to ensure hygiene. Most of these uses are fairly constant in demand. However, the off-take to make PVC is large, and fluctuates significantly both seasonally and year to year with global economic activity. This is because it is used in many products for the construction industry. So when construction is buoyant, a strong demand arises for chlorine and also a high co-production of caustic soda. However, when construction is weak, caustic-soda production is low. Not only does this cyclic effect bear on production, it is multiplied in caustic soda's price as is seen in the section on market and price trends below.

Caustic Soda Characteristics, Specifications, and Analysis

Caustic soda is a commodity chemical, and as such is sold to an essentially common specification worldwide. The common commercial-grade specification has only four basic components. These are:

1. Sodium hydroxide—usually 47 wt%, because this concentration has a minimum in the freezing-point curve at 5.5°C but rising to 15°C at either 38 or 52 wt%. (Also a lower minimum exists of –27°C at 18 wt%.) The exact NaOH content is not particularly important because the price paid is usually for dry weight (i.e., equivalent to 100% of NaOH). Titration can do the analysis against any standardized acid.

2. Chloride—usually less than 0.1 wt%, but can be as low as 0.02 wt%. This is a measure of the leakage of brine within the electrolytic cell, and is much more likely in the diaphragm and membrane processes. It is important because the chloride contents of about 1 wt% can have a seriously corrosive effect on austenitic stainless steels. For soap making, the level of chloride is not a big issue, as long as it is low and, more importantly, is constant over time. This is because the chloride introduced from this source adds to the total electrolyte in the soap curd in the saponification reaction, and thus dictates its phase structure and thus, reaction time. One can do the analysis potentiometrically against silver nitrate.

3. Chlorate—usually less than 0.12 wt%, but a low-chlorate specification is at less than 0.01 by wt. Chlorate is again a measure of the leakage of brine within the electrolytic cell, and is much more likely in the diaphragm and membrane processes. The higher chlorate specification is suitable for DFA and glycerine soap making, but when the glycerine is recovered after neutralization, the low chlorate should be chosen. This is important because chlorate can be concentrated during glycerine recovery and collect in the glycerine-still bottoms, potentially causing fires and even explosions. Keeping the chlorate concentration in the crude glycerine feed to the still below 100 ppm by weight has been shown to prevent these hazards. One can do the analysis by titration against iodine/sodium thiosulfate.

4. Heavy and transition metals—usually below 20 ppm of total heavy metals evaluated as lead (Pb). This is important because soaps are used on the skin, and particularly chromium (Cr), nickel (Ni), and cobalt (Co) can lead to sensitization. These metals, except mercury (Hg) in the mercury-cell process, originate in the feedstock brine, and typically easily meet the standard, particularly since caustic is only used at about 10 wt% to fatty matter. Consequently, these metal contents are not measured frequently by the buyer, but are audited from records kept by the supplier backed up by retained samples of deliveries. One can do the analysis by precipitation with sulfate for heavy metals and/or individually for all trace metals by atomic absorption or plasma emission spectroscopy.

Caustic Soda Markets and Price Trends

No market exchange is available for caustic soda like for agricultural products such as TG oils. This is because relatively few producers exist worldwide, and the production of the chlorine/caustic complex can, within reason, be turned up and down at will to suit major seasonal demand fluctuations. However, market-intelligence organizations confidentially obtain negotiated prices for recent large contracts and from traders who sell to smaller users. From this data they produce a price index which is used as the equivalent of the market price. The quarterly prices shown in Fig. 13.4 below are a composite from various sources showing the last 20 years.

The first thing to note is that the prices in different parts of the world tend to move together. This is to be expected because caustic soda can be shipped fairly easily and quickly. So, traders arbitrage the price differences by filling their port-side storage tanks from different sources.

The caustic price shows a similar volatility to TG oils, but no correspondence exists between the peaks and troughs. As mentioned above, the price of caustic reflects its by-product status to chlorine, and so is counter-cyclical to general economic activity.

In the period from 2005–2007, a rising trend occurred caused by the general increase in fuel costs which reflect onto the electricity used in the production processes.

Finally, from early 2008, a large price rise occurred in the United States, but it was caused by the effect of the credit crunch on the construction industry's reduced off-take of PVC and not, as in the case of TG oils, on the arrival of biofuels. Europe and Asia followed this trend later in 2008. At present, caustic-soda demand is higher than that for chlorine, and so caustic has become the primary product. This situation has not occurred for many decades, but now caustic soda must bear a larger proportion of the overhead costs of the electrolysis operation. This has increased its price even more than would be expected from normal: supply-and-demand fluctuations. As a consequence, the prices have moved to a new all-time high.

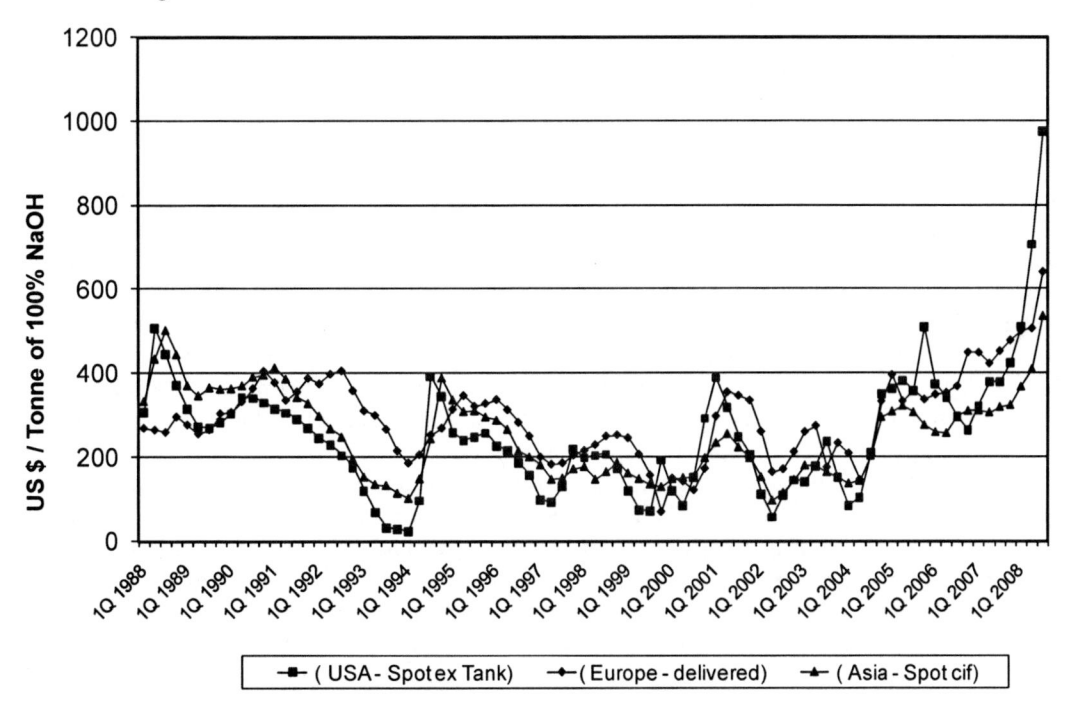

Fig. 13.4. Caustic Soda Quarterly Price Trends.

Caustic Soda Transport, Storage, and Handling

Most soap-making plants receive their caustic soda in bulk in road tankers or rail cars. Caustic soda is corrosive, and so operating staff must be protected, usually wearing full personal protection and following strict safety procedures and permits when working with it during deliveries or when maintaining any equipment.

In general, the only test on receipt is a specific-gravity measurement backed up by a sample titrated against a standardized acid. This combination is usually sufficient to ensure that the material received is indeed caustic soda and that it is at the correct composition. Chloride, chlorate, and trace-metal content usually use the supplier's certificate of analysis or conformance backed up by random audits and a comparison of retained received samples against the suppliers' own retained samples and analysis results.

The storage tanks are usually vertical cylinders made from carbon steel. The tanks are bunded to prevent any spillage flowing directly to storm drains. Any rainwater from the bunds must be sent to alkali drains and from there to effluent treatment for pH correction before off-site discharge.

In locations where ambient is likely to fall below about 10°C, the tanks are lagged and fitted with heating coils in their bases to prevent the freezing of the contents. The heating fluid is preferably temperature-controlled hot water because if the welding of the tank plates is to be subjected to temperatures greater than 60°C, then the whole tank must be annealed at above 650°C to prevent embrittlement. Similarly, any external pipelines need to be hot water or electrically traced.

Diluting 47 wt% of caustic soda with water liberates a considerable amount of heat. Therefore, to separately feed the full-strength caustic soda and the dilution water into the saponification reactor is usual. This way, no dilution vessel is required, and the heat of dilution can heat the reacting mix and thus reduce feed pre-heating.

Glycerine

Glycerine Production, Sources, and Users

For many readers of this book, glycerine is seen as a by-product from soap making by the TG route, and so their interest is in its sale. However, for those with interest in DFA soap making, they are buyers of glycerine because almost all soap formulations include some of it. So the following sections of this chapter cover both aspects of this material.

Glycerine is mostly produced as a by-product from soap making and oleochemical production, the latter being principally fatty acids but more recently including biodiesel. The dilute glycerine from these processes is usually treated to remove gross contaminants, and then concentrated by evaporation to about 80 wt% for soap making or 90 wt% for oleochemical production. The difference between the two is that the soaper's crude contains about 10 wt% of salt which originated in the soap-washing step. Both of these crudes are traded internationally, but eventually nearly all are refined by distillation and adsorption on ion-exchange resin and/or activated carbon to a set of grades which contain greater than 99.5 wt% of glycerine. The remainder of this refined product is mostly water.

All the >99.5 wt% of glycerine is essentially to the same chemical specification because the plants are set up to meet U.S. and European pharmacopoeia standards (i.e., USP and EP). All standards are almost identical; only their methods of analysis are different. However, a producer can elect, or not, to carry the extra expense of certification, and thus sell to users demanding these certificates. The producers who elect not to obtain certification can still sell the uncertified product stream, but at a cheaper price, to users who don't need the certificate for their products. An additional complication is that all these glycerines are also graded for cultural-acceptance reasons by the original TG feedstock [i.e., as of vegetable origin, porcine-free (Kosher and Halal), and of animal origin].

Until the advent of biodiesel, a synthetic route also made glycerine from petrochemical feedstock. In essence, it was made as a side-stream in epichlorohydrin plants by hydrolyzing any excess of the main product. However, these facilities are now all mothballed because synthetic glycerine could not compete on price once the biodiesel by-product glycerine came onto the market.

As is seen in Table 13.3, the amount of glycerine produced was rising slowly until the arrival of biodiesel which started to become significant in the last few years, and is now just over one-half the total. The glycerine arising from saponification via the TG process is down by about one-third over the period, mostly due to the switch to liquid personal-washing products, but those are more than compensated for by the rise from oleochemicals, as this industry expanded in Southeast Asia and China.

Looking at Fig. 13.5, one can see that the major user of glycerine is the toiletries and cosmetics industry, with many others each taking significant amounts. Over the last 10 years, these traditional users expanded their businesses, and essentially kept supply and demand in balance. However, with the massive expansion of biodiesel arising, glycerine prices were expected to fall markedly. Two things are now happening to mop up the excess. Users of other polyols, like sorbitol, propylene glycol, and monoethylene glycol which are used in toothpastes, foods, antifreeze, resins and paints, are substituting glycerine instead. In addition, new processes were developed to use cheap glycerine as a feedstock for producing epichlorohydrin, propylene glycol, and acrylic acid. Between them, these additional users can more than consume the excess glycerine from the biodiesel expansion. Unfortunately, the timings are not coordinated, and so likely a severe imbalance will occur over the next few years which will result in highly volatile market prices.

Table 13.3. Glycerine Production Over the Last 10 Years and the Traditional User Industries (Tonnes thousands at 100% glycerine equivalent).

Source	1998	2003	2008
Saponification	200	180	130
Oleochemicals*	430	460	680
Biodiesel	50	150	950
Synthetic	75	80	0
Others	45	50	25
Total	800	920	1785

* Fatty Acids + Alcohols

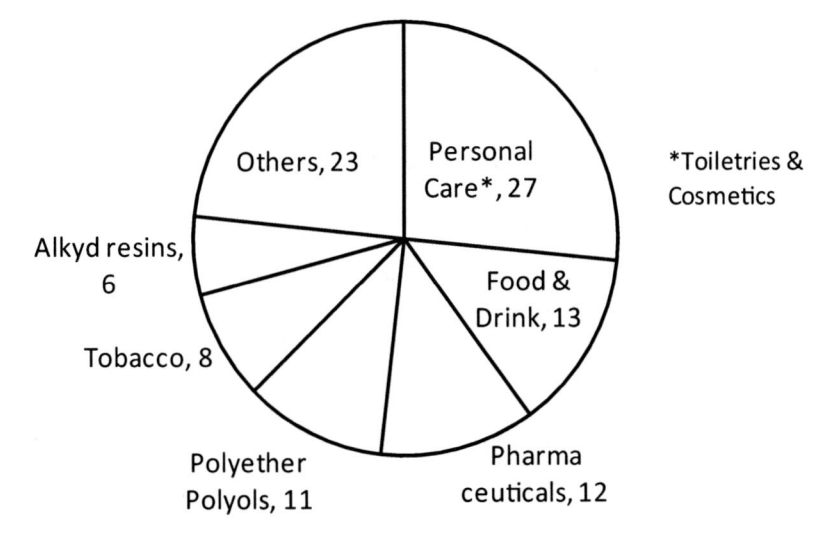

Pie chart values: Others, 23; Personal Care*, 27; Food & Drink, 13; Pharmaceuticals, 12; Polyether Polyols, 11; Tobacco, 8; Alkyd resins, 6. *Toiletries & Cosmetics

Fig. 13.5. Traditional Glycerine User Industries (%).

Glycerine Characteristics, Specifications, and Analysis

Glycerine is a commodity chemical, and as such it has essentially a common specification based on the U.S. or European pharmacopoeia requirements. All other grades are subsidiaries of this. The various pharmacopoeia specifications are detailed and are updated regularly. Therefore, one must refer to these original documents before making any serious decisions. Note also that the analytical methods have come down from earlier versions, and often better modern ways of analysis are available. Nevertheless, to obtain pharmacopoeia certification, one must use the specified method. The various specifications do differ in detail, but all have the following five main areas:

- Assay—a measure of the amount of glycerine in the sample. For USP this is based on specific gravity, while for EP refractive index is used. Note that USP allows up to 5 wt% of water and EP up to 2 wt%. However, the commercial standard is greater than 99.5 wt% of glycerine. The only thing to note here is that most users are set up for this high concentration, and so supplying a lower content, although meeting the "official" standard, would probably cause serious disruption in their operations and product performance.

- Ash—usually less than 0.01 wt%. This is a measure of inorganic contamination, and is very easily met by any modern process, which results in levels measured in ppm.

- Heavy metals—usually less than 15 ppm by wt evaluated as lead (Pb) by precipitation of sulfates. Again, this is easily met by any modern process.

- Diethylene glycol (known as substance A in EP) is measured by gas–liquid chromatography. No reason exists as to why this substance, which is moderately toxic, should be present in glycerine produced by any known process. So, the only conclusion is that it is a deliberately added adulterant. However, the latest versions of the pharmacopoeias call for its measurement to prevent the harmful effects of this adulterant.

- Color and odor are included, but any modern commercial process easily meets the requirements.

Other items, often written into individual buyer's specifications, are additional to and often more stringent than the pharmacopoeia standards. These are often a consequence of the more onerous requirements of toiletries and cosmetics regulations. Here the major one is

- Trace metals is aimed at picking up transition metals which are important for skin leave-on products because they can lead to sensitization, particularly with chromium (Cr), nickel (Ni), and cobalt (Co). Trace metals' contents are measured by atomic absorption or plasma emission spectroscopy. Either technique has the additional benefit of measuring the metals individually, and covers both the heavy and transition metals.

If sold, crude glycerine is usually analyzed only for its glycerine and salt content.

Glycerine Markets and Price Trends

Like caustic soda, no market exchange is available for glycerine as for agricultural products such as TG oils. This is because relatively few producers exist worldwide, and the production of the oleochemical/glycerine complex can, within reason, be turned up and down at will to suit major seasonal fluctuations in demand. However, market-intelligence organizations confidentially obtain negotiated prices for recent large contracts and from traders who sell to smaller users. From this data, they produce a price index which is used as the equivalent of the market price. The quarterly prices shown in Fig. 13.6 below are a composite from various sources showing the last 20 years.

As one can see, a major volatility in prices is present, but the peaks and troughs do not correspond to those of either the TG oils or caustic soda. This is because the traditional users of glycerine tend to be

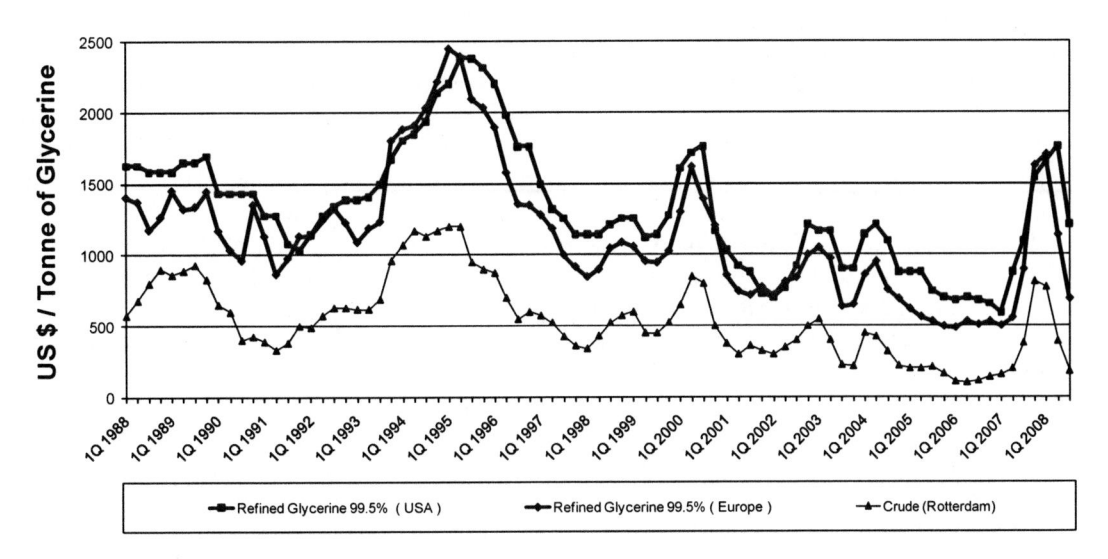

Fig. 13.6. Glycerine Quarterly Price Trends.

industries whose sales do not follow general economic cycles. However, they do respond to supply-and-demand imbalances.

The two upper prices are for refined glycerine in the United States, a major importer, and the European Union (EU), a significant exporter. As would be expected, the prices follow each other closely because to ship glycerine across the Atlantic only costs about $100–200/tonne.

The lower line is for crude glycerine, and, as would be expected, it has a reasonably constant discount to refined glycerine which covers the refining costs and margins.

From about 2005 onward, prices did respond to the advent of biodiesel glycerine. Until about early 2007, the expected down movement occurred. However, at that point, two competing forces met market sentiment! First, a large number of governments across the globe encouraged biodiesel production with promises of subsidies, and so plants were scheduled for construction. The market was to be flooded with glycerine. However, at the same time, the prices of TG oils started a rapid increase, thus neutralizing any chance of profitable biodiesel production.

Furthermore, in anticipation of large amounts of cheap glycerine, many petrochemical producers developed a process to use glycerine as a raw material (e.g., for propylene glycol, epichlorohydrin, and as a substitute for ethylene glycol). The market saw a deficit of supply and so the price sky-rocketed. After about a year, everyone involved realized that all the promises and schedules were just that, and prices have again fallen back to the long-term trend.

All these issues are still in the air, essentially hinging on biodiesel. As was stated above in the prospects for TG-oil prices, all the decisions are essentially grounded in politics. So, all we can do is form business strategies and have selling tactics which accommodate both sides of these political decisions.

Glycerine Transport, Storage, and Handling

The purpose of this chapter is not to cover the details of glycerine recovery, but a brief description is useful of the processes used in soap making and fatty-acid and biodiesel production as background for soap makers to compare their competitors in supplying the glycerine market.

In soap making, glycerine is transferred from the soap curd into the lye stream during the soap-washing step. The spent lye, typically at 20–30 wt% of glycerine, is first treated to remove excess alkali and fatty matter by adding acid, usually hydrochloric, and ferrous chloride to precipitate iron soaps. Coagulants are added and then the stream is filtered. Multi-effect evaporation is then used to remove the water to give crude glycerine which contains about 80 wt% of glycerine with about 10 wt% of water and 10 wt% of salt. This concentration is chosen because above this the elevation of the boiling point increases very rapidly. The salt level is at saturation and exceeds that entering with the lye stream. So, salt crystallizes during evaporation and is removed, and then returned to soap washing. One can sell crude glycerine into the merchant market at this stage; it is often referred to as soaper's crude glycerine.

The crude glycerine is distilled typically to give a product of about 99.9 wt% of glycerine. However, this material still has some color and odor which is removed by refining with ion exchange and/or carbon-bed adsorption. The product here is refined glycerine.

In fatty-acid making, the glycerine is liberated during the TG-splitting process and passes from the hydrolysis column as sweet water containing about 18 wt% of glycerine. This is flashed to atmospheric pressure, giving about 30 wt% of glycerine, and then multi-effect evaporated to 90 wt% of glycerine and 10 wt% of water. This concentration is chosen because, again, above this value, the elevation of the boiling point increases very rapidly. One can sell this crude glycerine into the merchant market; it is often called splitter's crude glycerine.

Most biodiesel plants have to separate their glycerine from methanol. This is done by distillation to produce biodiesel crude glycerine. Again, one can sell this into the merchant market.

As with soaper's crude, the oleochemical plants can use very similar processes of distillation and adsorption to yield refined glycerine.

The cost of producing crude tends to be slightly more expensive for soap makers than for oleochemical plants, and it contains less glycerine because of its salt content. Similarly, the cost tends to be higher for refining soaper's crude.

Crude glycerine tends to be handled in carbon-steel equipment, and refined glycerine in stainless steel.

Refined glycerine freezes at about 15°C, splitter's crude at about −2°C, and soaper's crude at −20°C or below. In fact, glycerine rarely freezes in normal plant conditions even in cold climates because it tends to supercool. Nevertheless, if concern arises in any cold location, then the storage tanks can be hot-water-jacketed and lagged.

Although glycerine is flammable, its flash and auto-ignition temperatures at 193 and 400°C, respectively, are far higher than operating and storage temperatures. So, no special precautions need to be made. However, glycerine has a high oxygen demand, and so its storage tanks should be bunded and any collected rainwater should be sent to effluent treatment rather than discharged to the storm drains.

Crude glycerine is almost always dispatched in bulk, but refined can also be packed in 200-liter resin-coated drums and/or in 500- to 1000-liter polyethylene containers.

Because soap-making plants using the TG process are selling glycerine, they will have to provide Certificates of Analysis and/or Conformance for each dispatch. This means they need the appropriate analytical facilities with a backed-up database of analytical results. The database must allow reference to a store of samples retained for at least one and often three years after dispatch. These are most important because they may be called on during the random inspections of supplier's Certificates of Analysis and/or Conformance, retained samples, and lab analyses by their buyer's auditors.

Acknowledgments and Disclaimer

The information given in this chapter is a composite from my own experiences supported over many years by private communications with a large number of my colleagues at Unilever, at equipment manufacturers, at material suppliers, at trade bodies, etcetera. I acknowledge a debt of gratitude to them all.

The volume and price data are again composites of information gained from various national governments and international organizations, trade bodies, market exchanges, and commercial index providers.

In compiling the information for this chapter, I used my own judgment on what to include or omit from my many sources. Also, where the sources differ, I have used my experience to select what I consider to be the most appropriate.

The author believes the information to be correct, but it is given in good faith and is only to be used for guidance. Before making any decisions, the reader must check the latest information from their own sources, and choose which is the most relevant for their own applications.

14

Analysis of Soap and Related Materials

Thomas E. Wood
Consultant, 305 Coweta Court, Loudon, Tennessee 37774, USA

Introduction

The analytical methods available to the soap chemist for evaluating raw material and soaps include both the classical physical and chemical analytical procedures that are found in standard reference manuals of analytical methods, and the modern instrumental methods including gas chromatography and high-performance liquid chromatography. The two most important reference sources for analytical methods for soap and soap raw materials include the *Official Methods and Recommended Practices of the American Oil Chemists' Society* (AOCS, 2009) and the *Annual Book of ASTM Standards* of the American Society for Testing and Materials, Volume 15.04 (ASTM, 2008). The instrumental methods are generally found in publications like the *Journal of the American Oil Chemists' Society*, the *Journal of Chromatography*, the *Journal of Liquid Chromatography*, and other similar publications. Please refer to Table 14.1 for a concise list of references to selected AOCS Official Test Methods for soap and soap raw materials.

Chemical and Physical Characteristics of Soap Raw Materials

Among the chemical properties of fats, oils, and fatty acids that are important to the soap chemist, three of the most important are acid value, saponification value, and iodine value. Additionally, the *trans*-isomer content can be an important consideration whenever hydrogenated stocks are involved. Other important characteristics include the titer, free fatty acid, raw color, bleached color, moisture, insoluble impurities, and unsaponifiable matter.

Acid Value

Acid value is defined as the number of milligrams of potassium hydroxide (KOH) required to neutralize the free acids in 1 gram of sample.

For samples of fats and oils, the determination of the acid value involves the simple titration, in alcohol, of an appropriately sized sample to the phenolphthalein end point with essentially no sample preparation required. In this case, the acid value is virtually a measure of the percentage of free fatty acid present, and can be expressed as such by simple mathematical conversion. Typically, in the case of fats and oils, the level of free fatty acid should be very low, ranging from near zero to several tenths of a unit for higher grade material to a few whole units for lower grade oils. Please refer to AOCS Official Method Cd 3d-63 for more information on acid-value determination for fats and oils.

For samples of fatty acid stock, the acid-value determination is also a direct titration of the sample in alcohol to the phenolphthalein end point without any special sample preparation. Since fatty acid stocks, by their nature, are essentially all free fatty acid, the acid value will range above two hundred units for most of the common fatty acid blends used in soap making. For fatty acid stocks, the acid value will approach the saponification value. Please refer to AOCS Official Method Te 1a-64 for the details for acid-value determination of fatty acid blends.

For soap products, the acid value is determined on the total fatty acids present including the free acids, which may range from none to several percent, plus the major portion that is combined with the cation as soap. Consequently, the acid-value determination for soap requires an initial sample

Table 14.1. Selected AOCS Official Test Methods

Test/AOCS Method for:	Fats & Oils	Fatty Acids
Acid value	Cd 3d-63	Te 1a-64
Free fatty acid	Ca 5a-40	
Insoluble impurities	Ca 3a-46	
Iodine value	Cd 1d-92	Tg 1a-64
Moisture/volatiles, 130°C	Ca 2c-25	
Colorimetric color, Wesson method	Cc 13b-45	
Colorimetric color, Lovibond method	Cc 13e-92	
Photometric color	Cc 13c-50	Td 2a-64
Alcoholic saponification color	Cc 13g-94*	
Refined & bleached color	Cc 8d-55	
R&B; alcoholic saponification color	Cc 13f-94*	
Saponification value	Cd 3-25	
Titer	Cc 14-59	Tr 1a-64
Unsaponifiable matter	Ca 6a-40	Tk 1a-64
Water, Karl Fischer method	Ca 2e-84	Tb 2-64
Test/AOCS Method for:	Soap	Soap with Detergent
Acid value	Da 14-48	
Anhydrous soap content	Da 8-48	Db 6-48
Chlorides	Da 9-48	Db 7-48
Free fatty acid/free alkali	Da 4a-48	Db 3-48
Glycerin	Da 23-56	
Iodine value	Da 15-48	
Moisture & volatiles	Da 2a-48	Db 1-48
Saponification value	Da 16-48	Db 8-48
Titer	Da 13-48	

*AOCS Recommended Practice

preparation where the combined fatty acids are liberated by acidulation with an excess of sulfuric acid, recovered, and then dried. Following this sample preparation, the acid value is determined on the fatty acids as described above for fatty acid blends. Again, an acid value more than two hundred units would be expected for the fatty acids associated with most ordinary soaps. Please see AOCS Official Method Da 14-48 or ASTM Standard Method D 460, Sections 48 and 49 for more details on acid-value determination of soap.

The acid value is a key indicator of the fatty acid composition of soap since it is inversely and linearly related to the average molecular weight and chain length of the fatty acids in the blend. For example, tallow fatty acid has an approximate acid value of 204, while coconut fatty acid has an approximate acid value of 268. The usefulness of the acid-value determination lies in helping to establish the composition soap fatty acid blends. In the case of a fatty acid blend that is derived from an 80:20 tallow/coconut oil blend, for example, the expected acid value should be about 216 based on the weighted average of the acid values for the blend. If the ratio were to shift toward higher tallow content and lower coconut oil content, a lower acid value would result. For example, a 90:10 tallow/coconut oil blend would result in an approximate acid value of 210. Conversely, a lower tallow-to-coconut oil ratio would result in a higher acid value. A 70:30 tallow/coconut oil blend would have an acid value near 223.

The acid value is frequently included in specifications for soap with only a lower limit indicated. This is done for reasons of both quality and economics since tallow, with the lower acid value, is the lower cost commodity with the poorer performance characteristics.

Saponification Value

Saponification value is defined as the number of milligrams of KOH required to saponify 1 gram of sample.

The procedure is carried out directly on a sample of the stock for both commercial whole oils and fatty acid blends. In the case of soap samples, the fatty acids must first be prepared by acidulating a sample of the soap, recovering the liberated fatty acids, and drying the recovered fatty acids. The saponification-value determination is carried out by refluxing the sample of oil or fatty acid with an excess of KOH in an alcoholic solution for 30 to 60 minutes. Along with each sample, or set of samples, a blank is also run. After the reaction is completed and cooled, the excess KOH is titrated with standardized 0.5 N of hydrochloric acid. With an appropriate calculation, the difference between the titrations of the blank and the sample is then reported as the saponification value. For more information, see AOCS Official Methods Cd 3-25 for fats and oils, Tl 1a-64 for fatty acids, and Da 16-48 for soap and soap products.

Note that any difference between the results of the saponification value and the acid value on a given sample is called the ester value. Not unusually, the saponification value can be one or two units higher than the acid value on a given sample of soap fatty acid. This results from the reaction of trace amounts of naturally occurring ester-like compounds that are reactive under the conditions of the saponification-value determination, but not under the milder conditions of the acid- value determination.

Iodine Value

Iodine value is defined as the number of centigrams of iodine absorbed by 1 gram of sample.

For samples of fats, oils, and fatty acids, the iodine-value determination is performed directly on an appropriately sized sample of the material that was liquefied by melting and filtered to remove any trace impurities including moisture. For samples of soap, the fatty acids must first be liberated with sulfuric acid, recovered, and then dried. The iodine value is then determined on the prepared fatty acids.

The procedure is carried out both by reacting the sample of oil or fatty acid with an excess of an iodine monochloride solution (Wij's solution), and by titrating the excess iodine with a standardized sodium-thiosulfate solution by using a starch indicator for the end-point determination. Each sample, or set of samples, is done with a blank to determine the quantity of iodine consumed by the sample. The difference between the blank and the sample is attributable to the iodine absorption by the sample. With appropriate calculation, the difference is reported as the iodine value for the sample. See AOCS Official Methods Cd 1d-92 for fats and oils, Tg 1a-64 for fatty acids, and Da 15-48 for soap and soap products for detailed information on the procedures. Also, for soap and soap products, refer to ASTM Standard Method D 460, Sections 50 to 52, for detailed procedures.

The iodine value is directly related to the degree of unsaturation present in the fat, oil, or fatty acid. The iodine value also serves as an indicator of the relative hardness of fats and of the derived fatty acids and soap product for otherwise comparable fat stocks. In general, higher levels of unsaturation, as indicated by higher iodine values, indicate a tendency toward softer fat stocks and softer soap product. This relationship is clearly evident when comparing natural animal fats such as tallow and grease that have roughly comparable average chain-length distributions and average molecular weights, as indicated by their saponification values, but which have markedly different amounts of unsaturated fatty acids. Tallow, with its lower iodine value indicating less unsaturation, will melt at a higher temperature and will result in firmer soap compared with grease.

The underlying basis for the physical effects of lower melting point and softer material associated with natural animal fats that have a higher degree of unsaturation is found in the predominance of *cis*-unsaturated fatty acids in naturally occurring fats. Due to the geometry around the *cis* double bonds, the molecules with the *cis* configuration have a distinctive bend in the carbon chain at the site of the carbon-to-carbon double bond. This results in the looser packing of molecules in the solid, causing reduced intermolecular forces, and consequently lower, melting point. The effects on the properties of fatty acids and soap resulting from varying the *cis*- and *trans*-isomer content are discussed in a later section.

Titer

Titer is defined as the temperature, expressed in degrees centigrade, at which fatty acids solidify.

Generally, titers of fatty acids will vary inversely with iodine values. Titer is frequently used as a quality-control and process-control measure in soap making since it is a good indicator of the processing characteristics of the resultant soap at the bar-finishing stage of manufacturing. Soap made from higher titer fatty acid blends tends to be firmer, and vice versa. The effects of *trans*-isomer content on this relationship are discussed in a later section.

The typical inverse relationship between titer and iodine value for otherwise comparable fats is illustrated in Table 14.2. Both lard and tallow are similar in average molecular weights, but lard, with the significantly higher iodine value, has a much lower titer and will form a softer soap

For tallow, Grompone (1984) reported that titer is fundamentally dependent on the stearic/oleic acid ratio. The relationship between titer of tallow and stearic-acid content is direct and linear. The relationship between titer and oleic-acid concentration is inverse and linear. Apparently, the titer is not influenced significantly by the varying levels of palmitic acid.

The titer test is always performed on the sample in the form of free fatty acids. For fatty acid blends, the procedure requires no special sample preparation. For fats and oils, a sample of the stock to be evaluated must first be saponified, followed by acidulation, and then the recovery and drying of the fatty acids. For soap samples, the sample needs to be acidulated with subsequent recovery and drying of the fatty acids. Please refer to AOCS Official Methods Cc 14-59 for fats and oils, Tr 1a-64 for fatty acids, and Da 13-48 for soap and soap products. Also, for soap and soap products, see ASTM Standard Method D 460, Sections 46 and 47.

Table 14.2. Comparative Data for Lard and Tallow

Property	Lard	Tallow
Saponification Value	190–202	190–200
Iodine Value	53–77	35–48
Titer (°C)	32–43	40–46

Effects of Trans Isomers on Fatty Acid and Soap Properties

The variation in the *trans*-isomer content of fatty acid blends impacts the titer, soap characteristics, and the relationship between titer and iodine value for otherwise similar fatty acid stocks. While the iodine value can serve as a useful indicator of hardness for naturally occurring fats where *cis* unsaturation is prevalent and *trans* unsaturation is low and consistent, this reliability can significantly diminish whenever hydrogenation is involved. If hydrogenation is carried out under conditions of higher-than-normal temperature, lower-than-normal hydrogen feed, or with a poor catalyst, a significant amount of rearrangement from *cis* to *trans* isomers can occur rather than the reduction of the double bonds. During the hydrogenation of fatty acids, the reduction of *cis*-unsaturated fatty acids (with their characteristically nonlinear structure) to saturated fatty acids (that are essentially linear molecules) will predictably result in the formation of harder fatty acid with a lower iodine value. However, the rearrangement of *cis* to *trans* isomers, insofar as it occurs, will not contribute to the lowering of the iodine value, but will result in harder fatty acid stock with a higher melting point.

The relative amount of *cis* and *trans* isomers not only has an impact on the melting point and hardness of the fatty acid, but also affects the hardness and plasticity of the resulting soap. The *trans*-isomer content will influence titer independent of the degree of saturation. Again, this phenomenon can become important when fatty acids used in soap making are subjected to hydrogenation. The conversion of a significant amount of *cis* to *trans* isomers will result in harder-than-expected soap for a given iodine value. Under these conditions, titer may be a more useful process-control measure for the soap maker than iodine value.

Some interesting relationships between melting points and iodine values are seen in Table 14.3 (Unichema Chemicals, 1987), where analogous sets of fatty acids that have varying degrees of unsaturation and *cis/trans* ratios are listed. For the four groups of isomers listed, both the iodine values and melting points are given. In the first pair, both isomers have a double bond at the number-6 carbon atom: the first is a *cis* structure, and the second is a *trans* structure. Both have an iodine value of 90. Note, however, the difference in melting points, with the *trans* isomer having a melting point of 21°C higher than the *cis* isomer. The other two pairs of isomers listed here show the same kind of relationship between isomeric structure and melting point. In the last set, each of the three fatty acid isomers contains three double bonds. Note that in the first structure shown, all three double bonds are *cis;* in the second case, one is *cis* and the other two are *trans;* and in the third case, all three are *trans.* All three of these isomers have the same iodine value of 274. However, a dramatic increase occurs in the melting point with increasing *trans*-isomer content.

Table 14.3. Comparative Iodine Values and Melting Points

Systematic Name	Formula	IV	MP(°C)
6*c*-octadecenoic acid	$C_{18}H_{34}O_2$	90	33
6*t*-octadecenoic acid	$C_{18}H_{34}O_2$	90	54
9*c*-octadecenoic acid	$C_{18}H_{34}O_2$	90	16.3
9*t*-octadecenoic acid	$C_{18}H_{34}O_2$	90	45
9*c*,12*c*-octadecadienoic acid	$C_{18}H_{32}O_2$	181	-5.0
9*t*,12*t*-octadecadienoic acid	$C_{18}H_{32}O_2$	181	28–29
9*c*,12*c*,15*c*-octadecatrienoic acid	$C_{18}H_{30}O_2$	274	-11
9*c*,11*t*,13*t*-octadecatrienoic acid	$C_{18}H_{30}O_2$	274	49
9*t*,11*t*,13*t*-octadecatrienoic acid	$C_{18}H_{30}O_2$	274	71.5

Trans *Isomer Measurement*

The classical method of analysis for iodine value will establish the relative amounts of saturated and unsaturated fatty acids present, but will not differentiate between the *cis* and *trans* isomers. Infrared spectroscopy can quantify the *trans*-acid content of fatty acid blends. The method is based on the absorption at 966 cm^{-1} by the *trans* double bonds in the sample. The absorption is directly proportional to the concentration of *trans* double bonds present in the sample. The absorption is thought to be independent of the position and number of *trans* double bonds present. Conjugated *trans* double bonds in excess of 5% may interfere with this method. The presence of such interference is indicated by absorption at 987 cm^{-1}. The details of this method are found in AOCS Official Method Cd 14-95. One may achieve a complete analysis of *cis/trans* isomers by capillary gas chromatography by using AOCS Official Method Ce 1h-05.

Color Evaluation of Soap Raw Materials

In the following sections, we review the various methods for evaluating the color of soap raw materials and of the resulting neat-soap product. The color of raw fats and oils stock is an indicator of the degree of abuse and degradation to which the material was subjected. To that extent, it is a measure of the degree of additional processing that is required to produce white soap and of the potential for the loss of fat in the soap-making process. In the case of fatty acid stocks, the color is also an indicator of the quality of the material, and will have a direct bearing on the color of the neat soap made from the fatty acid stock. If the color is high, it may also be an indicator of degradation that will affect the odor of the resulting soap.

Raw Color of Fats and Oils by Colorimeter

Probably the most widely used method of color measurement for fats and oils in the United States is the Wesson method that uses the Lovibond AOCS Tintometer, Model AF710, that is manufactured by The Tintometer, Ltd., Amesbury, Wiltshire, United Kingdom (www.tintometer.com).

This method determines the color of fats, oils, and fatty acids by the comparison of a column of the oil with standard glass slides under specific controlled conditions. The procedure is also commonly used for commercial fatty acid blends where, in the absence of any turbidity, the material can be evaluated while molten.

In the Lovibond method, a 5.25-inch column of material is placed in a standard color tube that is specially designed for use with the AOCS Tintometer. The tube is inserted into the port of the lit cabinet where the color of the material in the sample tube can be compared with the standard red- and yellow-glass slides that have numerical scale values. Usually, both the red and yellow values are specified. Frequently, only the red value is specified and reported, particularly for tallow and coconut oil. The details of this method are found in AOCS Official Method Cc 13b-45.

The accepted international standard based on BS 684 is ISO 15305, the Lovibond method. The Model F (BS 684) Lovibond Tintometer that is also made by The Tintometer, Ltd. is suitable for this method. Please refer to the AOCS Official Method Cc 13e-92 for details of this method.

Color of Fats, Oils, and Fatty Acids by Photometric Measurement

Another common method of measuring the color of fats and oils utilizes visible spectrophotometry. The photometric measurement of the color of fats and oils measures the absorbance at 460, 550, 620, and 670 nm. The photometric color is expressed as follows:

$$\text{Photometric color} = 1.29A_{460} + 69.7A_{550} + 41.2A_{620} - 56.4A_{670} \tag{Eq.14.1}$$

Please refer to AOCS Official Method Cc 13c-50 for the details of this method.

Frequently, the color of fatty acids is also determined by photometric measurement. Either the absorbance or transmission is measured at 440 nm and 550 nm. The results are reported either as "% transmission at 440/550 nm" or in terms of the "photometric index" that is expressed as: $100 \times A_{440}$ and $100 \times A_{550}$. Please refer to AOCS Official Method Td 2a-64 for details of this method.

Refined and Bleached Color of Fats and Oils

One can use the refined and bleached color test to predict the color after the processing of intermediate and lower grades of tallow that will require pretreatment before use in soap making.

The tallow sample is refined by neutralizing the free fatty acids with caustic soda. The neutral fat is then separated by filtration through cheesecloth or filter paper to remove any soap formed during the refining step. The neutral fat is then treated with activated bleaching earth. The mixture is then filtered, followed by the reading of the Lovibond color of the clear filtered fat. Please refer to AOCS Official Method Cc 8d-55 for the details of this procedure.

Direct Bleach Test for Fats and Oils

The direct bleach test for tallow is useful when working with higher grades of inedible tallow that are low in free-fatty acid content. The method uses 300 grams of fat that is heated to 110°C, followed by the addition of 15 grams of activated bleaching earth. The mixture is then agitated for 10 minutes at 80°C or higher, followed by filtration through a heated funnel at 70°C. After the bleaching and filtration, the Lovibond color is determined. All supplies used for this procedure are the same as those specified for the refined and bleach test above. The results of this test indicate the potential for making white soap from fats that will be pretreated by bleaching without refining.

Alcoholic Saponification Color for Fats and Oils

In AOCS Recommended Practice Cc 13g-94, a method is presented for saponification color determination of high-quality tallow and coconut oil intended for use in soap making. In this method, a sample of the untreated oil is reacted with an excess of alcoholic KOH solution. Upon completion of the saponification reaction, a sample of the soap solution is placed in a 5.25-inch tube color evaluation using the Wesson method for Lovibond-color measurement found in AOCS Official Method Cc 13b-45.

In AOCS Recommended Practice Cc 13f-94, methods are presented for refined and bleached color and for saponification color of lower grade tallows and greases intended for use in soap making. First, the sample is subjected to a caustic refining and bleaching operation. Following the refining and bleaching, a portion of the sample is reacted with an excess of an alcoholic KOH solution. After saponification is completed, a sample of the soap solution is transferred to a 5.25-inch tube for color reading using the procedure in AOCS Official Method Cc 13b-45.

Saponification Color of Fatty Acids

The color of fatty acids for soap making can be evaluated by measuring the alcoholic saponification color that determines the color of a potassium soap solution. This method is applicable to normal-soap fatty acid blends with a saponification value in the range of 214 to 220.

In this procedure, a sample of the fatty acid is dissolved in ethanol, followed by neutralization with an aqueous 39% of a KOH solution. The percentage of transmission at 440 and 550 nm is determined for the potassium soap solution. This method serves as a good indicator for the expected color of finished soap to be made from the fatty acid blend. Due to the relatively short period of time required to run this procedure, it can be useful as a receiving quality-control method for bulk fatty acids.

Approximate Conversions for Various Color Scales:

Figure 14.1 gives the approximate equivalence of various visual color scales commonly used for fats and oils, including Gardner, APHA, F.A.C, Lovibond, and percent transmission. Additional correlations for several methods of measuring the color of fats and oils are given in AOCS Official Method Cc 13b-45. The American Oil Chemists' Society (AOCS) does not encourage the use of conversion charts for color. They should be used with caution for an estimation or an approximation purpose only.

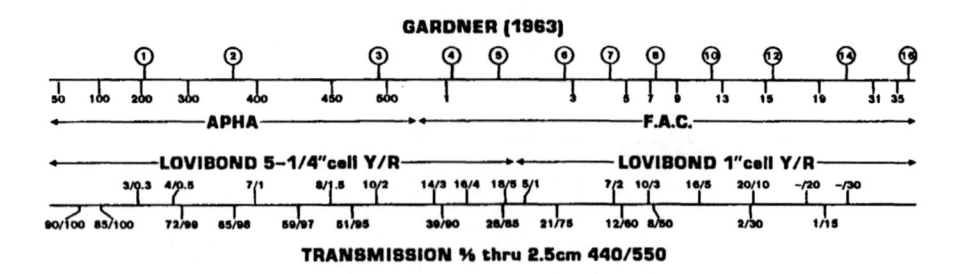

Fig. 14.1. Approximate equivalents of various visual color scales.

Free Fatty Acid Content of Fats and Oils

For determining the free fatty acid content of fats and oils, a weighed sample is dissolved in ethanol, followed by titration with a standardized NaOH solution to the phenolphthalein end point. The result is calculated and conventionally reported as percentage of oleic acid for most samples. Please refer to AOCS Official Method Ca 5a-40 for details of this method.

The presence of elevated levels of free fatty acid can result from the storage of fats and oils at elevated temperatures in the presence of water, causing hydrolysis of a portion of the triglycerides. Also, prolonged enzymatic activity in the case of animal fats can result in undesirable levels of free fatty acid. The presence of excessive free fatty acid will indicate a decreased level of available glycerine for recovery and an increased potential for color and odor degradation of the stock.

Moisture Content of Fats, Oils, and Fatty Acids

The moisture content of tallow and other nonlauric oils is generally determined by the 130°C air-oven method found in the AOCS Method Ca 2c-25. The air-oven method is not recommended for coconut and palm kernel oils. For the standard Karl Fischer titration method that is recommended for lauric oils and other fat stocks, please see AOCS Method Ca 2e-84. For the Karl Fischer method for fatty acids, see AOCS Method Tb 2-64.

Insoluble Impurities in Fats and Oils

This procedure determines dirt, bone meal, and other insoluble impurities present, under the conditions of the test, in all commercial fats and oils.

A properly sized sample of the fat is dried to constant weight in an oven; one can use the residue from the standard air-oven moisture determination if available. The dried sample is dissolved in a warm 50-mL portion of kerosene, and filtered through a dried, tared Gooch crucible with a glass-fiber filter. The filter is washed with five 10-mL portions of hot kerosene, followed by a thorough washing with petroleum ether. The crucible and contents are dried to constant weight in a 101°C air oven, cooled under desiccation, and weighed. The weight of residue is reported as the percentage of insoluble impurities. See AOCS Official Method Ca 3a-46 for more information.

Unsaponifiable Matter in Fats and Oils

Unsaponifiable matter can be composed of such things as cholesterol, fatty alcohols, and denaturants in fats. These materials do not react with NaOH to form soap and are extractable with fat solvents. The level of unsaponifiable material in fats and oils can range as high as a few tenths of a percentage for better grade materials. Apart from the obvious economic consideration associated with the purchasing of fats for soap making, the unsaponifiable matter can also have a negative impact on soap performance. One unit of unsaponifiable matter in soap can reduce the detergency action of at least three times its weight in soap (The Procter & Gamble Co., 1967).

The unsaponifiable-matter content of fats is determined by vigorously reacting a sample with KOH, followed by petroleum-ether extraction of the unreactive organic matter. Several petroleum-ether extractions are collected and dried, with the amount of residue determined gravimetrically. Unsaponifiable-matter content can be determined by AOCS Method Ca 6a-40 in fats and oils, Tk 1a-64 in fatty acids, and Da 11-42 in soap. Also, for soap and soap products, see ASTM Standard Method D 460, Sections 36 to 38.

MIU Content of Fats and Oils

The sum of the moisture, insoluble matter, and unsaponifiable matter present in commercial fats and oils is typically part of the material specification, the MIU, and may serve as the basis for a price adjustment if the total exceeds specified limits.

Analysis of Soap and Minor Ingredients

In the following sections, we review the principal test methods used for analyzing soap and soap products. Included here are methods for anhydrous soap, moisture, glycerine, chlorides, free fatty acid or alkalinity, and certain functional ingredients.

Anhydrous-Soap Content

In the anhydrous-soap determination, a precisely weighed soap sample is treated with mineral acid to liberate the fatty acids. The fatty acids are recovered and reacted with a NaOH solution to form soap. The resulting soap is dried and weighed to establish the anhydrous-soap content of the original sample. Please refer to either AOCS Official Method Da 8-48 or ASTM Method D 460, Sections 24 to 25, for details of this procedure.

Moisture Content of Soap

The moisture content of ordinary sodium soap can be reliably determined by the 105°C air-oven method as described in the AOCS Method Da 2a-48 and by the ASTM Method D 460. This method is not reliable in the presence of elevated glycerine and free-fatty acid levels.

Another useful method that is specific for water is the Karl Fischer titration. This method provides for a rapid determination of water content, and is especially useful when water is present in small quantities. It may be used on glycerine, fats and oils, fatty acids, soap, and soap product. Among the automated systems available for a Karl Fischer moisture analysis is the Photovolt Aquatest 2010, manufactured by Photovolt Instruments Inc. (www.photovolt.com), that generates a Karl Fischer reagent on demand based on the sample. Since a Karl Fischer reagent is generated electrolytically, no volumetric measurement of reagent is required, and no standardization of the solution is needed. Instead, the water content of the sample is determined and computed by the instrument based on the equivalence of one coulomb of electricity to 186.53 micrograms of water.

Free-Glycerine Content of Soap

This method determines the free-glycerine content of soap by way of the oxidation of the glycerine with periodic acid.

The sample of soap containing glycerine is mixed with chloroform and glacial acetic acid, and then quantitatively transferred to a volumetric flask. Distilled water is added, and the sample is dissolved with heating, if necessary, to completely dissolve the sample. The flask is filled to volume with distilled water and stoppered, followed by agitation to effect thorough mixing. The water and chloroform layers are then allowed to separate. An aliquot of the aqueous layer containing the glycerine is withdrawn and added to a beaker containing an excess of periodic-acid reagent where the oxidation–reduction reaction is allowed to proceed for 30 minutes. Two blanks are prepared for each batch of samples being analyzed.

After the reaction period has passed, potassium iodide is added to each beaker to liberate the excess iodine. Each sample and blank are then diluted with deionized water and titrated with standardized 0.1 N of a sodium–thiosulfate solution. The titration difference between the blank and the sample is converted with appropriate calculation and reported as a percentage of free glycerine in the soap sample. See AOCS Official Method Da 23-56 and ASTM Standard Method D 460, Sections 82 to 84.

Another useful procedure for the determination of glycerine content of soap is a modification of the sodium metaperiodate method in the AOCS Official Method Ea 6-51. In this procedure, a 40.0g ± 0.001 g sample (or other sample size based on the expected results found in the table in AOCS Method Ea 6-51) is weighed into a 250-mL Erlenmeyer flask. Add 100 mL of deionized water and heat on a steam bath to dissolve. Once dissolved, the solution is acidified to a slight excess with 1:4 sulfuric acid using a methyl-orange indicator. The flask is covered with a watch glass and heated to clarify the fatty acid layer. The solution is filtered through a qualitative filter paper into a 400-mL beaker, retaining the fatty acids on the filter paper. The fatty acids are washed with several small portions of hot deionized water until the washings are neutral to methyl orange. Cool and neutralize to methyl orange with 50% of NaOH very carefully. Adjust the pH of the solution so that it is definitely acid to methyl orange. Add a few glass beads and boil for 5 minutes to expel any CO_2. Boil long enough to reduce the volume to 75 mL. Some glycerine may be volatilized if the solution volume goes below 50 mL. Cool to room temperature. Prepare a blank of distilled water and process with the sample solutions in an identical manner. Buffer the pH meter using pH 7.0 and pH 10.0 buffer solutions. Neutralize the sample and blank to a pH of 8.1 ± 0.1 with NaOH. (The strength of NaOH used depends on the acidity of the sample.) Final adjustment should be made with 0.125 N or weaker of NaOH or with 0.1 N or weaker of H_2SO_4. The volume of the pH-adjusted solution should not be more than 100 mL. Pipet 50 mL of sodium metaperiodate solution to the pH-adjusted samples. Swirl the beaker to ensure thorough mixing, cover with a watch glass, and immediately place in a dark cupboard at room temperature for 30 minutes. Remove from the cupboard, add 10 mL of 50% of an ethylene-glycol solution by graduated cylinder, swirl gently, and allow to stand for 20 minutes. Titrate by using 0.125 N of NaOH to a pH of 8.1 ± 0.1. A dropwise addition of NaOH should be made as the 8.1 pH is approached. The results are calculated as follows, where A = mLs of NaOH used for sample titration and B = mLs of NaOH used for blank titration.

$$\% \text{ Glycerin} = \frac{(A - B)(N)(9.209)}{\text{Sample weight in grams}}$$

(Eq. 14.2)

Chlorides in Soap

The soap sample, usually 5 grams, is dissolved in about 300 mL of chloride-free deionized water, with boiling as needed. The soap is then reacted with an excess of magnesium nitrate to form insoluble magnesium soaps, filtered, and washed with chloride-free deionized water. The filtrate is then titrated with a standardized 0.1 N of a silver-nitrate solution with a potassium-chromate indicator. The result is calculated and usually reported as a percentage of sodium chloride. See AOCS Official Method Da 9-48 and ASTM Standard Method D 460, Sections 53 to 55.

Free Fatty Acid and Free Alkalinity in Soap

The free fatty acid or free alkalinity in soap is determined by titration, with standard alkali or acid as appropriate, to the phenolphthalein endpoint. For most soaps that contain no significant amount of alkaline salts, the procedure involves the titration of the sample, usually 10 or 20 grams dissolved in neutralized ethanol, with either a standardized NaOH solution (0.1 N or 0.25 N) for acidic samples or sulfuric acid (0.1 N or 0.25 N) for alkaline samples to the phenolphthalein end point. The results for alkaline soaps are usually reported as either a percentage of NaOH or Na_2O for sodium soaps and a percentage of KOH or K_2O for potassium soaps. For acidic soaps, the results are typically reported as a percentage of oleic acid, coconut acid, or lauric acid.

Note that the titration is always performed in neutralized ethanol, rather than water, due to the hydrolysis of soap in water that would buffer the solution and interfere with the endpoint determination. See AOCS Official Method Da 4a-48 and ASTM Standard Method D 460, Section 21 for more detail.

Triclocarban and Triclosan in Soap by Ultraviolet Absorbance

Two important and widely used additives found as the active ingredients in many antibacterial soaps and as the functional ingredients in many deodorant soaps are triclocarban and triclosan. The chemical structures of these two compounds make them amenable to quantitation by ultraviolet (UV) absorbance spectroscopy methods. Triclocarban absorbs at 265 nm, and triclosan at 282 nm.

Note that interference may be encountered due to absorbance by fragrance components or other soap ingredients at these same wavelengths. This interference can be nullified by preparing a calibration curve for each formulation of soap product. By preparing stock solutions of the product matrix without the active ingredient and a separate stock solution of the active ingredient, various concentrations of the active ingredient over the concentration range of interest can be prepared for UV-absorbance measurement.

Listed in Table 14.4 are the UV-absorbance values obtained for triclocarban concentrations over a range from 0.0 to 1.0 mg/100 mL in a typical soap formulation. Since the UV absorbance is a direct linear function of the analyte's concentration, the relationship between UV absorbance and concentration can be expressed by an equation in the form of a straight line:

$$A_{CORR} = mC + b$$

(Eq. 14.3)

where A_{CORR} is the UV absorbance and C is the triclocarban (or triclosan) concentration in mg/100 mL. The slope of the line, m, and the y-intercept, b, can be derived by the treatment of the concentration and absorbance data using linear least-squares equations (Arnold & Ford, 1972) as follows:

$$m = \frac{N\sum C_i A_i - (\sum C_i)(\sum A_i)}{N\sum C_i^2 - (\sum C_i)^2} = \frac{(5)(3.228) - (2.8)(4.27)}{(5)(2.16) - (7.84)} = 1.41$$

(Eq. 14.4)

$$b = \frac{(\sum A_i)(\sum C_i^2) - (\sum C_i)(\sum C_i A_i)}{N\sum C_i^2 - (\sum C_i)^2} = \frac{(4.27)(2.16) - (2.8)(3.228)}{(5)(2.16) - (7.84)} = 0.0624$$

(Eq. 14.5)

Thus, the linear equation for the data in this example would be:

$$A_{CORR} = (1.41)(C) + 0.0624$$

(Eq. 14.6)

Upon rearrangement, the equation becomes:

$$C = \frac{A_{CORR} - 0.0624}{1.41}$$

(Eq. 14.7)

The equation obtained in this manner can be applied at any time in the future during a routine analysis of the product of this same formulation. By preparing a test solution of the product and determining its UV absorbance at the specified wavelength, the triclocarban (or triclosan) concentration in the test solution can readily be determined. By dividing the triclocarban (or triclosan) concentration obtained from the soap sample by the soap-sample concentration in the test solution, the analyte's concentration in the product can then be expressed. The y-intercept value, 0.0624 in this example, represents the correction for interference from other formula components such as fragrance in the product matrix. In Fig. 14.2, the equation that was derived from the data in Table 14.4 is represented graphically.

$$A_{corr} = 1.41(C) + 0.0624$$

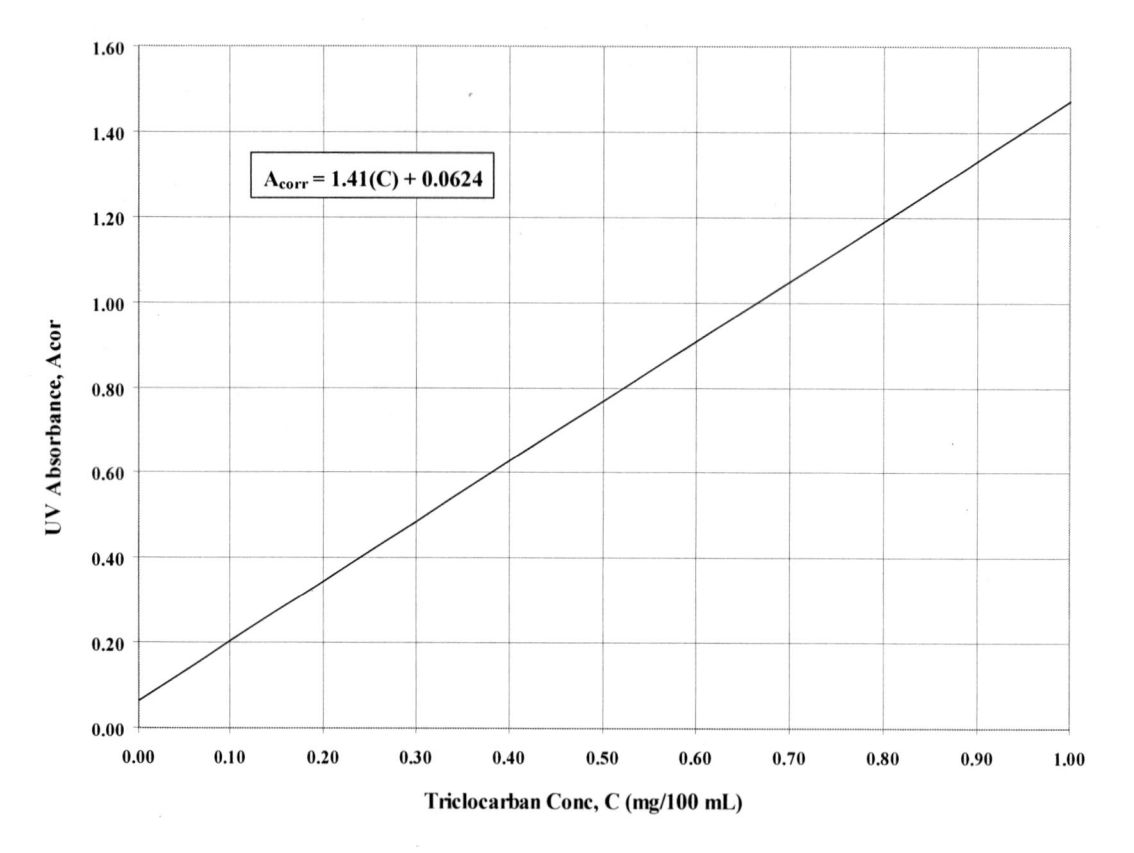

Fig. 14.2. UV Absorbance vs. Triclocarban Conc. for a Typcial Antibacterial Bar Soap

Table 14.4. UV Absorbance vs. Triclocarban Conc. for a Typical Bar Soap

Conc (mg/100 ml)	A265	A320	ACORR
0.0	0.17	0.11	0.06
0.4	0.73	0.12	0.61
0.6	1.03	0.10	0.93
0.8	1.33	0.11	1.22
1.0	1.49	0.14	1.45

Chromatographic Methods

Triglyceride Analysis by HPLC

Some work is reported in the literature for the separation of triglycerides using high-performance liquid chromatography (HPLC). Plattner et al. (1977) performed triglyceride separation by chain length and degree of unsaturation by using a C18 μ-Bondapak column with an acetonitrile–acetone mobile phase and a differential refractometer detector. Waters Associates has a specialty column for triglyceride analysis (Waters, #84346) which is used with a 50:50 acetone–tetrahydrofuran mobile phase (Waters Division of Millipore Corp., 1986). Supelco's reversed-phase "Supelcosil" LC-8 and LC-18 columns, using acetone–acetonitrile (63.6:36.4) mobile phase, are also reported to effect triglyceride separations (Supelco, 1980). Triglyceride composition may be determined using AOCS Official Methods Ce 5b-89 and Ce 5c-93.

Derivatized Fatty Acid Analysis by HPLC

Scholfield (1975) reported the use of a C18/Corasil column and an aqueous acetonitrile mobile phase for the separation of fatty acid methyl esters (FAMEs) by unsaturation and chain length. Work by Warthen (1975) achieved the analytical separation of FAMEs, including *cis* and *trans* isomers, by using a μ-Bondapak C18 column and methanol–water mobile phase.

Work was reported by Jordi (1978) for the separation of the phenacyl and naphthacyl derivatives of fatty acids by using a μ-Bondapak column and the Fatty Acid Analysis column (Waters) with an acetonitrile–water gradient. The HPLC of geometric isomers of the fatty acids of coconut oil and other seed oils was also reported by Wood & Lee (1983). This method also utilized the phenacyl and naphthacyl derivatives chromatographed on a C18 reversed-phase column by using an acetonitrile–water gradient.

Free Fatty Acid Analysis by HPLC

Analysis of free fatty acid, without derivatization, by HPLC was described by King et al. (1982). In this method, a Free Fatty Acid column (Waters) was used with varying combinations of a ternary mobile phase consisting of tetrahydrofuran, acetonitrile, and water at a reduced pH, and employing a refractive-index detector. Using this system, the identification and semi-quantification of fatty acid mixtures derived from industrial oils and alkyd resins are accomplished in about 15 minutes.

Additional work was reported by George (1994)— quantitation of fatty acids from triglycerides and soap without derivatization. The triglycerides are first saponified, and then acidulated to free the fatty acids. The fatty acids are then dissolved in a methanol or a tetrahydrofuran–methanol solution for injection. The mobile phase consists of 45% of acetonitrile, 20% of tetrahydrofuran, 34.5% of water, and 0.5% of glacial acetic acid at a flow rate of 1.1 mL/minute on a Waters fatty acid stainless-steel column. Work was also done by using radial-compressed stainless-steel columns. Soaps are dissolved in methanol, and then injected directly into the HPLC where they are acidulated on the column by the acidic mobile phase. A refractive-index detector is used in conjunction with the isocratic reverse-phase chromatography.

Derivatized Fatty Acid Analysis by Capillary GC

Capillary gas chromatography (GC) on wall-coated open-tubular fused silica columns has greatly enhanced the accuracy and speed of analysis due to the development of technology to attach very polar stationary phases to highly deactivated fused silica. Slover & Lanza (1979) used capillary glass columns coated with SP2340 for quantitation of FAMEs. The columns were used for extended periods, and, with up to 1900 samples, analyzed on a single column during an 11-month period with little deterioration of the column. The separation included the resolution of *cis* and *trans* isomers. Capillary GC analysis of *cis* and *trans* FAMEs is also described in Supelco's literature (1985) in which most of the C_{18} isomers are resolved. In this method, the FAMEs are injected into a GC equipped with a glass split-injection port (split ratio 100:1), SP2560 column (Supelco, #2-4056) with 100 m × 0.25 mm i.d., 0.20 μm film, and a flame-ionization detector (FID). By using an oven temperature of 175°C and helium as the carrier gas, an excellent resolution of all the C_{12} through C_{22} FAMEs is obtained including partial separation of most of the $C_{18:1}$ *cis*- and *trans*-positional isomers, but run times can exceed 60 minutes.

Lanza et al. (1980) reported the use of short glass capillary columns coated with SP2340 for the rapid analysis of fatty acids that provides some resolution of geometric isomers. Sampugna et al. (1982) also reported the rapid analysis of *trans* fatty acids by using an SP2340-coated glass capillary column.

The official AOCS method for FAME preparation (AOCS, 1994a) uses a boron trifluoride–methanol reagent. The official AOCS method for FAME analysis by GC (AOCS, 1994b) also utilizes packed-column technology. A complete analysis of *cis/trans* isomers may be achieved by capillary GC by using AOCS Official Method Ce 1h-05.

Glycerine Content of Soap by HPLC

Glycerine determination in soap base was accomplished by using HPLC as described by George and Acquaro (1982). A Carbohydrate Analysis column (Waters) is used with an acetonitrile–water (92.5:7.5) mobile phase, a 1.0 mL/minute flow rate, with a differential refractometer (Waters, Model 401). Sample analysis time is about 30 minutes, including sample preparation time under the above conditions.

Triclocarban and Triclosan in Soap by HPLC

George et al. (1980) also developed an HPLC method for the determination of triclocarban and triclosan. This method uses a radially compressed C18 column (Waters, Radial Pak A). The sample preparation involves using C18 solid-phase-extraction cartridges to remove most of the soap and fragrance components from the sample. The mobile phase for the separation consists of a tetrahydrofuran/water mixture (58:42) at a flow rate of 2 mL/minute with a UV detector at 280 nm. Sample preparation time is reported to be 15 minutes with a 15-minute analysis time.

Chelating Agents in Soap by HPLC

Identification and quantification of aminocarboxylate chelating agents used in bar soaps— such as ethylenediaminetetraacetic acid (EDTA), *N*-hydroxyethylenediaminetriacetic acid (HEDTA), and diethylenetriamine pentaacetic acid (DTPA) by HPLC—are described by Goldstein and Lok (1988). This method uses cupric sulfate to precipitate the soap and to form a water-soluble aminocarboxylate—copper complex. The copper complex is then isolated and chromatographed on an anion-exchange column (Wescan, Anion R Analytical) with a mobile phase consisting of 0.003 M of sulfuric acid at a flow rate of 1.5 mL/minute. A UV detector was used at a wavelength of 254 nm. Under the above conditions, the retention times for the copper complexes were 3.5 minutes for HEDTA, 6.2 minutes for DTPA, and 6.9 minutes for EDTA.

BHT in Soap by Capillary GC

A method for the determination of 2,6-di-*tert*-butyl-methylphenol (BHT) in soap products is offered by Goldstein et al. (1982). The authors present a method that can quantitate BHT in fragranced bar soap without employing the cumbersome standard addition method that was previously reported (Sedea & Toninelli, 1981). After blending the sample with dimethylformamide and adding 2,4-di-*tert*-butylphenol (DTBP) as an internal standard, *bis*-trimethylsilyltrifluoroacetamide (BSTFA) is added to a filtered aliquot to convert the BHT and DTBP to their silyl derivatives. The sample is then introduced into a GC equipped with a glass split-injection port (split ratio 200:1), methyl silicone column (Hewlett-Packard, #19091-60010) with a 12 m × 0.2 mm i.d., and an FID. By using helium as the carrier gas, the column temperature is held at 100°C for 2 minutes and then programmed up to 144°C at 2°C/minute. The program rate is then changed to 30°C/minute up to 240°C and held for 5 minutes. Under the above conditions, the reported retention times of the silylated derivatives of DTBP and BHT were 15.5 minutes and 22.4 minutes, respectively. One can also apply this method to neat soaps, pellets, and fatty acids as well.

Evaluation of Soap Color and Translucency

Visual Color Comparisons

Bar-soap color evaluation was traditionally performed by a visual comparison of fresh product against various standards such as paper or plastic color chips and retained standard product samples. Visual color comparisons are prone to be subjective, and individual differences in color perception and background-lighting conditions can affect them. All physical standards, including both color chips and retained product standards, must be carefully handled and protected to avoid fading. All such standards will change with time and must be periodically updated.

For visual color comparisons, a controlled lighting environment is a necessity. The use of a standardized light source and a working environment, such as that which a Pantone Color Viewing Light can provide, can be helpful in providing such standardized conditions. The Pantone Color Viewing Light is available from Pantone, Inc. (www.pantone.com). The Pantone Color Viewing Light provides three different light sources—including artificial daylight, fluorescent store light, and incandescent home light—that one can use independently or in combination. The interior of the cabinet has a matte neutral-gray finish that minimizes reflection and glare.

Soap Color by Hunter Reflectance Color Measurement

The instrumental approach to color evaluation can provide many advantages over visual comparison techniques. The instrument in common use by many soap manufacturers is the Hunter D25LT Colorimeter that is manufactured, sold, and serviced by Hunter Associates Laboratory, Inc. (www.hunterlab.com). Among the advantages of the instrumental approach are the objective basis for color comparison to standard, a defined target that is not subject to deterioration, and quantifiable data that can be retained indefinitely. Variables such as light source, background lighting, and differences in individual color perception are eliminated. Also, the nature of the instrumental approach easily lends itself to interfactory and intercompany color control.

The Hunter Opponent Color Scale (Hunter & Harold, 1987a) expresses color in terms of three values: "*L*" expresses "lightness," and ranges from 0 to 100 units; "*a*" expresses "redness" when positive, "gray" when zero, and "greenness" when negative; and "*b*" expresses "yellowness" when positive, "gray" when zero, and "blueness" when negative. The Hunter *L*, *a*, and *b* Color Solid is illustrated in Fig. 14.3.

The *L*, *a*, and *b* values are determined from fresh-product samples that were cut to a planar surface. The values obtained for each sample are then compared with previously established target values. The differences between sample and target *L*, *a*, and *b* values can then be interpreted individually, or they can be resolved into the total color difference (Hunter & Harold, 1987b) or delta-*E* (ΔE) value that serves as a very useful tool for color control at the production level. The total color difference or ΔE is expressed by the following equation:

$$\Delta E = \sqrt{\Delta L^2 + \Delta a^2 + \Delta b^2}$$

(Eq. 14.8)

This equation resolves the three component differences, ΔL, Δa and Δb, into the direct difference, in Hunter units, between the target and the sample. The ΔE serves as a very useful tool for routine quality control of bar-soap color. For example, typical ΔE limits might allow:

- Product with a ΔE below 2.0 units to be packed without qualification,
- Product with a ΔE between 2.0 and 4.0 units to be packed with an ongoing effort to bring the product back below 2.0 units, and
- Product with ΔE above 4.0 units not to be packed.

The Hunter colorimeter can also measure neat-soap whiteness. The color can be reported in terms of L, a and b values, or can be expressed in terms of the Whiteness Index (Hunter & Harold, 1987c). The perception of whiteness, in terms of the Hunter Opponent Color Scale values, is favored by a higher L value (lightness) and a more negative b value (higher degree of blueness). The Hunter Whiteness Index is expressed as:

$$WI = L - 3b.$$

(Eq. 14.9)

The human eye can consistently perceive the differences of 1.5 units of Hunter Whiteness Index (Appleby & Halloran, 1990).

In performing a color evaluation on neat soap, one must wait until the neat soap has thoroughly cooled to room temperature before measuring the whiteness. The sample needs to be cut to a flat surface before performing the Hunter readings. As recommended, one must read the sample six times, turning the sample about 60 degrees for each reading. One should then report the average of the six readings.

In a similar manner, the color of soap to be commercially made from fatty acid blends can be reliably estimated by quantitatively preparing a sample of sodium soap in the laboratory. For an 80:20 tallow–coconut oil blend, for example, 130 grams of fatty acid would be melted and weighed. A solution of NaOH would be prepared in a 600-mL beaker. The fatty acid would be added slowly with constant mixing to the caustic solution until the soap is formed. Soap would then be packed into a 125-mm Petri dish, and let stand for about an hour to cool before covering. The soap would then set until the following day when the color would be evaluated. If the color is to be measured with the Hunter colorimeter, six readings should be taken and averaged. This procedure is not very practical as a routine quality-control procedure, but can be very useful in qualifying a vendor's material.

Soap Translucency by Hunter Opacity Measurement

This technique has proven to be useful in quantifying the translucency of bar soap using the Hunter D25LT Colorimeter. The method is an adaptation of the "contrast ratio method" for measuring opacity (Hunter & Harold, 1987d). The procedure for performing the opacity reading is outlined in the Hunter instruction manual that accompanies those instrument models, such as the D25LT, that are designed to perform this measurement.

The application of this procedure for measuring the translucency of bar soap involves preparing the soap sample for reading by cutting a flat slice to a uniform thickness of .25 inch. Then the standard routine for opacity measurement, as outlined in the Hunter manual, is performed. This procedure involves calibrating the instrument, followed by two simple operations with the meter. In the first step, the translucent slab is backed by the uncalibrated white tile to achieve maximal reflectance through the sample, followed by the second step where the sample is backed up by the black tile for minimal reflectance through the sample. The processor in the instrument then converts this data into an opacity value. The opacity value provides an inverse indicator of translucency. A reasonable scale of values for interpreting these results might rate an opacity value below 10% as "excellent," between 10 and 25%

as "good," between 25 and 40% as "fair," between 40 and 50% as "poor," and greater than 50% as essentially a nontranslucent product. Action steps in the plant can then be set for these various grades.

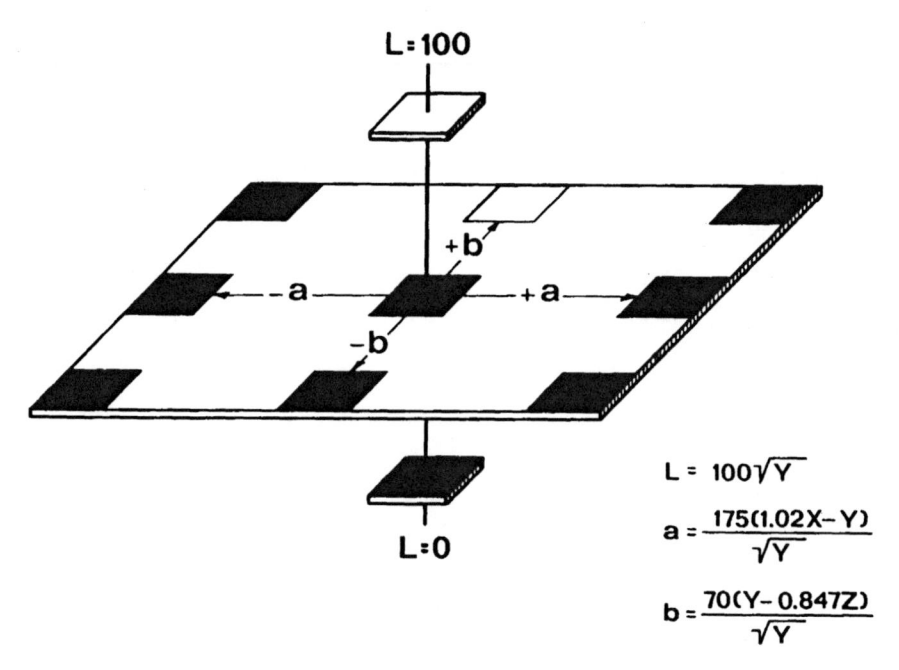

$$L = 100\sqrt{Y}$$

$$a = \frac{175(1.02X - Y)}{\sqrt{Y}}$$

$$b = \frac{70(Y - 0.847Z)}{\sqrt{Y}}$$

Fig. 14.3. Hunter *L, a, b* Color solid.

Foreign Particulate Matter in Soap

A convenient method for quantifying the presence of any particulate foreign matter in the soap base or finished bars utilizes a Model J Bulk Tank Sediment Tester (used in the dairy industry) which is available from the Clark Dairy Equipment Company of Greenwood, Indiana.

The procedure involves finely dividing 100 grams of sample and dissolving them in 1 to 2 L of hot deionized water on a steam bath with agitation. The resulting soap solution is then filtered through a specially designed 3.5-cm filter disk that is designed to be used with the sediment-tester device. The sediment tester is mounted on a large filter flask that is connected to an aspirator for suction. The particulate matter trapped on the filter disk can then be visually compared with a standard set of photographs to provide a grade for the sample.

Acknowledgments

The author was formerly Vice President/Director of Technical Services at Valley Products Company, Memphis, TN, USA, and The Hewitt Soap Company, Inc., Dayton, OH, USA.

References

American Oil Chemists' Society (AOCS). *Official Methods and Recommended Practices of the American Oil Chemists' Society*, 6th ed.; R.C. Walker, Ed.; Champaign, IL, 2009a; Method Ce 2-66.

Ibid., 2009b; Method Ce 1-62.

Ibid., 6th ed.; 2009c.

American Society for Testing and Materials (ASTM). *Annual Book of ASTM Standards;* : West Conshohocken, PA, 2008; Vol. 15.04.

Appleby, D.B.; K.A. Halloran. *Soap Technology for the 1990s;* L. Spitz, Ed.; American Oil Chemists' Society: Champaign, IL, 1990; p. 104.

Arnold, J.G.; R.A. Ford. *The Chemist's Companion: A Handbook of Practical Data, Techniques, and References;* John Wiley & Sons: New York, 1972; p. 188.

George, E.D. *J. Am. Oil Chem. Soc.* **1994**, *71,* 789.

George, E.D.; J.A. Acquaro. *J. Liq. Chromatogr.* **1982**, *5,* 927.

George, E.D.; E.J. Hillier; S. Krishnan. *J. Am. Oil Chem. Soc.* **1980**, *57,* 131.

Goldstein, M.M.; W.P. Lok. Ibid. **1988**, *65,*1350.

Goldstein, M.M.; K. Molever; W.P. Lok. Ibid. **1982**, *59,* 579.

Grompone, M.A. Ibid. **1984**, *61,* 788.

Hunter, R.S.; R.W. Harold. *The Measurement of Appearance,* 2nd ed.; John Wiley & Sons: New York, 1987a; pp. 173–174.

Ibid., 1987b; pp. 174–175.

Ibid., 1987c; pp. 206–207.

Ibid., 1987d; p. 90.

Jordi, H.C. *J. Liq. Chromatogr.* **1978**, *1,* 215.

King, J.W.; E.C. Adams; B.A. Bidlingmeyer. Ibid. **1982**, *5,* 275.

Lanza, E.; J. Zyren; H.T. Slover. *J. Agric. Food Chem.* **1980**, *28,* 1182.

Plattner, R.D.; G.F. Spencer; R. Kleiman. *J. Am. Oil Chem. Soc.* **1977**, *54,*511.

Sampugna, J.; L.A. Pallansch; M.G. Enig; M. Keeney. *J. Chromatogr.* **1982**, *249,* 245.

Scholfield, C.R. *J. Am. Oil Chem. Soc.* **1975**, *52,* 36.

Sedea, L.; G. Toninelli. *J. Chromatogr. Sci.* **1981**, *19,* 290.

Slover, H.T.; E. Lanza. *J. Am. Oil Chem. Soc.* **1979**, *56,* 933.

Supelco, Inc. *One-Step Triglyceride Separation;* HPLC Bulletin 787B, Bellefonte, PA, 1980.

Supelco, Inc. *Capillary Analyses of Positional Cis/Trans Fatty Acid Methyl Ester Isomers;* GC Bulletin 822, Bellefonte, PA, 1985.

The Procter & Gamble Company. *Better Rendering;* Cincinnati, OH, 1967; p. 17.

Unichema Chemicals, Inc. *Fatty Acid Data Book,* 2nd ed.; Chicago, IL, 1987; pp. 4–5.

Warthen, J.D. Ibid. **1975**, *52,* 151.

Waters Division of Millipore Corp. *Waters Sourcebook for Chromatography Columns and Supplies;* Milford, PA, 1986; p. 43.

Wood, R.; T. Lee. *J. Chromatogr.* **1983**, *254,* 237.

·• 15 •·

Soap Bar Performance Evaluation Methods

Yury Yarovoy and Albert J. Post
Unilever Research, Trumbull, Connecticut, USA

This chapter surveys the assessment procedures that are practiced in soap bar factories or in development laboratories around the world to ensure that manufacturing standards are met and, ultimately, that consumers will be satisfied with the cleansing bars they purchase. In principle, the appraisal methods include analytical chemistry tests to evaluate raw materials and finished product, methods of materials science to assess product structure, and straightforward tests of the consumer-observable attributes like lather, wet-bar feel, and wear rate, either in the laboratory or by consumer panel assessment. The main focus in this chapter concerns the finished product, although characterization (rheology) of partially manufactured soap is also considered.

Wood has provided a thorough review of many of the common practices of soap manufacturers and of producers of fats and fatty acids for soap makers (Wood, 1990, 1996). Since these methods remain essentially the same, some replication was inevitable.

When discussing evaluation procedures and quality control for soap bars, it should be noted that formal governmental standards for product performance and for methods of testing soap bars exist only in a few countries, namely India, Bangladesh, and Kenya. In these countries, soap is included in the list of items covered by a mandatory certification scheme, which had been introduced with the intention of protecting consumers against substandard, low-quality products on the market. In the rest of the world, consumer acceptance is the only criterion for judging the performance of marketed bars and their commercial success. Consequently, a variety of evaluation procedures are used and have been described in the literature, and each company adopts techniques that meet its particular quality requirements. In the following sections, we survey some of the more common procedures and, where applicable, include for comparison the standard tests of bar performance mandated by the Bureau of Standards in one of the above-mentioned three countries where such standards exist.

Lather Evaluation

An adequate instrumental evaluation of the lather or foaming properties of soaps is not simple, due to the complex nature of the lathering process. Many lather attributes, such as speed of generation, volume, bubble size, and stability, contribute to the overall consumer perception of lather, and these attributes are also greatly affected by the manner in which the lather is generated. Numerous evaluation techniques and devices to generate lather have been described in the literature to date. One of the techniques, proposed by J. Ross and G. Miles in 1941 (Ross & Miles, 1941) and now commonly referred to as the Ross-Miles foam test, measures the height of foam developed by pouring. It eventually became a standard test method for foaming properties of surface active agents (ASTM International, 2007). Though quite suitable for detergent solutions, this method does not differentiate well between soap bar products or provide a reliable correlation with the consumer perception of lather; consequently, many soap manufacturers developed a variety of other evaluation techniques.

The amount of lather generated by a soap bar is an important parameter affecting consumer acceptance and preference. According to the Bureau of Indian Standards, the lather requirement is intended to ensure the presence of an adequate level of surfactants in the bar for cleaning. The Indian Standards prescribe the following procedure for the determination of lather (Bureau of Indian Standards, 1983, 1992): a sample of 5 g uniformly grated (to about 0.5 to 2 mm size) soap is added to 100 ml of 300 ppm hard water in the blending jar of a kitchen food blender. The jar is covered and the blender is set on low speed for exactly 60 seconds. Then the lather is poured quickly into a graduated cylinder and measured immediately after leveling off the top surface of the foam. This method must be calibrated (by adjusting the speed of the blender) so that when the blender is operated for 60 seconds, it delivers 600±100 ml of lather from 100 ml of a 1% sodium lauryl sulfate solution at an ambient temperature of 27±2°C. The measurements should be performed on three samples with three duplicates, and the result is reported as the mean x (the sum of test results divided by the number of test results) and the range r (the difference between the maximum and the minimum values of the test results). In India, bars are deemed to conform to the standard requirements of a particular grade if the expression (x − $0.6r$) is greater than or equal to 280 ml for Grade 1, 240 ml for Grade 2, and 200 ml for Grade 3, respectively.

Another common lather evaluation technique often described in the patent literature is the conventional sudsing test (Wood, 1990), also referred to as a cylinder test, which is conducted by dissolving a specified amount of soap in de-ionized water, placing an aliquot of this solution into a graduated cylinder with a stopper, and shaking the cylinder. The net volume of the foam can be read directly, and also the stability of the foam can be measured by monitoring the foam height as a function of time. Typical specifications for this procedure are as follows: the soap solution (suspension) concentration is 1% wt.; the cylinder volume is 250 ml, wherein 50 ml of the solution is placed; the cylinder shaking is performed by inversion via rotating cylinders at several revolutions per minute for a total of about 30 inversions. A variety of modifications of this test, with adjustments to meet particular requirements, may be used. Figure 15.1 shows a simplified version of a cylinder test, conducted by hand-shaking 100-ml cylinders filled with 10 ml 1% solution of two different soap bars. This provides a direct comparison of the foam volume (height), and one can also monitor the rate of foam drain by measuring the volume of the drained liquor at the bottom of the cylinder.

Fig. 15.1. Foam generated in a cylinder hand-shake test from a syndet (left) and a regular soap (right) bars: (a) initial volume; (b) after 2 hours.

Recently, a new automated mechanical foam tester, SITA Foam Tester R2000, has been introduced to the market (SITA Messtechnik GmbH, 2008), which offers easy and repeatable testing and monitoring of the foaming characteristics for foaming *liquids*. It provides fully automated control of the test algorithm (measuring, cleaning, and refilling), reproducible test results, various settings for the test parameters, and automatic self-cleaning using tap water after completing an experiment.

A common feature of all tests described above is that a prepared solution or suspension of a soap bar is used to generate lather. The step of manipulating a soap bar in the hands (at the sink or in the shower), which is an essential factor in the process of soap dissolution and lather generation, is missing, and, therefore, these tests seem to be less suitable for bars than for liquid detergents. Consequently, in order to better simulate the in-use conditions, and achieve a better correlation of in-lab appraisals with the consumer perception of lather in the shower, a number of tests have been developed involving generation of lather in the hands by a trained technician.

In order to obtain an objective comparison of different soap formulations (and also good reproducibility and consistency between different appraisers), the lather tests are conducted using a standardized protocol under a set of strict conditions. A subjective assessment of lather creaminess is made by a technician during the generation of the lather.

The lather appraisal procedure is typically as follows (Yarovoy, 2003):

- Tablet pretreatment: Wearing disposable gloves, worn inside out and well-washed in plain soap, wash down all test tablets for at least 1 minute before starting the test sequence. This is best done by twisting them about 20 times under running water.
- Place about 5 liters of water of known hardness and at a specified temperature (typically 20–40°C) in a bowl. Change the water after each bar of soap has been tested.
- Take up the tablet, dip it in the water and remove it. Rotate the tablet 15 times between the hands. Replace the tablet on the soap dish.

The lather is generated from the soap remaining on the gloves:

- Stage 1: Rub the tips of the fingers of one hand (either hand) on the palm of the other hand 10 times.
- State 2: Grip the right hand with the left, or vice versa, and force the lather to the tips of the fingers. Repeat with the hands reversed. This operation is repeated five times with each hand.

Repeat Stages 1 and 2.

- Place the lather in the calibrated beaker.
- Repeat the whole procedure of lather generation twice more, combining all the lather in the beaker.
- Stir the combined lather gently to release large pockets of air. Read and record the volume.

Another common evaluation technique involving generation of lather in the hands is usually referred to as an inverted-funnel method (Farrell & Nunn, 2005). It requires a large measuring funnel and two large sinks. The measuring funnel is constructed by fitting a 10½-inch-diameter plastic funnel to a graduated cylinder that has had the bottom cleanly removed. The graduated cylinder should be at least 100 cc. The fit between the funnel and the graduated cylinder should be snug and secure. Before evaluations proceed, place the measuring funnel into one of the sinks and fill the sink with water until the 0 cc mark is reached on the graduated cylinder. The procedure involves the following steps, some of which are illustrated in Fig. 15.2:

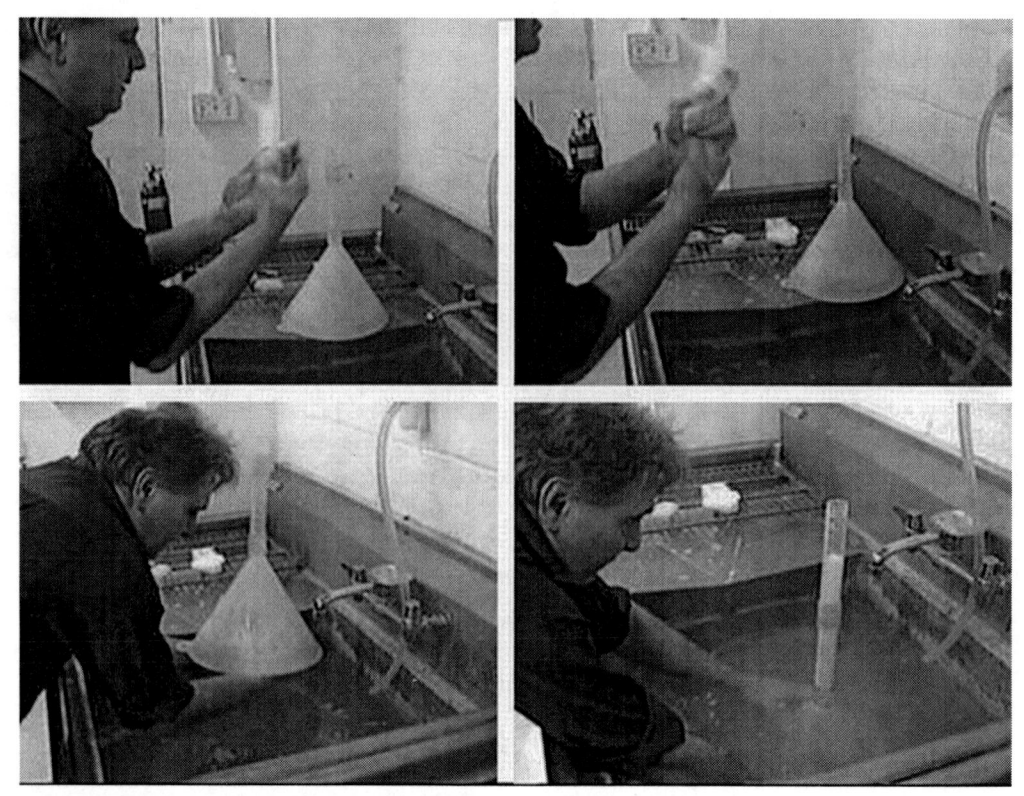

Fig. 15.2. Foam appraised by a technician using an inverted-funnel test.

- Run the faucet in the second sink and set the temperature to 95°F (35°C).
- Holding the bar between both hands under running water, rotate the bar for ten (10) half-turns.
- Remove hands and bar from under the running water.
- Rotate the bar fifteen (15) half-turns and lay the bar aside.
- Work up lather for ten (10) seconds.
- Place funnel over hands.
- Lower hands and funnel into the first sink.
- Once hands are fully immersed, slide out from under funnel.
- Lower the funnel to the bottom of the sink.
- Read the lather volume.
- Remove the funnel with lather from the first sink and rinse in the second sink.

The test should be performed on several bars, and the volume should be reported as an average.

When the goal of an appraisal is to compare lather of a new bar prototype with the control, a paired-comparison protocol is used for the data analysis, which involves calculating a value of least significant difference (LSD). Typically, six results for each bar are averaged, and paired comparisons are carried out between the averaged results for each bar. If the lather volume differs by more than the LSD, then the products are said to produce "significantly different amounts of lather."

Rate of Wear and Mush

After lather, the rate of wear (ROW) is one of the most important soap bar properties, especially in developing countries, where a purchase decision is often strongly influenced by a soap bar economy or a value-per-dollar consideration. A laboratory evaluation of the ROW is a relatively straightforward test performed by washing a bar multiple times and measuring the bar weight loss after the final wash. The ROW is usually reported as the bar mass loss (in grams) per wash. When comparing the ROW of bars having different weight, the ROW values should be normalized to a 100 g bar weight, or expressed as a percentage of the bar weight. The ROW often correlates strongly with mush—high mush scores usually signal a high ROW.

Even though this procedure appears to be straightforward, a variety of nuances have been introduced in order to better correlate the laboratory data with consumer perceptions. This is illustrated below by three test descriptions, randomly selected from the patent literature, which are presented in the order of increasing complexity.

Version 1. Wet a pre-weighed bar in running water and rotate 15 times while in the tester's hands. Place on a support stand. Repeat 10 times at half-hour intervals. Weigh the washed bar after allowing to dry at room temperature for 16 hours. The weight change multiplied by 100 and divided by the initial weight of the bar denotes the percent rate of wear of the bar (Abbas & Hui, 2004).

Version 2. The evaluation is carried out over a 4-day period in order to simulate at-home usage (Subramanyan et al., 1996). The initial weights of the bars are recorded. A few different individuals wash the bars for 10-second intervals in warm tap water, 32–38°C. The soap bars are placed in a soap dish with a grid to allow drainage of water. The bars are allowed to dry for at least a 30-minute interval between washings. The soap bars undergo a total of 20 washes of 10-second duration, and then are dried for 24 hours prior to reweighing. The results are reported as a weight loss per 100-gram bar per one use.

Version 3. In this test, water is added to a soap dish to induce the formation of mush. Weigh the bar to be tested. Set up an 8-liter bucket with continuous water running through it at 40.5°C. Immerse the bar into water, remove, and then rotate it in the hands 20 times. Repeat. Immerse the bar again to remove adhering lather, place in a dish, and dry in the air at 25°C and approximately 50% RH. Repeat every two hours over an 8-hour span. Let dry for 12 hours at 25°C and approximately 50% RH in a dish, and repeat for another 8-hour span. Add 10 g of de-ionized water to the dish between immersions and while the bar is resting in the dish. This should be additive over the 8-hour span: after the first 2 hours, 10 g water are added to the dish; after 4 hours, add 10 g more, totaling 20 g water in the dish; after 6 hours, add 10 g more, totaling 30 g water in the dish; and so on, for the 8-hour period. Then dry for 12 hours. The weight of the bar after 12 hours is recorded, and the wear rate is the percent weight loss of the bar (Brennan et al., 2008).

Closely related to the rate of wear is the propensity of most soap bars to form a soft, loose layer on the surface when in contact with water for a prolonged period of time (several hours). During subsequent use, this layer is readily washed down, contributing to both the generation of lather and the bar wear. In the patent literature, one finds different terminology used to describe this phenomenon. For example, Cussons and Unilever denote it as "mush" and "mushing," Procter & Gamble uses the term "smear," and Colgate refers to it as "slough." Excessive mush is disadvantageous because it represents a loss of soap from the bar and typically leads to greater rates of wear.

There are two somewhat different ways to generate and appraise mush in a laboratory environment: (I) soaking a bar in a dish and (II) immersing a suspended bar in water.

Procedure I. In the first procedure, a bar is placed in a small dish, 30 g of water are added, and the bar is soaked for 24 hours. Then the mush layer is gently scraped with a blunt blade. The weight W_m of the mush layer is measured and divided by the initial weight W_i of the bar prior to soaking to obtain a mush weight fraction, $x_m = W_m/W_i$. The final weight of the bar W_f after the mush layer has been scraped off is also measured. The water uptake weight fraction x_u can be calculated as $x_u = (W_m + W_f - W_i)/W_i$. Three bar samples of a formulation are typically evaluated in this manner, and the average x_m and x_u are reported (Post et al., 1998).

Procedure II. In the second procedure, a bar is shaved to the dimensions 7 cm × 4 cm × 2 cm, and a line is carved halfway down the center of the bar (at the 3.5 cm mark). Then half of the bar (to the line) is suspended in de-ionized water for 2 hours at a temperature of 25°C. After this time, the bar is removed, water is drained for 30 seconds, and the bar is weighed. This is the weight of the bar, the mush, and the absorbed water. After weighing, the mush is scraped from the bar, and the bar is dried for 12 hours. The difference in weight between the initial dry bar and the final dry bar, calculated for the 50 cm² bar surface area, is the amount of mush (grams). The difference in weight of the soaked bar and the initial dry bar is the amount of water absorbed (Brennan et al., 2008).

A variety of modifications of these techniques exist. For example, the soft layer may be removed by hand rather than spatula, the soaking time (in dishes) decreased to 17 hours, and the temperature of soaking increased to 35–40°C (Subramanyan et al., 1996; Colwell & Pflug, 1991). The weight loss due to mush formation in this test is reported as the loss per 100 grams. The Kenya Bureau of Standards recommends a modified version of the first procedure (tested bars are placed on wet fabric instead of in dishes with water), whereas the second procedure is adopted in India as the standard.

The mush (or smear) can be also graded using a subjective scale by a trained technician who grades soap bar smear by fingering the bar and taking into account both types of smear and amount of smear. Such a test is typically carried out as follows: (1) place a soap bar on a perch in a 1400-mm-diameter circular dish; (2) add 200 ml of room-temperature water to the dish such that the bottom 3 mm of the bar is submerged in water; (3) let the bar soak overnight (15 hours); (4) turn the bar over and grade qualitatively for the combined amount, characteristics, and depth of smear on a 1 to 10 scale, where 10 equals no smear, 8.0–9.5 equals low smear amount, 5.0–7.5 equals moderate smear similar to most marketed bars, and 4.5 or less equals very poor smear (Kacher, 1995).

Wet Cracking

Wet cracking, or cracking of soap bars during repeated usage, is a well-documented phenomenon that occurs when faults in the macro- and microstructure are stressed during bar/water interactions. Many factors affect the development of cracking: the composition of the soap, the efficiency of the soap finishing line (the degree of compaction during extrusion), and, last but not the least, the bar shape.

A generic procedure for evaluating wet cracking comprises the following steps: (i) soaking a bar by full immersion in water for a prescribed period of time at a constant temperature, (ii) draining the water and drying for 16–24 hours at the soaking temperature, (iii) visual evaluation by comparing to a set of photographs or illustrations (Marchesani, 1979). Figure 15.3 illustrates a possible set of photographs for rating both face and end cracking on a six-point scale (from 0 to 5) (Geoffrey, 2005), with zero being no cracking and five representing very severe and unacceptable cracking.

Various modifications of this generic procedure exist. For example, more recent Colgate patents (Colwell & Pflug, 1991) provide the following details for the wet cracking test: soap bars were suspended in tap water at room temperature (24°C) for 4 hours, then allowed to dry for 24 hours prior to being evaluated. Any resulting cracks on the bar surface were rated numerically using a scale of 0 (none) to 5 (severe), and then summed. For example, a soap bar having 5 cracks of severity 1, plus 2 cracks of severity 4, has a total rating of 13. A total rating of more than 25 is considered unacceptable. Shorter immersion times, such as ½ hour or 1 hour, may be also used, and the resulting cracking may be reported on a scale from 0 to 5 (Sonenstein, 1981).

The number of cycles of wetting and drying may also be changed, and a new step, rinsing, may be introduced to simulate in-use conditions more closely. For instance, Hyeon (Hyeon, 2005) describes a test in which a bar was dipped in tap water at 30°C for 2 hours. After soaking for 2 hours, the swelling part was removed from the soaps and slightly washed with flowing cold water. Then, the bar was dried at room temperature for one hour. After this sequence was repeated 3 times, the bar was dried for 24 hours in a thermostat at 30°C and rated for cracking.

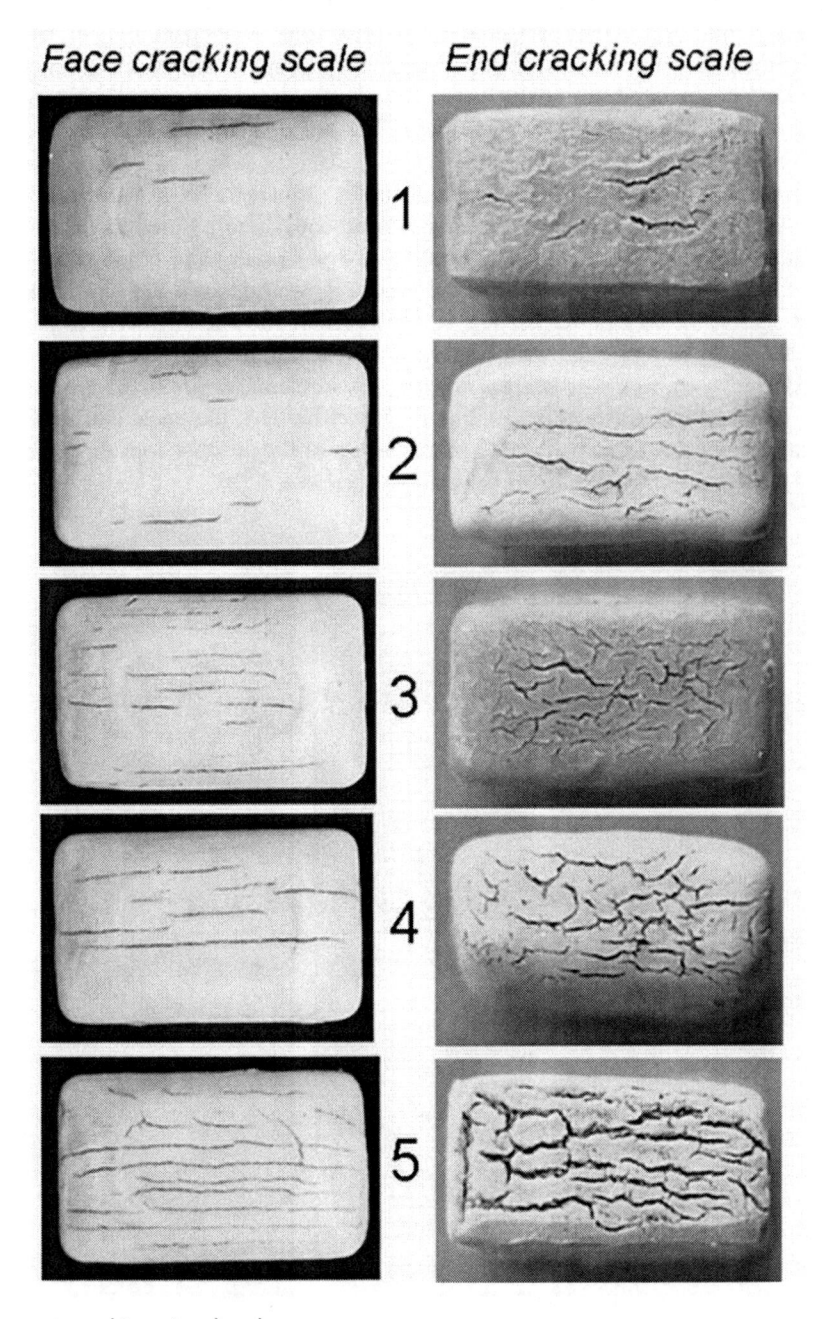

Fig. 15.3. Bar wet-cracking visual scale.

Wet Bar Feel

Wet-bar feel is what the consumer would experience during use of the bar at the sink or in the shower. A bar washdown test is intended to uncover any defects on the bar surface that would feel like grit, sand, or roughness. The average consumer is quite sensitive to bar surface quality and reacts negatively to even minor defects. The wet-bar feel test, or determination of grittiness, is a mandatory test that bars should pass to be certified by the Indian Bureau of Standards, which mandates the following test procedure. Hold the bar under running water at 30°C and rub the two sides of the bar gently on the palm for one minute. The bar shall show no rough surface and shall feel smooth to the touch. Allow this bar to dry in the open air for 4 hours and examine the surface. Bars pass the test if no gritty particles are visible on the surface. Figure 15.4 illustrates the visual grades of bar surface grittiness that can be appraised on a scale from 0–5.

Because the wet-bar feel is a tactile (rather than visual) property, a set of bars molded from plastic having various levels of surface grittiness and/or sandiness could also be used to calibrate the grading scale by touch. Such a reference scale is useful in product development and factory quality control.

Another wet-bar property that could be evaluated during washdown is bar slip (alternatively called drag). In materials science terms, wet-bar feel is a measure of the friction between the product and the skin, and this could be evaluated with friction testing equipment already used for skin tribology (Sivamani, 2003). However, we are unaware of any practical implementation of this type of measurement involving soap bars, and it is reasonable to assume that wet-bar feel, like most consumer traits, is best assessed by asking consumers directly if they find this aspect of the product appealing. Consumer panel assessments are discussed in the User Panel Evaluations section.

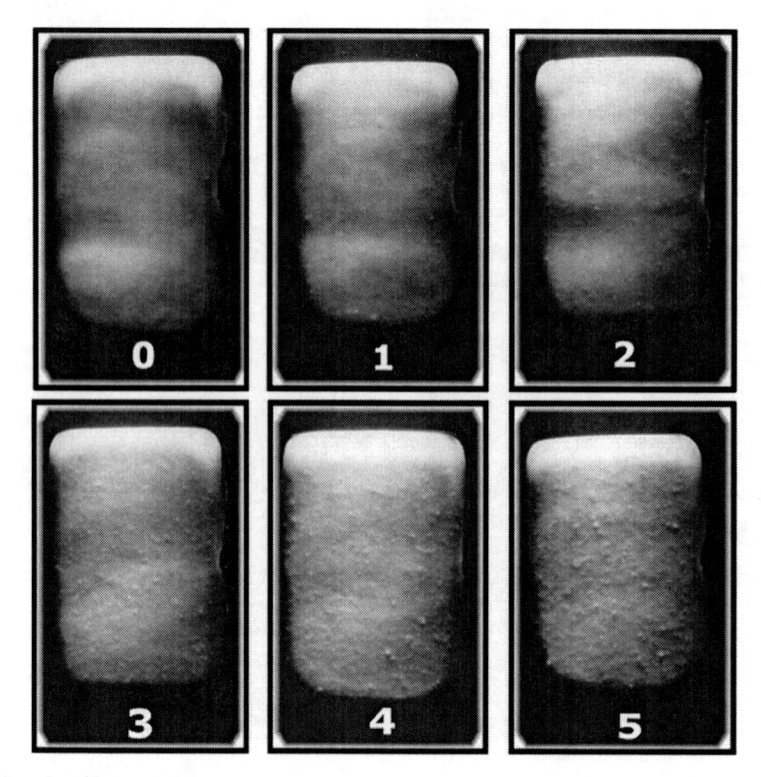

Fig. 15.4. Visual scale of bar grittiness.

Mildness to Skin

It is well known that frequent use of a regular soap (as well as other cleansing products) may result in increased skin dryness and tautness, especially when applied to the face during the cold and dry season and in individuals with dry, sensitive skin. The full potential of a cleansing product to dry and irritate the skin can be assessed in vivo in various clinical tests and in vitro using a number of laboratory assays. Since the clinical tests are usually expensive, it is a good strategy to initially screen the products using in vitro methods.

One of the most common mildness assays is the zein test, which evaluates the effect of soap on proteins by measuring the solubility of the water-insoluble corn protein, zein, in a solution of the cleansing base. Goette(Goette, 1967) and Schwuger (Schwuger, 1969) have shown that a surfactant's ability to solubilize zein correlates well with the surfactant's irritation potential. The lower the zein score, the milder the product is considered to be. A typical procedure involves measuring the percent of dissolved zein as follows: (i) prepare 30 g of 1% cleansing base solution; (ii) add 1.5 g zein and mix for 1 hour; (iii) centrifuge for 30 minutes at 3000 rpm; (iv) extract the pellet, wash with water, and dry in vacuum for 24 hours or to a constant weight; (v) measure the weight of the dry pellet. The percent of zein solubilized is calculated using the following equation (Post et al., 1998; Fujiwara, 1999):

$$\% \text{ zein solubilized} = 100 \times [1 - (\text{weight of dried pellet} / 1.5)]$$

Final assessment of the mildness of the soap may be conducted using the forearm controlled application technique (FCAT), which is an industry standard method for estimating the relative irritation potential of personal cleansers (Ertel, 1995). This test is usually conducted by independent, certified laboratories. The FCAT uses an exposure protocol that is based on consumer washing habits and a number of instrumental techniques (such as skin conductivity and transepidermal water loss), coupled with visual observations of dryness and redness (erythema), in order to assess the condition of the skin before, during, and after washing.

A detailed review of the effect of cleansers on skin has recently appeared in a special journal issue (*Dermatologic Therapy*, 2004).

Fragrance

Although perfume does not contribute to the cleansing function of soap, it is an important factor in attracting consumers, reinforcing a product image, and reinforcing subjective judgments of product performance (Ho, 2000). One important function of perfume is to mask the base odor that comes from the raw materials. This function can be assessed only by a human expert panel (usually in fragrance houses). Another function is to provide a pleasant olfactory experience during and after soap use. This is closely linked to the release of perfume from soap during the shower, perfume retention in a soap bar over time, and the perfume's ability to deposit on the skin. All of these can be evaluated using headspace analysis, in which vapors are trapped and then analyzed using conventional gas chromatography–mass spectrometry (GC-MS) analysis. A recent technique that has achieved widespread use in analytical laboratories is solid phase micro-extraction (SPME) that eliminates the need for organic solvents and complicated apparatus. In this technique, vapors are trapped on a fine fiber before being directly injected into GC-MS. More details on this technique can be found elsewhere (Smith, 2004).

A typical procedure for evaluating the fragrance headspace using the SPME technique is as follows (Zhou, 2006). A soap bar is placed into a jar and covered by a polyethylene film. Following the incubation period, a preconditioned 100 μm PDMS fiber is inserted into the jar to carry out the headspace extraction for 30 minutes. After extraction the analysis is performed using GC-MS. Details on the GC column and desorption temperature depend on the type of the GC-MS instrument used.

This procedure can be applied to measure the headspace over the whole bar, a bar slurry, or diluted solutions (to evaluate the release of fragrance during wash), or over the skin to assess the perfume substantivity after wash. Figure 15.5 illustrates the latter test, in which a microfiber absorbing the perfume is seen inside a glass container covering the area of the forearm skin washed with the bar being tested.

User Panel Evaluations

Product testing with trained (expert) or untrained panelists is also relevant to our discussion of bar soap appraisal. Major consumer products companies use consumer panels or expert panels of appraisers at one or more stages of product development. Panels comprising several hundred people may be used to confirm that a new prototype meets the criteria for launch in the market. Expert or untrained panel appraisals at early and intermediate stages often help direct formula development, and their use is worthy of discussion here because it is often employed to assess one or more of a product's sensory characteristics that are not easily quantified by an instrumental technique.

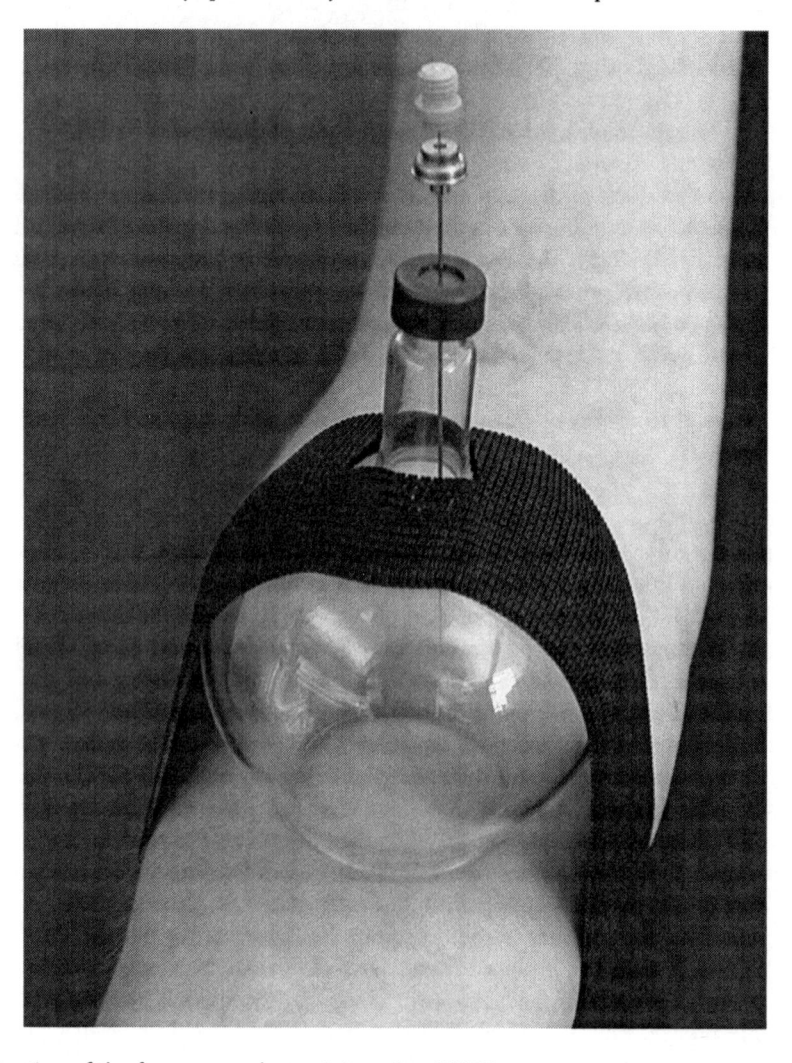

Fig. 15.5. Evaluation of the fragrance substantivity using SPME.

Characteristics like lather volume, mush, wet-bar feel, and skin feel after wash can be quantified with instruments (with varying degrees of difficulty) or by trained technicians, but sometimes a human evaluation may be the best way to assess a consumer preference or sensation. Consider fragrance assessment for a bar soap. A gas chromatograph can certainly identify and quantify the mass transport of molecules from the product into the atmosphere, but the assessment of the fragrance's likeability is a qualitative judgment made by a person. Thus, an expert or an untrained consumer panel might be used to help select an appropriate fragrance for a product, and the qualitative judgments of the panelists can be mapped onto a scale to provide a quantitative measure if necessary. In the case of lather, where volume or generation speed can certainly be quantified by an instrument or an expert evaluator, the sensory experience for the consumer remains the characteristic that will influence product acceptance. An instrumental measure may not correlate adequately with consumer preference or may distinguish subtleties that are beyond the average consumer's perception. In such cases, a panel test provides the most relevant product appraisal, and familiarity with some of the basics of consumer testing is useful for the product development scientist.

Moskowitz has reviewed consumer testing procedures (Moskowitz,1984). We note several testing methodologies for use with small panels that could help guide product development at the early stages. Forced-choice tests, like the triangle test or the same-different test with pairs of prototypes, can be used with relatively small numbers of panelists. Panelists are asked to distinguish the different product in a set of three choices, or they are asked to judge whether a pair of products are the same or different and, if different, to state what the difference is. There are a number of systematic errors to avoid. Expectation error arises when panelists have too much information. This information may include spurious hints, as when a label (such as "1" or "a") is assigned to a prototype that has nothing to do with real properties of the prototype, but triggers a general bias that people have(for example, paying more attention to objects in a set labeled "1" or "a"). This type of expectation error can be avoided by assigning random three-digit codes to the prototypes. Another systematic error is positional error. The triples of samples in the triangle test should not be arranged in a line due to people's bias toward selecting the middle object in this arrangement; instead, the prototypes should be arranged in a triangle, which gives the procedure its name. Also, the arrangement of samples needs to be randomized from panelist to panelist in any test where panelists compare multiple prototypes. Standards have been written to implement various kinds of sensory product tests so that these errors are avoided. (See ASTM E1885-04 concerning the triangle test.)

Bar Hardness

Cleansing bar hardness is not a product trait that generally determines consumer acceptance, since most marketed products show little variation as far as the consumer can assess. All products meet some threshold level of hardness at room temperature, and whatever differences might be revealed by an instrumental test are usually not discernable to the touch. Thus, the basic consumer requirement regarding hardness is that the soap bar not be malleable when it is gripped. However, a bar must be malleable to some degree, so that when a larger stress is imposed quickly, as when the user drops a bar, the product dents rather than shatters. This behavior is ensured by the phase structure of cleansing bars, which we discuss below, and a room-temperature hardness test for a cleansing bar is merely used to reject clearly inadequate formulations during development.

Hardness tests are much more important on the production floor, because the optimum factory production rate and product quality are achieved over a certain range of product hardness. Soap manufactured by extrusion and stamping will have numerous stamping defects if it is too soft, whereas soap that is too hard can lead to unsteady extrusion rates or reach a limiting extrusion rate that falls below the desired production rate. Soap bars manufactured by casting may also suffer from handling defects if they are too soft, but extrusion and stamping are more sensitive to material hardness, so our summary of assessment methods primarily concerns extruded bars. Different batches of the same nominal formulation may exhibit different hardness on the factory floor due to variation in characteristics of the raw materials, like fatty chain length distribution or moisture, and environmental factors, such as different ambient conditions in the factory or different storage conditions, either of which will result in variation in the moisture level and temperature of the product as it enters the extrusion operation.

The most practical hardness characterization methods for cleansing bars in the factory environment involve indentation or penetration experiments using a needle, cone, or wire. The manufacturers of these devices may refer to the measurement as a hardness test, or a measure of yield stress or consistency. We will define these terms more rigorously in the next section. Consider a penetration experiment, in which a cone is driven into a soap sample by its weight, and the movement of the cone is arrested by the resistance of the soap. Penetration eventually stops when enough of the cone has entered the soap so that the material resistance to deformation balances the load. The penetration depth serves as a measure of product softness. Alternatively stated, a force balance when the cone penetration stops can be used to determine the hardness H, resulting in the following definition:

$$H = \frac{W}{A_c} = \frac{W}{\pi d^2 \tan^2 \alpha} \qquad [1]$$

where W is the total load (weight) that causes the deformation, and A_c is the contact area, which is expressed in terms of the depth d and angle α of the cone (see Fig. 15.6). This formula allows results of tests with different cones and weights to be compared, though it may be best to fix the cone and weight and simply track the depth of penetration.

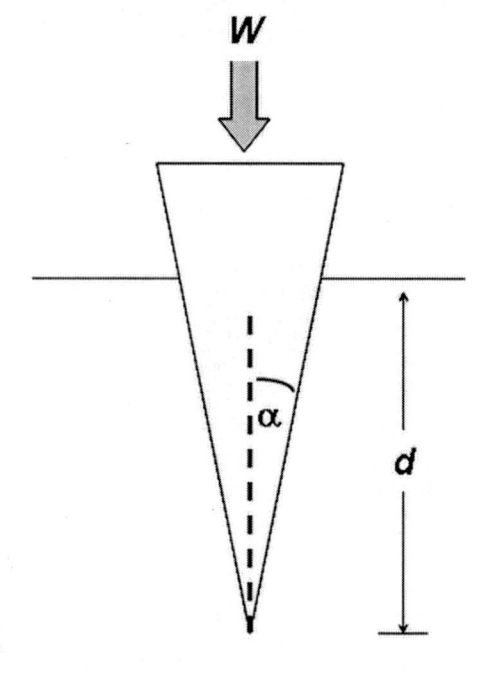

Fig. 15.6. Cone penetrometer.

The hardness H defined in Equation [1] has units of stress, and this may be associated with the yield stress (that is, is the minimum stress required to cause the material to flow), because the cone has, for practical purposes, stopped moving, and the material is resisting deformation at the stress H. Analysis of this test in terms of fundamental material parameters is difficult because two kinds of stress are involved: extensional stress (in the direction of the applied force) and frictional stress (tangent to the cone surface). Separation of these effects requires more detailed knowledge of material properties for a soft solid like soap. A more rigorous discussion of indention tests and their relation to yield stress can be found elsewhere (Adams et al., 1996). Another difficulty with the test is that a soap continues to deform under any load, so the choice of time point when the cone appears to stop is arbitrary. If we wait twice as long, or ten times longer, the depth of the cone will be greater by some amount—we are merely approaching a zero penetration rate. Thus, implementation of the cone penetrometer test and similar tests requires selection of consistent procedures, including specification of the weight, the geometry of the penetrator, and the contact time. Some standard methods can be used with minor modifications, such as the standard ASTM test for cone penetration of petrolatum, or needle penetration in bituminous materials, or petroleum waxes (ASTM International, 2004–2007).

A less common penetrometer that uses a wire to slice through soap is shown in Fig. 15.7. Note how a cross section of soap is positioned in the holder. As in the case of the cone penetrator, the wire comes to rest when it has penetrated enough soap that material resistance balances the weight. For this geometry, hardness is defined by force over projected area:

$$H = \frac{W}{A_c} = \frac{W}{L\,d}$$

[2]

where L is the length of soap supporting the weight when the wire comes to rest, and d is the wire diameter. Again, a contact time should be arbitrarily selected (a minute is appropriate), because the wire will continue to penetrate the soap, albeit very slowly.

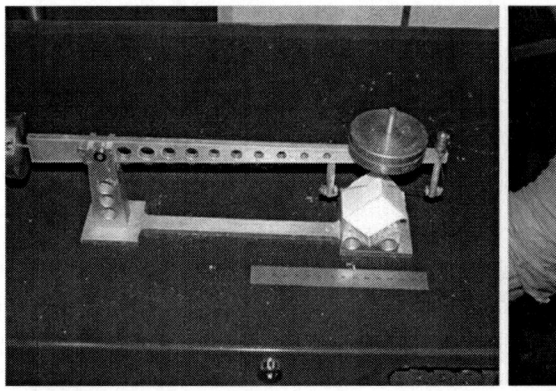

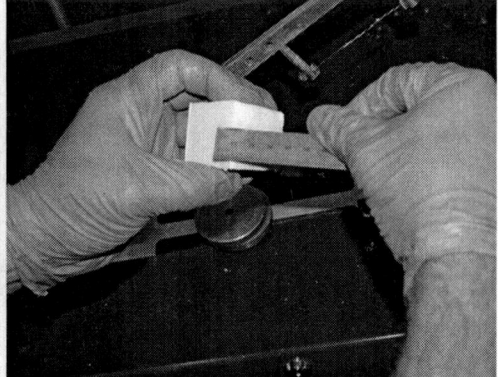

Fig. 15.7. Wire penetrometer.

Another type of experiment involves subjecting soap to a constant strain rate (constant rate of probe movement) and measuring the force required to maintain the strain rate. The force or stress associated with a moving probe is sometimes referred to as consistency. Barnes described (Barnes, 1980) some of the earliest reported hardness tests on soap by Bowen and Thomas and by Vold and Lyon (Bowen and Thomas, 1935; Vold and Lyon, 1945). These researchers used sectilometry, that is, the movement of a wire through a material. Barnes repeated the experiments of the original studies and cast his results in terms of tensile stress σ versus extensional strain rate $\dot{\varepsilon}$, defined as follows:

$$\sigma = K_1 \frac{F}{2\ell R} \qquad [3]$$

$$\dot{\varepsilon} = K_2 \frac{2V}{R} \qquad [4]$$

where ℓ is the length of wire moving through the soap, R is the wire radius, V is the velocity of the wire, and K_1 and K_2 are dimensionless constants. He suggested that K_1 and K_2 can both be taken as unity. The original workers reported the force required for a given V, or in the limit as V approached zero. Barnes plotted tensile stress as a function of $\dot{\varepsilon}$ and extrapolated to $\dot{\varepsilon} = 0$, which provides the effective yield stress for extensional flow. He found yield stress to be in the range of $2-5 \times 10^5$ Pa for typical toilet soap.

Instruments are available that perform constant stress or constant strain rate experiments with various probes. A universal penetrometer is a simple device that releases a weighted needle or cone into a material and measures the maximum depth of penetration; the Humboldt Manufacturing Company offers one example (Humboldt Mfg. Co., 2008). This is a constant stress experiment, and the device reports a penetration depth, but does not report a hardness such as the one defined by Equation [1]. ASTM test D5-05a for hardness of bituminous materials, noted in (ASTM International), describes how this penetrometer is used. A simple handheld device called a Green Hardness Tester (473 "B" Scale model), made by Dietert for the foundry industry to measure the hardness of green sand molds, would be appropriate to evaluate soap on the production line (Dietert Foundry Testing Equipment, Inc., 2008). This device uses a spring-loaded ball-point-pen-like penetrator and measures the penetration depth. The so-called texture analyzer, marketed by Stable Micro Systems, Ltd., is a more sophisticated penetrometer, which finds extensive application in the food industry (Stable Micro Systems, Ltd., 2008). An operator can choose from a variety of penetration tools. The maximum force is recorded for settings of penetration velocity and maximum penetration depth. The Chatillon mechanical tester, MT 150 series, is another example of a sophisticated penetrometer that provides force as a function of penetration rate (Chatillon Brand Products, 2008).

Rheological Characterization

Rheology is the study of the flow and deformation behavior of materials (Macosko, 1994). Measurement of soap hardness is only one aspect of its rheological behavior. A better understanding of soap behavior during extrusion and stamping operations requires a more complete characterization of soap rheology. A thorough review from 1980 (Barnes, 1980) remains the most comprehensive and relevant discussion of the subject in the open literature. In the years since his review, the rheology of bar soap materials has rarely been presented in peer-reviewed scientific literature, and published work has largely been confined to the patent art.

The material structure of soap at usual extrusion processing temperatures (38–45°C) on the colloidal length scale (≈0.001–1 micron) is a concentrated suspension of various solids in a viscous fluid, and has been likened to "bricks and mortar" (Hill & Moaddel, 2004). The solid bricks comprise, in large part, C_{16} and longer-chain saturated fatty soap crystals, which are embedded in a continuous mortar phase comprising C_{12} and shorter saturated soaps, C_{18} unsaturated soaps, water, and salts. Other solids, such as talc, clays, and starches, may be dispersed in the mortar phase. The bricks, or long-chain fatty soap crystals, represent about 42% by weight, judging from known chain-length distributions for a soap bar composed of only 80/20 tallow/coconut soap at 15% moisture. The mortar is a mixture of solid and liquid whose relative amounts depend on the temperature. We expect only about half of the mortar to be solid in the processing temperature range.

The phase structure of soap controls its rheological properties, which in turn define the limits for extrusion and stamping. Extrusion processing lines are frequently designed to be controlled by the stamping machine. A stamping rate is set, and a control loop will adjust the extruder to deliver soap at the required rate, so the extruder screw speed will adjust accordingly. A soap formula usually increases in temperature as it passes through the extrusion process, which may consist of one or more refiners and a final extruder. The temperature increase of the soap effectively adjusts the ratio of liquid in the mortar phase to allow the soap to extrude at the specified rate. There is a screw speed associated with maximum throughput for a given soap material, above which slower throughput occurs because the soap churns, that is, rotates on the screw. Churning is caused by excessive formation of liquid phase, which reduces friction at the extruder barrel wall. The phase structure and resulting rheology of a soap formula should allow it to extrude at the rate demanded by the stamper, but too large a rise in temperature causes excessive softening, which leads to a high occurrence of stamping defects. Thus, characterization of soap rheology, especially as a function of temperature, provides insight into the processing limits of a formulation.

The term "soft solid" is an appropriate description of soap material. A soft solid is a material that behaves as an elastic solid when subjected to small strains for a short time, deforms and does not recover its original shape when subjected to large strains, and can be made to flow like a fluid when stress exceeds a critical value, which is the yield stress. A concise scientific term for such a material is a viscoelasto-plastic fluid (Adams et al., 1996). The relation between stress σ and strain ε for soap has been described by Hooke's law below the yield stress σ_y,

$$\sigma = E\varepsilon \quad \text{for} \quad \sigma \leq \sigma_y \tag{5}$$

where E is the elastic modulus, and the relation between stress and strain rate $\dot{\varepsilon}$ above the yield stress is given by the Hershal-Bulkley equation:

$$\sigma = \sigma_y + k\dot{\varepsilon}^n \quad \text{for} \quad \sigma > \sigma_y \tag{6}$$

where k is called the flow consistency, and n is the flow index. The behavior, as indicated by Equations [5] and [6], is depicted by the dashed line in Fig. 15.8. The determination of yield stress depends on the experimental method used, and the sharp distinction between the regimes governed by Equation [5] and Equation [6] is an idealization. Barnes has argued that no soft solid has a true yield stress, since nearly all soft-solid materials will flow or deform under a finite stress if we wait long enough (Barnes, 1999). In spite of this observation, Equation [6] remains a meaningful description for the flow behavior of soft solids like soap.

Ram extrusion through an orifice can be used to make a practical estimate of the yield stress of soft solids, and to obtain the flow consistency and flow index as well. This experiment can be performed with a capillary rheometer device using an orifice fitting. The soap is forced by a ram of diameter D_r through an orifice of diameter D located at the bottom of a cylinder (cylinder diameter $\cong D_r$), and the force F required to move the ram at speed V_r is recorded by a force transducer. Benbow and colleagues (Benbow & Bridgewater, 1993) derived a semi-empirical relation, known as the Benbow-Bridgewater equation, for the pressure drop of paste forced through an orifice:

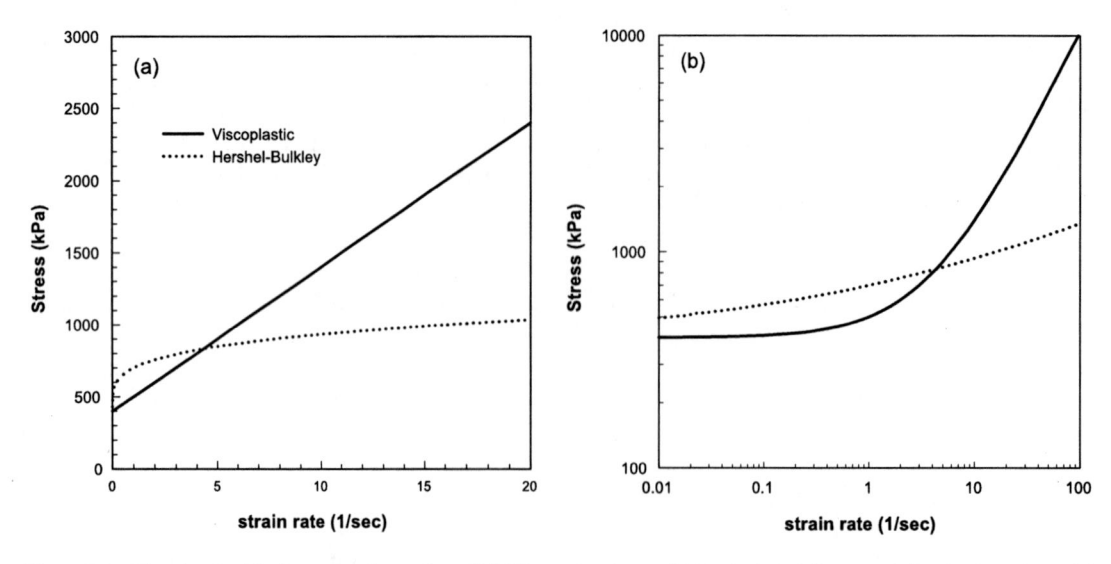

Fig. 15.8. Rheological behavior of a soft solid. Viscoplastic and Hershel-Bulkley models depicted on (a) linear scale and (b) log-log scale.

$$P = 2(\sigma_y + c V^n) \ln \frac{D_r}{D}$$

[7]

where V is the exit velocity of the material, and P is the pressure drop. The measured force F on the ram is related to P by $P = \frac{F}{\pi D_r^2}$, and the exit velocity V is related to the ram velocity V_r by $V = V_r(D_r/D)^2$. The σ_y and n of Equation [7] are the yield stress and flow index of the Hershel-Bulkley relation, Equation [6], and c is a constant. The yield stress is found from experiment by plotting pressure drop versus velocity and determining the intercept of the curve on the pressure axis, which is $2\sigma_y \ln(D_r/D)$. If desired, the flow index can be determined as the slope of the asymptote of $\log(P)$ versus $\log(V)$ for large V.

Basterfield and coworkers have shown that all three Hershel-Bulkley parameters can be obtained through a more rigorous analysis of orifice extrusion (Basterfield, 2005). They obtained an expression for pressure drop similar to the Benbow-Bridgewater equation:

$$P = 2\sigma_y \ln(D/D_r) + Ak\left(\frac{2V}{D}\right)^n \left(1 - \left(\frac{D}{D_r}\right)^{3n}\right)$$

[8]

where the quantities have the same definitions as above, and A is a parameter related to the geometry and the flow index n, which the authors suggest can be approximated as

$$A = \frac{2}{3n}[\sin\phi\,(1+\cos\phi)]^n$$

The ϕ above is an angle associated with the funneling of material as it passes through an orifice. Basterfield and coworkers indicate it can be approximated by 45°. We refer the interested reader to

(Basterfield et al., 2005) for more detail. Extracting all three Hershel-Bulkley parameters from Equation [8] is slightly more complicated than extracting the set of parameters from Equation [7], but the yield stress is determined in the same way.

The values of yield stress and the other Hershel-Bulkley parameters depend on temperature and formulation and especially on moisture level. For superfatted toilet soap at room temperature, Barnes has provided order-of-magnitude estimates of

$$\sigma_y \sim 5 \times 10^5 \, \text{Pa}$$
$$k \sim 3 \times 10^5 \, \text{Pa}$$
$$n \sim 0.2$$

(Barnes, 1980). (Note: The text in this reference indicates a different order of magnitude for the first two parameters, but the figures suggest the values shown here.)

It would be convenient if factory operation could be modified in response to hardness measurements made online automatically or by an operator, but few adjustments on the factory floor can be made to alter product rheology. Addition of small amounts of water to the pellet mixer can be used to soften material, and a soap can be made harder by doing less mechanical work on it. One simple mechanical change can be implemented by replacing a screen in a refiner with one having a larger or smaller mesh size. A larger mesh, or no screen at all, will result in less work done on the soap, and should lower the extrusion temperature, making the soap slightly harder. The disadvantage of this approach is that a screen of a certain mesh size may be required to adequately blend a dye into the soap.

Ideally, a formulation that is too soft or too hard to be manufactured with the factory equipment would be excluded during the product development process, or the equipment would be modified to accommodate the formulation. However, measurement of product hardness in the factory remains a useful exercise to determine how the production line performance correlates with variation in factors like soap composition (within the bounds of a formulation or for different formulations), ambient temperature and humidity, storage history, and mechanical settings.

Conclusion

Manufactured soap bars have existed for centuries, and appraisal methods have evolved together with technological understanding of the product and consideration of consumer needs. This chapter has provided an introduction to current appraisal methods used to evaluate properties of personal wash bars relevant to the consumer, and to the rheology of bar materials relevant to processing. Typically, individual companies describe appraisal methods in internal documentation, which is generally not accessible to the public, so we have attempted to provide the reader with references from the open literature as much as possible, for example from the patent art and from scientific or engineering publications. Therefore, the chapter may also serve as an entry point for more advanced study.

References

Abbas, S.H.; R. Hui. Skin Cleansing Bars with High Level of Liquid Emollient. U.S. Patent 6,680,285, 2004.

Adams, M. J.; B.J. Briscoe; S.K. Sinha. An Indention Study of an Elastoviscoplastic Material. *Philos. Mag. A* **1996**, *74*, 1225–1233.

ASTM International. *Cone Penetration of Petrolatum*; ASTM D 937, 2007. *Penetration in Bituminous Materials*; ASTM D5, 2006. *Needle Penetration of Petroleum Waxes*; ASTM D5, 2005, ASTM D 1321, 2004.

ASTM International. *Standard Test Method for Foaming Properties of Surface-Active Agents*; ASTM D1173-07; 1 October 2007.

Barnes, H. Detergents. In *Rheometry: Industrial Applications*; Walters, K., Ed.; Research Studies Press: Letchworth, England, 1980; pp. 81–118.

Vold, R. D.; L.L. Lyon. A New Cutting Wire Plastometer—Application to Viscous and Plastic Materials. *Ind. Eng. Chem—Analytical Edition* **1945**, *17*, 585–590.

Barnes, H.J. The Yield Stress—A Review, Or 'pi alpha nu tau alpha epsilon iota'—Everything Flows. *J. Non-Newtonian Fluid Mech.* **1999**, *81*, 133–178.

Benbow, J.J.; J. Bridgewater. *Paste Flow and Extrusion*; Clarendon Press: Oxford, 1993.

Basterfield, R.A.; C.J. Lawrence; M.J. Adams. On the Interpretation of Orifice Extrusion Data for Viscoplastic Materials. *Chem. Eng. Sci.* **2005**, *60*, 2599–2607.

Bowen, J.L.; R. Thomas. Properties of Solid Soaps. *Trans. Fara. Soc.* **1935**, *31*, 164–182.

Brennan, M.A.; M. Massaro; S.H. Abbas; Y. Yarovoy. Mild Acyl Isethionate Ca/Mg Salts and Toilet Bar Composition. U.S. Pat. Appl. Publ., U.S. 2008058237, 2008.

Bureau of Indian Standards *Specifications for Toilet Soaps* (second revision); Indian Standards IS 2888; 1983.

Bureau of Indian Standards. *Bathing Bar—Specification*; Indian Standards IS 13498; New Delhi, 1997.

Chatillon Brand Products. MT-150 Series Mechanical Tester. Available at http://www.chatillon.com/. accessed 2008.

Colwell, D.J.; J.J. Pflug. Bar Soap Having Improved Resistance to Cracking. U.S. Patent 5,017,302, 1991.

Dietert Foundry Testing Equipment, Inc. No. 473 Green Hardness Tester. Available at http://www.dietertlab.com/ (accessed 2008).

Ertel, K. D.; B. H. Keswick; P.B. Bryant. A Forearm Controlled Application Technique for Estimating the Relative Mildness of Personal Cleansing Products. *Journal of the Society of Cosmetic Chemists*, **1995**, *46* (2), 67–76.

Dermatologic Therapy **2004**, *17* (Supplement 1).

Farrell, T.; C.C. Nunn. Bar with Good User Properties Comprising Acid-Soap Complex as Structurant and Low Levels of Synthetic Surfactants. U.S. Patent 6,849,585, 2005.

Fujiwara, M.; C. Vincent; K. Ananthapadmanabhan; M. Aronson. Soap Bars Having Quick Kill Capacity and Methods of Enhancing Such Capacity. U.S. Patent 6,007,831, 1999.

Duncalf G. Evaluating Tallow- and Palm Oil-based Soaps. Savonnerie Anglaise: Chesham, UK, 2005. http://www.rendermagazine.com/October2005/EvaluatingTallow.pdf (accessed 2008).

Goette, E. Skin Compatibility of Surfactants, Based on Zein Solubility. Chem. Phys. Appl. Surface Active Subst., Proc. Int. Congr., 4th (1967), Meeting Date 1964, 3, 83–90.

Hill, M.; T. Moaddel. Soap Structure and Phase Behavior. In *SODEOPEC: Soaps, Detergents, Oleochemcials, and Personal Care Products*; Spitz, L., Ed.; AOCS Press: Champaign, IL, 2004; pp. 73–95.

Ho, L.L.T. *Formulating Detergents and Personal Care Products: A Complete Guide to Product Development*; AOCS Press: Champaign, IL, 2000.

Humboldt Mfg. Co. Universal Penetrometer H-1200. Available at http://www.humboldtmfg.com/ (accessed Sept. 2008).

Hyeon, K.; P. Kim; S. Kim. Toilet Soap Composition. Int. Patent WO/2005/017087, 2005.

Kacher, M.L.; N.W. Geary; M.W. Evams; S.K. Hedges; J.A. Ehrhard, Jr.; J.R. Schwartz; D.J. Weisgerber. Combined Skin Moisturizing and Cleansing Bar Composition. Int. Patent WO 1995/026710, 1995.

Macosko, C.W. *Rheology, Principles, Measurements and Applications*; VCH Publishers: New York, 1994.

Marchesani, C. Wet Crack Test Method for Soap Bars. U.S. Patent 4,147,053, 1979.

Moskowitz, H.R. *Cosmetic Product Testing. A Modern Psychophysical Approach*; (Cosmetic Science and Technology Series, Vol 3) Marcel Dekker, Inc.: New York, 1984.

Post, A.J.; E. Van Gunst; M. He; M. Fair; M. Massaro. Bar Comprising Copolymer Mildness Additives. U.S. Patent 5,786,312, 1998.

Ross, J.; G. Miles, G. An Apparatus for Comparison of Foaming Properties of Soaps and Detergents. *Oil & Soap* **1941**, *18*, 99–103.

Schwuger, M.J. Interaction of Proteins and Detergents Studied with the Model Substance Zein. *Kolloid Zeitschrift &*

Zeitschrift für Polymere **1969**, *233* (1–2), 898–905.

Sonenstein, G.G. Soap Bar. U.S. Patent 4,265,778, 1981.

Subramanyan, R.; S. H. Abbas; S.K. Chopra. Soap Composition Containing Sodium Pyrophosphate. U.S. Patent 5,571,287, 1996.

SITA Messtechnik GmbH. http://www.online-tensiometer.com/produkte/frame_products_foamtester.html. (accessed Sept. 2008).

Sivamani, R. K; J. Goodman; N.V. Gitis; H.I. Maibach. Friction Coefficient of Skin in Real Time. *Skin Research and Technology* **2003**, *9*, 235–239.

Smith, L.C. Technical Aspects of Perfumery. In *SODEOPEC: Soaps, Detergents, Oleochemcials, and Personal Care Products*; Spitz, L., Ed.; AOCS Press: Champaign, IL, 2004; pp. 444–453.

Stable Micro Systems, Ltd. Texture Analyzer TA-XT. Available at http://www.stablemicrosystems.com/ (accessed Sept. 2008).

Wood, T. Analytical Methods, Evaluation Techniques, and Regulatory Requirements. In *Soap Technology for the 1990s*; L. Spitz, Ed.; AOCS Press: Champaign, IL, 1990; pp. 260–291.

Wood, T. Quality Control and Evaluation of Soap and Related Materials. In *Soaps and Detergents: A Theoretical and Practical Review*; L. Spitz, Ed.; AOCS Press: Champaign, IL, 1996; pp. 46–74.

Yarovoy, Y.; M. Massaro; R. Patel. Soap Bars Comprising High Levels of Specific Alkoxylated Triglycerides. Int. Patent WO/2003/031553, 2003.

Zhou, M. Rapid Study of Fragrance Loss from Commercial Soap after Use by Solid Phase Microextraction–GC/MS and Olfactory Evaluation. *Anal. Sci.* **2006**, *22* (September), 1249 and references therein.

16

Soap Calculations, Glossary, and Fats, Oils, and Fatty Acid Specifications

Luis Spitz
L. Spitz, Inc., Highland Park, Illinois, USA

Soap Calculations

Two Routes to Soap Making

Two primary routes exist for soap making:

1. *Triglycerides*

$$\begin{array}{c}RCOOCH_2\\|\\RCOOCH\\|\\RCOOCH_2\end{array} + 3NaOH \rightarrow 3RCOONa + \begin{array}{c}CH_2OH\\|\\CHOH\\|\\CH_2OH\end{array}$$

Triglyceride + caustic soda → soap + glycerine

2. *Fatty acids*

$RCOOH + NaOH \rightarrow RCOONa + H_2O$

Fatty acid + caustic soda → soap + water

Fat Blend Calculations

The SV, IV, and Titer are all additive of the proportional values of fat blend components. The SV, IV, and Titer of an 80/20 blend of tallow (SV 197, IV 45, Titer 41) and coconut oil (SV 257, IV 10, Titer 22) can be calculated as shown below:

Calculation of SV, IV, and Titer of an 80/20 Tallow/Coconut Oil Blend

Fat Component	(%)	SV	Titer	IV
Tallow	80	197 × 80%	41 × 80%	45 × 80%
Coconut oil	20	257 × 20%	22 × 20%	10 × 20%
Blend	100	209	37.2	38

Caustic Soda Requirement Calculations

The saponification of a fat blend results in the formation of soap and glycerine (Eq. 16.1).

$$triglyceride + 3\ NaOH \longrightarrow 3\ RCOONa + glycerine$$

(Eq. 16.1)

The fat blend and caustic soda are mixed in a nearly stoichiometric ratio, with ~0.1–0.5% excess of alkali. The molecular weight of the fat blend is calculated per Eq. 16.2.

$$MW_{fat} = \frac{56.1}{SV} \times 3 \times 1000 = \frac{168,300}{SV}$$

(Eq. 16.2)

Problem 1.
Calculate the MW of an 80/20% blend of tallow (SV 197) and coconut oil (SV 258).

$$MW_{tallow} = \frac{168,300}{197} = 854$$

$$MW_{coco} = \frac{168,300}{258} = 652$$

$$MW_{80/20\,T/C} = (854 \times 80\%) + (652 \times 20\%) = 813.6$$

The amount of caustic soda required to saponify a fat blend of known SV can be calculated *via* Eq. 16.3.

$$\frac{SV}{1000} \times \frac{MW_{NaOH}}{MW_{KOH}} = \frac{SV}{1000} \times \frac{40}{56.1} = SV \times 0.000713_{g/g}$$

(Eq. 16.3)

Problem 2.
Calculate the amount of caustic soda (50%) required to saponify 500 lb of an 80/20% (SV 209) tallow/coconut oil blend.

$$NaOH_{50\%} = SV \times \frac{0.000713}{50} \times 100$$

$$= 209 \times \frac{0.000713}{50} \times 100_{g/g}$$

$$= 0.298_{g/g} \times 500\,lb = 149\,lb$$

Saponification Products—Quantity Calculations
The Quantity of Soap
The amount of soap produced in a saponification reaction can be calculated from Eq.s 16.4, 16.5, and 16.6.

$$\text{triglyceride} + 3\,\text{NaOH} = 3\,\text{soap} + \text{glycerine}$$
$$1\,\text{mol} + 3 \times 40 = 120 \quad 3\,\text{moles} + \quad 92$$

$$\text{Soap}_{(wt)} = MW_{fat} + 120 - 92 = MW_{fat} + 28 \tag{Eq. 16.4}$$

$$= \left(\frac{\text{fat}_{wt}}{\text{fat}_{MW}}\right) \times \left(\text{fat}_{MW} + 28\right) \tag{Eq. 16.5}$$

$$= \left(\frac{\text{fat}_{wt} \times SV}{168,300}\right) \times \left(\frac{168,300}{SV} + 28\right) \tag{Eq. 16.6}$$

The Quantity of Glycerine

Glycerine liberated in a saponification mixture can be calculated from Eq. 16.7.

Since triglyceride + 3 KOH = soap + glycerol,

$$\text{glycerine \%} = \frac{SV}{1000} \times \frac{92}{168.3} \times 100$$

$$= SV \times 0.0547$$

$$\text{glycerine}_{wt} = \frac{SV \times 0.0547 \times \text{fat}_{wt}}{100} \tag{Eq. 16.7}$$

For fat blends with high levels of FFA, the following calculation will give the glycerine content of the saponification mixtures (Eq. 16.8):

$$\text{fat}_{\text{triglyceride}}, (TG) = SV - AV$$

$$\text{glycerine \%} = \frac{TG}{1000} \times \frac{92}{168.3} \times 100$$

$$\text{glycerine \%} = TG \times 0.0547 \tag{Eq. 16.8}$$

Problem 3.

Calculate the amounts of soap and glycerine produced in the saponification of a blend of 250 lb of tallow (SV 197) and 250 lb of coconut fatty acid (AV 260; SV 260).

In this case, $\text{fat}_{wt} = 500$ lb, and

	SV	AV
Tallow	$197 \times 50\% = 98.5$	
Coconut fatty acid	$260 \times 50\% = 130$	$260 \times 50\% = 130$

$$\begin{aligned} TG &= SV - AV \\ &= (SV\ [tallow] + SV\ [coconut\ fatty\ acid]) - AV \\ &= (98.5 + 130) - 130 \\ &= 228.5 - 130 \\ &= 98.5 \end{aligned}$$

$$\begin{aligned} \text{glycerine}_{wt} &= TG \times 0.0547 \\ &= 98.5 \times 0.0547 \\ &= 5.39\% \\ &= 5.39 \times 500 / 100 \\ &= 26.95\ \text{lb (from Eq. 8)} \end{aligned}$$

$$\text{soap}_{wt} = \left(\frac{500}{168,300} \times 228.5\right) \times \left(\frac{168,300}{228.5} + 28\right)$$

$$= 0.679 \times 76$$
$$= 519\ \text{lb}$$

The amount of caustic soda required for this reaction can be calculated per Eq. 16.3.

$$\begin{aligned} \text{NaOH}_{wt} &= 228.5 \times 0.000713 \times 500 \\ &= 81.46\ \text{lb (100\% NaOH)} \end{aligned}$$

Total Fatty Matter (TFM) Calculation

For product specifications purposes, TFM of a fat mixture is obtained from triglyceride (TG) mass, less its glycerine content, plus water (Eq. 16.9).

$$\begin{array}{ccc} \text{triglyceride} + 3\ H_2O & = 3\ RCOOH & + \text{glycerine} \\ 3 \times 18 = 54 & \text{fatty acid} & 92 \end{array}$$

$$TFM = (TG + 54) - 92$$

$$= TG - 38$$

$$TG_{eq.wt} = \frac{56.1}{SV} \times 1000$$

$$FFA_{eq.wt} = TG - (38/3)$$

$$= \left(\frac{56,100}{SV}\right) - 12.66$$

$$TFN\% = \left(\frac{FFA_{eq.wt}}{TG_{eq.wt}}\right)$$

(Eq. 16.9)

Fatty Acid Blend Calculations

The following three methods are utilized in the calculation of fatty acid and alkali reactants:
- the molecular weight (MW) method,
- the gram-mole (G-Mole) method,
- and the acid value (AV) method.

In the molecular weight method, the fatty acid and alkali are blended in the ratio of their molecular weights. The G-Mole method is a variant of the molecular weight method; the reactants are mixed in their grams per mole ratio. The acid value method permits the blending of fatty acids and alkalis on the basis of the acid value of the fatty acid utilized in the neutralization reaction.

$$NaOH_{wt} = FA_{wt} \times AV \times 0.713 \times NaOH\,(\%)$$

Since molecular weight and acid value are interrelated: MW = 56.1/AV x 1000, the molecular weight method will be described in more detail in this chapter.

For fatty acids, the acid value (AV), Titer, and saponification value (SV) are all additives of the partial moieties present in the blend. Thus, for a blend of tallow and coconut fatty acids in a 80/20% ratio, the AV of the blend is 218.

Calculation of the Acid Value of a Fatty Acid Blend

Fatty acid component	%	AV
Tallow (AV 205)	80	205 × 80% = 164
Coconut (AV 270)	20	270 × 20% = 54
		Total = 218

Caustic Soda Requirement Calculations

For the reaction, RCOOH + NaOH = RCOONa + H$_2$O, use Eq. 10.

$$FA_{MW} \quad\quad 40 \quad\quad (FA_{MW} + 22) \quad\quad 18$$

$$FA_{MW} = \frac{56.1}{AV} \times 1000 = \frac{56,100}{AV}$$

$$NaOH_{wt} = \left(\frac{FA_{wt}}{FA_{MW}}\right) \times 40$$

$$= \left(\frac{FA_{wt}}{(56,100\,/\,AV)}\right) \times 40$$

$$= \frac{FA_{wt} \times AV \times 40}{56,100}$$

(Eq. 16.10)

Fatty Acid Neutralization Products
The Quantity of Soap Produced

Soap produced in the neutralization reaction can be calculated per Eq.s 16.11 and 16.12.

$$\text{soap}_{wt} = \left(\frac{FA_{wt}}{FA_{wt}}\right) \times (FA_{wt} + 22) \quad \text{(Eq. 16.11)}$$

$$= \left(\frac{FA_{wt}}{FA_{wt}}\right) \times \left(\frac{56,100}{AV} - 22\right) \quad \text{(Eq. 16.12)}$$

The Amount of Water Produced
This is calculated as per Eq. 16.13.

$$\text{water}_{wt} = \left(\frac{FA_{wt}}{FA_{MW}}\right) \times 18 \quad \text{(Eq. 16.13)}$$

Formula Adjustments
Occasionally, the fatty acid blends or neat soap mixtures require an adjustment of blend composition due to a weighing or calculation error. This section describes practical approaches to handling such manufacturing problems.

Fatty Acid Blend Molecular Weight Adjustment
This adjustment usually requires the addition of a fatty acid of molecular weight lower or higher than the molecular weight of the blend to be adjusted. Eq. 16.14 can be used for this purpose.

Let x = portion of fatty acid to be added.

$(1 - x)$ = portion of fatty acid blend to be adjusted.

$$x\,(FA\text{ added})_{MW} + \{(1 - x) \times [FA\text{ (initial blend)}_{MW}]\} = FA\text{ (final blend)}_{MW} \quad \text{(Eq. 16.14)}$$

Problem 4.
You have a tallow/coconut fatty acid blend of MW 244. How much tallow fatty acid (MW 274) should be added to it to convert it into a blend of MW 255?

From Eq. 16.14, let x = portion of tallow fatty acid (MW 274) to be added; $(1 - x)$ = portion of initial blend (MW 244).

$$x(274) + [(1-x)244] = 255$$
$$(274x) + (244 - 244x) = 255$$
$$(274x) - (244x) = 255 - 244$$
$$30x = 11$$
$$x = 0.37; \text{ or } 37\%$$
$$(1-x) = 100 - 37 = 63\%$$

Thus, fatty acid blend (initial) = MW 244 × 63% = 153.7
tallow fatty acid added = MW 274 × 37% = 101.3
final blend = MW 255

An alternative to this calculation is described in Equation 15.

Let initial fatty acid weight = Wt_1; MW = MW_1
final fatty acid weight = Wt_2; MW = MW_2; and
fatty acid added weight = x; MW = MW_3

$$FA\ added_{wt} = \left(\frac{Wt_1}{x}\right) - Wt_1$$

$$x = \frac{MW_3 - MW_2}{MW_3 - MW_1}$$

(Eq. 16.15)

Problem 5.
You have a 100 lb blend of fatty acid, MW 244. How much of a fatty acid of MW 274 should be added to it to make a final blend of MW 255?

$$x = \frac{274 - 255}{274 - 244} = \frac{19}{30} = 0.633$$

$$FA\ added_{wt} = \left(\frac{100}{0.633}\right) - 100 = 58.0\ \text{lb}$$

Thus, $FA\ initial_{wt}$ = 100 lb (63%) $Fa\ initial_{MW}$: 244 × 63% = 153.7
$FA\ added_{wt}$ = 58 lb (37%) $Fa\ added_{MW}$: 274 × 37% = 101.3
$FA\ final\ blend_{wt}$ = 158 lb $FA\ final\ blend_{MW}$: = 255

Alkalinity/Acidity Adjustment

In cases of the downward adjustment of the acidity of superfatted formulas, Eq. 16.16 can be used, where FFA refers to the free fatty acid to be neutralized.

$$FFA_{wt} = soap_{wt} \times FFA_{\%}$$

$$NaOH_{added} = \left(\frac{FFA_{wt}}{FFA_{MW}}\right) \times 40$$

$$= \left(\frac{FFA_{wt}}{FFA_{MW}}\right) \times \left(\frac{40}{NaOH_{\%}} \times 100\right)$$

(Eq. 16.16)

Problem 6.
A neat soap batch (400 lb) was found to contain 3% FFA (MW 280). How much NaOH (50%) should be added to it to make it neutral?

$$FFA_{wt} = 400 \times 3\% = 12 \text{ lb}$$

$$NaOH_{added} = \left(\frac{12}{280}\right) \times \left(\frac{40}{50} \times 100\right) = 3.4 \text{ lb}$$

The upward adjustment of the FFA level of a neat soap blend is done *via* Eq. 16.17.

$$\text{conversion factor, } CF = \frac{soap_{wt} - (soap_{wt} \times FFA \text{ initial}_{\%})}{soap_{wt} - (soap_{wt} \times FFA \text{ final}_{\%})}$$

$$FFA \text{ added}_{wt} = (CF \times soap_{wt}) - soap_{wt}$$

(Eq. 16.17)

Problem 7.
You have a 400 lb batch of neat soap with FFA (MW 280) of 1%. How much FFA (MW 280) should be added to it for a final FFA (MW 280) content of 2% in the neat soap?

$$CF = \frac{400 - (400 \times 1\%)}{400 - (400 \times 2\%)} = \frac{400 - 4}{400 - 8} = 1.01$$

$$FA \text{ added}_{wt} = (1.01 \times 400) - 400 = 4.0 \text{ lb}$$

The adjustment of a formula of high alkalinity, *via* the addition of a fatty acid, is performed by Eq. 16.18.

$$\text{alkali}_{wt} = \text{soap}_{wt} \times \text{alkalinity}_{\%}$$

$$\text{FA added}_{wt} = \frac{\text{alkali}_{wt} \times \text{FA}_{MW}}{\text{alkali}_{MW}}$$

(Eq. 16.18)

Problem 8.
A 400 lb batch of neat soap has an alkalinity of 1.6% (as NaOH). How much of a fatty acid of MW 280 should be added to it to make the net soap neutral?

$$\text{alkali}_{wt} = 400 \times 1.6\% = 6.4 \text{ lb as NaOH}$$

$$\text{FA added}_{wt} = \frac{64 \times 280}{40} = 4418 \text{ lb}$$

The molecular weight interconversion of fatty acids can be calculated *via* Eq. 16.19.

$$\frac{\text{FFA}_A}{\text{FFA}_B} = \frac{\text{MW}_A}{\text{MW}_B}$$

(Eq. 16.19)

Problem 9.
A soap bar sample contains 2% coconut fatty acid (MW 207) as the superfat. Convert this and express it as tallow fatty acid (MW 273) superfat value.

$$\frac{2}{\text{FFA}_B} = \frac{207}{273}$$

$$\text{FFA}_B = \frac{2 \times 273}{207} = 2.63\%$$

Problem 10.
A 400 lb batch of soap contains 2% coconut fatty acid (MW 208) as superfat. How much additional coconut fatty acid should be added to this batch to contain a total of 4% superfat, expressed as stearic acid (MW 274)?

In this example, we need to determine the amount of coconut fatty acid present initially in the soap, the above quantity expressed as stearic acid, the additional amount of coconut fatty acid required, and that quantity expressed as stearic acid.

From Eq. 16.19,

$$\text{stearic FA}_{\text{initial}} = \frac{208}{274} \times \frac{274 \times 2}{208} = 2.63\%$$

$$\text{FFA}_{\text{added}} = \text{FFA}_{\text{calculated}} - \text{FFA}_{\text{Iinitial}}$$

$$= 4.0 - 2.63$$

$$= 1.37\%$$

The above quantity of stearic acid to be added should then be converted into the coconut fatty acid equivalent, as per Eq. 16.19.

$$\frac{\text{FFA}_A}{\text{FFA}_B} = \frac{\text{MW}_A}{\text{MW}_B}$$

$$\frac{\text{FFA}_A}{1.37} = \frac{208}{274}$$

$$\text{FFA}_A = \frac{1.37 \times 208}{274} = 1.04\% = (400 \times 1.04\%) = 4.16 \text{ lb}$$

The batch contains now a total of 8.0 + 4.16 = 12.16 lb (3.01%) of coconut fatty acid. This is equivalent to 4% stearic acid, as per Eq. 16.19.

The calculation for the adjustment of alkalinity follows:

- The alkalinity of a formula containing FFA can be increased as per Eq. 16.20.
- First, calculate the amount of alkali needed to bring the formula to neutrality *via* Eq. 16.16. Then, the amount of additional alkali needed to reach the desired alkali level is calculated.

$$\text{alkalinity adjustment, AA} = \frac{\text{alkali final}_\%}{100} - \frac{\text{alkali initial}_\%}{100}$$

$$\text{alkali}_{\text{wt}} = \frac{\text{AA} \times \text{soap}_{\text{wt}}}{\text{alkali}_\%}$$

Problem 11.
For a 400 lb batch of neat soap containing 2% FFA (MW 270), how much NaOH should be added to increase the alkalinity (as NaOH) to 0.1?

From Equation 16, $FFA_{wt} = 400 \times 2\% = 8\,lb$

$$NaOH\ added = \frac{8}{270} \times 40 = 1.18\,lb$$

From Equation 20, $AA = \frac{0.1}{100} - \frac{0}{100} = 0.001$

$$NaOH_{wt} = \frac{0.001 \times 400}{100} \times 100 = 0.4\,lb$$

$$NaOH\ total\ added = 1.18 + 0.4\,lb = 1.58\,lb \times 100$$

(Eq. 16.20)

The alkalinity of a formula, which is already alkaline, can be increased further by Eq. 16.20; the alkalinity of a formula can be decreased by the addition of a fatty acid *via* Eq. 16.18.

Problem 12.
A 400 lb batch of neat soap contains 2% FFA (MW 270). How much of a 30% NaOH solution should be added to it to bring the FFA level to 1% (MW 208)?
A combination of Eq.s will be used for this calculation.

$$FFA_{wt} = 400 \times 2\% = 8\,lb$$

From Equation 19,

$$\frac{8}{FFA_B} = \frac{270}{208}$$

$$FFA_B = \frac{8 \times 208}{270} = 6.16\,lb$$

Now, the batch requires 400 x 1% = 4 lb FFA (MW 208). Thus, excess FFA (MW 208) = 6.16 − 4 = 2.16 lb. To neutralize 2.16 lb of fatty acid with 30% NaOH, Eq. 16 is utilized:

$$30\%\ NaOH\ added = \frac{2.16}{208} \times \frac{40}{30} \times 1.38\,lb$$

To summarize in molar equivalents, the formulas containing free fatty acids will have the following weight ratios:

$$coconut\ fatty\ acid_{wt} < stearic\ acid_{wt}$$

To illustrate, 5% coconut fatty acid, MW 208 (as FFA) = 6.6% stearic acid, MW 274 (as FFA). Therefore, it will take a greater quantity of stearic acid than coconut fatty acid to neutralize a given quantity of alkali.

Glossary

Acid Value (AV):
The Acid Value is the number of milligrams of potassium hydroxide (KOH) necessary to neutralize the fatty acids in 1 gram of sample.

$$RCOOH + KOH \rightarrow RCOOK + H_2O$$

- Higher AV materials allow faster appearing but less stable suds creation.
- Lower AV materials allow slower to appear but more stable suds formation.
- Lower acid value means more cleansing (detergency).
- AV is used only on fatty acids to provide an estimate of SV. The AV for fatty acids is very close to the SV. AV usually runs about 2 points lower than the SV.

Commercially used Color Scales
APHA:
For light colored liquids; compares light absorption relative to Pt-Co standards

Gardner:
Measures in step; compares sample color to standards of specified colors with associated "Gardner numbers."

Lovibond Color:
Color measurement of the fats, oils and fatty acids determined with a Lovibond Tintometer. A 5¼-inch glass cell containing the sample is compared with Lovibond glass red (R) and yellow (Y) color standards and the colors are recorded in R and Y units. The R value is the color controlling value. The results are cell dependent.

- R value of 1.0 or les is preferred for the production of white soaps.
- R value above 2.5 will result in less off white (darker color) soaps.

Spectrophotometric:
Continuous scale; measures % transmission at 440 nm and 550 nm.

Fatty Acids:
Fatty acids are linear, mostly even carbon-numbered long chain hydrocarbons with a terminal carboxyl group. Unsaturated fatty acids are those with one or more double bonds in their carbon chain structure.

Notes on Fatty Acid Profiles

- Long chain C_{14} to C_{20} fatty acids provide best cleaning but minimal sudsing.
- Short chain C_6 to C_{12} fatty acids provide faster but less stable suds and lower cleansing than long chain fatty acids.
- C_{12} provides best sudsing.
- C_{18} provides best cleaning. Among long chain fatty acids, unsaturated fatty acids provide better cleansing.

- Saturated fatty acids are more stable to oxidation, discoloration and rancidity.
- Extra long chain fatty acids (C_{22} and higher) do not contribute much to either sudsing or cleansing but do contribute to bar integrity.
- High sudsing does not mean high cleansing. However most consumers do not think that way.

Mixing Oils & Fatty Acids for Soap Making

- Mixtures of high and low SV stocks provide a very desirable lever of foaming and cleansing.
- Typically, 10 to 30% of high SV and 90% to 70% of low SV triglyceride oils are used.
- The average AV or SV is calculated, based on the % weight ratio of fatty acids or oils. The amount of sodium hydroxide used is based on this average value.
- High foaming does not mean high cleansing. However, most consumers think that way.
- Examples of high SV oil ; Coconut oil, Palm Kernel oil, Babaçu oil
- Examples of low SV oils; Tallow, Palm oil, Canola (rapeseed) oil, Rice bran oil.
- For detailed analytical testing procedures please consult the official AOCS Methods.

Free glycerine:
The amount of free glycerine present in the sample expressed as percentage weight of the total sample.

- Perceived as a moisturizer
- Glycerine levels up to 2% will harden soap
- Glycerine levels above 2% gives softer and stickier soap
- Used as a processing aid for low moisture, high titer products.

Free Alkalinity (Free Caustic):
Free alkalinity is the amount of alkali content present in a sample expressed as percentage weight of free sodium hydroxide (NaOH).

- The higher free alkalinity the greater the skin irritation from soap.
- The higher the alkalinity the greater the cleaning power of soap.
- For toilet soaps, lower alkalinity is preferred for a better product stability with respect to color and odor.
- For laundry soaps, higher alkalinity is preferred for better cleansing properties of the product.

Free Fatty Acid (FFA):
It is the free fatty acid content present in a sample commonly expressed as oleic acid but it can also be expressed as palmitic acid or stearic acid.

Foam Volume:
It is the measure of the foamability of a cleansing product.
    The foam volumes (mL) listed in the tables were determined using a 0.1% soap solution placed in a measuring cylinder and agitated with a perforated paddle stirrer for thirty strokes. The initial measurement was made after agitating the solution and final after 5 minutes standing.

There are different foam test protocol methods. The Ross Miles method is the most widely used. All the methods give relative volumes and are used for lather comparison. There is no standard method giving absolute values.

Iodine Value (IV):
The Iodine Value is a measure of the unsaturation (double bonds) in fats, oils and fatty acids. IV is expressed in terms of the number of grams of iodine absorbed by 100 grams of sample (% iodine absorbed)

$$\text{"-CH=CH- } + I_2 \rightarrow \text{ -CHI-CHI-"}$$

The higher the IV, the higher the degree of unsaturation and the greater the vulnerability for rancidity. As the IV level increases, soaps become softer and stickier.

The foam and cleansing increase as the iodine value increases and decrease as the IV decreases in higher chain saturated fatty acids.

Coconut Oil (7 –12 IV range) is an exception. It produces the hardest soap, the fastest sudsing, but lacks suds stability.

The analytical determination is actually carried out using Wijs solution (ICl in acetic anhydride) as opposed to iodine per se.

Melting Point:
The temperature expressed in °C at which a triglyceride or fatty acid liquefies.

Moisture Content:
It is the amount of volatile materials present in a sample expressed in percentage weight.

Penetration Value:
Is a measure of bar soap hardness. It is expressed as the distance (depth) in millimeters a penetrometer needle penetrates a bar of soap when subjected to a 50 gram weight. The deeper the needle travels, the softer is the soap.

Saponification Value (SV):
The Saponification Value is defined as the number of milligrams of potassium hydroxide (KOH) required to saponify 1 gram of sample.

SV is used to determine the average molecular weight (MW) of fats and oils being saponified, using the formula: $MW = 56,100 / SV$.

Mixtures of high and low SV stocks provide very desirable levels of sudsing and cleansing. Typically, 10 to 30% of high SV and 90 to 70% of low SV triglycerides are used.

Sodium Chloride:
The amount of sodium chloride (NaCl) present in a sample expressed in percentage weight of sodium chloride or simply chloride. Sodium chloride is one of the most critical ingredients for soap processing and product attributes.

Sodium chloride hardens the soap but high levels can create "cracking" and decrease sudsing. In standard non-superfatted low coco soaps the level should not exceed 0.5%.

The chloride level increases in non-superfatted soaps depending upon the level of coco/ palm kernel soap in the product. It can be as high as 2-3% in high coco containing soaps.

Superfatted soaps can have up to 1.5 % sodium chloride without detrimental effects.

Titer:
It is the measure of the solidification point of fatty matter (fats, oils and fatty acids) measured in °C. A peaked or plateau temperature point in the temperature curve profile characterizing the temperature rise in a crystallizing sample due to the release of heat of crystallization.

- Higher titer provides harder soap.
- Lower titer provides better cleansing (with longer chain fatty acids or triglycerides).

Total Fatty Matter (TFM):
Total Fatty Matter is expressed as the fatty acids obtained from soap. and is the sum of the free fatty acid, the fatty acid obtained from soap and the unsaponifiables. The test method used for determination of TFM requires splitting of soap by using mineral acids and then the extraction of the fatty matter by using petroleum ether. Total fatty matter does not include the fatty matter generated by non-soapy synthetic actives.

- The TFM of triglyceride is the amount of fatty acids produced by splitting the oil.
- The TFM of fatty acids is the total weight of fatty acids. Fatty acids are 100% TFM.

Unsaponified & Unsaponifiable Matter (U&U):
The unsaponified matter consists of neutral unreacted fat, which is not saponified. The unsaponifiable matter includes substances frequently found dissolved in fats and oils that cannot be saponified with caustic alkalies but are soluble in ordinary fat solvents. U&U it is the amount of substances soluble in petroleum ether present in the sample and is expressed in percentage weight.

- High unsaponifiable content makes soap sticky and can lead to discoloration.
- Unsaponifiables contribute to the emolliency and skin feel attributes of soap bars. They basically act as "superfatting" agents and are a part of the TFM.

Fats, Oils, and Fatty Acid Specifications
Tables 16.1–8.

Table 16.1. Fats, Oils ,and Fatty Acids Specifications

Fats and oils	Saponification value	Iodine value	Titer (°C)	Melting point (°C)	Glycerine (%)
Coconut oil	250–264	7–12	20–24	23.0–26.0	13
Tallow	192–202	48–52	40–47	40.0–47.0	10
Palm kernel oil	245–255	14–19	20–28	24.0–26.0	12
Palm kernel olein	231–244 [a]	25–31	—	21.8–26.0 [a]	—
Palm oil	196–202	50–55	40–47	27.0–50.0	10
Palm stearin	193–206 [b]	48 (max)	20–26	24.0–26.0	—

Sources: Witco Fats and Oils Brochure (1); [a]Tang (3); [b]Tang (2).

Table 16.2. Distilled Fatty Acids for Soaps

Distilled fatty acids	Saponification value	Acid value	Iodine value	Titer (°C)	Unsaponifiable matter (%)	Lovibond color 5.25" cell red-yellow
Coconut	266–278	265–277	5.0–10.0	22–27	0.5	0.8R–8.0Y
Stripped coconut	251–263	252–260	5.0–12.0	27–30	0.5	0.3R–3.0Y
Palm kernel[b]	255–267	255–265	14.0–19.0	22–26	1.0[a]	0.5R–5.0Y
Stripped palm kernel[b]	248–260	248–258	16.0–22.0	24–28	1.0[a]	0.5R–5.0Y
Tallow	—	198–207	36.0–65.0	39–49	1.5[a]	0.2R–0.9R 0.3Y–4.3Y
Palm oil[b]	205–214	205–212	46.0–57.0	45–49	1.0[a]	1.0R–4.0Y
Palm stearin[b]	206–218	206–216	28.0–38.0	48–54	1.0[a]	1.0R–4.0Y
Soap blend (T/C)	—	214–222	36.0–42.0	40[a]	1.0[a]	0.4R–4.0Y
Soap blend[c] (40PO/40POs/20PKO)	—	215–225	37.0–45.0	41–46	1.0[a]	0.3R–3.0Y

[a]maximum.
Abbreviations: Tallow, T; coconut, C; palm oil, PO; palm stearin, POs; palm kernel oil, PKO.
Source: Witco (1), [b]Cognis (4), [c]Cognis (personal communication, 2003).

Table 16.3. Distilled Fatty Acid Soap Blends

Distilled fatty acid soap blends	Acid value	Total fatty matter (%)	Moisture (%)	Iodine value	Titer (°C)	Free caustic (%)	Sodium chloride (%)	Penetration value (mm)	Foam volume (mL)[a]
90PO:10PKO	214	84	10	54.5	43.0	0.06	0.4	53	530/365
85PO:15PKO	217	81	11	52.2	42.2	0.06	0.4	52	520/355
80PO:20PKO	219	81	10	45.2	41.3	0.06	0.4	51	505/345
75PO:25PKO	221	82	11	44.5	38.4	0.04	0.4	50	513/343
70PO:30PKO	225	83	10	43.5	37.5	0.06	0.4	50	520/340
60PO:40PKO	229	80	11	38.2	37.5	0.03	0.3	50	510/350
40PO:40PS:20PKO	224	81	11	37.0	46.0	0.10	0.2	48	470/306
80POs:20PKO	219	83	12	27.0	48.0	0.10	0.5	48	510/345
70POs:30PKO	225	82	11	29.0	47.0	0.10	0.5	48	500/325
65POs:35PKO	228	82	10	25.2	45.0	0.10	0.5	48	525/355
60POs:40PKO	229	83	11	22.0	43.8	0.10	0.3	48	500/368

[a]See Glossary for foam volume test details for all tables that list foam volumes.
Abbreviations: See Table 16.2.

Table 16.4. Fats and Oils Composition

Fats and oils composition

Fats and oils	Caprylic C8:0	Capric C10:0	Lauric C12:0	Myristic C14:0	Palmitic C16:0	Stearic C18:0	Oleic C18:1	Linoleic C18:2	Other
Tallow	—	—	—	3.4	26.3	22.4	43.1	1.4	3.4
Palm oil	—	—	0.3	1.1	43.1	4.6	39.3	10.7	0.9
Palm stearin	—	—	0.7	1.5	55.7	4.8	29.5	7.2	0.6
Coconut oil	7.6	7.3	48.2	16.6	9.0	3.8	5.0	2.5	—
Palm kernel oil	1.4	2.9	50.9	18.4	9.7	1.9	14.6	1.2	—
Palm kernel olein	4.3	3.7	42.6	12.4	8.4	2.5	22.3	3.4	0.4

Source: Witco (1).

Table 16.5. Fatty Acid Composition of Distilled Fatty Acid Blends

Fatty acid composition

Fatty acid soap blends	Caprylic C8:0	Capric C10:0	Lauric C12:0	Myristic C14:0	Palmitic C16:0	Stearic C18:0	Oleic C18:1	Linoleic C18:2	Other
90PO:10PKO	0.07	0.27	6.05	2.01	40.00	4.79	38.41	8.12	0.28
85PO:15PKO	0.30	0.56	8.69	3.71	31.27	4.21	37.25	7.03	0.51
80PO:20PKO	0.98	0.86	12.82	5.80	35.75	3.53	32.89	7.04	0.33
75PO:25PKO	0.97	0.85	13.55	6.56	33.58	3.69	33.43	7.05	0.42
70PO:30PKO	0.16	1.05	17.38	6.19	29.78	3.87	33.24	7.73	0.58
60PO:40PKO	1.09	1.23	19.48	7.40	33.18	3.39	29.97	6.07	0.16
40PO:40PS:20PKO	0.30	0.10	11.80	4.70	45.80	3.80	29.40	3.70	0.33
80POs:20PKO	0.30	0.60	9.80	4.50	53.10	3.70	22.45	4.70	0.82
70POs:30PKO	0.20	1.00	9.22	4.21	54.13	4.65	22.42	3.71	0.73
65POs:35PKO	0.50	1.75	14.12	5.63	49.65	3.58	20.97	3.07	0.67
60POs:40PKO	0.60	2.20	19.50	7.40	44.30	3.30	19.40	2.50	0.61

Abbreviations: See Table 16.2.

Table 16.6. Distilled Tallow and Vegetable Fatty Acid Soap Blends Composition

Tallow and vegetable fatty acid soap blends	Caprylic C8:0	Capric C10:0	Lauric C12:0	Myristic C14:0	Palmitic C16:0	Palmitoleic C16:1	Stearic C18:0	Oleic C18:1	Linoleic C18:2	Other
80T:20PKO	0.3	0.6	9.3	6.7	22.3	1.0	17.7	32.4	2.9	6.5
40T:40PS:20PKO	0.5	0.6	9.0	4.8	32.6	1.1	9.2	24.2	3.0	3.0
60T:20PS:20PKO	0.5	0.7	9.7	5.7	30.9	1.4	1.0	28.9	2.8	5.5

Abbreviations: See Table 16.2.

Table 16.7. Fatty Acid Composition of Vegetable Oil Soap Blends

Vegetable oil soap blends	Caprylic C8:0	Capric C10:0	Lauric C12:0	Myristic C14:0	Palmitic C16:0	Palmitoleic C16:1	Stearic C18:0	Oleic C18:1	Linoleic C18:2	Other
90PS:10PKOo	0.3	0.4	4.8	2.7	60.3	0.2	3.9	22.2	4.4	0.7
80PS:20PKOo	0.6	0.7	8.5	3.7	53.5	0.1	4.5	22.5	4.1	1.8
70PS:30PKOo	1.1	1.0	13.2	5.1	48.8	0.1	4.0	22.3	4.1	0.5
60PS:40PKOo	1.4	1.3	16.3	6.0	44.0	0.1	3.9	22.4	3.9	0.6
50PS:50PKOo	1.7	1.5	20.6	7.6	39.1	0.1	3.8	21.5	3.4	0.5

Abbreviations: See Tables 16.2.

Table 16.8 Soap Bases Superfatted with Distilled Soap Fatty

Superfatted soap blends with distilled fatty acids (DFA)	Total fatty matter (%)	Free fatty acid (%)	Moisture (%)	Free caustic (%)	Sodium chloride (%)	Penetration value (mm)	Foam volume (mL)[a]
70PO:30PKO	76	—	16	0.16	0.5	50	415/230
with 2% (DFA)	77	1.3	17	—	0.5	51	340/205
with 4% (DFA)	78	1.3	14	—	0.5	51	240/107
with 6% (DFA)	77	3.7	16	—	0.5	62	270/112
40POs:40PS:20PKO	76	—	14	0.15	0.5	46	470/305
with 2% (DFA)	80	1.3	12	—	0.5	51	490/335
with 4% (DFA)	78	3.2	15	—	0.5	52	355/185
with 6% (DFA)	83	6.8	10	—	0.5	74	360/175

Abbreviations: See Table 16.2. [a]See Glossary.

References

Gupta, S., Chemistry, Chemical, and Physical Properties & Raw Materials, in *Soap Technology for the 1990's*, ed. L. Spitz, AOCS Press, Champaign, Illinois, 1991, pp. 48–93.

Ainie Kuntom and Luis Spitz Comparison of Palm-and Tallow-Based Soaps, in *SODEOPEC* ed. L. Spitz, AOCS Press, Champaign, Illinois, 2004, pp. 114-146

Index

A

Absolute pressure, 300–301
Ace Detersivo Marsiglia, 5
Acid value (AV), 381, 399–401, 448
Acne deterrents as additives, 148
Additive-base interactions in soap cleansing systems, 135–139
Additives
    anti-acne compounds, 148
    anti-irritants, 148
    antimicrobial compounds, 147
    CTFA Cosmetic Handbook, 143
    dermabrasive/exfoliating agents, 144, 146
    drug components, 146
    emollients, 144–145
    humectants/moisturizers, 144, 146
    miscellaneous, 149–150
    occlusive agents, 144, 147
    OTC active ingredients, 146
    secondary surfactants, 148–149
    syndet bars and, 170–171
Advertising, 34. *See also* Marketing
Alcoholic saponification color, 405
Alfa-Laval dryer, 268
Alkali-free cleansing bars. *See* Syndet bars
Alkyl polyglucosides (APG) and foaming, 168
"Alphabet--Pretty pictures and truism about children's friend Wool Soap", 15–16
Amalgamator with open-arm sigma blades, bar soap finishing and, 308–309
American Family brand, 21, 76–77
Amphoteric surfactants, 154–155, 160–161. *See also* Surfactants
Andrew Jergens Company (1880), 22, 30
Angled bar soap stamping, 373

Animal sacrifices and soap origins, 3
Anionic surfactants, 154–158. *See also* Surfactants
Annual Book of ASTM Standards, 399
Anti-acne compounds as additives, 148
Anti-irritants as additives, 148
Antibacterial/ingredient/deodorant/moisturizing soaps (1994-2009)
    history
        Dove (2001-2008), 56
        Fa (2002-2008), 57
        Ivory (2004-2008), 58–59
        Ivory Moisture Care (1997), 52
        Lever 2000 (2000-2009), 55
        Lifebuoy (2003-2008), 58
        Lux (2000), 53–54
        Oil of Olay (1994-2008), 49–50
        Palmolive Bars (1999-2008), 50–51
        Pears (2004-2008), 58
        Safeguard (1999-2008), 52
Antimicrobial compounds as additives, 147
AOCS Official Test Methods, 399–400
APHA color scale, 448
Ariel, 21
Armour & Company (1867), 21–22
Armour Soap Works, 11
Arrow Borax Soap, 16
Art in marketing/advertising, 5–8, 28, 34, 72
Ash and glycerine, 394
Assay of glycerine, 394
Atlantic and Pacific (1969)
    history, 46
    manufacture of, 350
Atmospheric systems and semi-boiled soap batch process of, 252–253

compared with pressurized system, 260
continuous process of, 254
semi-continuous process of, 253–254
Autoclaves and semi-boiled soap, 254–255
Automation of plants, 290–292

B

Bar hardness, 427–430
Bar smear/slough, 170
Bar soap finishing
    amalgamator with open-arm sigma blades and, 308–309
    chiller sizing, 331
    cooling units, 331
    cutters, 323–326
    direct noncontact transfers for cartoner/
      wrapper interface
        Binacchi direct product transfer (DPT) system, 328
        flashstamping presses, 330
        Mazzoni LB direct transfer (DTS) system, 328, 343–344
    finishing line types/selection, 333–337
    history, 61, 66
    low-temperature glycol chillers, 331, 333
    mixers
      amalgamator with open-arm sigma blades, 308–309
      double-arm mixers with sigma blades, 309–310
      mixer-kneaders, 309–310
    mottled laundry soap manufacturing system, 374
    packaging
      ACMA 330 soap cartoner with YT/YV infeeds, 340
      ACMA 7250/7350 soap wrappers, 339
      ACMA 770 soap cartoner with TH non-contact infeed/transfer, 340
      ACMA 771 TH-non-contact soap infeed/transfer, 339
      ACMA rotary infeed/transfer systems, 338
      ACMA TH-non-contact soap infeed/transfer, 338
      Binacchi systems, 343–344
      CAM soap caroner infeed/transfer systems, 341
      carton blank terminology, 345
      carton styles, 346
      definitions/conversion factors, 346
      Doboy IL3 three belt contact feeder, 342
      Doboy IL4 four belt contact feeder, 342
      general information, 337
      Guerze 1000HS horizontal flow wrapper, 341
      soap cartoners, 345
      soap wrappers, 344
      styles, 337
    plodders
      extrusion group, 315–316
      fiinishing line, 336–337
      Mazzoni LB multirefining plodder (MRP), 319, 321
      packaged water chillers, 331–332
      refining/pelletizing group, 315–316
      SAS Transavon Duplex Vacuum Plodder, 322–323
      short-/long-Lc/D-ratio, 319
      types, 314
      worms, 317–320
    presses
      flashstamping presses, 327
      soap shapes, 326
    processing steps for, 303–308
    roll mills, 311–313, 331–332
    temperature control units, 332–333
Barometric condensers, 285
Barrat, Thomas J., 7–8
Bathing as religious/cleanliness activity, 2–3
Beauty soaps. *See* Purity/beauty/health soaps (1872-1947)
Beecham Group, 6
BHT chromatographic analysis and, 413
Binacchi "CHBS" process, 244–245, 256–259
Binacchi "CSFA" process, 244–245
Binacchi "CSWE-3" system, 237–238, 241, 243
Binacchi direct product transfer (DPT) system, 328
Binacchi SDE dryer system, 281–282
Binders, syndet bars and, 169
Biofuels, 383, 393

B.J. Johnson Soap Company, 34
Bleachability, 382
Blumenschien, Mary Green, 36
Boiling. *See* Kettle saponification
Borit, 2
Brooks, Benjamin, 31
Bubbles (Millais), 8
Buerger nomenclature for solid soap phases, 118, 194
Bulk of lye, 248
Bumetrizole as additive, 149
A Busy Day (Humphrey), 28
Buteth-3 as additive, 149

C

Calculations
  alkalinity/acidity adjustments, 444–447
  caustic soda requirements, 437–438, 441
  FA blend molecular weight adjustment, 442–443
  FA blends, 441
  FA neutralization products, 442
  fat blend, 437
  kettle saponification
    finishing/fitting kettle, 219–220
    fluctuating fat ratios, 217
    kettle settling, 218–219
    loading kettle, 214–217
  primary soap making routes, 437
  saponification products - quantities, 438–440
  total fatty matter (TFM), 440
Calorimetry, 127–128
Camay (1926)
  history, 38–39
Caress (1972-2009)
  history, 47
  ingredient label compositions of, 173
Casely, Robert E., 40
Cashmere Bouquet (1872), 24–26
Cast-melt process and transparency, 200
Cast transparent bars, 131
Cast transparent/translucent bars, 170
Castilla, Spain and soap making, 3
Cataphil, ingredient label compositions of, 174
Cationic surfactants, 154–155. *See also* Surfactants

Caustic soda
  characteristics/specifications/analysis of, 390
  markets/prices and, 391
  production/sources/users of, 388–390
  requirement calculations, 437–438, 441
  saponification and, 224
  transport/storage/handling of, 392
Cavity Transfer Mixer, 198–199
Centrifugal pump metering system, 227–228
Centrifuged lye, 248
Certified colorants, 139, 142
Chardin, Jean Siméon, 78
Charles H. Geilfus & Company, 22
Chase, Helen, 39
Chelating agents, chromotographic analysis and, 412–413
Chemistry of soap structures. *See also* Physical characteristics of soap raw materials
  AOCS Official Test Methods, 399–400
  basic, 83–84
  chromotographic analysis
    BHT by capillary GC, 413
    chelating agents by HPLC, 412–413
    drivatized FA analysis by capillary GC, 412
    drivatized FA analysis by HPLC, 411
    FFA analysis by HPLC, 411–412
    glycerine content by HPLC, 412
    triclocarban/triclosan by HPLC, 412
    triglyceride analysis by HPLC, 410
  continuous saponification and, 228–230, 233–234
  and effect of *trans* isomers on FA/soap properties, 402–403
  FAs, 437
  glycerides and, 84–85
  history, 223
  iodine value (IV), 401–402
  minor ingredients
    anhydrous-soap content, 407
    chlorides, 408–409
    FFA/free alkalinity, 409
    free-glycerine content, 408
    moisture content, 407
    triclocarban/triclosan by UV absorbance, 409–411
  molecular structure and, 115–116

● L. Spitz

mush/cracking
   formulation implications, 108–112
   relationship, 93
   structural implications, 103–108
   structure weakness, 98–103
oleate:laurate eutectic mixture and, 85–87
performance models
   macro model, 87–89
   molecular model, 89–92
saponification value (SV), 401
soap/water interactions, 92–93
structure weakness
   consequences, 98–103
   crystal orientation during plodding, 95–96, 112–114
   overall effects, 97–98
   pressure effects during plodding, 94–95
   structure fundamentals, 93–94
   structure visualization in plodder, 96–97
syndet bars and, 162–167
titer and, 402
and *trans* isomer measurement, 403
transparent/translucent soaps and, 191–195
triglyceride oil structure, 379–380
triglycerides, 437
Cherry Ripe (Millais), 8
Chevreul, Michel Eugène, 4
Chicago Family, 13
A Child's World (Millais), 8
Chill rolls/flakers, 268
Chiller sizing, bar soap finishing and, 331
Chlorate and caustic soda, 390
Chloride and caustic soda, 390
Chromotographic analysis
   BHT by capillary GC, 413
   chelating agents by HPLC, 412–413
   drivatized FA analysis by capillary GC, 412
   drivatized FA analysis by HPLC, 411
   FFA analysis by HPLC, 411–412
   glycerine content by HPLC, 412
   triclocarban/triclosan by HPLC, 412
   triglyceride analysis by HPLC, 410
Clarette, 13
Cleansing bar formulation, syndet bars and, 159
Cleansing grains
Coast (1974-2008)
   history, 48
   manufacture of, 350
Coates, F. Graham, 36
Coconut oil (CNO), 84, 224, 378, 437, 451–453
Colbert, Jean Batiste, 4
Colgate & Company
   Cashmere Bouquet (1872) of, 24–25
   Colgate Palmolive Company (1806), 17–21
   Global Handwashing Day and, 33
   history, 60
Colgate clock, 19–21
Colgate, William, 17
Color
   alcoholic saponification color, 405
   analysis by colorimeter, 404
   analysis by Hunter reflectance measurement, 413–414
   analysis by photometric measurement, 404
   color scale conversions, 405–406
   commercially used scales, 448
   direct bleach test and, 405
   evaluation of raw materials, 404
   fatty matter and, 381
   Hunter reflectance measurement and, 413–414
   refined/bleached, 404–405
   saponification color of FAs, 405
   scale conversions of, 405–406
   visual color comparisons, 413
Colorants
   categories of, 139–141
   certified, 139, 142
   natural, 143
   noncertifiable, 141–142
   noncertified, 142
Combo bars. *See also* Syndet bars
   ingredient label compositions of, 173–175
   surfactant advantage and, 153
   surfactant composition in, 161
Computer-controlled plants, 290–292
Condensate temperature, 300–301
Cone, Fairfax M., 41
Continuous saponification
   commercially available systems, 241–243
   continuous FA neutralization, 241–243

defined, 247
history, 223
metering/dosing
    critical metered streams, 227
    metering equipment, 227–228
    raw materials, 224, 227
neutralization, 240–241
    equipment, 243–245
    reaction, 243
overview of, 223–226
saponification
    chemical reaction, 228–229
    critical reaction factors, 230
    equipment, 230–232
    rate of reaction, 229–230
soap washing/extraction
    cooling and spent lye removal, 237–238
    general information, 232
    glycerine washing efficiency, 239–240
    glycerine washing equipment, 237
    limit lye concentration, 233–234
    lye and neat soap separation, 239–240
    soap grain, 233, 235
    soap phase chemistry/diagram, 233–234
    washing and half spent lye removal, 237–239
    Wigner's model, 235–236
Cooling tower systems
    soap drying systems, 289
Cooling units
    bar soap finishing and, 331
Copco, 13
Countercurrents. *See* Kettle saponification
Countway, D. L., 32
Coupons, in marketing/advertising, 19, 76–77
Cracking
    evaluation of, 422–423
    influences on, 108–109, 112
    mush relationship with, 93
    structural implications and, 103–108
    structure weaknesses and, 98–103
Cream soaps, 150
Cries of London (Wheatley), 5–6
Critical micelle concentration (CMC), 123–126
Crosfield, Joseph, 22
Crude soap, 248

Crutcher and semi-boiled soap, 252–254
Crystallization, 191–195. *See also* Chemistry of soap structures
CTFA Cosmetic Handbook, 143, 149–150
Curd phase, 118
Curd soap, 248
Curtis Davis Company, 31
Cutters
    bar soap finishing and, 323–326

D

Danforth, Philip, 21
Delta phase of molecular model, 92, 99–100
Density, defined, 293
Deodorant/skin care soaps (1948-1967). *See also* Antibacterial/ingredient/deodorant/moisturizing soaps (1994-2009); Freshness/deodorant/skin care soaps (1968-1993)
    history
        Dial (1948-2008), 40–41
        Dove (1955), 44
        Jergens (1951-2008), 41–42
        Safeguard (1963-2008), 44
        Tone (1968), 45
        Vel (1948-2008), 41
        Zest (1952-2008), 42–43
Dermabrasive/exfoliating agents as additives, 144, 146
Detergents. *See* Laundry powder detergents
Deupree, Richard R., 38
Dial, 11, 22, 60
Dial (1948-2008), 40–41
Die Frische Fa. *See* Fa (1968)
Diethylene glycol and glycerine, 394
Differential Scanning Calorimetry, 127–128
Direct noncontact transfers for cartoner/wrapper interface
    bar soap finishing, 328, 330, 343–344
Discontinued soaps
    history
        Colgate, 60
        Dial, 60
        Jergens, 60
        Lever, 60
        Proctor & Gamble, 60
Distilled fatty acid (DFA) process, 377, 452–454

Double-arm mixers with sigma blades, bar soap finishing and, 309–310
Dove (1955), 44
Dove (2001-2008), 56
Dove Beauty Bar design, 162, 173, 176
Dove Nutrium (2000), 351
Dreft (1933), 63–64
Drug components as additives, 146
Drying systems. *See* Soap drying systems
Dual cyclone fines recovery system, 283–284
Dutch Margarine Union, 23

E

Economics. *See* Marketing
Edict of Colbert, 4
Electrolytes and soap production, 102, 111, 205–207, 247, 250–251
Emollients as additives, 144–145
Equipment. *See also* Soap drying systems
   continuous saponification, 230–232
      glycerine washing equipment, 226, 237
   continuous saponification and, 237
   glycerine washing, 237
   neutralization, 240–241
Ethylene propylene diene monomer (EPDM), 279
Eubose, ingredient label compositions of, 175
European use of bar soaps, syndet bars and, 183–185
Eutectic mixture, 85–87
Evaluation methods
   bar hardness, 427–430
   fragrance, 425–426
   lather evaluation, 417–420
   mush, 422
   rate of wear (ROW), 421–422
   rheological characterization, 430–433
   skin mildness, 425
   user panel, 426–427
   wet bar feel, 424
   wet cracking, 422–423
Expansion dryers, 268–269
Extraction
   continuous saponification
      cooling and spent lye removal, 237–238
      general information, 232
      glycerine washing efficiency, 239–240
      glycerine washing equipment, 237
      limit lye concentration, 233–234
      lye and neat soap separation, 239–240
      soap grain, 233, 235
      soap phase chemistry/diagram, 233–234
      washing and half spent lye removal, 237–239
      Wigner's model, 235–236
   RDC countercurrent extraction column, 237–239
Extrusion
   bar soap finishing and, 315–316
   soap production and, 104–106, 362–369
   translucent bars and, 131

F

Fa (1968), 45, 349
Fa (2002-2008), 57
Fairbank, Nathaniel Kellogg, 13
Fairy Soap, 13–15
Fatty acids (FA)
   calculations for blends, 441
   chromatographic analysis and, 411–412
   defined, 448–449
   fats/oils/FA specifications, 451–453
   saponification color of, 405
   surfactant synthesis and, 156–158
Ferguson nomenclature for solid soap phases, 118, 194
Fewa (1932), 63
Fillers, syndet bars, 171
Flashstamping presses, 327, 330
Flex wash mildness test, 178
Flower Sellers Group (Wheatley), 5–6
Foam volume, defined, 449–450
Foaming. *See* Lather evaluation
Focardi, G., 7
Forearm controlled application technique (FCAT), 425
Foreign particulate matter, 415
Formulation, 162–167, 200. *See also* Chemistry of soap structures
Fragrances, 89, 110–111, 143, 425–426
Franklin, Benjamin, 5
Free alkalinity, defined, 449

Free fatty acids (FFA)
   chromatographic analysis and, 411–412
   defined, 449
   molecular performance model and, 91–92
   soap structure and, 89
   superfatting and, 111, 129–130
Free glycerine, defined, 449
Frémy, Edmond, 61
Freshness/deodorant/skin care soaps (1968-1993)
   history
      Atlantic and Pacific (1969), 46
      Caress (1972-2009), 47
      Coast (1974-2008), 48
      Fa (1968), 45
      Irish Spring (1972-2008), 46
      Lever 2000 (1987), 49
      Pure & Natural (1985-2008), 48
Frosch and Kligman toxicity ladder, 177

G

Gain, 21
Gamble, James, 21
Gardner color scale, 448
Geilfus, Charles H., 22
Gibbs' Phase Rule, 125
The Gillette Company
   history, 21
Givaduan-Delawaana, Inc., 40
Global Handwashing Day, 33
Global Public-Private Partnership for
   Handwashing with Soap (PPPHW), 33
Global use
   of caustic soda, 391
   of glycerine, 394–395
   syndet bars and, 183–185
   of triglyceride fatty matter, 382–384
Glossary, 448–451
Glycerides, soap structure and, 83–85
Glycerine
   characteristics/specifications/analysis of, 394
   chromatographic analysis and, 412
   defined, 449
   fats/oils/FA specifications, 451–452
   markets/prices and, 394–395
   production/sources/users, 392–393
   quantity calculations, 439–440

   semi-boiled soap production and, 249
   transport/storage/handling of, 395–396
Glycerol and soap production, 102, 110
Gold Dust Twins, 13, 15
Gold Dust Washing/Scouring Powder, 13–15
Good News Bible, 2
Grain, 233, 235, 247
Graining efficiency, 206
Gross, Henry, 31
Gross, Sydney, 31
Gump, William, 40

H

Half spent lye, 237–239, 248
Handcrafted artisan/specialty soaps, 60–61
Handcrafted Soap Makers Guild (HSMG), 352
Hardness of soap bar, 427–430
Harworth, W. T., 22
Health soaps. *See* Purity/beauty/health soaps
   (1872-1947)
Healy, Tim, 76
Heat transfer rate, 299–300
Heavy metals and glycerine, 394
Heavy/transitional metals and caustic soda, 390
Heckler & Company, 15
Hemoglobin denaturation (haemolysis) test for
   mildness, 178
Henkel & Cie, 61
Herbal extracts as additives, 149–150
Hexagonal liquid crystal phase, 120–122, 124
"Hints to Intending Advertisers" (Smith), 71
Historia Naturalis (Pliny the Elder), 3
History of soaps/detergents
   animal sacrifices and, 3
   antibacterial/ingredient/deodorant/
      moisturizing soaps (1994-2009)
         Dove (2001-2008), 56
         Fa (2002-2008), 57
         Ivory (2004-2008), 58–59
         Ivory Moisture Care (1997), 52
         Lever 2000 (2000-2009), 55
         Lifebuoy (2003-2008), 58
         Lux (2000), 53–54
         Oil of Olay (1994-2008), 49–50
         Palmolive Bars (1999-2008), 50–51
         Pears (2004-2008), 58

Safeguard (1999-2008), 52
bar soap categories and, 24
colonial America and, 5
current brands
   Andrew Jergens Company (1880), 22
   Armour & Company (1867), 21–22
   Colgate Palmolive Company (1806), 17–21
   Lever Brothers Company (1884), 22–23
   Pears (1789), 7–10
   Procter & Gamble Company (1837), 21
   Yardley (1770), 5–7
deodorant/skin care soaps (1948-1967)
   Dial (1948-2008), 40–41
   Dove (1955), 44
   Jergens (1951-2008), 41–42
   Safeguard (1963-2008), 44
   Tone (1968), 45
   Vel (1948-2008), 41
   Zest (1952-2008), 42–43
discontinued soaps
   Colgate, 60
   Dial, 60
   Jergens, 60
   Lever, 60
   Proctor & Gamble, 60
first industrial production, 4
freshness/deodorant/skin care soaps (1968-1993)
   Atlantic and Pacific (1969), 46
   Caress (1972-2009), 47
   Coast (1974-2008), 48
   Fa (1968), 45
   Irish Spring (1972-2008), 46
   Lever 2000 (1987), 49
   Pure & Natural (1985-2008), 48
handcrafted artisan/specialty soaps, 60–61, 60–61
handcrafted artisn/specialty soaps, 60–61, 60–61
laundry powder detergents
   bar soaps, 61, 66
   Dreft (1933), 63–64, 63–64
   Fewa (1932), 63
   liquid detergents, 69–70
   Persil (1907), 61–62, 61–62
   powder detergents, 68
   soap powders, 61, 67
   Tide (1946), 64–65, 64–65
   unit dosing products, 69
laundry washing products, 61
marketing
   coupons, 76–77
   Ivory stamp club, 76
   magazines, 71–72
   premiums, 75
   product booklets, 75–76
   radio and soap operas, 72–74
   selling of soap, 71
   slogans/jingles, 77–78
   soap posters, 75
   soap sculpting, 76
   television and soap operas, 74
   trade/trading cards, 74
   trolly car signs, 75
Mount Sapo and, 223
obsolete brands
   J. S. Kirk & Company (Chicago 1859), 10–13
   N. K. Fairbank Company (Chicago 1865), 13–15
   other companies, 17
   Swift & Company (Chicago 1892), 15–16
plagues and Romans, 3
purity/beauty/health soaps (1872-1947)
   Camay (1926), 38–39
   Cashmere Bouquet (1872), 24–26
   Ivory (1879), 26–30
   Jergens (1893), 30
   Jergens (1947-2005), 39
   Lifebuoy (1887), 31–33
   Lux (1925), 36–38
   Palmolive (1898), 34–36
   Woodbury (1897), 33
   Woodbury (1899), 36
soap making in Marseilles, France, 3–5
Sumerian references to soap (2500 BCE), 1–2
transparent/translucent soaps and, 191
Hollow disc type filtration pumps, 274
Hopkins, Samuel, 5
Hot-air cabinet dryers, 267
Hot-air spray towers, 267
Hudson, R. S., 22

Humectants/moisturizers as additives, 144, 146
Humphrey, Maude, 28
Hunter opacity measurement, 415
Hunter reflectance measurement, 413–414

I

I'm Forever Blowing Bubbles (Kenbrovin & Kellette), 78–79
Indirect cooling water system (ICS), 286
Ingredient soaps. *See* Antibacterial/ingredient/deodorant/moisturizing soaps (1994-2009)
International Maritime Organization (IMO), 384
Inverted-funnel method of lather evaluation, 419–420
Iodine value (IV), 84, 110, 381
   calculation in 80/20 tallow/coconut oil blend, 437
   defined, 450
   fats/oils/FA specifications, 451–452
   soap analysis and, 401–402
Irish Spring (1972-2008)
   history, 46
   manufacture of, 350
Isotropic soap solution, 123–128
Ivorette, 13
Ivory (1879), 26–30
Ivory (2004-2008), 58–59
Ivory Moisture Care (1997), 52
Ivory stamp club, 76

J

J. S. Kirk & Company (Chicago 1859), 10–13, 21, 76–77
Jergens, 60
Jergens (1893), 30
Jergens (1947-2005), 39
Jergens (1951-2008), 41–42
Jergens, Andrew, 22, 33, 36
John H. Woodbury Dermatological Institute, 33
John Hopkins University School of Public Health, 33
Johnson, Burdette, 34
Journal of Chromatography, 399
Journal of Liquid Chromatography, 399
Journal of the American Oil Chemists' Society, 399
Jovan, 6

K

Kappa phase of molecular model, 90, 92, 98–100
   transparent/translucent soaps and, 196
Kellette, John William, 78–79
Kenbrovin, Jaan, 78–79
Kettle saponification
   kettle soap boiling
      countercurrent/full boil, 206–207, 213–214
      finishing/fitting kettle, 212–213
      finishing/fitting kettle (mathematics), 219–220
      fluctuating fat ratios (mathematics), 217
      graining/brine addition, 209–210
      kettle settling (mathematics), 218–219
      kettle washing, 211–212
      loading kettle, 207–208
      loading kettle (mathematics), 214–217
      lyeless, 207
      oil finish, 207
      semi boiled, 207
      settling/spent lye removal, 210–211
      settling/wash lye removal, 212
      variations, 213
   phase diagram theory and, 205–207
   secondary water considerations of, 213
   terminology/properties of, 203–205
Kettle soap boiling. *See* Kettle saponification
Kimball, Alonso, 36
Kirk, J. S., 11
Kirk's brands. *See* J. S. Kirk & Company (Chicago 1859)
Knight, John, 22
Krafft temperature, 86, 125–126

L

Lamellar liquid crystal phase, 119–122
Lanosan, ingredient label compositions of, 175
Latent heat of vaporization, defined, 294
Lather evaluation, 417–420, 449–450
   lathering, of syndet bars, 168

Laundry powder detergents
  history
    bar soaps, 61, 66
    Dreft (1933), 63–64
    Fewa (1932), 63
    liquid detergents, 69–70
    Persil (1907), 61–62
    powder detergents, 68
    soap powders, 61, 67
    Tide (1946), 64–65
    unit dosing products, 69
Laundry soap dryers
  feed pumps for, 276–278
  filtration pumps/filters for, 274, 276–277
  heat exchangers for, 279–280
  processing steps for, 272–275
Laundry tablets. *See* Unit dosing products
Laundry washing products
  history of, 61
Lava Chemical Resolvent Soap, 21
Leglanc, Nicholas, 4
Levenhulme, Lord, 22
Lever, 10, 15, 60
Lever 2000 (1987), 49
Lever 2000 (2000-2009), 55, 174
Lever Brothers Company (1884), 22–23, 73
Lever, William Hesketh, 22–23
Lifebuoy (1887), 31–33
Lifebuoy (2003-2008), 58
Light and transparency of soap, 191–192
Limit lye concentration (LLC), 247
Liquid crystalline soap, 118–119, 118–124
Liquid detergents, 69–70
Lobe pumps, 277–278
L'Oreal Cosmetics, 6
Lovibond color scale, 448
Lux (1925), 36–38
Lux (2000), 53–54, 73–74
Lye. *See* Spent lye; Washing lye
Lyotropic phases of liquid crystalline soap, 118–124

M

Macro model of soap structure, 87–89
Magazines in marketing/advertising, 11, 25–26, 29, 33, 35–36, 71–72

Marbleized soaps, 351, 353–359, 362–363, 366–367
Marketing
  caustic soda and, 391
  glycerine and, 394–395
  history of soaps/detergents
    art, 5–8, 28, 34, 72
    coupons, 19, 76–77
    Ivory stamp club, 76
    magazines, 11, 25–26, 29, 33, 35–36, 71–72
    movie stars, 37–38
    premiums, 75
    product booklets, 75–76
    radio and soap operas, 72–74
    selling of soap, 71
    slogans/jingles, 5, 9, 33, 38, 77–78. *See also specific soap brands, i.e. Dial*
    soap posters, 75
    soap sculpting, 76
    television and soap operas, 74
    trade/trading cards, 74
    trolly car signs, 12, 75
  mildness and global trends, 181–185
  Palmolive soap and, 34
  and selling of soap, 71
  triglyceride fatty matter and, 382–384
Marseilles, France and soap making, 3–5, 51
Mascot, 13
Material balances, 294–298
Mathematics. *See* Calculations
Maxine Elliott Complexion Soap, 16
Maybelline Company, 6
Mazzoni
  Mazzoni LB direct transfer (DTS) system, 328, 343–344
  Mazzoni LB multirefining plodder (MRP), 319, 321
  Mazzoni LB "SCNT" system, 230, 232, 237, 242–243, 353
  Mazzoni LB SCT-SSCT process, 243–244, 256–259
McElroy, Neil Mosley, 38
Melting point
  defined, 450
  fats/oils/FA specifications, 451–452

Metering/dosing and saponification
  critical metered streams and, 227
  metering equipment and, 227–228
  raw materials and, 224, 227
Miag "Double Expansion Drying" system, 268
Micelles, 123–126
Mildness concept, 425
  comparisons, 180–183
  global marketing trends and, 181–185
  red blood cell (RBC) test, 178–180
  skin barrier destruction (SBD) test, 178, 180
  soap chamber test (SCT), 176–177
  syndet bars and, 171–178
  syndet bars and the future, 186
  in vivo/in vitro test, 177–178
  Zein test, 178–179, 181–182
Miles, G., 417
Millais, John Everett, 8
Minerals and soap production, 89
Mixer-kneaders, bar soap finishing and, 308–310
Moisture content, defined, 450
Moisture, impurities, and unsaponifiables (MIU), 381, 406–407
Moisturizing soaps. *See* Antibacterial/ingredient/deodorant/moisturizing soaps (1994-2009)
Molecular model of soap structure, 89–92
Molecular structure. *See* Chemistry of soap structures
Molecular weight adjustments of FA blends, 442–443
Mono screw pumps, 277–278
Mottled laundry bar soaps, 374
Movie stars, in marketing/advertising, 37–38
Mucha, Alphonse, 22
Multicolored/multicomponent soap production
  handcrafted artisan soaps, 352
  history/general information, 349–351
  manufacturing systems/methods, 353
  mottled laundry soap manufacturing system, 374
  recycling methods and, 369–372
  solid/liquid systems for, 353–358
  solid/solid co-extrusion systems for, 362–369
  solid/solid systems for marbleized soaps, 359, 362–363, 366–367
  solid/solid systems for speckled soaps, 361
  solid/solid systems for striped soaps, 362, 364–369
  solid/solid systems for two-tone soaps, 360, 362, 368–369
  stamping options, 373
  types of, 351–353
Mush
  annealing and, 102
  bar moisture content and, 102–103
  defined, 98
  as delta phase, 100
  evaluation of, 422
  formulation implications, 108–112
  liquid phase of, 99–100
  mush/cracking relationship, 93
  solid phase of, 98–99
  structural implications, 103–108
  structure weakness and, 98–103
Musk Oil, 5–6

N

N. K. Fairbank Company (Chicago 1865), 10, 13–15
Natural colorants, 143
Neat soap, 239, 248, 251, 291
Neca (Israel), 183
Neutralization
  calculations for FA products of, 442
  continuous saponification
    equipment, 243–245
    reaction, 243
  defined, 247
  neat soap and, 248
  semi-boiled soap production and, 260–264
Nivea Milk Bar, 175
Non-ionic surfactants, 154–155, 160–161. *See also* Surfactants
Non-occlusive moisturizers, syndet bars and, 172
Noncertifiable colorants, 141–142
Noncertified colorants, 142
Nut oils and soap production, 85, 88, 91, 110

O

Occlusive agents as additives, 144, 147, 172
Octagon products, 19

Odor, 381, 425–426. *See also* Frangrances
Official Methods and Recommended Practices of the American Oil Chemists' Society, 399
Oil bleaching, 385–388
Oil finish (OF) kettle, 211–213, 219–220
Oil of Olay (1994-2008)
   history, 21, 49–50
   ingredient label compositions of, 174
   surfactants and, 164
Oils and soap production, 84–85, 115–116, 150. *See also* Soap phase structure/behavior
Oleate:laurate eutectic mixture, 85–87
Omega phase of molecular model, 90
One Soap for the Whole Family, 16
Operating vacuum (absolute pressure), 300–301
OTC active ingredients as additives, 146
Oxydol, 21

P

Palm kernel oil (PKO), 84–85, 224, 378, 383, 451–453
Palm oils (PO), 85, 378–379, 384, 451–453
Palmolive (1898), 34–36
Palmolive Bars (1999-2008), 50–51
Palmolive Building, 18–19
Parkson Atmospheric Dryer, 268
Pears (1789), 7–10, 200
Pears (2004-2008)
   history, 58
Pears, Andrew, 7
Pears' Annuals, 8
Pears' Shilling Cyclopaedia, 9
Peet Brothers, 71
Penetration value, 450
Perfume and soap production, 89, 110–111, 143
Persil (1907), 61–62
pH
   calculations for alkalinity/acidity adjustments, 444–447
   cast transparent/translucent bars and, 170
   soaps after manufacturing and, 154
   soaps during manufacturing and, 153
Phase diagram theory of kettle saponification, 205–207
Physical characteristics
   alcoholic saponification color, 405
   color by colorimeter, 404
   color by Hunter reflectance measurement, 413–414
   color by photometric measurement, 404
   color evaluation of raw materials, 404
   color, refined/bleached, 404–405
   color scale conversions, 405–406
   direct bleach test, 405
   FFA content, 405–406
   foreign particulate matter, 415
   insoluble impurities, 406
   MIU content, 406–407
   moisture content, 406
   saponification color of FAs, 405
   translucency by Hunter opacity measurement, 415
   unsaponifiable matter, 407
   visual color comparisons, 413
Physical characteristics of soap raw materials. *See* Raw materials for soap making
Pierce, Charles S., 34
Piggly Wiggly, 71
Pink Palace Museum of Arts and Industry (Memphis), 71
Plant automation, 290–292
Plasticizers/binders, syndet bars, 169
Plate-and-frame heat exchangers (PAF), 279–280
Pliny the Elder, 3
Plodders
   bar soap finishing
      extrusion group, 315–316
      Mazzoni LB multirefining plodder (MRP), 319, 321
      packaged water chillers, 331–332
      refining/pelletizing group, 315–316
      SAS Transavon Duplex Vacuum Plodder, 322–323
      short-/long-Lc/D-ratio, 319
      types, 314
      worms, 317–320
Plodding and soap production
   crystal orientation during, 95–96
   pressure effects during, 94–95
   soap structures in plodder, 96–97
Polyalkylene glycol and binding, 169
Positive displacement metering system, 227–228

Posters in marketing/advertising, 75
Powder detergents, history, 68
Premiums in marketing/advertising, 75
Presses, bar soap finishing and, 326–327
Pressurized systems and semi-boiled soap
  combination (integrated) saponification/
    drying plants and, 256, 258–259
  compared with atmospheric system, 260
  high shear mixer/reactor of, 255–256
  semiconcentrated saponification systems and,
    256–257
  stirred vessel reactor (autoclave) of, 254–255
Pride Cleanser, 16
Pride Soap, 16
Pride Washing Powder, 15–16
Primary base, defined, 351
Processing of syndet bars, 159–160
Procter & Gamble Company
  Ace Detersivo Marsiglia, 5
  Global Handwashing Day and, 33
  history, 6, 11
    discontinued soaps, 60
    Oxydol's Own Ma Perkins, 73–74
    The Songs of the City, 74
  mildness and, 183
  Procter & Gamble Company (1837), 21
  Richardson-Vick and, 50
  surfactants and, 164
Procter, Cooper, 38
Procter, Harley, 26–27
Procter, William, 21
Product booklets in marketing/advertising,
  75–76
Production. See Chemistry of soap structures;
  Plodding and soap production
Pummo Glycerine Pumice Soap, 13
Pumps, 287. See also Laundry soap dryers
Pure & Natural (1985-2008)
  history, 48
Purity/beauty/health soaps (1872-1947)
  history
    Camay (1926), 38–39
    Cashmere Bouquet (1872), 24–26
    Ivory (1879), 26–30
    Jergens (1893), 30
    Jergens (1947-2005), 39

Lifebuoy (1887), 31–33
Lux (1925), 36–38
Palmolive (1898), 34–36
soap campaigns, 32–33
Woodbury (1897), 33
Woodbury (1899), 36

Q
Quick Naphtha Chips, 16

R
Rackham, Arthur, 25
Radio and soap operas in marketing/advertising,
  72–74
Rate of wear (ROW), 421–422
Raw materials for soap making
  caustic soda
    characteristics/specifications/analysis, 390
    markets/prices, 391
    production/sources/users, 388–390
    transport/storage/handling, 392
  glycerine
    characteristics/specifications/analysis, 394
    markets/prices, 394–395
    production/sources/users, 392–393
    transport/storage/handling, 395–396
  physical characteristics
    alcoholic saponification color, 405
    color by colorimeter, 404
    color by Hunter reflectance measurement,
      413–414
    color by photometric measurement, 404
    color evaluation of raw materials, 404
    color, refined/bleached, 404–405
    color scale conversions, 405–406
    direct bleach test, 405
    FFA content, 405–406
    foreign particulate matter, 415
    insoluble impurities, 406
    MIU content, 407
    moisture content, 406
    saponification color of FAs, 405
    translucency by Hunter opacity
      measurement, 415
    unsaponifiable matter, 407

visual color comparisons, 413
triglyceride fatty matter
    characteristics/specifications/analysis, 379–382
    as largest constituent, 377
    markets/prices, 382–384
    oil bleaching, 385–388
    sources/grades/users, 378–379
    transport/storage/handling, 384–385
RDC countercurrent extraction column, 237–239
Recycling methods, multicolored/multicomponent soap production and, 369–372
Red blood cell (RBC) test for mildness, 178–180
Refining/pelletizing group, bar soap finishing and, 315–316
Religious influence on cleanliness/soap, 2–3
Rheological characterization, 430–433, 430–433
Richardson-Vick, 49–50
Rider, Ed, 26
Roll mills, bar soap finishing and, 311–313, 331–332
Ross, J., 417
Ross-Miles foam test, 417

S

Safeguard (1963-2008), 44
Safeguard (1999-2008), 52
Santa Claus, 13
Saponification, 438–439. *See also* Continuous saponification; Continuous saponification; Kettle saponification; Semi-boiled soap production
Saponification color of FAs, 405
Saponification value (SV), 381, 401, 450
    calculation in 80/20 tallow/coconut oil blend, 437
    fats/oils/FA specifications, 451–452
SAS Transavon Duplex Vacuum Plodder, 322–323
Satina, 175
Saturated fatty acids, soap structure and, 83–84
Saunders, Clarence, 71
Savon De Marseille, 4
Savona, Italy and soap making, 3

Sculpting in marketing/advertising, 76
Sebamed, 175
Secondary base, defined, 351
Secondary surfactants as additives, 148–149
SELA drying plants, 268–269
Sela GmbH--KVN Plant, 243
Semi-boiled soap production
    atmospheric systems
        batch process, 252–253
        compared with pressurized system, 260
        continuous process, 254
        semi-continuous process, 253–254
    continuous saponification plant use
        main components, 260–261
        neutralizer route - atmospheric system, 261–264
        reactor route - pressurized system, 261–262
    crutcher-mixer/reactor step, 252
    definitions/calculations, 264–266
    general information, 249–251
    pressurized systems
        combination (integrated) saponification/drying plants, 256, 258–259
        compared with atmospheric system, 260
        high shear mixer/reactor, 255–256
        semiconcentrated saponification systems, 256–257
        stirred vessel reactor (autoclave), 254–255
Sensible heat, defined, 294
Shell-and-tube heat exchangers, 279–280
Short-/long-Lc/D-ratio, 319
Silver Dust, 13
Single cyclone fines recovery system, 284
Sinking Ivory promotion, 30
SITA Foam Tester R2000, 418
Skin barrier destruction (SBD) test for mildness, 178, 180
Skin care soaps. *See* Deodorant/skin care soaps (1948-1967); Freshness/deodorant/skin care soaps (1968-1993)
Slogans/jingles in marketing/advertising, 5, 9, 33, 38, 77–78. *See also specific soap brands, i.e. Dial*
Smith, Thomas, 71
Snow Boy Washing Powder, 16
Soap Bubbles (Chardin), 78
Soap calculations. *See* Calculations

Soap chamber test (SCT) for mildness, 176–177
Soap cleansing systems
    additive-base interactions in, 135–139
    additives
        anti-acne compounds, 148
        anti-irritants, 148
        antimicrobial compounds, 147
        CTFA Cosmetic Handbook, 143
        dermabrasive/exfoliating agents, 144, 146
        drug components, 146
        emollients, 144–145
        humectants/moisturizers, 144, 146
        miscellaneous, 149–150
        occlusive agents, 144, 147
        OTC active ingredients, 146
        secondary surfactants, 148–149
    colorants
        categories of, 139–141
        certified, 139, 142
        natural, 143
        noncertifiable, 141–142
        noncertified, 142
    fragrances, 143
    overview, 135
Soap drying systems
    chill rolls/flakers, 268
    cooling tower systems, 289
    definitions/terminology, 293–294, 296–297
    expansion dryers, 268–269
    heat transfer rate for, 299–300
    hot-air cabinet dryers, 267
    hot-air spray towers, 267
    laundry soap dryers
        feed pumps, 276–278
        filtration pumps/filters, 274, 276–277
        heat exchangers, 279–280
        processing steps, 272–275
    material balances, 294–295, 297–298
    operating vacuum (absolute pressure)/condensate temperature and, 300–301
    plant automation, 290–292
    soap fines recovery system, 283–284
    solid/liquid additive systems, 290
    steam consumption by, 288
    and TFM/moisture content of soaps, 300–301
    toilet soap dryers, 268–272
    vacuum chambers, 282–283
    vacuum formation systems
        pumps, 287
        steam jet ejectors (boosters), 287
    vacuum producing systems
        barometric condensers, 285
        indirect cooling water system (ICS), 286
        surface condensers, 285
        vapor condensation equipment, 284–285
    vacuum spray drying, 267
    vapor-liquid separators, 279–282
Soap fines recovery system, 283–284
Soap finishing. *See* Bar soap finishing
Soap-less soaps. *See* Syndet bars
Soap making raw materials. *See* Raw materials for soap making
Soap operas, 74
Soap phase structure/behavior
    isotropic soap solution, 123–128
    liquid crystalline soap
        defined, 118–119
        lyotropic phases, 118–124
        thermotropic phases, 118–119
    molecular structure and, 115–116
    solid soap and, 115, 117–118
    superfatting and, 129–130
    transparent/translucent soaps and, 131
Soap powders, 61, 67
Soap shapes, 326
Soap structure. *See also* Chemistry of soap structures; Soap phase structure/behavior
    macro performance model, 87–89
    molecular performance model, 89–92
Sodium benzotriazolyl sulfonate as additive, 149
Sodium butylphenol as additive, 148
Sodium chloride, defined, 450
Sodium-distilled, topped cocoyl isethionate (STCI), 163
Sodium hydroxide and caustic soda, 390
Solid/liquid systems for multicolored/multicomponent soap production, 353–358
Solid phase micro-extraction (SPME), fragrance evaluation and, 425–426
Solid soap, 115, 117–118
Solid/solid co-extrusion systems, 362–369

Solid/solid systems for marbleized soaps, 359, 362–363, 366–367
Solid/solid systems for speckled soaps, 361
Solid/solid systems for striped soaps, 362, 364–369
Solid/solid systems for two-tone soaps, 360, 362, 368–369
Solubility, syndet bars and, 170–171
Solvay, Ernest, 4
Soybean oil, 378, 383
Specialty soaps. *See* Handcrafted artisan/specialty soaps
Specific gravity, defined, 293
Specific heat (Cp), defined, 294
Specifications for fats/oils/FA
    distilled FA blend compositions, 453
    distilled FAs for soap blends, 452
    distilled FAs for soaps, 452
    distilled tallow/vegetable FA blends compositions, 454
    FA composition of vegetable oil blends, 454
    fats/oils compositions, 453
    fats/oils/FA specifications, 451
    soap bases superfatted with distilled soap fatty, 454
Speckled soaps, 352, 361
Spent lye, 224, 234, 237–238, 237–238, 248
Stamping options, multicolored/multicomponent soap production, 373
Standard bar soap stamping, 373
Steam consumption by soap drying systems, 288
Steam jet ejectors (boosters), 287
Steichen, Edward, 33, 36
Storer, Doug, 76
Striped soaps, 351, 362, 364–369
Sudsing test, 418–419
Sunbrite Cleanser, 16
Sunlight Almanac, 23
Sunlight Flakes, 36
Sunlight label, 23
Sunlight Year Book, 23
Sunny Monday Laundry Soap, 13
Superfatting
    FFA and, 111
    soap phase structure/behavior and, 129–130
    syndet bars and, 171–172

Surface condensers, 285
Surfactants
    cleansing/rinsing efficiency and, 166–167
    importance of, 153
    sodium-distilled, topped cocoyl isethionate (STCI) and, 163
    syndet bars and, 154–158, 160–162
Swift & Company (Chicago 1892), 11, 15–16
Swift, Gustavus F., 15
Swift's Cleanser, 16
Syndet bars
    appearance-improving additives, 170–171
    cast transparent/translucent bars, 170
    cleansing bar formulation, 159
    cleansing/rinsing efficiency and, 166–167
    Dove Beauty Bar design and, 176
    fillers and, 171
    foaming/lathering of, 168
    formulation of, 162–167
    ingredient label compositions of, 173–175
    mildness concept and, 176–178
    mildness evaluation methods
        comparisons, 180–183
        global trends of mild soaps, 181–185
        red blood cell (RBC) test, 179–180
        skin barrier destruction (SBD) test, 178, 180
        soap chamber test (SCT), 176–177
        in vivo/in vitro test, 177–178
        Zein test, 178, 181–182
    mildness formulations in the future, 186
    mildness improvers/skin conditionsers/moisturizers and, 171–175
    performance-improving additives, 170–171
    plasticizers/binders and, 169
    processing of, 159–160
    raw materials/bases of, 165
    soap pH and, 154
    surfactant composition in, 161
    surfactants
        advantage, 153
        anionic surfactants, 155
        general information, 160–162
        synthesis, 156–158
        types of surfactants, 154–155

T

Talcum powder as filler, 171
Tallows and soap production
   and calculations of SV/titer/IV in 80/20 blend, 437
   continuous saponification and, 224
   fatty matter and, 378–379, 384
   soap structure and, 85
   specifications of, 451–454
Television and soap operas
   in marketing/advertising, 74
Temperature
   and batch process of semi-boiled soap production, 252
   condensate, 300–301
   cracking and, 108–109, 112
   distilled FA process and, 377
   fats/oils/FA specifications, 451–452
   Krafft temperature, 86, 125–126
   macro performance model and, 88
   molecular performance model and, 90, 92
   spent lye removal and, 237
   structure/mush and, 103
   thermotropic phase and, 118
Thermal conductivity (k), defined, 294
Thermotropic phase, 118
Tide (1946), 21, 64–65
Titer, 402
   calculation in 80/20 tallow/coconut oil blend, 437
   defined, 451
   fats/oils/FA specifications, 451–452
Toilet soap dryers, 268–272
Tom, Dick, and Harry, 13
Tone (1968), 45
Total fatty acid (TFA), defined, 247
Total fatty matter (TFM)
   calculations for, 440
   defined, 451
   and moisture content of toilet soaps, 300–301
   percentage of, 249
   toilet soap dryers and, 269–270, 272–273
Total fatty matter (TFM) and semi-boiled soap, 249
Toxicity ladder, 177
Trade/trading cards in marketing/advertising, 74

Trans isomers and FA/soap properties, 402–403
Transparent/translucenct soaps
   Hunter opacity measurement and, 415
Transparent/translucent soaps
   cast-melt process and, 200
   history, 191
   neat soap drying/finishing
      intensive mixing/refining, 198–200
      vacuum spray drying, 195–198
   phases/properties/characterization methods for, 192–195
   structure and development of, 131, 191–192
   syndet bars, 170
Tributyl citrate as additive, 149
Triclocarban/triclosan, chromatographic analysis and, 412
Triglyceride fatty matter
   characteristics/specifications/analysis of, 379–382
   as largest constituent of soap, 377
   markets/prices and, 382–384
   oil bleaching and, 385–388
   sources/grades/users of, 378–379
   transport/storage/handling of, 384–385
Triglycerides
   chromatographic analysis and, 410
   soap structure and, 83–85
Trolly car signs in marketing/advertising, 12
Tubular Drying System, 268
Two-tone soaps, 352, 362, 368–369

U

UNICEF, 33
Unilever, 33
Unit dosing products, 69. *See also* Metering/dosing
Unsaponified & unsaponifiable matter (U&U), 451
Unsaturated fatty acids, soap structure and, 83–84
U.S. Agency for International Development (USAID), 33
U.S. Centers for Disease Control and Prevention, 33
User panels and evaluations, 426–427

V

Vacuum chambers for drying systems, 282–283
Vacuum formation systems
    pumps and, 287
    steam jet ejectors (boosters) and, 287
Vacuum producing systems
    barometric condensers and, 285
    indirect cooling water system (ICS) and, 286
    surface condensers and, 285
    vapor condensation equipment and, 284–285
Vacuum spray drying, 267
Vanity Fair Beauty Soap, 16
Vapor condensation equipment, 284–285
Vapor-liquid separator (VLS), 269–272, 279, 281
Vapor-liquid separators, 279–282
Vapor pressure, defined, 293
Vel (1948-2008), 41
VEL, ingredient label compositions of, 173
Viscosity, defined, 293
VVF Limited, 22

W

Wahl Brothers Glue Works, 22
Wash mildness tests, 178
Washing, 247
Washing lye, 247
    continuous saponification
        cooling and spent lye removal, 237–238
        general information, 225, 232, 234
        glycerine washing efficiency, 239–240
        glycerine washing equipment, 237
        limit lye concentration, 233–234
        lye and neat soap separation, 239–240
        soap grain, 233, 235
        soap phase chemistry/diagram, 233–234
        washing and half spent lye removal, 237–239
        Wigner's model, 235–236
    and half spent lye removal, 237–239
    semi-boiled soap production and, 250
Washing ratio, 248
Water and Sanitation Program, 33
Water and soap production
    hard/salt water, 153
    soap structure and, 91–93
    water penetration and, 101–102
Water content of soap
    hard/salt water and lathering, 168
    soap-to-water ratio, 109
Water Supply and Sanitation Collaborative Council, 33
Wella Company, 6
Wesley, John, 2
Western Soap Company, 22
Wet bar feel, 424
Wet cracking, 422–423
Wheatley, Francis, 5–6
Whitefield, George, 2
Wigner's model, continuous saponification and, 235–236
William Waltke company, 21
Woodbury (1897), 33
Woodbury (1899), 36
Woodbury, John H., 33, 36
Wool Soap Chips Borated, 16
Woolworth, Frank, 26
World Bank, 33
Worms, bar soap finishing and, 317–320
Wrigley, William, Jr., 11

X

X-ray pattern of soap, 127

Y

Yardley (1770), 5–7
Yardley, William, 5
You Dirty Boy (Focardi), 7

Z

Zein test for mildness, 178–179, 181–182
Zest (1952-2008)
    history, 42–43
    ingredient label compositions of, 174
    surfactants and, 164
Zeta phase of molecular model, 91–92, 99–100, 196